Computational Methods for Electron–Molecule Collisions

Computational Methods for Electron–Molecule Collisions

Edited by

Winifred M. Huo

NASA Ames Research Center
Moffet Field, California

and

Franco A. Gianturco

University of Rome, City of Rome
Rome, Italy

Springer Science+Business Media, LLC

Library of Congress Cataloging-in-Publication Data

On file

Proceedings of the Workshop on Comparative Study of Current Methodologies in Electron–Molecule Scattering, held March 11–13, 1993, in Cambridge, Massachusetts

ISBN 978-1-4757-9799-2 ISBN 978-1-4757-9797-8 (eBook)
DOI 10.1007/978-1-4757-9797-8

Originally published by Plenum Press, New York in 1995
Softcover reprint of the hardcover 1st edition 1995

10 9 8 7 6 5 4 3 2 1

PREFACE

The collision of electrons with molecules and molecular ions is a fundamental process in atomic and molecular physics and in chemistry. At high incident electron energies, electron-molecule collisions are used to deduce molecular geometries, oscillator strengths for optically allowed transitions, and in the case of electron-impact ionization, to probe the momentum distribution of the molecule itself. When the incident electron energy is comparable to or below those of the molecular valence electrons, the physics involved is particularly rich. Correlation and exchange effects necessary to describe such collision processes bear a close resemblance to similar effects in the theory of electronic structure in molecules. Compound state formations, in the form of resonances and virtual states, manifest themselves in experimental observables which provide details of the electron-molecule interactions. Ro-vibrational excitations by low-energy electron collisions exemplify energy transfer between the electronic and nuclear motion. The role of nonadiabatic interaction is raised here. When the final vibrational state is in the continuum, molecular dissociation occurs. Dissociative recombination and dissociative attachment are examples of such fragmentation processes.

In addition to its fundamental nature, the study of electron-molecule collisions is also motivated by its relation to other fields of study and by its technological applications. The study of planetary atmospheres and the interstellar medium necessarily involve collision processes of electrons with molecules and molecular ions. For example, the fine-structure changing transitions of the oxygen atom by electron impact and the resonant electron-impact vibrational excitation of N_2 are two major cooling mechanisms of electrons in the earth's ionosphere. Electron-atom and electron-molecule collisions in intense laser fields are used to probe the properties of atoms and molecules under high field conditions. Electron scattering is also employed as a research tool in material science to investigate the properties of condensed matter, from simple crystals to molecules physisorbed or chemisorbed on the surface of a metal or molecular crystal. The difference between the condensed phase versus gas phase spectrum provides information such as the change of bonding in the condensed environment. Recently, this technique was employed in the study of new materials formed by clusters of atoms or molecules.

Electron-molecule collisional cross sections are required to input data for the design and modeling of plasma processes, including plasma etching, chemical vapor deposition, and advanced laser developments. The operating condition of these processes can be optimized, and new processes devised, based on the knowledge of electron excitation and fragmentation rates. Such data are also employed to model the weakly ionized flow field experienced by high-speed space vehicles upon re-entry into the planetary atmosphere. Here the actual environment in space is difficult to duplicate in a laboratory and much of the input is from theoretical studies.

The combined effort of theory and experiment contributes to advances in the field of electron-molecule collisions. While the theory of high-energy electron collisions is well understood and cross section calculations using the Born approximation are readily available, the situation in the low-energy regime has lagged behind. Much of the earlier development in this area was devoted to qualitative predictions involving symmetry selections, categorizing resonances, and descriptions of threshold behavior. The development of *ab initio* methods for low-energy electron-molecule collision came rather late.

Stimulated by the availability of modern-day high-speed computers, the advance in molecular electronic structural theory, and access to accurate experimental data, activities in the development of *ab initio* computational methods for low-energy electron-molecules collisions have increased significantly in the last ten to fifteen years. At this point the computed cross sections for small polyatomic molecules can be of comparable accuracy as the best experimental data. In some cases, theoretical studies actually were carried out before experiment.

While the starting point of all *ab initio* methods is the Schrödinger equation or Lippmann-Schwinger equation, each method approaches the problem with different emphasis on the physics. Also, as a method is refined, it is frequently found that certain modification in its implementation may expedite the convergence of the calculation. In view of the growth in this area, it was deemed timely for a gathering of the practitioners of the field to discuss the interplay between fundamental theory and practical implementation, to compare how each method treats various physical aspects of the scattering process, and also the effect of approximate treatments. Under the sponsorship of the Institute for Theoretical Atomic and Molecular Physics (ITAMP) at the Harvard-Smithsonian Center for Astrophysics, a workshop on "Comparative Study of Current Methodologies in Electron-Molecule Scattering" was held March 11-13, 1993, at ITAMP in Cambridge, Massachusetts.

The last day of the workshop, March 13, happened to be the time of the great blizzard of '93 in the eastern part of United States. However, the wind and snow outdoors did not diminish the heated discussion inside. During one of the discussion sections, the participants agreed to prepare a book based on the proceedings of the workshop, with the specification that the book should be useful not only to practitioners in the field, but also should serve as an introduction to graduate students and postdoctoral researchers new to the field. Such a book should also offer opportunities for cross fertilization for researchers from different disciplines.

The present volume is a result of that effort. It covers the time-independent *ab initio* methods for low-energy electron-molecule collisions currently in use. Electron impact ionization is not considered here. Care is taken throughout the book to demonstrate how these methods are implemented in actual computations. While this book is concerned with technical aspects of *ab initio* methods, it is not designed as a review book for cross section data. Other conference proceedings, such as the biennial ICPEAC and its associated satellite meetings, are good sources for such data.

The book is organized according to the methods. Some methods are described in a single chapter, and some have multiple contributions. The numbering of equations starts afresh in each chapter except the R-matrix chapters where the authors number their equations sequentially for easy references. The description of each method is self-contained. The reader is assumed to have a general background in scattering theory, but needs not be well versed in electron-molecule collisions.

We thank Prof. Alex Dalgarno and Dr. Kate Kirby of ITAMP, for their generous support and help in organizing the workshop and for their hospitality to those participants stranded in Cambridge over the snowy weekend. We are grateful to Ms. Valerie Sorenson and Ms. Verity Parris, for arranging housing for the participants from overseas and for their secretarial help. Special mention should be made of a conscientious caterer, who delivered our coffee and cookies under weather conditions that stopped the Boston subway from running. We also appreciate the help from Professor Jonathan Tennyson, who edited the R-matrix section of the book, and the editorial assistance of Dr. Helmar Thümmel.

Winifred M. Huo

Franco A. Gianturco

September 1994

CONTENTS

THE COMPLEX KOHN VARIATIONAL METHOD

Chapter 1. *THE COMPLEX KOHN VARIATIONAL METHOD*

T.N. Rescigno, C.W. McCurdy, A.E. Orel, and B.H. Lengsfield, III

THE LINEAR ALGEBRAIC METHOD

Chapter 2. *THE LINEAR ALGEBRAIC METHOD FOR ELECTRON-MOLECULE COLLISIONS*

Lee A. Collins and Barry I. Schneider

THE MULTICHANNEL QUANTUM DEFECT METHOD

METHOD BASED ON SINGLE-CENTER EXPANSION OF THE TARGET

Chapter 5. *A STUDY OF THE PORTING ON SIMD AND MIMD MACHINES OF A SINGLE CENTRE EXPANSION CODE TO TREAT ELECTRON SCATTERING FROM POLYATOMIC MOLECULES*

F.A. Gianturco, N. Sanna, and R. Sarno

ROTATIONAL AND VIBRATIONAL CLOSE COUPLING

Chapter 6. *HOW TO CALCULATE ROTATIONAL AND VIBRATIONAL CROSS SECTIONS FOR LOW-ENERGY ELECTRON SCATTERING FROM DIATOMIC MOLECULES USING CLOSE-COUPLING TECHNIQUES*

Michael A. Morrison and Weiguo Sun

THE PARTIAL DIFFERENTIAL EQUATION METHOD

Chapter 7. *THE (NON-ITERATIVE) PARTIAL DIFFERENTIAL EQUATION METHOD: APPLICATION TO ELECTRON-MOLECULE SCATTERING*

A. Temkin and C.A. Weatherford

THE R-MATRIX METHOD

Chapter 11. *ELECTRON COLLISIONS WITH THE* He_2^+ *CATION*

Brendan M. McLaughlin and Charles J. Gillan

Chapter 12. *ROVIBRATIONAL EXCITATION BY ELECTRON IMPACT*

Helmar T. Thümmel, Thomas Grimm-Bosbach, Robert K. Nesbet and Sigrid D. Peyerimhoff

Chapter 13. *TAILORING THE R-MATRIX APPROACH FOR APPLICATION TO POLYATOMIC MOLECULES*

Kurt Pfingst, Bernd M. Nestmann and Sigrid D. Peyerimhoff

THE SCHWINGER VARIATIONAL METHOD

THE COMPLEX KOHN VARIATIONAL METHOD

T.N. Rescigno[1], C.W. McCurdy[2], A.E. Orel[3] and B.H. Lengsfield, III[4]

[1]Physics Department, Lawrence Livermore National Laboratory
[2]National Energy Research Supercomputer Center
[3]Department of Applied Science, University of California Davis
Livermore, CA 94550
[4]IBM Almaden Research Center
San Jose, CA 95120

INTRODUCTION

Though substantial progress has been made in the theoretical study of electron collisions with molecules and molecular ions, most work has been restricted to diatomic or linear targets. Electron- and photon-molecule collision cross sections are needed in such diverse areas such as advanced laser development, pollution control, the design of high-speed space re-entry vehicles, the manufacture of semiconductor devices and plasma driven chemical synthesis. For example, the photoionization of polyatomic radicals, which plays an important role in combustion[1], requires a description of electron scattering from a polyatomic molecular ion. Such studies are scarce. In the area of plasma enhanced chemical vapor deposition and etching[2], studies indicate a subtle interplay between the neutrals, ions, electrons, and the surface. A critical lack of fundamental cross sections is hindering our understanding of these processes. Reliable theoretical methods are exceptionally important because of the extreme difficulty of experiments in this area.

Dissociative recombination of ions[3],

$$e^- + AB^+ \rightarrow A + B$$

Computational Methods for Electron-Molecule Collisions
Edited by W.M. Huo and F.A. Gianturco, Plenum Press, New York, 1995

plays an essential role in plasma dynamics, especially in low temperature plasmas, and also plays a major role in the operation of many excimer lasers.[4,5] For example, in the interstellar medium, the dissociative recombination of H_3^+ is an important contributor to the formation of hydrogen atoms and molecules[6]. For this simplest polyatomic dissociative recombination process, there is no adequate theory for very low collision energies and experiments have yielded cross sections that vary by orders of magnitude[7]. For these problems, there is additional complexity due to the coupling of the nuclear motion with the electron scattering. For diatomics, where the majority of work has been done, there is only one nuclear degree of freedom; even for a triatomic, three must be considered.

In addition to the practical interest, electron-polyatomic molecule collisions are of fundamental theoretical importance. With the many vibrational degrees of freedom and opportunities for interaction between excited electronic surfaces in several dimensions, it is clear that polyatomic targets offer a richer and more complex area of study than that found with diatomic targets. We have therefore been interested in developing a technology that is robust enough to study such polyatomic systems.

The complex Kohn method is an algebraic variational technique which, over the last few years, has been developed into a powerful approach for studying both heavy-particle (reactive) collisions[8] and electron scattering problems. It is somewhat surprising that although algebraic variational methods based on the Kato identity[9], such as the Kohn variational method[10] have been (and continue to be) used with much success in electron-atom problems[11], they received virtually no attention in molecular applications during the decade of the 1970's and were not developed in this connection until the mid 1980's. Two fundamental difficulties contributed to this neglect. One factor was the longstanding problem associated with the occurance of anomalous singularities in the reactance or K-matrices that made these techniques difficult to apply in large scale calculations[12]. The second problem concerns computational difficulties associated with the evaluation of multi-center integrals involving continuum functions.

The first problem can be trivially solved by formulating the variational problem with physical, outgoing wave (complex) boundary conditions, rather than the usual (real) standing-wave boundary conditions. This fact was originally used in nuclear physics by Mito and Kamimura[13], but went largely unnoticed until it was introduced into atomic and molecular physics by Miller and Jansen op de Haar in reactive heavy-particle scattering problems[14]. The connection between Miller's original work, the Kohn method and Kapur-Peierls theory was pointed out by McCurdy, Rescigno and Schneider[15] in 1987 and the complex Kohn method for electron-molecule scattering was developed over the ensuing few years.

A practical solution to the second problem required further theoretical developments. These included the judicious use of separable approximations to transfer the major burden of the problem to the construction and manipulation of bound-bound matrix elements[16]. Moreover, because the Kohn variational method is based on a Hamiltonian formulation of the collision problem, the full arsenal of bound-state structure methodology can be exploited in this connection. In order to evaluate the continuum matrix elements that are needed, we rely on efficient 3-dimensional adaptive quadrature techniques[17,18]. This divorces the method from reliance on any specific analytic schemes and allows us considerable flexibility in choosing continuum basis functions.

The complex Kohn variational method is an efficient and flexible computational tool for studying electron-molecule scattering and molecular photoionization. We do not intend to give an exhaustive review of the many calculations which have been carried out with this method. Such a review can be found elsewhere[19]. Our purpose here is to provide a pedagical exposition of the technical aspects of the method. We will, however, site illustrative examples to highlight various aspects of the formulation. We should also point out that the fixed-nuclei electronic scattering problem, with which this review is mainly concerned, is only part of the full electron-molecule scattering picture. Much work centers on the nuclear motion problem, which is important in resonantly driven processes such as recombination and dissociative attachment[20], as well as near threshold vibrational and rotational excitation[21]. Meaningful (vibrationally and rotationally summed) cross sections can frequently be extracted solely from fixed-nuclei calculations. We will also show how the fixed-nuclei picture can be modified to provide threshold vibrational excitation cross sections. In many other cases the fixed-nuclei electronic problem can provide the input necessary to solve the full problem including the nuclear dynamics[22,23]. We will give one example of this for the case of dissociative recombination.

THEORETICAL FOUNDATION

The Kohn Variational Method for Neutral and Ionic Scattering

The basic concepts of the algebraic variational method are most easily illustrated for the case of scattering of a spinless particle by a spherically symmetric, short-range potential, V(r)[24]. We will therefore consider this problem in some detail before generalizing to the multichannel case and inelastic processes. Atomic units will be assumed throughout, unless otherwise stated. For the partial wave radial Schödinger equation, which we write as $Lu_\ell=0$, where

$$L = -\frac{1}{2}\frac{d^2}{dr^2} + \frac{\ell(\ell+1)}{2r^2} + \frac{Z}{r} + V(r) - \frac{k^2}{2} \quad , \tag{1}$$

we define the functional

$$L=\int_0^\infty u_\ell(r)Lu_\ell(r)dr \quad . \tag{2}$$

The boundary conditions assumed for u_ℓ are

$$u_\ell(0) = 0$$

$$u_\ell(r) \underset{r\to\infty}{\sim} F_\ell(kr) + \lambda G_\ell(kr) \qquad (3)$$

where F_ℓ and G_ℓ are two *linearly independent* solutions of the radial Schröedinger equation with V(r)=0. Their asymptotic normalization is, at this point, arbitrary. Note that Z/r is explicitly included in Eq. (1) so that we will be able to generalize to scattering by either neutral or ionic targets.

We now examine how L, which clearly vanishes when u_ℓ is an exact solution of the radial Schödinger equation, changes when the exact u_ℓ is replaced by a trial solution u_ℓ^t satisfying the same boundary conditions specified by Eq. (3), that is, with

$$\delta u_\ell(r) \equiv u_\ell^t(r) - u_\ell(r) \qquad (4)$$

and

$$\delta u_\ell(0) = 0$$
$$\delta u_\ell(r) \underset{r\to\infty}{\sim} \delta\lambda G_\ell(r) \qquad (5)$$

The result is

$$\delta L = \int_0^\infty \delta u_\ell \mathcal{L}\, u_\ell dr + \int_0^\infty u_\ell \mathcal{L}\, \delta u_\ell dr + \int_0^\infty \delta u_\ell \mathcal{L}\, \delta u_\ell dr \qquad (6)$$

Integrating the second term on the RHS of Eq. (6) by parts and making use of Eqs. (3) and (4) and the fact that $\mathcal{L}\, u_\ell = 0$ gives

$$\delta L = -\frac{k}{2} W \delta\lambda + \int_0^\infty \delta u_\ell(r) \mathcal{L}\, \delta u_\ell(r)\, dr \qquad (7)$$

where W is the Wronskian defined by

$$W = F_\ell(x)\frac{d}{dx}G_\ell(x) - G_\ell(x)\frac{d}{dx}F_\ell(x)$$
$$= \text{constant} \qquad (8)$$

Eq. (7), which is known as the Kato identity, expresses the exact value of λ in terms of the approximate value of the functional L and a term which is second order in the error in the wave function. The Kato identity thus provides a stationary principle for approximating λ:

$$\lambda^s = \lambda^t + \frac{2}{kW}\int_0^\infty u_\ell^t L\, u_\ell^t dr \quad . \tag{9}$$

Eq. (9) can be solved explicitly for a trial wave function that contains only linear trial coefficients. If we choose a trial function of the form

$$u_\ell^t = f_\ell(r) + \lambda^t g_\ell(r) + \sum_{i=1}^{n} c_i\varphi_i \tag{10}$$

where $\{\varphi_i\}$ are a set of square-integrable functions and

$$\begin{aligned} f_\ell(r) &\underset{r\to\infty}{\sim} F_\ell(kr) \\ g_\ell(r) &\underset{r\to\infty}{\sim} G_\ell(kr) \end{aligned} \quad , \tag{11}$$

then the coefficients λ^t and c_i are determined from the conditions:

$$\frac{\partial\lambda^s}{\partial c_i} = \frac{\partial\lambda^s}{\partial\lambda^t} = 0 \quad . \tag{12}$$

Substituting Eq. (10) into Eq. (9) and taking the derivative with respect to c_i gives

$$\int_0^\infty \varphi_i L\, u_\ell^t dr = 0 \quad , i = 1,,,n \quad . \tag{13}$$

Differentiation with respect to λ^t gives, after some rearrangement,

$$\int_0^\infty g_\ell L\, u_\ell^t dr = 0 \quad . \tag{14}$$

Eqs. (13) and (14) can be expressed in a compact form by relabeling the basis functions $g_\ell(r)$ and $\varphi_i(r)$ into a single set $\{\varphi_i\}$, i=0,,,n with $\varphi_0 \equiv g_\ell$ and denoting the linear parameters $(\lambda^t, c_1,,,c_n\}$ by the vector **c**. The result is:

$$\mathbf{c} = -\,\underline{\mathbf{M}}^{-1}\,\mathbf{s} \tag{15}$$

where $\underline{\mathbf{M}}$ is a matrix with elements

$$\mathbf{M}_{ij} = \int_0^\infty \varphi_i L\, \varphi_j dr \quad , i,j = 0,,,n \tag{16}$$

and **s** is a vector with elements

$$s_i = \int_0^\infty \varphi_i L\, f_\ell dr \quad , i = 0,,,n \quad . \tag{17}$$

The *stationary* value of λ^s is obtained by substituting Eq. (15) into Eq. (9) to obtain:

$$\lambda^s = \frac{2}{kW}\left[\int_0^\infty f_\ell L\, f_\ell dr - \mathbf{s}\mathbf{M}^{-1}\mathbf{s}\right] \quad . \tag{18}$$

We must now specify a form for the reference continuum functions F_ℓ and G_ℓ. We can take F_ℓ (and f_ℓ) to be the regular Coulomb function[25]. The traditional choice for G_ℓ is the irregular Coulomb function. The function g_ℓ, which must be regular at r=0, is usually obtained by multiplying G_ℓ by a suitable cut-off function that approaches unity for large r. With this choice, W=-1, λ is simply the tangent of the phase shift, $\tan\delta_\ell$, and Eq. (18) reduces to the familiar Kohn variational expression for $\tan\delta_\ell^s$. In the case of Coulomb scattering, δ_ℓ is of course the phase shift caused by the potential V(r) and does not include the Coulomb phase shift. When the continuum functions are so chosen, **M** is a real symmetric matrix and has zero determinant at the real values of E corresponding to the eigenvalues of H in the basis of φ_i plus g_ℓ. At these energies, Eq. (18) is singular; this is the origin of the so-called Kohn anomalies[12].

If g_ℓ is chosen to be an outgoing function $h_\ell^+(r)$, defined as

$$h_\ell^+(r) = i\,[\,F_\ell(kr) - i\,G_\ell(kr)]/\sqrt{k} \ , \tag{19}$$

then W=-1/k and Eq. (18) becomes an expression for the T-matrix, $\lambda = T_\ell = e^{i\delta_\ell}\sin\delta_\ell$:

$$T_\ell^s = -2\left[\int_0^\infty f_\ell L\, f_\ell dr - \mathbf{s}\mathbf{M}^{-1}\mathbf{s}\right] \quad . \tag{20}$$

M is now a complex symmetric matrix and *its inverse is generally nonsingular at real energies*[15]. We thus call this form of the variational equations the complex Kohn method. While $\mathbf{M}^{-1}$ can , in principle, be singular at a real energy, such an occurrence is improbable and is not a problem in practical calculations.

Continuum Basis Functions

The continuum basis functions used in a variational trial function must be regular at the origin. The traditional choice has been to use Ricatti-Bessel (or in the case of ionic targets, Coulomb-Bessel) functions, with a simple exponential cut-off applied to the irregular solution. This choice was historically motivated by the desire to use simple functions that would not overly complicate the analytic evaluation of free-free integrals[26]. However, in our implementation of the method, we do not rely on any specific analytic schemes and thus have considerable flexibility in choosing continuum basis functions. Moreover, since we found in early applications that there could be quite a bit of sensitivity to the traditional cut-off, we were led to seek to a more physical choice for regularizing the outgoing wave continuum basis functions[27]. The choice we made was inspired by the work of Sun et al.[28] on the Kohn and generalized Newton variational principles.

The outgoing continuum function we employ is defined as:

$$g_\ell(r) = A\,G_\ell^+\,VF_\ell \quad , \tag{21}$$

where G_ℓ^+ is the radial partial wave free particle Green's function,

$$G_\ell^+(r,r') = \begin{array}{ll} -2/kF_\ell(kr)h_\ell^+(kr') & ,\ r \le r' \\ -2/kF_\ell(kr')h_\ell^+(kr) & ,\ r > r' \end{array} , \tag{22}$$

V is any suitable "test potential' (we use a simple exponential) and the constant A is introduced in order to normalize g_ℓ (r) as in Eq. (19) and is thus given by

$$A = \left[2\int_0^\infty F_\ell V F_\ell dr\right]^{-1} \tag{23}$$

The function g_ℓ is easily calculated by noting that it satisfies the inhomogeneous differential equation

$$\left(\frac{d^2}{d\rho^2} - \frac{\ell(\ell+1)}{\rho^2} - \frac{2Z}{k\rho} + 1\right) g_\ell(\rho) = 2AVF_\ell(\rho) \tag{24}$$

It is a simple matter to construct the function required in Eq. (21) as a linear combination of a solution to Eq. (24) obtained by numerical integration and a solution of the corresponding homogeneous equation. This technique has the added advantage of not requiring the irregular function G_ℓ on the integration mesh[27].

Trial Wave Function for Electron-Molecule Scattering

To implement the complex Kohn method for electron-molecule scattering, we must generalize the simple trial function previously discussed to a form appropriate for multichannel scattering. We will use body-frame coordinates within the framework of the fixed-nuclei approximation. We write:[29]

$$\Psi_{\Gamma_0} = \sum_{\Gamma} A\left(\chi_\Gamma F_{\Gamma\Gamma_0}\right) + \sum_{\mu} d_\mu^{\Gamma_0}\Theta_\mu \quad . \tag{25}$$

The first sum runs over a set of energetically open N-electron target states, χ_Γ, and $F_{\Gamma\Gamma_0}$ is a one-electron function that describes the scattered electron. The index Γ_0 labels a particular degenerate solution corresponding to the initial state of the target and the operator A antisymmetrizes the orbital functions $F_{\Gamma\Gamma_0}$ into the functions χ_Γ. Note that we are using the symbol Γ to label all the quantum numbers needed to represent a physical state of the composite system, that is, the internal state of the target and the angular momentum of the scattered electron.The target functions χ_Γ may be single or multiconfiguration descriptions of the bound states of the molecule. It is assumed that $F_{\Gamma\Gamma_0}$ is, by construction, orthogonal to all the orbital functions used to construct χ_Γ. We use average natural orbital techniques to insure that the orbital space used to generate these states is kept manageably small. The functions Θ_m are a set of (N+1)-electron configuration state functions, (CSF's) orthogonal to the terms in the first sum of Eq. (25), constructed from square-integrable (Cartesian Gaussian) functions.

In general, the (N+1)-electron CSF's can be classified as one of three distinct types[30]. First, this sum contains the so-called 'penetration' terms, which relax any constraint implied by requiring $F_{\Gamma\Gamma_0}$ to be orthogonal to the orbitals used to build the χ_Γ[29]. Thus, these CSF's are built entirely from target orbitals. A second class of terms is possible in the case that multiconfiguration target states are used. Since the χ_Γ are then constructed as fixed linear combinations of CSF's, there will be an orthogonal complement of target states, presumed to be energetically closed, that uses the same CSF's in different linear combinations. These "CI relaxation terms" are built as the direct product of these other states and a square-integrable orbital. Finally, we can include other closed channels by constructing terms which are the direct product of a bound orbital and a disjoint set of target configurations (to be compared to CI-relaxation terms, which use the same set of CSF's as the open channel states). We refer to this latter set of CSF's as polarization terms. Both types of closed-channel contributions are necessary to describe the polarization and correlation effects that can dominate the scattering cross sections at low energy[31]. Note that these closed-channel contributions to the wave function fall off exponentially as any electron coordinate tends to infinity. This form of the trial function, which is often referred to as a close-coupling plus correlation (CCPC) expansion, is quite general and common to many closed-coupling formulations[32].

In algebraic variational methods such as the complex Kohn method, $F_{\Gamma\Gamma_0}$ is further expanded as

$$F_{\Gamma\Gamma_0} = \sum_{i} c_i^{\Gamma\Gamma_0}\varphi_i + \sum_{\ell m}[f_\ell^\Gamma(k_\Gamma r)\delta_{\ell\ell_0}\delta_{mm_0}\delta_{\Gamma\Gamma_0} + T_{\ell\ell_0 mm_0}^{\Gamma\Gamma_0} g_\ell^\Gamma(k_\Gamma r)]Y_{\ell m}(\hat{r})/r \tag{26}$$

where the φ_i are a set of square-integrable functions, $Y_{\ell m}$ is a normalized spherical harmonic and the functions f_ℓ^Γ and g_ℓ^Γ are the regular and outgoing-wave continuum functions discussed in the previous section. This is the generalization of Eq. (10) to the multichannel case. The channel momenta k_Γ are determined by energy conservation,

$$k_\Gamma^2 / 2 = E - E_\Gamma \quad , \tag{27}$$

where E is the total energy of the composite system and E_Γ is the energy of the target molecule corresponding to state χ_Γ. The coefficients $T^{\Gamma\Gamma_0}_{\ell\ell_0 m m_0}$ are elements of the T-matrix and are the fundamental dynamical quantities from which differential and integral cross sections are constructed. In order to simplify the notation in what follows, we will adopt the convention of using $f^\Gamma_{\ell m}$ (or $g^\Gamma_{\ell m}$) to denote the product of a radial function f^Γ_ℓ (or g^Γ_ℓ) and $Y_{\ell m}/r$.

In principle, the total wave function can be described to any desired level of accuracy with a trial wave function of the form just outlined. In practice, the open-channel expansion is usually truncated rather severely and the target states themselves are approximate; we will see that the consequences of these two facts can profoundly effect the quality of the computed cross sections and that it will be very important to balance correlation effects in the N- and (N+1)- electron systems in the process of deciding what types of terms are reasonable to include in the trial wave function. We will return to these physical considerations later, but now we continue with the mechanics of how the variational equations are constructed and solved.

Feshbach Partitioning

It is convenient to partition the total wave function into two parts, $P\Psi_{\Gamma_0}$ and $Q\Psi_{\Gamma_0}$, corresponding to the two sums defined in Eq. (25). As we previously stated, the channel orbitals $F_{\Gamma\Gamma_0}$ which appear in $P\Psi_{\Gamma_0}$ are expanded as linear combinations of bound and continuum functions. These basis functions can be mutually orthogonalized without changing the resulting T-matrix, a property known as *transfer invariance*[11]. We will show later how this property can be exploited to simply the calculation of certain matrix elements. We also follow the common practice of requiring the channel orbitals to be orthogonal to all the bound-state orbitals which are used to form the target wave functions, and have commented that these additional constraints, if necessary, can be relaxed by including appropriate penetration terms in $Q\Psi_{\Gamma_0}$[33,34].

$P\Psi_{\Gamma_0}$ contains the asymptotic parameters from which collision cross sections are determined; since $Q\Psi_{\Gamma_0}$ is square-integrable, it is not explicitly needed for the determination of scattering parameters. In calculations which include the effect of closed channels, the set of Q-space configurations can become quite large. For this reason, and because of the fact that the configuration state functions $\{\Theta m\}$ are built solely from square-integrable orbitals, there are cases where it desirable to use bound-state molecular structure methods to treat the Q-space portion of the problem. However, there are applications, such as photoionization[35], in which the entire wave function, $P\Psi_{\Gamma_0} + Q\Psi_{\Gamma_0}$, is required to calculated a desired property. In such cases, Feshbach partitioning may not be the most efficient way to proceed.

Feshbach partitioning[36] is accomplished by formally solving the Schrödinger equation for $Q\Psi_{\Gamma_0}$ and then substituting the result into the equation for $P\Psi_{\Gamma_0}$. Defining M as H-E, we can derive in the usual way a modified Hamiltonian that determines $P\Psi$:

$$\begin{aligned} H_{eff} &= H_{PP} - M_{PQ} M_{QQ}^{-1} M_{QP} \\ &= H_{PP} - V_{opt} \end{aligned} \tag{28}$$

where M_{QQ}^{-1} is the inverse of the Hamiltonian matrix spanned by the functions $\{\Theta m\}$. This allows us to drop the variational coefficients $d_\mu^{\Gamma_0}$ in Eq. (25) from further consideration. Note that the optical potential, V_{opt}, that we have defined is technically not the same as the Feshbach optical potential[36]. The difference lies in the fact that our $Q\Psi_{\Gamma_0}$ can contain orthogonality relaxing "penetration terms", in which case $Q\Psi_{\Gamma_0}$ will contain a portion of the open-channel wave function.

We can now proceed to use the effective Hamiltonian defined in Eq. (28) to define a functional, just as we did in the single-channel case, for the purpose of getting a stationary approximation to the T-matrix. The multichannel T-matrix can again be characterized as the stationary value of the Kohn functional

$$\mathbf{T}_s^{\Gamma\Gamma'} = \mathbf{T}_t^{\Gamma\Gamma'} - 2\int P\Psi_\Gamma (\, H_{eff} - E\,)\, P\Psi_{\Gamma'}\, dr_1 \,..\, dr_{N+1} \tag{29}$$

By following the same procedures outlined in the single-channel case, we obtain the generalization of Eq. (20):

$$\mathbf{T}_s = -2\left(\mathbf{M}_{00} - \mathbf{M}_{q0}^t\, \mathbf{M}_{qq}^{-1}\, \mathbf{M}_{q0}\right) \tag{30}$$

where we have used a condensed matrix notation, in which open-channel indices are suppressed. The index 0 denotes the subspace spanned by the functions $\left\{\chi_\Gamma f_{\ell m}^\Gamma\right\}$ and the index q refers to the subspace spanned by $\left\{\chi_\Gamma g_{\ell m}^\Gamma\right\}$ and $\{\chi_\Gamma \varphi_i\}$. M refers to the operator H_{eff}-E. For example, the elements of M_{00} are defined as

$$(\mathbf{M}_{00})_{\ell m \ell' m'}^{\Gamma\Gamma'} = \int A\left(\chi_\Gamma f_{\ell m}^{\Gamma}\right)(\, H_{eff} - E\,)\, A\left(\chi_{\Gamma'} f_{\ell' m'}^{\Gamma'}\right) dr_1 \,..\, dr_{N+1} \tag{31}$$

and M_{q0} and M_{qq} are defined similarly.

The matrix **M** is thus built from elements that can be classified as either free-free, bound-free or bound-bound, depending upon whether two, one or zero continuum orbitals appear in the expression. The bound-bound terms, which generally constitute the largest class of matrix elements, can be evaluated using bound-state structure methodology. The bound-free and free-free terms must be handled numerically. Since the methods employed in evaluating these various Hamiltonian matrix elements are quite different, it is critical that we have some way of guaranteeing a consistent set of definitions for defining and ordering CSF's and establishing phase conventions. We will discuss the bound-bound problem first.

The effective Hamiltonian defined by Eq. (28) is an (N+1)-electron operator in a product space labelled by open channel target states. Operationally, we must reduce this many-electron operator to a matrix of one-electron operators defined over the functions used to expand $F_{\Gamma\Gamma_0}$ by integrating out the N coordinates of the target. The construction of the optical potential also requires the manipulation of operators built from (N+1)-electron Q-space configurations which are not in general labeled by target channels. The Q-space configurations of the "penetration" type are built entirely from target orbitals, while the relaxation and polarization configurations can involve both target and scattering orbitals. The insurance of consistency between the N-electron and (N+1)-electron problems with respect to phase conventions, ordering of configurations etc. may be difficult with molecular structure codes which make provision for automatic generation of configurations since these conventions may not be transparent. To avoid any possible ambiguity, we determine the target CI coefficients from a pseudo-(N+1)-electron CI calculation in which a *single* orbital is used to represent the scattered electron. Let $\{\Phi_i\}$, i=1,,n denote the N-electron configurations used to expand the target eigenstates. We first set up an n×n Hamiltonian in the (N+1)-electron space where each configuration is a direct product of a Φ_i and the single square-integrable function, φ_s. However, before carrying out this CI calculation, all one- and two-electron integrals involving φ_s are set to zero. Because there is no interaction between φ_s and the remaining N electrons, the diagonalization of this Hamiltonian will produce the desired N-electron target states. This set of target vectors can serve as prototypes for building a new set of vectors for use in the full (N+1)-electron problem in which a Hamiltonian is built from repeated products of the target configurations $\{\Phi_i\}$ and all the scattering orbitals in succession, $\{\varphi_j\}$,j=1,,p. Penetration and polarization configurations, which are denoted by Θ_q, are appended to the end of this list. This prototyping scheme removes any possible inconsistencies between the N- and (N+1)-electron problems and preserves the phase conventions used in the original determination of the target states. The procedure can also be modified if we wish to take specific account of the symmetry of the various open-channel target states. In that case, we would first group the target CSF's into different symmetry classes and then use a different prototype orbital, φ_s^{Λ}, for each class in order to couple up to a specific total symmetry, Λ .

The Q-space configurations, from which the optical potential is built, consist of the penetration and polarization terms explicitly included as Θ_q, and the CI relaxation terms to which we now turn our attention. Since the P-space vectors are constructed as direct products of scattering orbitals and a *fixed* number, m, of eigenstates of an n × n Hamiltonian matrix, the complement of P-space consists of the direct product of the scattering orbitals and the remaining(n-m) target eigenstates, which are presumed to be energetically closed. However, the algorithm that we use to include the CI relaxation terms in Q-space does <u>not</u> require the determination of all of the eigenvectors of the target Hamiltonian. Rather, we employ projection operators to account for these terms and can thereby avoid the explicit construction of H_{QQ}. We express H_{QQ} in a configuration state function basis rather than in the basis of CI eigenfunctions[37]:

$$H_{QQ} = \begin{bmatrix} 1-|c_p><c_p| & 0 \\ 0 & 1 \end{bmatrix} \begin{bmatrix} H_{ij}^{kl} & H_{iq}^{k} \\ H_{iq}^{k\dagger} & H_{qq} \end{bmatrix} \begin{bmatrix} 1-|c_p><c_p| & 0 \\ 0 & 1 \end{bmatrix} \equiv \rho_Q H_{QQ}^{CSF} \rho_Q \tag{32}$$

where we have used the notation

$$H_{ij}^{kl} = \langle \Phi_i \varphi_k | H | \Phi_j \varphi_l \rangle \quad , \tag{33}$$

and c_p is a matrix built from the m open-channel eigenstates

$$c_p = \begin{bmatrix} c_1^1 \cdots c_1^m & & \\ & \ddots & 0 \\ c_n^1 \quad c_n^m & & \\ \vdots & \ddots & \\ & & c_1^1 \quad c_1^m \\ 0 & & \\ & & c_n^1 \quad c_n^m \end{bmatrix} \quad . \tag{34}$$

where the number of replications of the n×m matrix c_j^i, which specifies the target state expansion coefficients, is determined by the number of scattering orbitals.

In building the optical potential defined by Eq. (28), H_{QQ} is used to solve a set of linear equations,

$$(E - H_{QQ}) X_{QP} = H_{QP} \tag{35}$$

and

$$V_{PP}^{OPT} = H_{PQ} X_{QP} = H_{PQ} (E - H_{QQ})^{-1} H_{QP} \tag{36}$$

The linear equations, Eq. (35), are solved iteratively by successively multiplying (E-H_{QQ}) by a set of trial vectors, (X^i, i=1, ··r) and solving a small set of linear equations in the trial space until convergence is achieved. This type of algorithm has been discussed in a number of papers and a detailed description will not be given here[38]. We only note that neither the projection operator, ρ_Q, nor the Q-space Hamiltonian, H_{QQ}^{CSF} , need be explicitly constructed[37]. A direct-CI procedure can be used to perform the multiplication of $(E - H_{QQ}^{CSF})$ times the projected trial function. This is a standard procedure in most modern electronic structure packages and allows us to employ large Hamiltonian matrices in the scattering calculations[39]. In cases where Q-space is very large, the procedure just outlined for obtaining V_{opt} will be for more efficient than explicitly inverting (E-H_{QQ}).

APPROXIMATION SCHEMES

Primitive Separable Expansions

The techniques of bound-state electronic structure theory can be used to evaluate bound-bound matrix elements; the matrix elements involving continuum basis functions are more problematic and the principal difficulty of any molecular scattering calculation is their evaluation. There are two critical steps that make this problem tractable for polyatomics. The first step, which we will now discuss, uses separable approximations to eliminate an entire class of continuum exchange and optical potential matrix elements[16]. The second step, which we take up in the following section, is the development of an efficient numerical technique for evaluating the remaining continuum matrix elements.

There are several ways to achieve a separable representation of an operator. In this discussion, we confine ourselves to what is sometimes referred to as a primitive separable representation of an operator which is obtained by its projection onto a finite basis[40]:

$$H \approx \sum_{\alpha\beta} |\varphi_\alpha > H_{\alpha\beta} < \varphi_\beta| \quad . \tag{37}$$

With reference to the effective Hamiltonian defined in Eq. (28), this type of representation is invoked for the exchange components of H^{PP}, as well as V_{opt}. The basis we use for this separable representation is the set of target orbitals, along with the square-integrable functions used to expand the channel orbitals $F_{\Gamma\Gamma_0}$. To see the usefulness of such representations, consider the expansion of the channel functions $F_{\Gamma\Gamma_0}$ given in Eq. (26). If we were to add any multiple of the square-integrable basis functions φ_i to the continuum functions f_ℓ^Γ and g_ℓ^Γ, the asymptotic form of $F_{\Gamma\Gamma_0}$ would of course remain unchanged. The process simply transfers a square-integrable term from the first sum to the second. We have already mentioned that the Kohn variational equations have the property of *transfer invariance* which means that the T-matrix elements are invariant to such a transformation[11]. We have formulated the Kohn variational equations with a strong orthogonality condition between the continuum basis functions and both the scattering orbitals, φ_i, as well as the orbitals used to construct the target wave functions. (Remember, we add penetration terms to relax any constraints implied by "target orthogonality" [29,33,34].)

Now consider a continuum matrix element, for example, a free-free matrix element of the type defined in Eq. (31). It is useful to separate this integral into a direct component,

$$\begin{aligned} {}^{\text{Direct}}\left(\mathbf{M}_{00}^{PP}\right)_{\ell m \ell' m'}^{\Gamma\Gamma'} &\equiv \int \chi_\Gamma(\mathbf{r}_1..\mathbf{r}_N) f_{\ell m}^{\Gamma}(\mathbf{r}_{N+1}) \left(H_{pp} - E\right) \\ &\times \chi_{\Gamma'}(\mathbf{r}_1..\mathbf{r}_{N+1}) f_{\ell' m'}^{\Gamma'}(\mathbf{r}_{N+1}) \; d\mathbf{r}_1 \, .. d\mathbf{r}_{N+1} \end{aligned} \tag{38}$$

and an exchange component arising from the interchange of "bound" and "free" electron coordinates. The short-range character of the exchange operators allows for their expansion in a rapidly convergent series of separable terms[16,40]. In the present context, the separable expansion of exchange terms leads to a great simplification. Indeed, if the exchange operators are projected onto a finite set of (N+1)-electron configurations, Λ_m, which are made up from the entire set of target plus scattering square-integrable basis orbitals, then all free-free and bound-free exchange integrals will simply vanish. This is a consequence of the strong orthogonality condition we have imposed on the continuum functions, ie.

$$
\begin{aligned}
{}^{\text{Exchange}}\left(\mathbf{M}_{00}^{PP}\right)_{\ell m \ell' m'}^{\Gamma\Gamma'} \approx & \sum_{m,n} \left\langle A\left(\chi_\Gamma f_{\ell m}^{\Gamma}\right) \middle| \Lambda_m \right\rangle \\
& \times \left\langle \Lambda_m \middle| (H_{PP} - E) \middle| \Lambda_n \right\rangle \left\langle \Lambda_n \middle| A\left(\chi_{\Gamma'} f_{\ell' m'}^{\Gamma'}\right) \right\rangle \\
& = 0
\end{aligned}
\tag{39}
$$

since $\left\langle \Lambda_n | A\left(\chi_{\Gamma'} f_{\ell' m'}^{\Gamma'}\right) \right\rangle = 0$. The presumption is that since the exchange interactions are relatively short-ranged, the square-integrable basis can be made large enough to make this a good approximation. Similarly, matrix elements involving continuum functions and (N+1)-electron CSF's are also expanded in a square-integrable bases:

$$
\begin{aligned}
\left\langle A\left(\chi_\Gamma f_{\ell m}^{N}\right) \middle| (H-E) \middle| \Theta \right\rangle \approx & \\
& \sum_{m} \left\langle A\left(\chi_\Gamma f_{\ell m}^{\Gamma}\right) \middle| \Lambda_m \right\rangle \left\langle \Lambda_m \middle| (H-E) \middle| \Theta \right\rangle \\
& = 0
\end{aligned}
\tag{40}
$$

The consequence of these approximations is that no continuum matrix elements involving either exchange or optical potential interactions need be constructed and the entire effect of these interactions is carried in the bound-bound portions of the Hamiltonian. Whether or not this is a good approximation depends upon how many functions we include in the underlying L^2 basis and how close it comes to being complete for the purpose of representing the operators in question. In this context, one must bear in mind that the orbitals used in the determination of target wave functions must necessarily be excluded from the set of L^2 functions used to expand the channel scattering functions. There is no formal difficulty here, since these target orbitals, as we have stated, appear in appropriate (N+1)-electron "penetration" terms in Q-space. However, because the optical potential is only represented in separable form, the transference of terms from P-space to Q-space does involve an approximation. For this reason, we try to keep the target orbital space as compact as possible.

Adaptive Quadrature

We now turn our attention to the evaluation of the direct continuum integrals. Because the integrands can involve long-range multipole interactions, they are very difficult to converge by using separable expansions, so we handle these terms numerically. The first step is to write the direct integrals in terms of one-particle transition potentials. For example, we can write

$$\langle A(\chi_i f)|(H-E)|A(\chi_j g)\rangle_{direct} =$$
$$\delta_{ij}\langle f|-1/2\nabla^2 + E_i - E + V_{nuc}|g\rangle + \langle f|V_{ij}|g\rangle \tag{41}$$

where V_{ij} is the direct potential associated with the i→j transition:

$$V_{ij}(\mathbf{r}) = \sum_{q=1,N} \int \chi_i(\hat{\mathbf{r}}_1..\hat{\mathbf{r}}_N)\frac{1}{|\mathbf{r}-\mathbf{r}_q|}\chi_j(\hat{\mathbf{r}}_1..\hat{\mathbf{r}}_N)d\mathbf{r}_1...d\mathbf{r}_N \tag{42}$$

The direct integrals may thus be reduced to three-dimensional quadrature.

Numerical quadrature is only used to evaluate bound-free and free-free matrix elements of the kinetic energy, nuclear attraction and direct transition potential operators. We require a set of points and weights in the space $\vec{r} = (x,y,z)$. The prerequisite for any quadrature is that we be able to construct the integrand at points in three-space. For example, for a bound-free potential matrix element, $\langle \varphi_k|V_{nuc} + V_{ij}|g^{\Gamma'}_{\ell' m'}\rangle$, we need Gaussian basis functions which contribute to φ_k, a continuum function and a spherical harmonic evaluated at the points in $\vec{r}$-space, as well as the potential $V_{ij}(\vec{r})$. The Gaussian functions and nuclear attraction potential, V_{nuc}, are of course given by simple formulas, the spherical harmonic functions can be evaluated by efficient recursion schemes and the continuum functions by the procedures described earlier. For the moment, let us assume we also have a way to generate the potential $V_{ij}(\vec{r})$; we will return to this point momentarily. We now turn our attention to the construction of the points and weights of the grid.

Three-dimensional quadrature of the integrals typically required in molecular problems are difficult to converge because the integrands are dominated by cusps at the atomic nuclei. It is near these integrable singularities that we must use the greatest care to avoid quadrature errors. Some time ago[17], we developed an adaptive 3-D quadrature scheme to address this problem. In this scheme, we begin with a grid which is separable in spherical coordinates consisting of shells of points chosen from standard 1-D quadratures. This grid is adapted to the molecular problem by choosing a denser spacing of points in those subshells which contain the nuclei. The grid is then further refined by transforming the points under an analytic mapping which has the effect of rearranging those points in the vicinity of each nucleus so that they are drawn in closer to the nuclei. Since the mapping is analytic, the Jacobian of the transformation can be easily evaluated. This procedure has been used in most previous applications of the complex Kohn method.

More recently, we have implemented a multi-center numerical integration scheme originally proposed by Becke[18] and which is currently being used in applications of density functional theory[41]. The idea is to decompose a single molecular integral into a sum of one-center, atomic-like integrations. Specifically, we write a general integral corresponding to a molecule with N atomic centers

$$I = \int f(\mathbf{r})d^3r \tag{43}$$

$$I = \sum_{n=1}^{N} \int w_n(\mathbf{r}) f(\mathbf{r}) d^3r \equiv \sum_n I_n \tag{44}$$

where w_n is a *relative weight* function which has the property that it is unity in the vicinity of nucleus n and vanishes in a well-behaved fashion near any other nucleus. The normalization must be such that, for any **r**,

$$\sum_{n=1}^{N} w_n(\mathbf{r}) = 1 \quad . \tag{45}$$

Since the function $w_n(\mathbf{r})F(\mathbf{r})$ is well behaved near any nucleus $i \neq n$, the integral I_n should be easy to converge by numerical quadrature in a spherical polar coordinate system about the center n. Even though the integral I is decomposed into N separate integrals, the total number of quadrature points should be far less than what would be required to converge the integral in any single-centered coordinate system.

The weighting function proposed by Becke is based on a partitioning of coordinate space into the so-called Voronoi polyhedra associated with each nucleus, the ith of which is simply the locus of points closer to nucleus i than to any other nucleus. The Voronoi polyhedra have a convenient analytical definition in terms of the coordinates

$$\mu_{ij} = \frac{r_i - r_j}{R_{ij}} \tag{46}$$

where r_i and r_j denote distances from the point **r** to nuclei i and j, respectively, and R_{ij} is the internuclear separation. If we define the step function

$$s(\mu_{ij}) = \begin{cases} 1, & -1 < \mu_{ij} \leq 0 \\ 0, & 0 < \mu_{ij} \leq 1 \end{cases} \tag{47}$$

then the Voronoi polyhedron associated with nucleus i is defined by the product:

$$P_i(\mathbf{r}) = \prod_{j \neq i} s(\mu_{ij}(\mathbf{r})) \quad . \tag{48}$$

This product is unity if **r** lies within the polyhedron and vanishes outside the cell. Choosing

$$w_n(\mathbf{r}) = \frac{P_n(\mathbf{r})}{\sum_i P_i(\mathbf{r})} \tag{49}$$

clearly satisfies the normalization condition of Eq. (45) and provides a function which vanishes near all nuclei $i \neq n$; unfortunately, it does not provide a suitable function for numerical integration because of the step discontinuities at $\mu_{ij}=0$. We can get a suitable function by retaining the definitions in Eq. (49), but replacing the step function $s(\mu_{ij})$ with an analytic function that passes continuously from one to zero as μ_{ij} passes through zero. Becke proposes the function

$$s(\mu) = \tfrac{1}{2}(1 - f_q(\mu)) \tag{50}$$

where

$$\begin{aligned} f_1(\mu)&=g(\mu) \\ f_2(\mu)&=g(g(\mu)) \\ f_3(\mu)&=g(g(g(\mu))) \;,\;,\;, \end{aligned} \tag{51}$$

and

$$g(\mu) = \tfrac{3}{2}\mu - \mu^3 \quad . \tag{52}$$

Successively sharper switching functions are generated as q increases. Like Becke, we have found that q=3 generates the best results.

In summary then, the procedure we use to evaluate continuum matrix elements is to first choose a set of quadrature points independently about each atom. Or course, we can make use of any local symmetry to reduce the number of points required. In a case like H_2O, for example, we would place points about only one hydrogen atom and double their associated weights. We then sum over all points, multiplying the integrand at each point by the product of its quadrature weight and the relative weight defined by Eq. (49). We have found that this quadrature scheme provides more accurate results with fewer that half the number of points required by our original scheme.

Before closing this section, we take up the question of how to evaluate the direct transition potential defined in Eq. (42). This potential can be written in terms of a one-electron density matrix (or transition density matrix if $\Gamma \neq \Gamma'$), $\rho^{\Gamma\Gamma'}$:

$$V_{\Gamma\Gamma'}(\mathbf{r}) = \sum_{kk'} \rho_{kk'}^{\Gamma\Gamma'} \int \frac{\varphi_k(\mathbf{r}') \quad \varphi_{k'}(\mathbf{r}')d^3\mathbf{r}'}{|\mathbf{r} - \mathbf{r}'|} \tag{53}$$

The density matrices are constructed with the target state vectors obtained from the pseudo-(N+1)-electron CI previously described, to assure consistency between the numerically and analytically obtained matrix elements. We will drop the $\Gamma\Gamma'$ notation below with the understanding that we are considering one such potential. The bound functions, φ_k, appearing in Eq. (53) are simply sums of Gaussians so the integrals appearing in this equation are identical to nuclear attraction integrals with a nucleus at the position $\bar{r}$. These integrals are one-electron integrals which appear in any quantum chemistry computer code. An algorithm can be constructed which computes these for the $\vec{r}$ points in the quadrature grid. Although this algorithm can easily be constructed to make use of vectoriztion on vector architecture supercomputers and provides an exact evaluation of the potential, it can become quite time consuming , especially for large molecules, since it can typically require the computation of millions of one-electron integrals.

Alternatively, we can opt for a numerical determination of the direct potential. We begin by decomposing the density with the local weight function defined above:

$$\begin{aligned}\rho(\mathbf{r}) &\equiv \sum_{\alpha,\beta} \rho_{\alpha\beta}\varphi_\alpha(\mathbf{r})\varphi_\beta(\mathbf{r}) = \sum_{n} w_n(\mathbf{r})\rho(\mathbf{r}) \\ &\equiv \sum_{n} \rho_n(\mathbf{r})\end{aligned} \tag{54}$$

We next expand each component, ρ_n, of the density in spherical harmonics about its respective center:

$$\rho_n(\mathbf{r}) \equiv \sum_{\ell,m} Y_{\ell m}(\hat{\mathbf{r}}_n)\rho_n^{\ell m}(r_n) \tag{55}$$

where

$$\rho_n^{\ell m}(r) = \int d\hat{\mathbf{r}} Y_{\ell m}^*(\hat{\mathbf{r}})\rho_n(\mathbf{r}) \tag{56}$$

We use the index n on the angular and radial variables in Eq. (55) to underscore the fact that these quantities are defined in the local coordinate system attached to atom n. By using the familiar expansion of $1/|\mathbf{r}-\mathbf{r}'|$ in spherical harmonics[42], the direct potential defined by Eq. (53) is expressed as

$$V(\mathbf{r}) = \sum_{n}\sum_{\ell,m} \frac{4\pi}{2\ell+1} Y_{\ell m}(\hat{\mathbf{r}}_n) f_n^{\ell m}(r_n) \tag{57}$$

with

$$f_n^{\ell m}(r) = \left[\frac{1}{r^{\ell+1}}\int_0^r r'^{\ell} + r^{\ell}\int_r^\infty \frac{1}{r'^{\ell+1}}\right]\rho_n^{\ell m}(r')r'^2 dr' \tag{58}$$

The essential point to note is that the decompostion of the density into atomic components via Eq. (54) leads to very rapid convergence of the spherical harmonic expansions. This is in marked contrast to a single-center expansion of the potential which, for an arbitrary polyatomic molecule, can be quite slowly convergent.

The radial components of the density, $\rho_n^{\ell m}(r)$, are obtained by an angular quadrature on a mesh of radial points and the functions $f_n^{\ell m}(r)$ are generated at the mesh points using, for example, Simpson's rule integration. The spherical harmonics can be analytically evaluated in any coordinate system by recursion. Since the function $f_n^{\ell m}(r)$ is only computed on a mesh of radial points with respect to its own center, it must be *interpolated* to provide values at the radial points with respect to other centers. We have found that this procedure can provide substantial computational speed over analytic evaluation of the direct potentials. We have also found this approach to offer significant advantages over a technique described by Becke and Dickson based on numerical solution of Poisson's equation, which is simply the differential form of Eq. (53)[43].

Pseudoresonances and Intermediate Energy Scattering

The foregoing discussion, which has concentrated on the mechanics of how the variational equations are constructed and solved, shows that the complex Kohn formulation admits the use of multi-configuration target wave functions and large-scale optical potentials and that modern electronic structure techniques can be used to carry out most of the manipulations required quite efficiently. We now wish to turn our attention to the physical considerations of the underlying approximations that determine the partitioning of P-space and Q-space terms and the criteria we use for selecting the appropriate configurations to include in the trial wave function.

In applying the close-coupling formalism, we are confronted with two problems. The first problem is that, with the exception of single-electron atoms and ions, exact target states are not known and approximate wave functions must be used. These are generally expressed as linear combinations of CSF's built from a specified list of "target" orbitals. Moreover, since it is not practical to employ non-orthogonal functions in many-electron applications, we have imposed a strong orthogonality constraint between the scattering functions and the target orbitals[44-46]. It is well known that this constraint can lead to physically incorrect results if it is not relaxed, as was seen in early calculations on electron impact dissociation of H_2[47,48]. The standard remedy for removing this constraint, which is to include all (N+1)-electron penetration terms that arise by taking the direct product of the target orbitals and the various open-shell N-electron target configurations, can lead to an enormous number of terms as the expansion of the target wave functions is increased. In addition, anomolous pseudoresonances can arise at intermediate energies as a consequence of these terms[32].

Second, there is the obvious difficulty that, as the incident energy in a collision is increased toward the ionization energy, the number of open channels approaches infinity. Practical considerations thus generally limit the number of target states that can be included in a close-coupling calculation. Even if one is only interested in cross sections for excitation to low-lying excited states, these cross sections may be needed at energies well above the thresholds of other (bound and continuum) states. It is, therefore, natural to inquire about the effect of neglected channels at intermediate collision energies and to examine whether there is anything about the close-coupling formulation that might lead to pathological results. As we shall see, both of the difficulties just mentioned can lead to problems that are symptomatic of the same disease.

We can illustrate the kind of problem that may arise with the simple example of elastic scattering of an electron from H_2 in $^2\Sigma_g^+$ symmetry. If we use a simple configuration-interaction wave function of the form ($C_1\ 1\sigma_g^2 + C_2\ 1\sigma_u^2$), with $C_1 >> C_2$, for the ground-state $X^1\Sigma_g^+$ wave function, the formalism we have outlined would dictate that we include a single penetration term $1\sigma_u^2\ 1\sigma_g$ in the trial wave function to relax the orthogonality constraint on the scattering wave function with respect to the $1\sigma_g$ orbital. The elastic cross section so calculated is shown in Fig. 1. There is evidently a broad resonance near 16 eV. Elastic scattering cross sections obtained with a simple self-consistent field (SCF) target wave

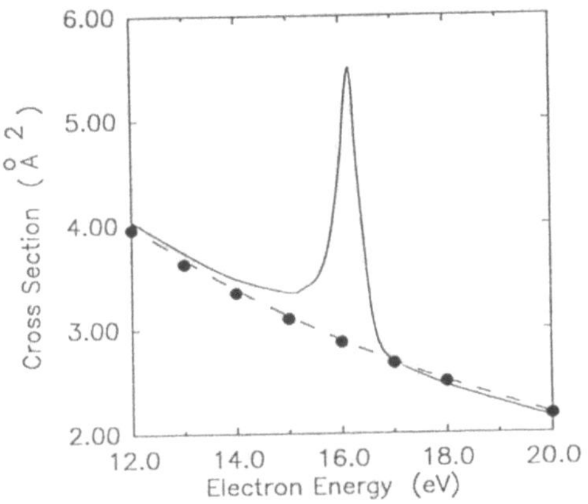

Figure 1. $e^- + H_2\ \Sigma_g^+$ elastic cross section. Solid curve: static-exchange plus penetration term; broken curve: static-exchange without penetration term; solid circles: 2-channel calculation including penetration term.

function, which differs only slightly from the target CI wave function we used, show no such structure. Moreover, if the calculation is performed without the $1\sigma_u^2\ 1\sigma_g$ term, the resonance disappears. This result is also shown in Fig. 1.

The $1\sigma_u^2\ 1\sigma_g$ term can be thought of as a doubly excited negative ion term whose parent is the ($1\sigma_g\ 1\sigma_u$), $b^3\Sigma_u^+$ state of H_2. This state is physically open at energies above 10 eV, but is not properly included in the single channel calculation. To demonstrate this, we performed a third calculation in which the $X^1\Sigma_g^+$ and $b\ ^3\Sigma_u^+$ states were both included in the open channel portion of the variational trial wave function and the single penetration term was again retained in Q-space. The resulting elastic cross section, which is shown in Figure 1, is evidently quite close to the smooth result obtained from the single channel calculations in which the penetration term was dropped. This simple example serves to show how excited states, which are not explicitly included in the close-coupling expansion, can enter the problem in an indirect way when correlated target wave functions are used. In this simple case, it is possible to easily identify the cause of the trouble and remove the term responsible for this behavior from the single-channel trial function.

Another thing to note about the formalism as outlined above is that, in the case of multi-configuration target states, it provides more flexibility than the minimum needed to compensate for any orthogonality constraints, since the coefficients of the *individual* penetration terms are chosen variationally. However, all that is required to remove the target

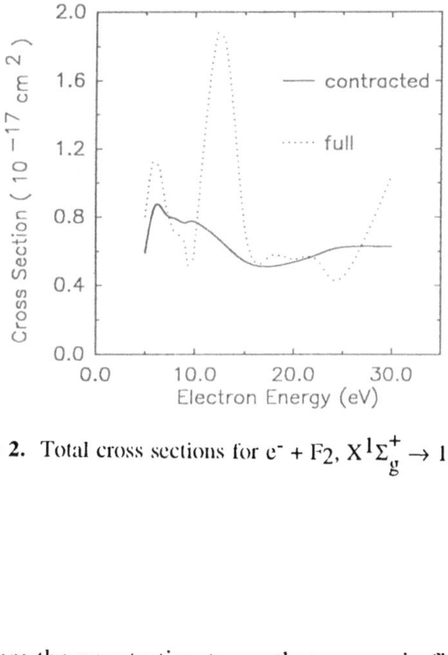

Figure 2. Total cross sections for $e^- + F_2$, $X^1\Sigma_g^+ \rightarrow 1^3\Pi_u$.

orthogonality constraints are the penetration terms that appear in fixed linear combinations that are consistent with the approximate target states that are being employed. We have shown elsewhere that it is possible to develop an algorithm for incorporating the penetration terms in these fixed linear combinations[32]. When elaborate CI target wave functions are used, this procedure can exclude many of the individual penetration terms built from N-electron configurations that contribute weakly to the target states explicitly chosen. Moreover, this contraction of the Hilbert space component of the total wave function makes it easier to identify and eliminate the terms in the trial wave function that will give rise to pseudo-resonances. To illustrate the power of this new formalism, we show in Fig. 2 integral cross sections for electron impact excitation of the dissociative $1^3\Pi_u$ state of F_2[32]. The cross sections were obtained from three-state ($X^1\Sigma_g^+$, $1^3\Pi_u$, and $1^3\Pi_u$) close-coupling calculations using correlated target states and illustrate the unphysical pseudoresonances that occur as a result of including uncontracted penetration terms in the trial function. An approximate, but simpler, alternative to the procedure just outlined is to only include those penetration terms that arise from the dominant configuration for each target state[49-51]. Either of these techniques will allow one to meaningfully apply the close-coupling method at intermediate energies even when elaborate multi-reference target states are employed.

A note about the target orbitals bears further comment. The previous discussion has presupposed the existence on an orbital set in terms of which the target wave functions have a compact representation so that it is easy to identify a principal configuration for each target state. For this reason, we have emphasized the use of averaged natural orbital techniques in many of our studies[30,32,49-52], since this often provides an ideal way to achieve a compact representation for a number of excited states without overly compromising on the accuracy.

APPLICATIONS AND EXTENSIONS

Off-Shell Extension to Threshold Vibrational Excitation

Much of the theoretical literature on electron-molecule scattering has focused on the development of *ab initio* methods for solving the electronic fixed-nuclei problem; this review is no exception. The complex Kohn technique, like other sophisticated methods presently available for tackling the low-energy electron-molecule scattering problem, has been implemented within the framework of the adiabatic-nuclei approximation[53]. In many cases, meaningful cross sections can be extracted solely from fixed-nuclei calculations, while in other cases the fixed-nuclei electronic problem can provide the input necessary to solve the full problem including the nuclear dynamics[22]. However, there are many interesting cases where the adiabatic-nuclei approximation breaks down.

The assumption that underlies the adiabatic-nuclei approximation is that the target molecule is effectively fixed during the course of the collision; this approximation consequently provides an accurate description of electron-molecule scattering only when the collision time is short compared with the rotational and vibrational periods of the molecule. This assumption of course breaks down in near-threshold regions where cross sections computed with the adiabatic-nuclei approximation do not vanish at the appropriate thresholds[54]. The breakdown of the adiabatic-nuclei approximation can be particularly severe for light molecules with large vibrational energy spacings (eg. ~.5 eV in H_2).

Laboratory-frame close-coupling, in which the total wave function is expanded in terms of the rotational, vibrational and electronic states of the target along with angular momentum states of the projectile[55], provides the most rigorous approach to electron-molecule collisions. Unfortunately, it becomes unwieldy for all but the simplest targets and, to date, has proved to be impossible for systems other than $e^- + H_2$. Body-frame vibrational close-coupling[56], in which only the rotational motion is treated adiabatically, is somewhat easier to apply; however, even this approach has relied on model potentials for all or part of the electron-target interaction and has yet to be carried out in a completely *ab initio* context.

In 1975, Shugard and Hazi[57], in a paper that examined the underpinnings of the adiabatic-nuclei approximation, suggested that an off-the-energy-shell generalization of the T-matrix (scattering amplitude) could extend the range of validity of the adiabatic-nuclei approximation while retaining much of its inherent simplicity. Their basic idea was to employ the approximate expressions for scattering amplitudes given by the adiabatic-nuclei approximation, but to use the proper energies in the entrance and exit channels. A numerical verification of soundness of these early ideas was provided some years later by Morrison and coworkers[58] in calculations of $e^- + H_2$ vibrational excitation carried out with model potentials. The off-shell treatment, which they termed the first-order nondegenerate adiabatic (FONDA) approximation, gave cross sections that were far more accurate than those given by conventional adiabatic-nuclei theory, and almost identical to those obtained from a full close-coupling treatment.

The complex Kohn method can be modified to provide a computational tool for variational evaluation of the off-shell T-matrix in a completely general *ab initio* context. We begin by noting that the T-matrix, whose elements we have defined by the asymptotic behavior of a trial wave function, also has a formal definition[59] as the matrix element

$$T^{el}{}_{i\to f} \;=\; \left\langle \Psi'_o \middle| V \middle| \Psi^+ \right\rangle \tag{59}$$

where Ψ^+ is the full scattering wave function for the N+1 electron system, Ψ'_o is the corresponding unperturbed wave function and V is that part of the full Hamiltonian that describes the electron-target interaction. We wish to evaluate this quantity "off the energy shell" and so the prime on Ψ'_o indicates that it is not evaluated at the same total energy as is Ψ^+.

Ψ^+ is of course antisymmetric under interchange of any pair of electron coordinates, but Ψ'_o is not and V is not a symmetric operator. Hence, Eq. (59) is not a convenient expression for many-body applications. We can of course also write the T-matrix as

$$T^{el}{}_{i\to f} \;=\; \left\langle \Psi^- \middle| H - E' \middle| \Psi'_o \right\rangle \tag{60}$$

since

$$(H_o - E')\Psi'_o \;=\; 0 \tag{61}$$

where H_0 is the unperturbed Hamiltonian and $H = H_0 + V$. Since $H - E'$ is a symmetric operator, we can now antisymmetrize Ψ'_o in Eq. (60) without changing the resulting value of $T^{el}{}_{i\to f}$.

It is possible to develop a variational principle for $T^{el}{}_{i\to f}$ by following the procedure given by Gerjouy, Rau and Spruch[60]. Defining $W \equiv (H - E')\Psi'_o$, a stationary principle for $T^{el}{}_{i\to f}$ is given by:

$$T^{s}{}_{i\to f} \;=\; \langle \Psi^{-t} | W \rangle + \langle f^{t} | H - E | \Psi^{+t} \rangle \tag{62}$$

where the auxiliary trial function f^t satisfies the equation

$$(H - E)|f^{t}\rangle + |W\rangle \;=\; 0 \tag{63}$$

along with the boundary condition

$$\langle f^{t} | H - E | \delta\Psi^{+t} \rangle - \langle \delta\Psi^{-t} | H - E | f^{t} \rangle = 0 \tag{64}$$

where

$$\delta\Psi^{\pm t} \;=\; \Psi^{\pm t} - \Psi^{\pm}. \tag{65}$$

We now choose a trial scattering wave function of the same form given in Eqs. (25-26) but, for notational simplicity, we assume there is only one electronically open channel:

$$\Psi^{+} = A\left(\chi\, F_{\Gamma_0}\right) + \sum_{\mu} d_{\mu}^{\Gamma_0} \Theta_{\mu} \quad , \tag{66}$$

with

$$F_{\Gamma_0} = \sum_{i} c_i\, \varphi_i + \sum_{\ell m} [f_{\ell m}(k\, r)\delta_{\ell\ell_0}\, \delta_{mm_0} + T^{t}_{\ell\ell_0 mm_0} g_{\ell m}(k_\Gamma r)] \tag{67}$$

where the $T^{t}_{\ell\ell_0 mm_0}$ are now to be regarded simply as a set of trial parameters. The boundary condition specified by Eq. (64) is satisfied by expanding f^t as[61]

$$f^{t} = \sum_{i} a_i\, \varphi_i + \sum_{\ell m} [b_{\ell\ell_0 mm_0} g_{\ell m}(k_\Gamma r)] \tag{68}$$

The derivation proceeds along the same lines we outlined in our earlier discussion of potential scattering; by carrying out the variation with respect to the linear coefficients in Eqs. (66) and (67), we get a stationary expression for the off-shell T-matrix elements:

$$T^{s}_{\ell\ell_0 m m_0} = \left\langle A\chi f_{\ell m} \middle| H - E' \middle| A\chi j'_{\ell_0 m_0} \right\rangle - \sum_{\alpha,\beta} \left\langle A\chi f_{\ell m} \middle| H - E \middle| \alpha \right\rangle \left\langle \alpha \middle| H - E \middle| \beta \right\rangle^{-1} \left\langle \beta \middle| H - E' \middle| A\chi j'_{\ell_0 m_0} \right\rangle \tag{69}$$

where α and β refer to the space spanned by the functions $A\chi\varphi_i$ and $A\chi g_{\ell m}$. We have again adopted the convention of using g_{lm} to denote the product of a radial function and Y_{lm}/r. Note that the term $j'_{\ell_0 m_0}$ on the RHS of Eq. (69) arises from the partial wave expansion of the plane-wave part of the unperturbed wave function Ψ'_0.

The variational expression for the off-shell T-matrix, Eq. (69), is clearly similar to the result we developed for the more familiar on-shell result, Eq. (30), but there are some key differences to note. Aside from the obvious fact that the continuum functions f_ℓ (and g_ℓ) are evaluated at a momentum that differs from the value used in j'_ℓ, there is a more subtle difference. The on-shell T-matrix has the property of *transfer invariance* [11], which means that it is not changed by any transformation which mixes an arbitrary component of the square-integrable functions φ_i into the continuum functions. Thus, in the on-shell case, the function j_ℓ could be replaced by f_ℓ without changing the result (recall that we constructed f_ℓ to be orthogonal to all the square-integrable basis functions). In the on-shell case, we used this fact to eliminate an entire class of bound-free and free-free exchange matrix elements[16] by using separable approximations of the form:

$$H^{ex} \approx \sum_{i,j} |i\rangle\langle i|H^{ex}|j\rangle\langle j| \quad . \tag{70}$$

The off-shell T-matrix, by contrast, is only transfer invariant with respect to the functions f_ℓ and g_ℓ, which appear in the expansion of the trial wave function Ψ^{+t}, but not the functions j_ℓ. The latter cannot be arbitrarily orthogonalized to the functions φ_i. Thus, continuum matrix elements involving j_ℓ can be simplified by using separable expansions, but not eliminated entirely.

To compute an amplitude for vibrational (or rotational) excitation, we evaluate the off-shell T-matrix for different nuclear geometries and use the adiabatic approximation [62]:

$$T(k_0 \to k_f) = \int dR\, v_0(R)\, T^{el}_{0\to f}(R)\, v_f(R) \tag{71}$$

where v_0 and v_f are the initial and final nuclear target wave functions and R denotes the appropriate nuclear coordinates. The wave vectors k_0 and k_f are simply chosen to conserve the total energy in the initial and final channels. Thus, aside from the use of an off-shell electronic T-matrix, the procedures for computing excitation cross sections are identical to those used in the adiabatic-nuclei approximation.

We carried out an *ab initio* treatment of low-energy $e^- + H_2$ scattering[63], using the off-shell formalism just described, in an effort to help resolve a longstanding dilemma posed by the discrepancy between crossed-beam measurements of the vibrational excitation cross section of H_2[64] and valued derived from an analysis of swarm data[65]. Previous theoretical treatments of this problem were based on a Hartree-Fock wave function for the target H_2 molecule and correlation effects were modelled with a local potential[66]. Because the vibrational excitation cross section is sensitive to the R-dependence of the electronic amplitudes, as well as to the dynamic distortion of the target molecule, we used a two-term $\left(c_1\, 1\sigma_g^2 + c_2\, 1\sigma_u^2\right)$ target wave function with natural orbitals extracted from a configuration-interaction expansion. The dynamic distortion of the target was treated by including in the Kohn trial function energetically closed electronic channels in the form of direct products of a bound orbital and a two-electron state formed by singly exciting the ground state target wave function into a large set of virtual orbitals.

Fig. 3 compares our calculated result for the integrated vibrational excitation cross section with previous theory[66] as well as available experimental data[64,65,67,68]. Our results are clearly in better agreement with the crossed-beam data than with the swarm derived cross sections. The agreement with the most recent experiments of Brunger et al.[64] is particularly good. There is also excellent agreement between our result and the theoretical results of Buckman et al.[66] (which used a model polarization potential in a full body-frame vibrational close coupling treatment) in the energy range below 1.6 eV, although our calculated cross section is somewhat smaller at higher energies. It is noteworthy that when we carried out our calculations with an uncorrelated target wave function, we obtain a result which is closer to that of Buckman et al. , which also used an SCF target wave function.

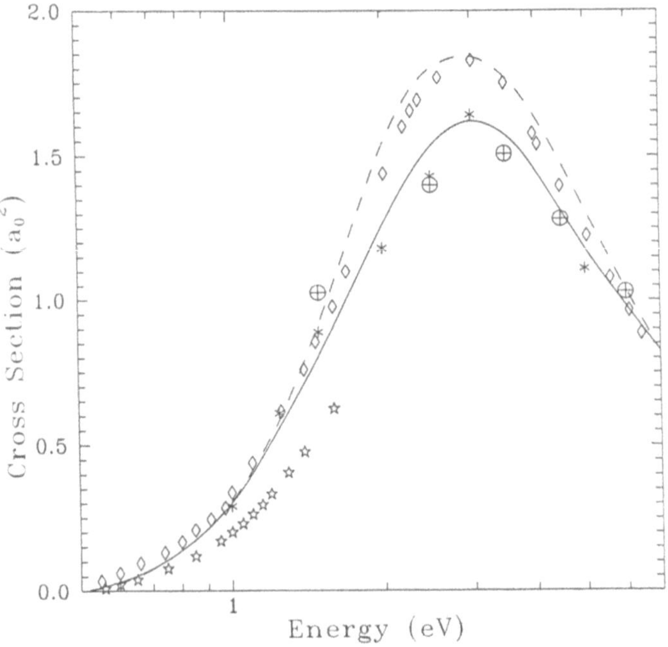

Figure 3. Integral cross sections for 0→1 vibrational excitation. Solid curve, off-shell Kohn result (ref. 63); dashed curve, theoretical result of Buckman et al. (re. 66); asterisks, Brunger et al. (ref. 64); circles, Linder and Schmidt (ref.67); diamonds, Ehrhardt et al. (ref. 68); stars, swarm results (ref. 65).

The vibrational excitation cross section is dominated by a broad, negative ion shape resonance of $^2\Sigma_u^+$ symmetry which shows a strong dependence on internuclear separation. The $^2\Sigma_u^+$ shape resonance is evidently softened when a correlated target function is employed, resulting in a smaller cross section. Our calculations, coupled with previous theoretical work, strongly support the validity of the crossed-beam measurements of vibrational excitation in H_2.

In cases where the threshold vibrational excitation cross section displays no anomolous behavior, the off-shell T-matrix formalism, which correctly conserves energy in the entrance and exit channels, provides a simple way of extending the validity of conventional adiabatic-nuclei theory. There are other cases, however, where vibrational excitation cross sections display prominent threshold features that cannot be explained on the basis of simple threshold laws[54]. Such is the case with the hydrogen halides[69], where the sharp threshold peaks that have been the subject of much theoretical study over the years[54] evidently correspond to nuclear-excited Feshbach resonances whose proper description requires more than a simple modification of adiabatic-nuclei theory[70]. It is interesting to question whether the simple off-shell theory is ever applicable in situations where threshold features are present. The CH_4 molecule provides an interesting example since it possesses no permanent dipole moment, but nevertheless displays vibrational excitation cross sections that rise steeply from threshold[71,72]. We have recently found that the present formalism can provide a qualitatively accurate description of the observed effects[73].

The first thing to note about e^-+CH_4 is that the *elastic* cross section shows a pronounced Ramsauer-Townsend minimum near .4 eV[74], a fact which, coupled with the absence of a stable negative ion, is consistent with the presence of a virtual state[54]. Accurate *ab initio* calculations carried out at the equilibrium geometry have reproduced the observed elastic cross section in the energy region of the Ramsauer-Townsend minimum[31]. One might expect the position of this minimum to be sensitive to nuclear geometry, in which case the off-shell T-matrix formalism, which is guaranteed to give a vibrational excitation cross section that vanishes at threshold, could give a qualitatively correct description of the observed results.

We have carried out complex Kohn calculations on this system at selected geometries using the off-shell formalism previously outlined. The methane molecule has nine normal modes of which, due to symmetry considerations, only four are energetically different. We studied the non-degenerate, Raman-active ν_1(symmetric stretch) and triply-degenerate, infrared-active ν_3 modes, which have threshold energies of 362 and 374 meV, respectively. The electron-energy-loss peaks corresponding to these transitions are not resolved in current experiments. The trial wave functions were constructed along the same lines that we found to be successful in our earlier studies of elastic scattering[31]. We used an SCF description of the ground-state target wave function and included single excitations of the target into a compact space of polarized orbitals.

For the symmetric stretch mode, the "fixed-nuclei" elastic cross sections computed from the off-shell amplitudes are plotted in Fig. 4 at five different values of the C-H bond length. These cross sections rise from the 362 meV threshold with vertical onset that is characteristic of an s-wave inelastic process. For the larger C-H bond values, the cross sections rise to a prominent peak before decreasing through a Ramsauer-Townsend minimum. This behavior is less pronounced for the shorter C-H values since the RT minimum moves below the vibrational threshold energy as the CH bond is squeezed below its equilibrium value. We also carried out variational calculations for several values of the normal coordinate corresponding to one of the three degenerate infrared-active ν_3 modes.

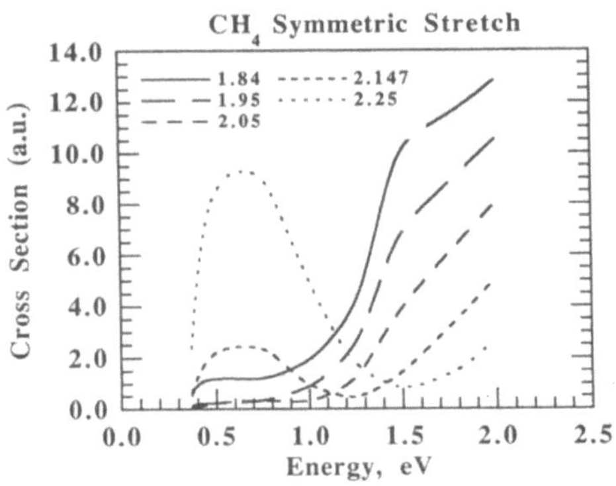

Figure 4. Fixed-nuclei elastic cross sections computed from the off-shell T-matrix for different C-H bond lengths along the symmetric stretch mode. The equilibrium CH distance in 2.05 Bohr.

Fig. 5 shows both the ν_1 and ν_3 contributions to the total vibrational excitation cross section, which were obtained by using off-shell amplitudes in Eq. (71) along with harmonic oscillator functions for the vibrational wave functions. It is interesting to note that the minimum at 1.25 eV, which reflects the geometry dependence of the fixed-nuclei Ramsauer-Townsend minimum, is less pronounced in the ν_3 cross section. This is to be expected, since the RT minimum is basically an s-wave phenomenon. The ν_3 mode, however, is infrared-active, which means that there is a transition dipole associated with motion in this mode. There is consequently a coupling between s-waves and p-waves that occurs in this mode which is not present in the symmetric-stretch amplitude. This coupling tends to wash out the effect of the RT minimum.

Polar Molecules

There are several complications one encounters when studying electron scattering by polar molecules. Because of the long-range character of the underlying electron-dipole interaction, which behaves asymptotically as $\mathbf{D} \cdot \mathbf{r} / r^3$, where $\mathbf{D}$ is the permanent dipole moment of the target, it turns out that many partial-wave components are needed to properly describe scattering for small deflection angles of the incident electron and to obtain converged total cross sections. One also has the additional *formal* difficulty that a fixed-nuclei treatment of electron scattering by a polar molecule gives infinite total cross sections and differential cross sections that diverge at zero scattering angle, which is a consequence of ignoring the

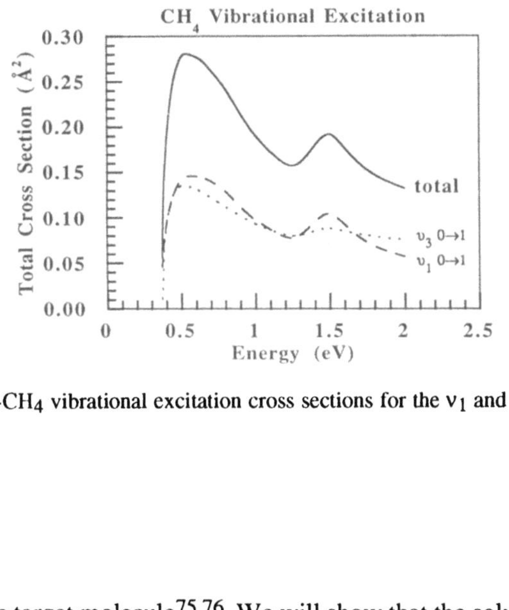

Figure 5. Electron-CH_4 vibrational excitation cross sections for the ν_1 and ν_3 normal modes.

rotational motion of the target molecule[75,76]. We will show that the solution to *both* problems is to exploit the fact that beyond a certain ℓ cutoff the partial-wave components of the scattering functions do not penetrate the interior part of the molecular target. This allows us to use a hybrid treatment in which only the low order partial-wave components of the T-matrix are determined from variational calculations and the higher order terms are included in the Born approximation via a closure formula. In the course of this discussion, we will also describe the numerical techniques we use to calculate differential cross sections, which differ from the procedures commonly used.

Let us first consider the case of a *non-polar* target. When the incident electron energy is large compared to the rotational energy spacings of the molecule, one is usually justified in ignoring the rotational motion of the target and treating the scattering as that produced by an ensemble of randomly oriented, motionless targets[77]. Under this assumption, the differential cross section for exciting a molecule from initial electronic state Γ and vibrational state ν to final states Γ',ν' can be expressed as:

$$\frac{d\sigma}{d\omega}(\Gamma\nu \rightarrow \Gamma'\nu') = \frac{(4\pi)^2}{k_\Gamma^2}\int \frac{d\alpha\, d\cos\beta d\gamma}{8\pi^2} |\langle \mathbf{k}_{\Gamma'}\nu'|\mathbf{T}^{\Gamma\Gamma'}|\mathbf{k}_\Gamma \nu\rangle|^2 \tag{72}$$

where α,β and γ are the three Euler angles that orient the initial and final wave vectors $\mathbf{k}_\Gamma$ and $\mathbf{k}_{\Gamma'}$ with respect to the target. The laboratory scattering angle ω is the angle between $\mathbf{k}_\Gamma$ and $\mathbf{k}_{\Gamma'}$. The body-frame **T**-matrix has the expansion:

$$\langle \mathbf{k}_{\Gamma'}\nu'|\mathbf{T}^{\Gamma\Gamma'}|\mathbf{k}_\Gamma \nu\rangle = \sum_{\substack{\ell\ell' \\ mm'}} i^{\ell-\ell'} Y_{\ell m}(\mathbf{k}_\Gamma)\, Y^*_{\ell' m'}(\mathbf{k}_{\Gamma'}) \times \int \chi_\nu(\mathbf{s}) T^{\Gamma\Gamma'}_{\ell m \ell' m'}(\mathbf{s}) \chi_{\nu'}(\mathbf{s}) d\mathbf{s} \tag{73}$$

where we have used **s** to denote the internal vibrational coordinates of the target and the vibrational wave functions are χ_ν and $\chi_{\nu'}$. If we are not interested in the individual vibrational levels and simply wish total cross sections associated with a particular electronic transition, we can sum Eq. (72) over the final vibrational levels of the target. If we assume that the electron kinetic energy is large compared to the spacing between vibrational levels, then we can use closure,

$$\sum_{\nu'} \chi_{\nu'}(\mathbf{s})\chi_{\nu'}(\mathbf{s}') = \delta(\mathbf{s}-\mathbf{s}') \tag{74}$$

to obtain

$$\frac{d\sigma}{d\omega}(\Gamma \rightarrow \Gamma')_{total} = \int \chi_\nu(\mathbf{s}) \frac{d\sigma^{\Gamma\Gamma'}}{d\omega}(\mathbf{s}) \chi_\nu(\mathbf{s}) d\mathbf{s} \tag{75}$$

Thus, the vibrationally summed cross section is expressed as the expectation value of the fixed-nuclei cross section with respect to the initial vibrational state of the target. One frequently approximates this quantity by the value of the fixed-nuclei cross section obtained at the equilibrium geometry of the target.

The usual procedure for carrying out the angular average required in Eq. (72) is to use Wigner rotation matrices to express the spherical harmonics that appear in Eq. (73) in laboratory frame coordinates and then carry out the averaging analytically[77]. Alternatively, we can carry out the integrations in Eq. (72) by numerical quadrature, keeping the angle between $\mathbf{k}_\Gamma$ and $\mathbf{k}_{\Gamma'}$ fixed. Starting with an initial pair of wave vectors $\mathbf{k}_{\Gamma_0}$ and $\mathbf{k}_{\Gamma'_0}$ in the body frame, we can express the wave vectors needed to evaluate the integrand for any set of Euler angles, from the expression[78]:

$$\mathbf{k}_\Gamma(\alpha,\beta,\gamma) = \begin{bmatrix} \cos(\alpha)\cos(\beta)\cos(\gamma)-\sin(\alpha)\sin(\gamma) & -\cos(\alpha)\cos(\beta)\sin(\gamma)-\sin(\alpha)\cos(\gamma) & \cos(\alpha)\sin(\beta) \\ \sin(\alpha)\cos(\beta)\cos(\gamma)+\cos(\alpha)\sin(\gamma) & -\sin(\alpha)\cos(\beta)\sin(\gamma)+\cos(\alpha)\cos(\gamma) & \sin(\alpha)\sin(\beta) \\ -\sin(\beta)\cos(\gamma) & \sin(\beta)\sin(\gamma) & \cos(\beta) \end{bmatrix} \mathbf{k}_{\Gamma 0} \tag{76}$$

This numerical approach allows us to avoid the complex angular momentum algebra necessitated by an analytic formulation and substantially reduces the complexity of the coding required. It also makes it fairly easy to carry out the modifications of the fixed-nuclei approach that are necessary in order to study electron collisions with *polar* molecules.

For elastic scattering by polar targets, a useful strategy is to expand the *difference* between the desired T-matrix and its value in the first Born approximation as a rapidly convergent partial-wave series and then to use a completion formula to obtain the final result. (Note that similar considerations apply to the case of inelastic, dipole-allowed allowed transitions.) The partial-wave Born approximation for the *electronically* elastic ($\Gamma \rightarrow \Gamma$) T-matrix element is given by

$$^{B}T^{\Gamma\Gamma}_{\ell m \ell' m'} = 2\sqrt{k_\Gamma k_\Gamma} \int j_{\ell'}(k_{\Gamma'} r)\, Y^*_{\ell' m'}(\hat{\mathbf{r}}) (\chi_\Gamma H \chi_\Gamma)\, j_\ell(k_\Gamma \mathbf{r})\, Y_{\ell m}(\hat{\mathbf{r}})\, d^3\mathbf{r} \tag{77}$$

Consistent with this approximation is the replacement of $(\chi_\Gamma H \chi_\Gamma)$ by its asymptotic form[79]:

$$\begin{aligned} V_{\Gamma\Gamma}(\mathbf{r}) &= (\chi_\Gamma H \chi_\Gamma) \\ &\equiv \int \chi_\Gamma(\mathbf{r}_1,,\mathbf{r}_N)\, H(\mathbf{r},\mathbf{r}_1,,\mathbf{r}_N)\, \chi_{\Gamma'}(\mathbf{r}_1,,\mathbf{r}_N) \\ &\sim \mathbf{D}\cdot\hat{\mathbf{r}}/r^2 \end{aligned} \tag{78}$$

where $\mathbf{D}$ is the dipole moment and $\hat{\mathbf{r}}$ is a unit vector in the body frame of the molecule. With this limiting form of the potential, the first Born approximation gives extremely simple formulas for scattering amplitudes and cross sections that can serve as a useful starting point for a more elaborate treatment of the dynamics. For example, the full body-frame, fixed-nuclei T-matrix is given as[75]:

$$\begin{aligned} \left\langle k_{\Gamma'} \middle| {}^{B}T^{\Gamma\Gamma} \middle| k_\Gamma \right\rangle &= -\frac{\sqrt{k_\Gamma k_{\Gamma'}}}{8\pi^2} \int e^{-i(\mathbf{k}_\Gamma - \mathbf{k}_{\Gamma'})\cdot\mathbf{r}} \frac{\mathbf{D}\cdot\hat{\mathbf{r}}}{r^2}\, d^3r \\ &= -\frac{\sqrt{k_\Gamma k_{\Gamma'}}\; i\,\mathbf{D}\cdot\widehat{\mathbf{k}'}}{2\pi\left|\mathbf{k}_\Gamma - \mathbf{k}_{\Gamma'}\right|} \end{aligned} \tag{79}$$

where $\widehat{\mathbf{k}'} = (\mathbf{k}_{\Gamma'} - \mathbf{k}_\Gamma)/|\mathbf{k}_{\Gamma'} - \mathbf{k}_\Gamma|$. We now write the partial wave expansion of the exact body frame T-matrix in the following form:

$$\left\langle \mathbf{k}_\Gamma \middle| T^{\Gamma\Gamma} \middle| \mathbf{k}_{\Gamma'} \right\rangle = -\frac{\sqrt{k_\Gamma k_{\Gamma'}}\, i\, \mathbf{D}\cdot \hat{\mathbf{k}}'}{2\pi^2 \left| \mathbf{k}_\Gamma - \mathbf{k}_{\Gamma'} \right|}$$

$$+ \sum_{\substack{\ell'\ell \\ mm'}} i^{\ell-\ell'} Y^*_{\ell' m'}\left(\hat{\mathbf{k}}_{\Gamma'}\right) Y_{\ell m}\left(\hat{\mathbf{k}}_\Gamma\right)\left(T^{\Gamma\Gamma}_{\ell m \ell' m'} - BT^{\Gamma\Gamma}_{\ell m \ell' m'}\right) \tag{80}$$

where we have merely added and subtracted the Born approximation to the partial wave expansion of the exact T-matrix. The point to note is that the sum in Eq. (80) now contains differences between exact and Born partial wave elements, which rapidly approach zero as $\ell(\ell')$ get large. To obtain converged cross sections, we need only compute variational T-matrix elements to high enough ℓ where the Born approximation is accurate. To obtain differential cross sections (away from zero scattering angle) corresponding to a random orientation of the target molecule with respect to an incident electron beam in the laboratory frame, we simply substitute the amplitude defined in Eq. (80) into Eq. (72) and carry out the average over molecular orientations numerically.

In the usual fixed-nuclei approximation, which does not treat the motion of the nuclei dynamically, the magnitudes of $\mathbf{k}_\Gamma$ and $\mathbf{k}_{\Gamma'}$ are identical and hence the Born amplitude in Eq. (80) diverges as the scattering angle approaches zero. Eq. (80) is nevertheless perfectly adequate for studying scattering out of the forward direction and for calculating momentum transfer cross sections, which are insensitive to the form of the differential cross section near zero degrees, as long as the electron energy is large compared to the spacing of the rotational levels[80].

To compute total cross sections for polar molecules, we must introduce the rotational motion of the target. However, as Padial and Norcross[81] have pointed out, this does not mean we have to carry out fully coupled laboratory frame calculations, which are slowly convergent and hence only feasible when model potentials are used. We can still use our hybrid formula, but following the suggestion of Padial and Norcross, we use *fixed-nuclei* quantities for the rapidly converging differences between partial-wave T-matrix elements and include rotational quantum numbers in the full Born amplitude, which removes the divergence in the forward direction. This is justified because, as Collins and Norcross[80] have demonstrated, for low angular momentum quantum numbers, the differences in the T-matrix elements obtained from lab-frame calculations and frame-transformed fixed-nuclei calculations are insignificant for polar targets.

We have carried out complex Kohn studies of elastic electron scattering by several polar molecules, including H_2O[82], NH_3[83] and, more recently, H_2S[84]. It is not surprising to find that at very low energies, where the cross sections are large, the results are largely determined by dipole Born approximation, while at high energies, the static-exchange approximation is adequate. In the energy range from 1-4 eV, however, polarization is found to be important. Fig. 6 shows results we obtained for H_2S, along with the experimental results of Gulley et al.[85]. The cross sections develop a significant backward peak as the energy increases, indicating a significant deviation from the Born approximation. This backward peaking, which we found in all the polar systems we studied, is not evident from experimental measurements, which rarely extend beyond 120°. It is noteworthy that this large angle region contribites significantly to the determination of the momentum transfer cross section and that extrapolation of measured cross sections to large angles can be dangerous.

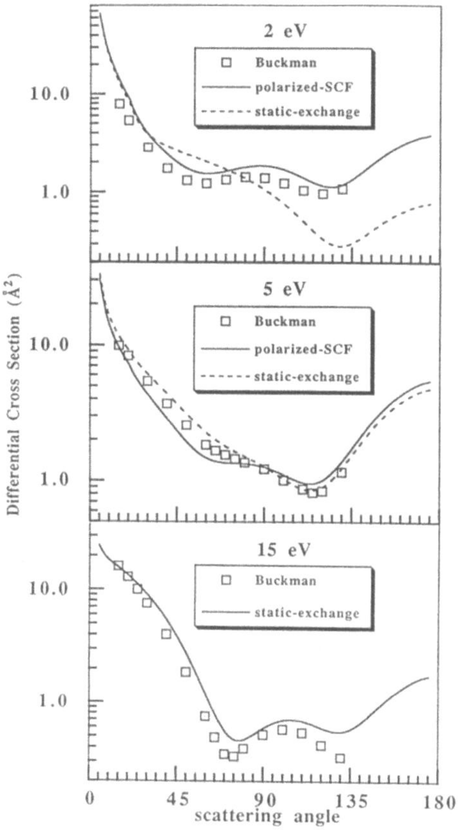

Figure 6. Differential cross sections for e^- + H_2S. Experimental points from ref. (85).

Photoionization

The complex Kohn method can be easily adapted to study molecular photoionization. We use a trial function of the same form as that given in Eq. (25), with incoming wave boundary conditions appropriate for scattering by an ionic target. The only difference between the photionization case and scattering from a molecular ion is the normalization of the contiuum basis function. This is chosen such that

$$h_\ell^-(kr) \underset{r\to\infty}{=} \frac{1}{\sqrt{k}} e^{-i(kr-\ell\pi/2+\ln(kr)/k+\delta_\ell)} \tag{81}$$

where δ_ℓ is the Coulomb phase shift, yielding purely incoming, rather than outgoing waves.

The matrix elements needed to construct photoionization sections can be expressed in terms of the body-frame amplitudes[86]:

$$\begin{aligned} I^\mu_{\Gamma_o} &\equiv \langle \Psi_o | \mu | \Psi^-_{\Gamma_o} \rangle \\ &= \sum_{i=1}^{N} \int \Psi_o(r_1,,r_N) r_i^\mu \Psi^-_{\Gamma_o}(r_1,,r_N) d^3r_1..d^3r_N \end{aligned} \tag{82}$$

where r^μ is the dipole operator which, in the length form, is defined as

$$r^\mu = \begin{cases} z & ,\mu = 0 \\ \mp(x \pm iy)/\sqrt{2} & ,\mu = \pm 1 \end{cases} \tag{83}$$

and Ψ_o is the wave function for the initial state of the target molecule. In order to construct an amplitude that represents an ejected photoelectron with momentum $\vec{k}_{\Gamma_o}$ associated with a particular ion channel, the matrix elements defined in Eq. (82) must be combined in a partial wave series

$$I_{\vec{k}_{\Gamma_o},\hat{\varepsilon}} = \sqrt{\frac{4\pi}{3}} \sum_{\ell_o m_o \mu} i^{\ell_o} e^{-i\delta_{\ell_o}} I^\mu_{\Gamma_o} Y^*_{1\mu}(\hat{\varepsilon}) Y^*_{\ell_o m_o}(\hat{k}_{\Gamma_o}) \tag{84}$$

where $\hat{k}_{\Gamma_o}$ specifies the direction of the ejected electron and $\hat{\varepsilon}$ is the direction of polarization of light. The doubly differential cross section for a hypothetical space-fixed target molecule is then given as

$$\frac{d^2\sigma^{\Gamma_o}}{d\Omega_{\hat{k}} d\Omega_{\hat{\varepsilon}}} = \frac{8\pi\omega}{3c} \left| I_{\vec{k}_{\Gamma_o},\hat{\varepsilon}} \right|^2 \tag{85}$$

where ω is the photon energy and c is the speed of light. In order to compute a cross section defined in the typical experiment where the target orientation is not resolved, the quantity defined in Eq. (85) must be averaged over all orientations of the target molecule in the laboratory frame. The resulting differential cross section has the form[87]

$$\frac{d\sigma^{\Gamma_o}}{d\Omega} = \left\langle \frac{d^2\sigma^{\Gamma_o}}{d\Omega_{\hat{k}} d\Omega_{\hat{\varepsilon}}} \right\rangle_{av} = \frac{\sigma^{\Gamma_o}}{4\pi}\left[1 + \beta^{\Gamma_o} P_2(\cos\theta)\right] \tag{86}$$

where P_2 is the Legendre polynomial of order 2, θ is the angle between $\hat{\varepsilon}$ and $\hat{k}$ and β^{Γ_o} is the so-called asymmetry parameter. We find it convenient to carry out the average required in Eq. (86) numerically, by a procedure analogous to the one we use to compute differential electron-molecule scattering cross sections. The quantity σ^{Γ_o} is the total photoionization cross section averaged over all polarizations and photoelectron directions and is given by

$$\sigma^{\Gamma_o} = \frac{8\pi\omega}{3c} \sum_{\ell_o m_o \mu} \left| I^{\mu}_{\Gamma_o} \right|^2 \tag{87}$$

We now turn our attention to the computation of the body-fixed amplitudes $I^{\mu}_{\Gamma_o}$. We choose a single set of molecular orbitals, $\{\varphi_i\}$, and partition them, as before, into a target set, in terms of which the initial state Ψ_0 and the ion states χ_Γ are expanded, and a complementary set of scattering orbitals for expanding the short-range part of the channel functions $F^{-}_{\Gamma_0\Gamma}$. It is clear that the amplitudes $I^{\mu}_{\Gamma_o}$ can be expressed as linear combinations of the bound-bound transition elements, $\langle\varphi_i|r^{\mu}|\varphi_j\rangle$, and bound-free elements $\langle\varphi_i|r^{\mu}|f_{\ell m}\rangle$ and $\langle\varphi_i|r^{\mu}|h^{-}_{\ell m}\rangle$. The former can be extracted from standard molecular property codes and the latter are evaluated by adaptive three-dimensional quadrature. Since the "scattering" orbitals are orthogonal to the target orbitals and the continuum basis functions $f_{\ell m}$ and $h^{-}_{\ell m}$ are orthogonalized to both, this constitutes a strong orthogonality condition between the channel functions $F^{-}_{\Gamma_0\Gamma}$ and the target orbitals; it is entirely equivalent to the Phillips-Kleinman pseudopotentials employed by Lucchese et al.[88] to enforce strong orthogonality.

The strong orthogonality constraint between the channel functions $F_{\Gamma_0\Gamma}$ and the target orbitals $\{\varphi_u\}$ can of course be relaxed by including in the trial function (Eq. (25)) appropriate N-electron penetration terms constructed solely from the target orbitals. The linear coefficients $d_i^{\Gamma_o}$ associated with these terms, or indeed with any other terms one might include in the set $\{\Theta_i\}$ to describe final-state correlation effects, are determined by the variational principle used to construct the Kohn trial function. There is a complication, however, that arises in cases where the target molecule does not possess a center of symmetry. In this case, the penetration terms can be identical to terms used to expand the initial target state Ψ_0. Inclusion of these terms in the expansion for $\Psi^{-}_{\Gamma_0}$ means that the N-electron final state is not necessarily orthogonal to the bound initial state Ψ_0 and that the transition dipole matrix elements we seek in Eq. (82) may be contaminated by introducing a component of the static dipole moment of the target. To remove these errors, we simply orthogonalize the scattering wave function $\Psi^{-}_{\Gamma_0}$ to Ψ_0 before computing the desired photoionization amplitudes. This simply corresponds to a redefinition of the coefficients $d_i^{\Gamma_o}$. This procedure guarantees orthogonality between the initial and final N-electron states and is an alternative to the strong orthogonality constraint implied by excluding any penetration terms in the expansion of the complex Kohn trial function.

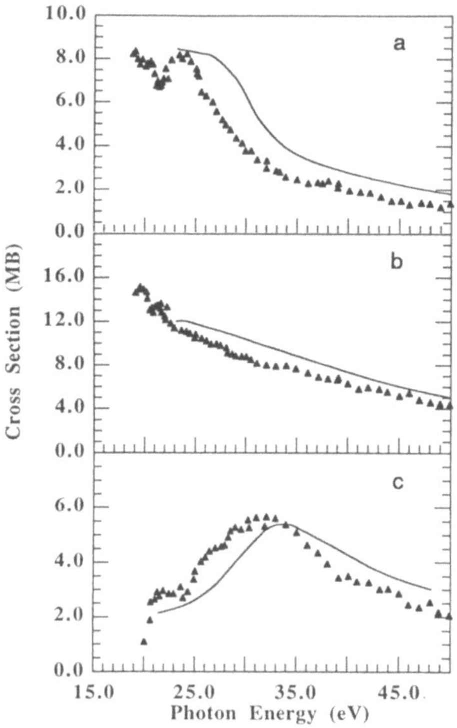

Figure 7. Comparison of present coupled-channel, CI partial cross sections for photoionization of CO with experimental results of Plummer et al. (ref. 89). Panel a: cross sections for production of $X^2\Sigma^+$ state of CO^+. Panel b: cross sections for production of $A^2\Pi$ state of CO^+. Panel c: cross sections for production of $B^2\Sigma^+$ state of CO^+.

We conclude this discussion with an explicit formula for the amplitudes $I^{\mu}_{\Gamma_o}$. We can once again characterize these amplitudes as the stationary value of a variational functional[61], as we did in the case of the off-shell T-matrix. This is another example of a specific application of a general class of supervariational principles discussed by Gerjouy et al.[60] The linear coefficients in the complex Kohn trial function (we use the vector $\underline{c}$ to denote the set of coefficients $\{c_i^{\Gamma_o}\}$, $\{d_i^{\Gamma_o}\}$ and $\{T^{\Gamma_o\Gamma}_{\ell_o m \ell_o m}\}$) are first determined by solving the set of equations.

$$\underline{\underline{\mathbf{M}}}_{cc}\underline{\mathbf{c}} = -\underline{\underline{\mathbf{M}}}_{c\Gamma_o} \tag{88}$$

where $\mathbf{M} \equiv H-E$, and E is the total energy, ie, $E=E_{target}+\omega$, $\underline{\underline{\mathbf{M}}}_{cc}$ is the matrix projection of $\mathbf{M}$ onto the space spanned by the functions $A\,(\chi_\Gamma \varphi_i)$, $A\,(\chi_\Gamma h^{-\Gamma}_{\ell m})$ and Θ_i, while Γ_o refers to the functions $A\,(\chi_{\Gamma_o} f_{\ell_o m_o})$. The techniques we use to assemble the various matrix elements required in Eq. (88), including the simplifying use of separable expansions to represent exchange and Q-space interactions, are identical to the procedures employed in our electron-molecule scattering studies. The complex Kohn expression for $I^{\mu}_{\Gamma_o}$ is thus given by

$$I^{\mu}_{\Gamma_o} = \langle \Psi_o | \mu | A\chi_{\Gamma_o} f_{\ell_o} Y_{\ell_o m_o} / r \rangle - \langle \Psi_o | \mu | c \rangle \underline{\underline{\mathbf{M}}}^{-1}_{cc} \underline{\underline{\mathbf{M}}}_{c\Gamma_o} \tag{89}$$

in cases where $\Psi^-_{\Gamma_o}$ and Ψ_o do not have the same symmetry. When Ψ_o and $\Psi^-_{\Gamma_o}$ do belong to the same irreducible representation, we employ the orthogonalized amplitude

$$\tilde{I}^{\mu}_{\Gamma_o} \equiv I^{\mu}_{\Gamma_o} - \langle \Psi_o | \mu | \Psi_o \rangle \langle \Psi_o | \Psi^-_{\Gamma_o} \rangle \tag{90}$$

as discussed above.

An an illustration of the above formalism, we show in Fig. 7 partial cross sections for photoionization of CO[35] computed with the complex Kohn method, along with available experimental values[89]. In contrast to previous theoretical studies on this system, these calculations included the effects of interchannel coupling and also employed a correlated ground-state wave function. The dominant effect of interchannel coupling is to remove a spurious $\pi \rightarrow \pi^*$ resonance feature from the continuum that appears at the frozen-core Hartree Fock level. It is also important to combine the effects of final channel coupling with a correlated initial target state to achieve quantitatively correct cross sections.

Dissociative Recombination

There are a number of resonant electron-molecule collisions processes in which the nuclear dynamics must be explicitly treated in order to make contact with experiment. The fixed-nuclei electronic problem can often provide the input necessary to solve the full problem including nuclear dynamics. We will discuss one such application, resonant dissociative recombination of H_3^+. In resonant or 'direct' DR, the electron is captured into a resonance state of the neutral molecule. After capture, the molecule begins to dissociate. During this time, the molecule can autoionize, leaving the ion in a vibrationally or rotationally excited state. Once the resonance curve crosses the ionic curve and becomes bound with respect to the emission of an electron, autoionization can no longer occur, and the system evolves into asymptotic final states. Thus to describe the direct DR process one must know the energies and widths of the resonant states and describe the dynamics of the dissociative process.

In the application to H_3^+[23], the nuclear dynamics were treated by a wave packet method which involves the direct integration of the time dependent Schrödinger equation:

$$i\hbar\frac{\partial \Psi}{\partial t} = H\Psi \qquad . \tag{91}$$

Our calculation was carried out in C_{2v} geometries, thereby restricting the problem to two degrees of freedom. The initial wave function is defined on an (r, R) grid, where r is the H_2 bond distance, and R is the distance from the remaining H to the center of mass of the H_2 bond. The angle between r and R was held fixed at 90°. In these coordinates, the Hamiltonian for the nuclear motion is given by:

$$H(r,R) = -\frac{1}{2\mu_1}\frac{\partial^2}{\partial r^2} - \frac{1}{2\mu_2}\frac{\partial^2}{\partial R^2} + V(r,R) \tag{92}$$

where μ_1 is the reduced mass of H_2, μ_2 is the reduced mass of $H + H_2$. The potential energy is:

$$V(r,R) = V_0(r,R) + \frac{i\Gamma(r,R)}{2} \tag{93}$$

where V_0 is the real part of the resonance energy, and $\Gamma(r,R)$ is the complex portion (the resonance width) that describes autoionization[90]. The resonant width and position are determined from fixed-nuclei electron scattering calculations from the H_3^+ ion.

Our treatment of dissociative recombination is a generalization of the time dependent treatment of photodissociation given by Heller[91]. At t=0 the wave packet is defined to be:

$$\begin{aligned}\Psi_0(r,R) &= \sqrt{\frac{\Gamma(r,R)}{2\Pi}}\Phi_0(r,R) \\ &= \sqrt{\frac{\Gamma(r,R)}{2\Pi}}\phi_i(Q_s)\phi_j(Q_b)\end{aligned} \tag{94}$$

where $\Phi_0(r,R)$ is the initial vibrational wave function on the resonant state surface, which we approximate as the product of harmonic oscillator wave functions, $\phi_i(Q_s)$ and $\phi_j(Q_b)$ in the symmetric stretch and bending normal modes of H_3^+ for each initial vibrational state of interest.

In order to determine the total dissociative recombination cross section, the final wave packet must be projected onto final states[92]. For example, when dissociation produces to an atom and a diatomic one can define:

$$S_2(E) = \sum S_i(E) \tag{95}$$

where the $S_i(E)$ are the final state probability distributions given by:

$$S_i(E) \propto \left| \iint drdR\chi_i(r)\phi_T(R)\Psi_o(r,R,t) \right|^2 \tag{96}$$

where $\chi_i(r)$ is a vibrational state eigenfunction of the diatomic, and $\phi_T(R)$ is the translation function describing the motion of the atom from the molecular center-of-mass; t is chosen

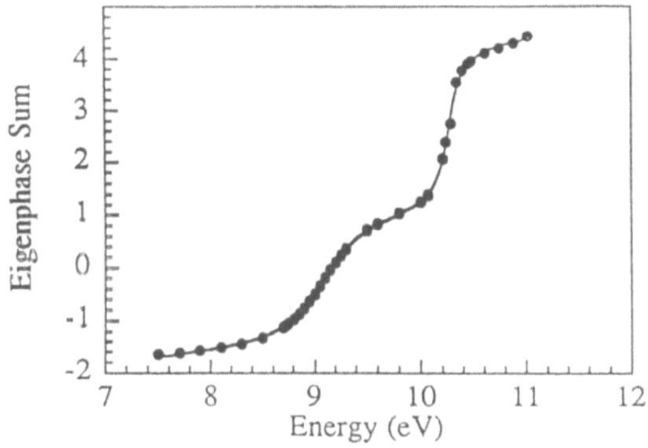

Figure 8. Eigenphase sums for electron H_3^+ scattering at the equilibrium geometry of the ion in A_1 symmetry. Solid circles are the calculated points and theline is the fit of the sum of two Breit-Wigner resonance lineshapes. Rapid jumps by pi in the eigenphase sum indicate resonances.

large enough that the dissociation is complete, i.e.-the fragments are no longer interacting. This expression, when summed over all bound vibrational final states, yields the branching ratio into the two-body channel, $S_2(E)$. Once the sum in Eq. (95) is extended to include the continuum H_2 states the result is the total dissociation cross section.

The ground state equilibrium geometry of H_3^+ is an equilateral triangle, (D_{3h} symmetry) with bond lengths 1.65 a_0[93]. The electronic configuration is $1a_1'^2$. The resonance states correspond to the capture of an electron into the low-lying doubly degenerate e'orbital, with the simultaneous promotion of one of the a_1' electrons to the same orbital:

$$H_3^+ (1a_1'^2) + e^- \rightarrow H_3 (1a_1'1e'^2).$$

Because of the degeneracy of the doubly occupied orbital, four distinct molecular configurations are possible. We carried out complex Kohn calculations to map out the four resonance potential energy surfaces from the Franck-Condon region to the points where they cross into the bound state manifolds. The eigenphase sums were fit to a Breit-Wigner form to obtain the resonance widths and positions. An example is shown in Fig. 8.

In Fig. 9 we show the individual contributions of the four resonance states and the total DR cross section as a function of impact energy compared to experiment. These results are in quantitative agreement, both for the position of the peak and its shape and magnitude, with the ~ 9 eV peak reported by high energy cross section measured by Larsson et al[94].

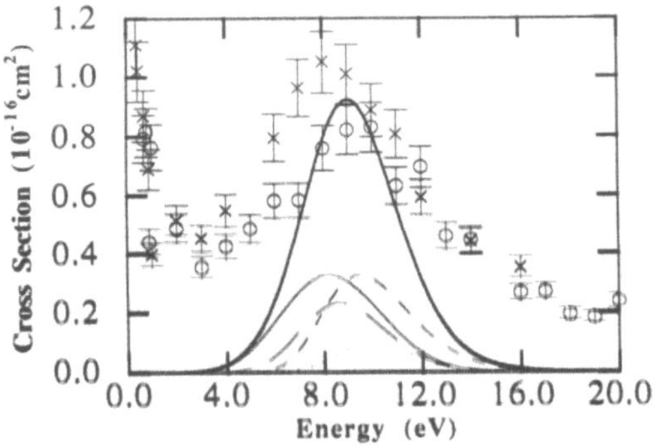

Figure 9. Total (heavy solid line and partial resonance DR cross sections as functions of incident electron energy: 1^2A_1 (solid line), 1^2B_2 (long dashed line), 2^2A_1 (medium dashed line) and 2^2B_2 (short dashed line). The experimental results are those of Larsson et al.[94]

Acknowledgement

This work was performed under the auspices of the U.S. Department of Energy by Lawrence Livermore National Laboratory under contract number W-7405-Eng-48. AEO wishes to acknowledge support from the National Science Foundation.

REFERENCES

1. A. B. Ehrhardt and W. D. Larger, *Collisional Processes of Hydrocarbons in Hydrogen Plasmas*, PPPL-2477 (Sept. 1987).
2. A. Gallagher, "Materials Issues in Amorphous-Semiconductor Technology Symposium," Palo Alto, 1986 (Mater. Res. Soc., Pittsburgh, PA1986), p. 3.
3. J. B. A. Mitchell, Phys. Repts. **186**, 216 (1990).
4. M. A. Biondi, in *Principles of Laser Plasmas*, ed. by G. Bekefi (Wiley, New York, 1976) p. 125.

5. See, for example, *Excimer Lasers,* ed. by C. K. Rhodes (Springer-Verlag, New York, 1979).
6. A. Dalgarno and J. H. Black, Rep. Prog. Phys., **39**, 573 (1976).
7. For a recent review, see D. R. Bates, M. F. Guest and R. A. Kendall, Planet. Space Sci. **41**, 9 (1993).
8. J. Z. H. Zhang and W. H. Miller, J. Chem. Phys. **91**, 1528 (1989).
9. T.Kato, Prog. Theo. Phys. **6**, 394 (1951).
10. W. Kohn, Phys. Rev. **74**, 1763 (1948).
11 R. K. Nesbet, *Variational Methods in Electron-Atom Scattering Theory* (Plenum, New York, 1980).
12. R. K. Nesbet, Phys. Rev. **175**, 134 (1968); **179**, 60 (1969).
13. Y. Mito and M. Kamimura, Prog. Theo. Phys. **56**, 583 (1976).
14. W. H. Miller and B. M. D. D. Jansen op de Haar, J. Chem Phys. **86**, 6213 (1987).
15. C. W. McCurdy, T. N. Rescigno and B. I. Schneider *Phys. Rev. A* **36**, 2061 (1987).
16. T. N. Rescigno and B. I. Schneider, Phys. Rev. A **37**, 1044 (1988).
17. C. W. McCurdy and T. N. Rescigno, Phys. Rev. A **39**, 4487 (1989).
18. A. D. Becke, J. Chem. Phys. **88**, 2547 (1988).
19. T. N. Rescigno, B. H. Lengsfield III and C. W. McCurdy, in *Modern Electronic Structure*, edited by D. Yarkony (World Scientific, 1994).
20. J. N. Bardsley, in *Electron-Molecule and Photon-Molecule Collisions,* edited by T. N. Rescigno, V. McKoy and B. I. Schneider (Plenum, New York, 1979), p. 267
21. M. A. Morrison, Adv. At. Molec. Phys. **24**, 51 (1988).
22. A. U. Hazi, T. N. Rescigno and M. Kurilla, Phys. Rev. A **23**, 1089 (1981); A. U. Hazi, A. E. Orel and T. N. Rescigno, Phys. Rev. Letts. **46**, 918 (1981).
23. A. E. Orel and K. C. Kulander, Phys. Rev. Letts. **71**, 4315 (1993).
24. P. G. Burke, *Potential Scattering in Atomic Physics,* (Plenum, New York, 1977).
25. M. Abramowitz and I. A. Stegun, *Handbook of Mathematical Functions*, (Dover, New York, 1965).
26. D. Lyons and R. K. Nesbet, J. Comp. Phys. **11**, 166 (1973);R. L. Smith and D. G. Truhlar, Comp. Phys. Commun. **5**, 80 (1973); J. Abdallah and D. G. Truhlar, *ibid.* **9**, 327 (1975).
27. T. N. Rescigno and A. E. Orel, Phys. Rev. A **43**, 1625 (1991).
28. Y. Sun, D. J. Kouri, D. G. Truhlar and D. W. Schwenke, Phys. Rev. A **41**, 4857 (1990).
29. B. I. Schneider and T. N. Rescigno, Phys. Rev. A **37**, 3749 (1988).
30. A. E. Orel. T. N. Rescigno and B. H. Lengsfield III, Phys. Rev. A **44**, 4328 (1991).
31. B. H. Lengsfield III, T. N. Rescigno and C. W. McCurdy, *Phys. Rev. A* **44**, 4296 (1991).

32. B. H. Lengsfield III and T. N. Rescigno, Phys. Rev. A **44**, 2913 (1991).
33. H. E. Saraph, M. J. Seaton and J. Shemming, Proc. Phys. Soc. London **89**, 27 (1966).
34. P. G. Burke, A. Hibbert and W. D. Robb, J. Phys. B **4**, 153 (1971).
35. T. N. Rescigno, B. H. Lengsfield III and A. E. Orel, J. Chem. Phys. **99**, 5097 (1993).
36. H. Feshbach, Ann. Phys. (N. Y.) **5**, 357 (1958); **19**, 287 (1962).
37. B. H. Lengsfield III and B. Liu, J. Chem. Phys. **75**, 478 (1981).
38. J. A. Pople, R. Krishnam, H. B. Schlegel and J. S. Binkley, Int. J. Quan. Chem. Symp. **13**, 225 (1979).
39. E. R. Davidson, J. Comp. Phys. **17**, 87 (1975); P. E. M. Siegbahn, J. Chem. Phys. **70**, 5391 (1980); **72**, 1647 (1980).
40. T. N. Rescigno and A. E. Orel, Phys. Rev. A **23**, 1134 (1981); **24,** 1267 (1981); B. Gyarmati, A. T. Kruppa and J. Revai, Nucl. Phys. **A 326**, 119 (1979).
41. C. W. Murray, N. C. Handy and G. J. Laming, Molec. Phys. **78**, 997 (1993).
42. J. D. Jackson, *Classical Electrodynamics,* (Wiley, New York, 1962), p. 69.
43. A. D. Becke and R. M. Dickson, J. Chem. Phys. **89**, 2993 (1988).
44. W. Eissner and M. J. Seaton, J. Phys. B **5**, 2187 (1972).
45. R. K. Nesbet, Comp. Phys. Comm. **6**, 275 (1973).
46. P. G. Burke, Comp. Phys. Comm. **6**, 288 (1973).
47. S. Chung and C. C. Lin, Phys. Rev. A **17**, 1874 (1978).
48. C. Weatherford, Phys. Rev. A **22**, 2519 (1980).
49. A. E. Orel, T. N. Rescigno and B. H. Lengsfield, Phys. Rev. A **42**, 5292 (1990).
50. T. N. Rescigno and B. I. Schneider, Phys. Rev. A **45**, 2894 (1992).
51. T. N. Rescigno, "Low energy electron collision processes in molecular chlorine", Phys. Rev. A **XX**,xxxx (1994)
52. T. J. Gil, C. W. McCurdy, T. N. Rescigno and B. H. Lengsfield III, Phys. Rev. A **47**, 255 (1993).
53. D. E. Golden, N. F. Lane, A. Temkin and E. Gerjuoy, Rev. Mod. Phys. **43** 642 (1971)
54. M. A. Morrison, Adv. Atom. Mol. Phys. **24**, 51 (1988)
55. A. M. Arthurs and A. Dalgarno, Proc. R. Soc. A **256**, 540 (1960).
56. N. Chandra and A. Temkin , Phys. Rev. A **13,** 188 (1976).
57. M. Shugard and A.U. Hazi, Phys. Rev. A **12,** 1895 (1975).
58. M.A. Morrison, J. Phys. B **19**, L707 (1986); M.A. Morrison, M. Abdolsalami and B. K. Elza, Phys. Rev. A **43**, 3440 (1991); B. K. Elza, University of Oklahoma, PhD Thesis, (1992,unpublished).
59. R. G. Newton, *Scattering Theory of Waves and Particles*, (McGraw-Hill, New York, 1966), p.188.
60. E. Gerjouy, A. Rau and L. Spruch, J. Math. Phys. **13,** 1797 (1972).

61. A. E. Orel and T. N. Rescigno, Phys. Rev. A **41,** 1695 (1990).
62. D. M. Chase, Phys. Rev. **104**, 838 (1956).
63. T. N. Rescigno, B. K. Elza and B. H. Lengsfield III, J. Phys. B **26**, L567 (1993).
64. M.J. Brunger, S.J. Buckman, D.S. Newman and D.T. Alle, J. Phys. B **24**, 1435 (1991).
65. J.P. England, M.T. Elford and R.W. Crompton, Aust. J. Phys. **41**, 573 (1988).
66. S.J. Buckman, M.J. Brunger, D.S. Newman, G. Snitchler, S. Alston, D.W. Norcross, M.A. Morrison, B.C. Saha, G. Danby and W.K. Trail, Phys. Rev. Letts. **65**, 3253 (1990).
67. F. Linder and H. Schmidt, Z. Naturforsch. **26a**, 1603 (1971).
68. H. Ehrhardt, L. Langhans, F. Linder and H.S. Taylor, Phys. Rev. **173**, 222 (1968).
69. K. Rohr and F. Linder, J. Phys. B **9,** 2521 (1976).
70. L. A. Morgan and P. G. Burke, J. Phys. B **21**, 2091 (1988); G. Snitchler, D. Norcross, A. Jain and S. Alston, Phys. Rev. A **42,** 671 (1990).
71. K. Rohr, J. Phys. B **13,** 4897 (1980).
72. W. Sohn, K. Jung and H. Ehrhardt, J. Phys. B **16,** 891 (1983).
73. T. N. Rescigno, A. E. Orel and C. W. McCurdy (unpublished).
74. W. Sohn, K. H. Kochem, K. M. Scheuerlein, K. Jung and H. Ehrhardt, J. Phys. B **19**, 3625 (1986).
75. O. H. Crawford, A. Dalgarno and P. B. Hays, Molec. Phys. **13**, 181 (1967).
76. W. R. Garrett, Molec. Phys. **24**, 465 (1972).
77. See, for example, N. F. Lane, Rev. Mod. Phys. **52**, 29 (1980).
78. M. E. Rose, *Elementary Theory of Angular Momentum* (Wiley, New York ,1967) pp. 62-65.
79. See, for example, M. J. Seaton, Proc. Phys. Soc. London **79,** 1105 (1962).
80. See, for example, L. A. Collins and D. W. Norcross, Phys. Rev. A **18**, 467 (1978). This paper gives a clear treatment of the subject and also contains many references to earlier work.
81. D. W. Norcross and N. T. Padial, Phys. Rev. A **25**, 226 (1982).
82. T. N. Rescigno and B. H. Lengsfield III, Z. Phys. D **24**, 117 (1992).
83. T. N. Rescigno, B. H. Lengsfield III, C. W. McCurdy and S. D. Parker, Phys. Rev. A **45**, 7800 (1992).
84. B. H. Lengsfield III and T. N. Rescigno, (unpublished).
85. R. J. Gulley, M. J. Brunger and S. A. Buckman, Phys. B **26**, 2913 (1993).
86. R. R. Lucchese and V. McKoy, Phys. Rev. A **24**, 770 (1981); R. R. Lucchese, G. Raseev and V. McKoy, Phys. Rev. A **25**, 2572 (1982).
87. J. C. Tully, R. S. Berry and B. J. Dalton, Phys. Rev. **176**, 95 (1968).
88. B. Basden and R. R. Lucchese, Phys. Rev. A **37,** 89 (1988).
89. E. W. Plummer, T. Gustafsson, W. Gudat and D. E. Eastman, Phys. Rev. A **15**, 2339 (1977).

90. C. W. McCurdy and J. L. Turner, J. Chem. Phys. **78**, 6773 (1983).
91. E. J. Heller, J. Chem. Phys. **68**, 3891 (1978).
92. K. C. Kulander and E. J. Heller, J. Chem. Phys. **69**, 2439 (1978).
93. G. Herzberg, *Electronic Spectra of Polyatomic Molecules*, (Van Nostrand Reinhold Company, New York), pg. 289 (1966).
94. M. Larssen, H. Danared, J. R. Mowat, P. Sigray, G. Sundström, L. Broström, A. Filevich, A. Källberg, S. Mannervik, K. G. Rensfelt and S. Datz, Phys. Rev. Lett.

THE LINEAR ALGEBRAIC METHOD FOR ELECTRON-MOLECULE COLLISIONS

Lee A. Collins[1] and Barry I. Schneider[2]

[1] Theoretical Division, Group T-4, Los Alamos National Laboratory, Los Alamos, NM 87545
[2] Physics Division, National Science Foundation, Arlington, VA 22230

1. BASIC CONCEPTS

In order to find numerical solutions to many problems in physics, chemistry and engineering it is necessary to place the equations of motion (classical or quantal) of the variables of dynamical interest on a discrete mesh. The formulation of scattering theory in quantum mechanics is no exception and leads to partial differential or integral equations which may only be solved on digital computers. Typical approaches introduce a numerical grid or basis set expansion of the scattering wavefunction in order to reduce the problem to the solution of a set of algebraic equations. Often it is more convenient to deal with the scattering matrix or phase amplitude rather than the wavefunction but the essential features of the numerics are unchanged.

In this section we will formulate the Linear Algebraic Method (LAM) for electron-atom/molecule scattering for a simple, one-dimensional radial potential.[1-2] This will illustrate the basic approach and enable the uninitiated reader to follow the subsequent discussion of the general, multi-channel, electron-molecule formulation without undue difficulty. We begin by writing the Schroedinger equation for the s-wave scattering of a structureless particle by a short-range, local potential.

$$-\frac{1}{2}\frac{d^2\Psi}{dr^2} + (V(r) - E)\Psi(r) = 0 \tag{1}$$

By re-writing this equation in its integral form,

$$\Psi(r) = \Psi_0(r) + \int dr' G(r \mid r')V(r')\Psi(r') \tag{2}$$

where

$$\begin{aligned} \Psi_0(r) &= \sin(\mathrm{k}r) \\ G(r \mid r') = \text{Green's function} &= -2\frac{\sin(\mathrm{k}r_<)\cos(\mathrm{k}r_>)}{\mathrm{k}} \\ \mathrm{k} &= \sqrt{2\mathrm{E}} \end{aligned}$$

it becomes straightforward to incorporate the two physical boundary conditions required of a standing wave solution to the scattering equation, regularity at the origin and incident free wave sin(kr) plus outgoing cos(kr) behavior at large distances from the scatterer. We now proceed by introducing a numerical quadrature with points r_i and weights w_i into the integral and then set the co-ordinate r to one of the quadrature points. This yields,

$$\Psi(r_i) = \Psi_0(r_i) + \sum_j G(r_i \mid r_j) w_j V(r_j) \Psi(r_j) \tag{3}$$

By defining vector and matrix elements as,

$$\begin{aligned} [\mathbf{\Psi}]_\mathbf{i} &= \Psi(r_i) \\ [\mathbf{\Psi_0}]_\mathbf{i} &= \Psi_0(r_i) \\ \left[\tilde{\mathbf{V}}\right]_\mathbf{ij} &= w_i V(r_i) \delta_{i,j} \\ [\mathbf{G}]_\mathbf{ij} &= G(r_i \mid r_j) \end{aligned}$$

it is possible to write Eq (3) in the matrix notation,

$$\begin{aligned} \mathbf{M\Psi} &= \mathbf{\Psi_0} \\ \mathbf{M} &= \mathbf{I} - \mathbf{G\tilde{V}} \end{aligned}$$

as a set of linear algebraic (LA) equations for the unknown vector $\mathbf{\Psi}$. These may be solved using Gaussian elimination or a number of other standard linear systems packages[3] which are readily available on most computers. If the matrix **GV** becomes too large to be held in central memory, there are iterative techniques[4] which may be employed to effect a solution of Eq (3). One such approach, the variation-iteration method, will be discussed in much greater detail in a subsequent section. The great advantage of the integral equation formulation, which becomes much more apparent in the multichannel scattering problem, is the numerical stability associated with the incorporation of the two point boundary conditions in the propagator, **G**. This allows weakly open and closed (exponentially decaying) channels to be calculated in a stable fashion, a much more difficult process to accomplish via the propagation of the solution of a (set of) differential equation(s).[5]

2. GENERAL FORMALISM

In this section we will formulate the general LAM[6-16] for a wavefunction which is expanded as a sum of a close-coupled plus correlation term (CCC).[17] In addition, we restrict ourselves to the fixed nuclei approximation; all rotational and vibrational degrees of freedom will be ignored and the molecule will be taken to be fixed in space. Rotational and vibrational motion away from thresholds and resonances may be treated using the adiabatic nuclei formalism. [19-20] When that is inappropriate, off-shell techniques or extensions of the close-coupled form of the equations to include rotations and vibrations may be invoked. These are treated in other chapters of this volume[17] and their inclusion here would add little to the present discussion. The CCC wavefunction may be expressed as,

$$\Psi(1,2,\cdots,n_{T+1}) = \sum_c \Theta_c(1,2,\cdots,n_T) f_c(n_{T+1}) + \sum_q \Xi_q(1,2,\cdots,n_{T+1}) D_q \tag{4}$$

where Θ_c are the channel wavefunctions and Ξ_q the correlation terms. The term channel wavefunction is used here to denote the product of an internal electronic state of the target times an overall spin eigenfunction of the composite system. This quite general form to the wavefunction allows us treat all desired open channels explicitly and to incorporate the effects of closed channels via an optical potential and/or by the explicit introduction of real or pseudo states as with the open channels. The optical potential results from using Feshbach[21] partitioning techniques to formally remove the second sum in favor of a non-local, energy-dependent potential. It will be assumed that the optical potential is capable of an N-term separable expansion in terms of a set of square integrable functions. Under these rather general assumptions it is possible to reduce the scattering problem to a set of coupled integro-differential equations of the form,

$$\left(-\frac{1}{2}\nabla^2 - \epsilon_c\right) f_c(r) + \sum_{c'} \int dr' V_{c,c'}(r,r') f_{c'}(r') = 0 \tag{5}$$

where the potential, $V_{c,c'}$, consists of a local(l) and non-local(nl) part, i.e.,

$$V_{c,c'}(r,r') = \delta(r-r') V^{l}_{c,c'}(r) + V^{nl}_{c,c'}(r,r') \tag{6}$$

and the non-local potential is expanded as,

$$V^{nl}_{c,c'}(r,r') = \sum_{\alpha} \sum_{\beta} \phi_{c,\alpha}(r) V_{c\alpha,c'\beta} \phi_{c',\beta}(r'). \tag{7}$$

The local potential includes the attractive interaction of the scattering electron with the nuclei and the direct ($c = c'$) or transition ($c \neq c'$) electronic potential arising from the repulsive electron-electron interactions. At large electron-molecule distances the local potential has a typical multi-polar form, reflecting the permanent and transition moments of the charge distributions of the target. The non-local interaction arises from the requirement of overall antisymmetry of the scattering wavefunction to exchange of the scattering electron with any of the bound electrons of the target and from the optical potential. Typical exchange interactions are short range and vanish exponentially at large electron-molecule separations. The optical potential reflects the moments induced by the scattering electron in the charge cloud of the target as well as shorter range, exchange terms. In the original LAM, Eq (5) is further expanded in a set of spherical harmonics and the three dimensional, coupled equations are replaced by an even larger set of coupled, radial equations. If we use the symbol γ to represent the composite label (c,l,m), where (l,m) are the quantum numbers of the spherical harmonic used in the expansion, it is possible to transform the set of coupled radial integro-differential equations into the following set of coupled integral equations by utilizing the Green's function for free-particle motion,

$$f^{\gamma''}_{\gamma}(r) = \delta_{\gamma,\gamma''} f^{0}_{\gamma}(r) + \sum_{\gamma'} \int dr' dr'' G^{0}_{\gamma}(r,r') V_{\gamma,\gamma'}(r',r'') f^{\gamma''}_{\gamma'}(r''). \tag{8}$$

where the label γ'' is used to denote the linearly independent solution having its incident wave in channel γ'' . By introducing a quadrature into the integrals over r and r', it is possible to convert the entire set of coupled integral equations into a larger set of linear algebraic equations defined by the composite index $\Gamma = (c,l,m,i)$, where $\gamma = (c,l,m)$ specifies the channel and i the quadrature point of the quadrature. Thus, assuming there are N_γ channels, any matrix element may be located using its γ and i index as,

$$\Gamma = (N_\gamma - 1)\gamma + i$$

This enables us to regard the vectors (matrices) as a set of supervectors (matrices) labelled by channel indices. The running index of each sub-block is the set of quadrature points. Proceeding in this fashion, we may define the elements of our vectors and matrices as,

$$[\mathbf{f}]_{\mathbf{\Gamma}} = f_\gamma(r_i) \qquad \left[\mathbf{f^0}\right]_{\mathbf{\Gamma}} = f^0_\gamma(r_i) \tag{9}$$

$$[\mathbf{M}]_{\mathbf{\Gamma},\mathbf{\Gamma}'} = \delta_{\gamma,\gamma'} - \sum_k G^0_\gamma(r_i, r_k) w_k V_{\gamma,\gamma'}(r_k, r_j) w_j \tag{10}$$

which enables us to write the entire set of equations in the compact notation,

$$\mathbf{Mf} = \mathbf{f^0} \tag{11}$$

Just as in the simple, one-dimensional example discussed earlier, the LA formulation has the advantage of numerical stability and conceptual simplicity. The disadvantage is that the practitioner must often deal with large sets of linear algebraic equations. The solution of these equations can be computationally and memory intensive. One useful approach to reduce the size of these equations is to formulate the LA problem inside a sphere sufficiently large to enclose the exchange/correlation region of the scatterer and to use more standard propagation techniques beyond that radius.[16] This philosophy, which is formally identical to the R-matrix method, allows us to treat the short-range, non-local interactions using the LAM inside the sphere and to integrate coupled, *differential* equations outside the sphere. The only modification required is to alter the boundary conditions imposed on the Green's function from outgoing wave to some (usually zero) log-derivative condition on the R-matrix surface. The form of the LA equations does not change. In essence, one is solving the integral equation for the full, interacting Green's function, which is equal to the R-matrix when evaluated on the surface of the sphere. This R-matrix may then be propagated to very large distances via R-matrix propagation. [22-23] continued fraction [24] or other approaches which are very efficient for coupled channel, *local* interactions.

3. NUMERICAL TECHNIQUES

3.1. General Remarks

Numerical techniques for the solution of the LA equations fall into two classes: 1) off-the-shelf methods which reduce the LA coefficient matrices to specialized forms (i.e. LU factorizations, Gaussian elimination, etc.) and 2) specialized iterative techniques. The off-the-shelf methods almost always work and are certainly to be preferred when the matrices are small enough to fit in core or are not particularly well behaved. The work to solve the LA equations using such techniques goes like N^3 where N is the size of the matrix. This is to be contrasted with iterative techniques which are dominated by matrix-vector multiplies and behave like N^2M where M is the number of iterations needed to achieve convergence. If M can be kept much smaller than N, the iterative approach can be quite practical and valuable. A further advantage of iterative techniques is that they may be easily developed to take advantage of any special structure to the matrices in performing the computationally intensive matrix-vector multiplies. For the problem under consideration, there are considerable simplifications in both computational effort and storage requirements resulting from developing a rapidly converging iterative approach. This has been done by the authors[11-12] and it is described in some detail in the next section.

3.2. Variation-Iteration Method

Iterative methods have a long history in numerical analysis. The Jacobi and Gauss-Seidel methods[4] are described in almost all textbooks dealing with iterative treatments of linear equations. In all of these methods one begins with some guess to the solution of the set of LA equations and then computes either a correction to that guess or what is equivalent, an updated solution. The process is continued until there is (essentially) no change in the next member of the iteration sequence. However, there is no guarantee that this process will converge. The question of convergence or the lack of it can often be traced to the eigenvalue spectrum of the iteration matrix. The iteration process is equivalent to expanding $(\mathbf{I} - \mathbf{GV})^{-1}$ in a power series in $\mathbf{GV}$ and is only justified if the eigenvalues of $\mathbf{GV}$ lie inside the unit circle. However, it is possible to combine iterative techniques with variational methods to produce convergent results even when there are eigenvalues outside the unit circle. In essence the variational step sums the geometric series by solving a "small" set of algebraic equations which results from the projection of the full set of LA equations into the space of the iterates. Stated somewhat differently, the iterates are used as a set of basis vectors for the expansion of the unknown solution. The expansion coefficients are determined either by invoking a variational principle or by projecting the full set of LA equations onto the iterate space. If the iterates are a good basis for the expansion of the unknown solution to the LA equations, convergence should be rapid. However, as long as one can develop N linearly independent vectors over the course of the iteration process, one will eventually have enough flexibility to span the space of the N dimensional vectors of the full LA equations. In practice, if convergence is slow, there is a reasonable probability that the iteration sequence will produce linear dependence of the expansion vectors and this can bring the entire procedure to a grinding halt. The variation-iteration process may be initiated using any starting vector. Typically we use the "normalized" Born approximation, defined as,

$$\begin{aligned} \mathbf{v^0} &= \frac{\mathbf{f^0}}{\langle \mathbf{f^0} \mid \mathbf{f^0} \rangle^{\frac{1}{2}}} \\ \langle \mathbf{f^0} \mid \mathbf{f^0} \rangle &= \sum_{\gamma,i} f^0_\gamma(r_i) w_i f^0_\gamma(r_i) \end{aligned} \tag{12}$$

to begin the process. We then initiate the following sequence:
1. At the nth step in the iteration process, compute an iterate via the formula,

$$\mathbf{f^n} = \mathbf{M}\mathbf{v^{n-1}} \tag{13}$$

2. Expand the solution to the LA equations as,

$$\mathbf{f} = \sum_{i=1}^{n} C_i \mathbf{v^{i-1}} \tag{14}$$

3. Determine the C_i by substituting the expansion of step 2 into Eq (11) and then projecting onto the set of $\mathbf{v}$. This yields,

$$C_i = \langle \mathbf{v^{i-1}} \mid \mathbf{f^0} \rangle + \sum_{j=1}^{n} \langle \mathbf{v^{i-1}} \mid \mathbf{f^j} \rangle C_j \tag{15}$$

a "small" set of equations which may be solved by standard techniques. Once the C_i have been calculated, it is possible to compute the RMS error to the set of LA equations. If the RMS error falls below the desired level, the process is stopped.

4. If that is not the case, a new vector is computed using the Gram-Schmidt process.

$$\begin{aligned} \mathbf{T^n} &= \mathbf{f^n} - \sum_{j=1}^{n} \mathbf{v^{j-1}} \langle \mathbf{v^{j-1}} \mid \mathbf{f^n} \rangle \\ \mathbf{v^n} &= \frac{\mathbf{T^n}}{\langle \mathbf{T^n} \mid \mathbf{T^n} \rangle^{\frac{1}{2}}} \end{aligned} \tag{16}$$

and then return to step 1. The vectors and iterates which form the basis of this procedure are called a Krylov sequence (KS). The KS is at the heart of most iterative approaches to the solution of large linear systems and eigenvalue problems. The differences between the various methods such as the Arnoldi, Generalized Minimum Residue, and Lanczos approaches is in how these quantities are manipulated to get the final information needed. The interested reader should consult the references for details.[18, 4] The main concern is how to make the calculation of the matrix-vector multiply as efficient as possible. Fortunately, the quasi-separable structure of the Lippmann-Schwinger kernel, facilitates some simplification. To see this in more detail, we specialize Eq (13) to the scattering problem.

$$f_\gamma(r_i) = \sum_{\gamma'} \sum_{j,k} G^0_\gamma(r_i, r_j) w_j V_{\gamma,\gamma'}(r_j, r_k) w_k v_{\gamma'}(r_k) \tag{17}$$

If we define N_c as the number of channels, N_b as the number of basis functions, N_p as the number of quadrature points, and the intermediate vector,

$$J_\gamma(r_i) = \sum_{\gamma'} \sum_{j} V_{\gamma,\gamma'}(r_i, r_j) w_j v_{\gamma'}(r_j) \tag{18}$$

the work required to construct J_γ goes like $N_\gamma^2 N_p$ for a local potential and as the greater of the products $N_\gamma N_b N_p$ or $N_\gamma^2 N_b^2$ for a non-local, separable potential. Once J_γ is formed we compute,

$$f_\gamma(r_i) = \sum_{j} G^0_\gamma(r_i, r_j) w_j J_\gamma(r_j) \tag{19}$$

Since the Green's function may be written as,

$$\begin{aligned} G^0_\gamma(r, r') &= -2\, \frac{\mathrm{j}(\mathrm{k}_\gamma r_<)\eta(\mathrm{k}_\gamma r_>)}{\mathrm{k}_\gamma} \\ \mathrm{k}_\gamma &= \sqrt{2\epsilon_\gamma} \\ \mathrm{j}(\mathrm{k}r) &= \text{Bessel Function} \qquad \eta(\mathrm{k}r) = \text{Neumann Function} \end{aligned} \tag{20}$$

it is possible to write the sum over the quadrature points as,

$$f_\gamma(r_i) = \eta_\gamma(\mathrm{k}_\gamma r_i) \sum_{j=1}^{i} \mathrm{j}_\gamma(\mathrm{k}_\gamma r_j) w_j J_\gamma(r_j) + \mathrm{j}_\gamma(\mathrm{k}_\gamma r_i) \sum_{j=i+1}^{n} \eta_\gamma(\mathrm{k}_\gamma r_j) w_j J_\gamma(r_j) \tag{21}$$

From the basic definitions, the following relationships arise,

$$\begin{aligned} \sum_{j=1}^{i} \mathrm{j}_\gamma(\mathrm{k}_\gamma r_j) w_j J_\gamma(r_j) &= \sum_{j=1}^{i-1} \mathrm{j}_\gamma(\mathrm{k}_\gamma r_j) w_j J_\gamma(r_j) + \mathrm{j}_\gamma(\mathrm{k}_\gamma r_i) w_i J_\gamma(r_i) \\ \sum_{j=i+1}^{n} \eta_\gamma(\mathrm{k}_\gamma r_j) w_j J_\gamma(r_j) &= \sum_{j=i+2}^{n} \eta_\gamma(\mathrm{k}_\gamma r_j) w_j J_\gamma(r_j) + \eta_\gamma(\mathrm{k}_\gamma r_{i+1}) w_{i+1} J_\gamma(r_{i+1}) \end{aligned} \tag{22}$$

Thus the forward and backward summations over quadrature points may be performed using recursion relations involving "one" matrix multiply per step. It should be noted that the effort required to perform the indicated recursions is identical to that of most propagation approaches. By performing the operation in the forward (backward) direction for the first (second) of Eq (22), we can insure the numerical stability of the recursion scheme. Standard propagation methods for the integration of the coupled differential or integral equations of scattering theory rely totally on either forward or backward integration. This is required since the solution must be known at some subset of prior points before it can "advanced". In contrast, the iterative approach always deals with an integrand which is known at all points in space. This enables the practitioner to integrate in both directions and to thus ensure numerical stability without elaborate stabilization procedures.

3.3. Further Developments

The current implementation of the LAM relies heavily on single-center expansions to reduce the three-dimensional equations to a larger set of one-dimensional equations. A serious disadvantage of this approach is the need to expand a singular, Coulomb potential away from its natural center. This often forces the spherical harmonic expansion to be quite slowly convergent, leading to linear systems of relatively large dimension. In addition, the idea of expanding an inherently multi-center charge distribution about a single co-ordinate system, is physically unappealing. Some recent developments, which can only be sketched here, will allow us to remove this limitation. At the core of the new approach is the recent multi-center integration scheme of Becke,[25] already being successfully applied in local density functional calculations in quantum chemistry. The central idea is to define local, i.e. atomic, numerical grids and to then weight these such that any three-dimension integral will be reproduced by the sum of the atomic integrals. The weighting function for each atomic grid is chosen to be unity at the atomic nucleus and to go smoothly to zero at any other nucleus. By experimentation, Becke found a practical functional form which achieves the desired purpose in an efficient fashion. In a subsequent paper, Becke[26] extended these ideas to the solution of the Poisson equation. Here he suggested the use of local spherical harmonic expansions combined using the weighting ideas of his first paper. Although the specifics of his numerical approach to the radial equations need not concern us here, it is worth pointing out that there are efficient, integral equation approaches to this problem which have been developed to solve scattering problems. We have already adapted some of these ideas to the Poisson equation and are now examining the inhomogeneous wave equation which forms the basis of our variation-iteration method. We are confident that we will be able to use this multi-center approach to effect a totally numerical solution to the electron-polyatomic scattering problem. Research along these lines is already in progress and we expect to report some preliminary results quite soon.

4. ELECTRON-ATOM SCATTERING IN INTENSE FIELDS

Physical phenomena abound for which the linear algebraic approach provides an effective means of explication. Those cases involving scattering of heavy or light particles are easily discerned. However, other classes of atomic and molecular processes lend themselves to the technique. One particular example on which we shall concentrate concerns an atom subjected to an intense electromagnetic radiation field. We demonstrate that the basic Schroedinger equation that describes this interaction can be casted in a form closely resembling electron-scattering from a vibrating ionic diatomic molecule, which in turn can be solved with the LA method. We shall only present a brief overview of the general concepts; an exhaustive treatment and comprehensive bibliographies appear in the book by Mittleman,[29] the review article by Gavrila, [32] and the papers by Csanak and Collins.[27, 28]

The time-dependent (TD) Schroedinger equation in atomic units in the $\mathbf{E}\cdot\mathbf{r}$ gauge that describes the interaction of an electron with an infinitely massive proton and with a temporally-varying electric field has the form[29, 31, 32]

$$[T(\mathbf{r}) + V_p(\mathbf{r}) + V_E(\mathbf{r}|t)]\psi(\mathbf{r}|t) = i\frac{\partial}{\partial t}\psi(\mathbf{r}|t), \tag{23}$$

where

$$T(\mathbf{r}) \equiv -\frac{1}{2}\nabla^2, \tag{24}$$

$$V_p(\mathbf{r}) \equiv -1/\mathbf{r}, \tag{25}$$

$$V_E(\mathbf{r}|t) \equiv \mathbf{r}\cdot\mathbf{E}(t). \tag{26}$$

The first term T represents the kinetic energy operator for the electron; the second term, the electrostatic interaction of the electron and proton. The final term, in which we have made the usual dipole approximation to the radiation field, gives the interaction of the electron with the electric field $\mathbf{E}(t)$. For the laser configurations and intensities under consideration, we may neglect the magnetic component and the spatial dependence of the electric field. We usually further simplify by taking the electric field as purely oscillatory

$$\mathbf{E}(t) = \mathbf{E}_0\cos(\omega t), \tag{27}$$

with $|\mathbf{E}_0|$ the amplitude and ω the frequency. This condition restricts attention to cases in which the laser pulse is long compared to the ramp times although this restriction can be lifted by treating multimode fields. With these caveats, the Schroedinger equation assumes a particularly familiar form, being the starting point for a TD perturbative treatment of photoionization of hydrogen. However, for our case, the electric field of the laser becomes comparable to or greater than the field binding the electron to the proton. A perturbative treatment of the radiation field becomes inappropriate. We therefore seek solutions to Eq.(23) directly.

Before embarking upon this daunting task, we explore certain limiting cases of Eq.(23). If we neglect V_E, the field terms, we recover the usual, but mundane, Schroedinger equation for an electron scattering from a proton or for an hydrogen atom. The cross section and various T-matrix elements show no structure for such collisional events. An analytical solution arises in terms of Coulomb functions for the continuum and hydrogenic functions for the bound states. On the other hand, if we drop the electrostatic term V_p, we obtain a Schroedinger equation describing the motion of a free electron in an oscillating electrical field. Somewhat surprisingly, this equation too has an analytical solution, termed a Klein-Volkov state.[34] If we form a well-defined gaussian wavepacket from these states, we find some rather interesting properties. First, the center of the packet moves as a classical electron in such an oscillating field

$$\mathbf{r}(t) = \boldsymbol{\alpha}(t) = \boldsymbol{\alpha}_0\cos(\omega t), \tag{28}$$

where $\boldsymbol{\alpha}_0$ is the classical displacement $|\boldsymbol{\alpha}_0| = E_0/\omega^2$. The packet quivers about the path of a freely-moving electron with an amplitude equal to $|\boldsymbol{\alpha}_0|$. Second, the packet spreads, a distinctly quantum mechanical property, as a free particle, unafffected by the field.

Having established the basic limiting cases, we now concentrate on a full solution of Eq.(23). Many approaches have been developed for directly solving the TD Schroedinger equation.[33] However, we take a different tact and convert to a time-independent(TI) form that can be solved by standard electron-molecule techniques. We begin by making a transformation due to Kramers and Henneberger[30] from the laboratory frame to a frame tied to the oscillating electron. In the frame of the electron, the proton appears to oscillate with frequency ω and amplitude $|\boldsymbol{\alpha}_0|$, and the effects of the field only appear implicitly. The interaction becomes simply electrostatic in terms of the instantaneous relative position of the two particles. In a more formal sense, the frame transformation is effected through a simple displacement operator

$$\Omega(\mathbf{r}|t) = \exp[-\boldsymbol{\alpha}(t) \cdot \nabla], \tag{29}$$

with ∇ the usual gradient operator and $\boldsymbol{\alpha}(t)$ given by Eq.(28). Applying this transformation to Eq.(23) yields a TD Schroedinger equation in the Kramers-Henneberger(KH) frame

$$[-\frac{1}{2}\nabla^2 + V_p(\mathbf{r} + \boldsymbol{\alpha}(t))]\Phi(\mathbf{r}|t) = i\frac{\partial}{\partial t}\Phi(\mathbf{r}|t), \tag{30}$$

where

$$V_p(\mathbf{r} + \boldsymbol{\alpha}(t)) = \frac{-1}{|\mathbf{r} + \boldsymbol{\alpha}(t)|} \tag{31}$$

The transformation, as advertised, has the effect of replacing $\mathbf{r}$ with $\mathbf{r} + \boldsymbol{\alpha}(t)$. We still have a TD equation; however, the field term has been subsumed into the potential. In the KH frame, an electron approaches a charge oscillating with the field frequency - reminiscent of scattering from a vibrating molecule.

We convert to a TI form by invoking the usual Floquet ansatz. For a purely periodic potential of frequency ω, the solution of Eq.(30) has the general form

$$\Phi(\mathbf{r}|t) = \exp[-iEt]\Phi_p(\mathbf{r}|t), \tag{32}$$

where E is a quasi-energy associated with the electron and Φ_p is a periodic function. Since Φ_p has periodic form, we can expand in terms of a Fourier series as

$$\Phi_p(\mathbf{r}|t) = \sum_{n=-\infty}^{\infty} \phi_n(\mathbf{r})e^{-in\omega t}. \tag{33}$$

Substituting Eqs (32-33) into (30), multiplying through by $\exp[in'\omega t]$, integrating over a period $[T = 2\pi/\omega]$ of the field, and using the relationship

$$T^{-1}\int_{-T/2}^{T/2} \exp[i(n-n')\omega t]dt = \delta_{nn'}, \tag{34}$$

we find the following TI equation:

$$[-\frac{1}{2}\nabla^2 - E_n]\phi_n(\mathbf{r}) = -\sum_{n'} V_{nm'}(\boldsymbol{\alpha}_0|\mathbf{r})\phi_n(\mathbf{r}), \tag{35}$$

where $\mathrm{E}_n \equiv \mathrm{E}+\mathrm{n}\omega$, and

$$V_{nn'}(\boldsymbol{\alpha}_0|\mathbf{r}) = T^{-1}\int_{-T/2}^{T/2} V(\mathbf{r} + \boldsymbol{\alpha}(t))\exp[i(n-n')\omega t]dt. \tag{36}$$

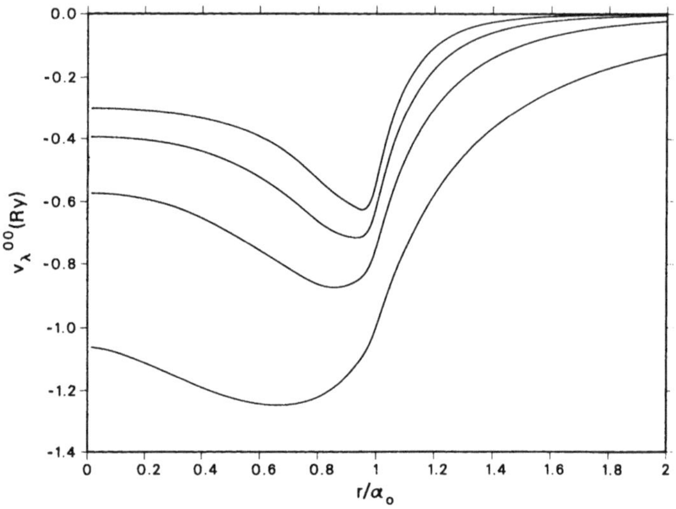

Figure 1. Single center radial expansion coefficients of V_{00}.

We have reduced the solution of the TD Schroedinger equation to a form [Eq.(35)] closely resembling a normal coupled channel scattering problem. We can further explicate this point by observing the boundary conditions of the channel functions of Eq.(33). As $\mathbf{r}$ becomes large, the potential tends to a simple Coulombic form since $|\boldsymbol{\alpha}_0|$ has a prescribed finite value

$$\lim_{\mathbf{r}\to\infty} |r + \boldsymbol{\alpha}_0 \cos(\omega t)|^{-1} \to \mathbf{r}^{-1}. \tag{37}$$

Therefore, the channel functions go to linear combinations of regular and irregular Coulomb functions just as in standard electron scattering from an ion. This characterizes one major advantage of the KH frame over the $\mathbf{E}\cdot\mathbf{r}$[Eq.(23)] or $\mathbf{p}\cdot\mathbf{A}$ formulations. For the latter two cases, the electric field lingers into the asymptotic realm making the boundary conditions more complicated. One other interesting point involves the potential $V_{nn'}(\boldsymbol{\alpha}_0|\mathbf{r})$, which has singularities at $\mathbf{r} = \pm\boldsymbol{\alpha}_0$. Although the order of the singularities differ, the form closely resembles the two-center potential encountered in electron scattering from a diatomic molecule. To reinforce this similarity, we display in Fig.1 the single-center radial expansion coefficients of the elastic term V_{00}. In this analogy, the classical displacement $\boldsymbol{\alpha}_0$ assumes the role of the internuclear separation R in the molecule. Finally, since the electric field has been treated classically, we have not mentioned photons. However, we intuitively identify and can formally associate the transition from the n to n′ Fourier states as the emission or absorption of n-n′ photons. The equal spacing between Fourier states again reminds us of vibrational excitation. All in all, we find this analogy between the the systems very strong. To solve Eq.(35), we can either apply a single-center expansion or a fully numerical solution by the grid techniques described above. As an example, we make a partial-wave expansion as

$$\phi_n(\mathbf{r}) = \sum_{\ell_n=0}^{\infty} \sum_{m_n=-\ell_n}^{\ell_n} r^{-1} f_{n\ell_n m_n}(r) Y_{\ell_n m_n}(\hat{r}), \tag{38}$$

with $Y_{\ell m}$ as spherical harmonic of order (ℓ, m). By substituting Eq.(38) into Eq.(35), multiplying through by $Y_{\ell_{n'} m_{n'}}(\hat{r})^*$, integrating over angles, and using the selection rules for spherical harmonics, we obtain

$$[\frac{d^2}{dr^2} - \frac{\ell_n(\ell_n+1)}{r^2} + k_\Gamma^2]f_\Gamma(r) = \sum_{\Gamma\Gamma'} U_{\Gamma\Gamma'}(r) f_{r'}(r), \tag{39}$$

where

$$\Gamma \equiv (n\ell_n, m_n), \tag{40}$$

$$k_\Gamma^2 \equiv 2E_n, \text{ and} \tag{41}$$

$$U_{\Gamma\Gamma'}(r) = \int Y_{\ell_{n'} m_{n'}}(\hat{r}) U_{nn'}(\alpha_o, \mathbf{r}) Y_{\ell_n m_n}(\hat{r}) d\hat{r} \tag{42}$$

We can now apply the usual linear-algebraic prescription directly to Eq.(39).

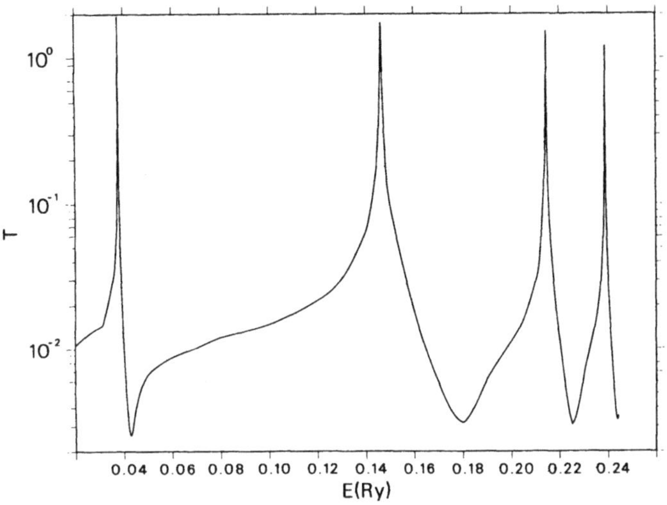

Figure 2. Multi-photon Feshbach resonances for the hydrogen atom in a laser field.

5. APPLICATIONS

5.1. Scattering in Intense Fields

As an example of an application of the LA technique to the realm of atoms interacting with intense lasers, we select electron scattering from a proton in the presence of a strong temporal electric field. The basic equations governing this process have been briefly reviewed in the previous section. In the absence of the laser field, the cross section and T-matrices as a function of electron energy exhibit no structure since the simplicity of the compound system prohibits Feshbach or shape resonances. However, as demonstrated in Fig.2, introducing the field leads to numerous Feshbach-like resonances in the elastic T-matrix element T(0,2|0,2) for a photon energy of 7.35eV (.27 au) and field strength of 0.0207 au($1.5\text{x}10^{13}$ W/cm^2). The origin of these capture-escape resonances stems from a multiphoton process. The electron emits n photons and falls to an energy level very close to a bound state of the composite system, the hydrogen atom. This compound resonance state has a much longer lifetime due to its temporary trapping in a bound state. In order to conserve energy, the electron must absorb n photons and return to the continuum. The lowest resonance state in Fig.2 corresponds to a two-photon process resonant with the 1s bound state of H. The higher resonances involve a single photon with the compound states of the n=2,3, and 4 levels. This second has a infinite number of members engaged with the hydrogenic Rydberg series. Inelastic processes such as Bremstralung also occur due to unequal absorptions and emissions. We can also extract multiphoton ionization rates by determining the resonance widths near the appropriate bound state(negative E).

5.2. Collisions of Electrons with H_2^+

As another illustration of the applicability of the LAM to a wide variety of phenomena, we describe the scattering of electrons from the hydrogen molecular ion H_2^+ in the regime strongly dominated by autoionizing resonances. While this seems a rather simple sytem, very large-scale calculations, involving many-state close-coupling formulations, are required to obtain accurate scattering information. We treat the basic collision process:

$$e^- + H_2^+ \rightarrow H_2^* \rightarrow e^- + H_2^+ \tag{43}$$

in which the incident electron becomes trapped in a two-electron excited state of the composite H_2 molecule and subsequently decays to the continuum. This process manifests itself as a strong signature in the basic collisonal cross section at an energy near an excited state of the compound system. Such resonances provide intricate tests of methods as well as considerable enhancement in the collisional process.

As an example, we shall concentrate on the energy region just below the first excitation threshold of the ion. The lowest lying states of H_2^+ have symmetries: $1\sigma_g$, $1\sigma_u$, $1\pi_u{}^+$, and $1\pi_u{}^-$. We use a generic symbol k for the channel energies, but we recognize that for a given total energy, the wavenumber for each channel will be different. For a choice of total symmetry of $^1\Pi_u$, the corresponding continuum states are: $k\pi_u^+$, $k\pi_g^+$, $k\sigma_g$, and $k\delta_g^+$ respectively. A series of bound states of the neutral molecule converge on the ionization thresholds of the ionic system. For example, as the principal quantum number n increases, the series $1\sigma_u n\pi_g^+$ has a limit of a bound electron in the $1\sigma_u$ state of H_2^+ and a continuum $k\pi_g^+$ electron. Similarly the series $1\pi_u^+ n\sigma_g$ and $1\pi_u^- n\delta_g^+$ converge on the $1\pi_u^\pm$. An electron impinging on the ion at an energy just below the $1\sigma_u$ threshold can easily be trapped in one of these two-electron excited states of H_2, producing resonance structure in the cross section. At large internuclear distances, these series remain well separated with those converging on the $\pi_u^\pm$ thresholds lying energetically above the $1\sigma_u$ ionic level. However, as we decrease R, the lowest resonance states of the $1\pi_u^+ n\sigma_g$ push below the $1\sigma_u$ threshold and begin to interact or mix with the higher-lying doubly-excited states of the $1\sigma_u n\pi_u^+$ series. This interference among the resonance series produces interesting effects in the positions and widths(lifetimes) of these levels. For the positions of the lowest few resonances, we note a behavior indicative of curve crossings in molecular systems. At the R value at which the the lowest-lying state of the higher series begins to cross resonance states of the lower series, we observe a distinct avoided crossing, indicative of the changing character of the resonance. Avoided crossings of electronic states of the same symmetry are well known for bound states but this marked the first quantitative study of the corresponding phenomenon for continuum levels. In Fig.3, we display the width as a function of R for the lowest three resonances in $^1\Pi_u$ symmetry . For large R, the third lowest resonance has distictly $1\sigma_u 3\pi_g^+$ character and has a small width. As the $1\pi_u^+ 2\sigma_g$ state begins to near this level, we note an enhanced feature in the width, characteristic of strong interfrence between the states. Once the intruder state has passed the $1\sigma_u 3\pi_g^+$ level, the width returns to its original character. We note similar behavior for the crossing of the $1\sigma_u 2\pi_g^+$ level. For very small internuclear distances, we observe a complete reversal with the first state of the upper series now becoming the lowest resonance. The actual interactions are more complicated, giving rise to the structure of the widths. To obtain reliable values for the resonance parameters, we had to employ an eight-state close-coupling expansion. For a single-center expansion of four partial waves per state and a mesh of one hundred radial points, the LA matrices reached orders of several thousand. Such large, non-sparse matrices were easily treated using the variation/iteration approach discussed in section 3.2. Whether these effects can be observed in synchrotron studies will depend on the development of an experimental capability to excite the initial H_2 to a vibrational level high enough to access the Franck-Condon region of the resonance. In addition, the observation of adiabatic as opposed to diabatic behavior of the electronic states involved in the crossing, can be influenced by other dynamic considerations. In any event, we hope this has demonstrated that even such simple systems as H_2 still have much to offer for theorists to contemplate.

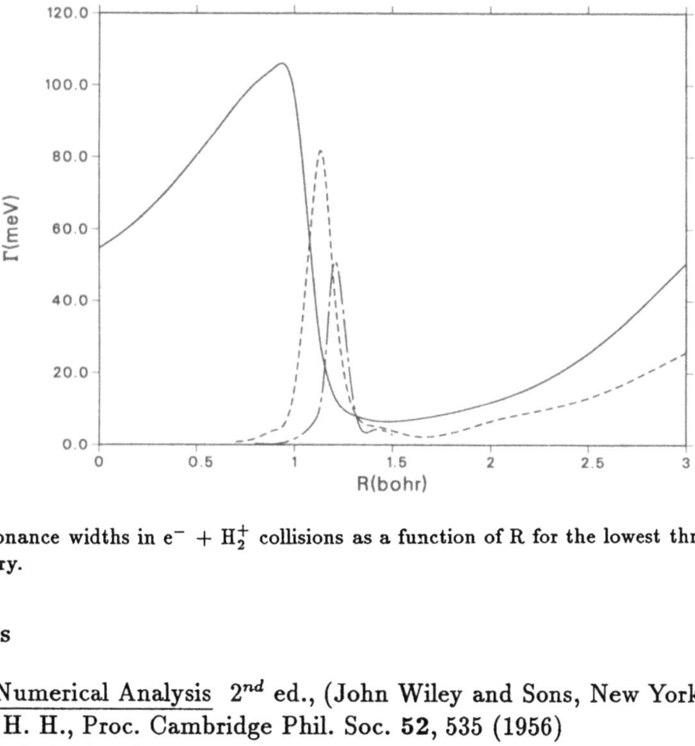

Figure 3. Resonance widths in $e^- + H_2^+$ collisions as a function of R for the lowest three resonances in $^1\Pi_u$ symmetry.

References

[1] Kopal, Z., Numerical Analysis 2nd ed., (John Wiley and Sons, New York, 1961).

[2] Robertson, H. H., Proc. Cambridge Phil. Soc. **52**, 535 (1956)

[3] Dongarra, J. J., Moler, C. B., Bunch, J. R. and Stewart, G. W., LINPACK Users' Guide, (SIAM, Philadelphia/1979)

[4] Faddeev, D. K. and Faddeeva, V. N., Computational Methods of Linear Algebra (W. H. Freeman and Co., San Francisco and London, 1963)

[5] Simply stated the problem is that the numerical integration process becomes contaminated by the exponentially growing, unwanted other solution to the second order set of equations. The undesired solutions may be eliminated during the integration process by matrix factorization procedures which take N^3 operations.

[6] Seaton, M. J., J. Phys. B **7**, 1817 (1974)

[7] Crees, M. A. and Moores, D. L., J. Phys. B **8**, L195 (1975)

[8] Schneider, B. I. and Collins, L. A., Phys. Rev. **A24**, 1264 (1981)

[9] Collins, L. A. and Schneider, B. I., Phys. Rev. **A24**, 2387 (1981)
[10] Schneider, B. I. and Collins, L. A., Phys. Rev. **A27**, 2847 (1983)
[11] Schneider, B. I. and Collins, L. A., Phys. Rev. **A33**, 2970 (1986)
[12] Schneider, B. I. and Collins, L. A., Comput. Phys. Comm. **53**, 381 (1989)
[13] Schneider, B. I. and Collins, L. A., Phys. Rev. **A33**, 2982 (1986)
[14] Collins, L. A. and Schneider, B. I., J. Phys. B **17**, L235 (1984)
[15] Collins, L. A. and Schneider, B. I., Phys. Rev. **A34**, 1564 (1986)
[16] Schneider, B. I. and Collins, L. A., Comp. Phys. Reports **10**, 51 (1989)
[17] See the articles on the R-matrix and Kohn variational methods in this volume.
[18] Comput. Phys. Comm. **53**, 381 (1989) was devoted entirely to the use of iterative methods in chemistry and physics. It is an excellent starting point and contains most of the important references.
[19] Chase, D. M., Phys. Rev. **104**, 838 (1956)
[20] Temkin, A. and Vasavada, K. V., Phys. Rev. **160**, 109 (1967)
[21] Feshbach, H, Ann. Phys. **5**, 357 (1958)
[22] Light, J. C. and Walker, R. B., J. Chem. Phys. **65** ,4272 (1976)
[23] Schneider, B. I. and Walker, R. B., J. Chem. Phys. **70**, 2466 (1979)
[24] Noble, C. J. and Nesbet, R. K., Comput. Phys. Comm. **33**, 399 (1984)
[25] Becke, A. D., J. Chem. Phys. **88**, 2547 (1988)
[26] Becke, A. D. and Dickson, R. M., J. Chem. Phys. **89**, 2993 (1988)
[27] Csanak, G and Collins, L. A. Phys. Rev. **47**, 3240 (1993)
[28] Collins, L. A. and Csanak, G., Phys. Rev. **44**, R5343 (1991)
[29] Mittleman, M. H., Introduction to the Theory of Laser-Atom Interactions, 2nd ed., (Plenum, New York, 1993)
[30] Henneberger, W. C., Phys. Rev. Lett. **21**, 838 (1968)
[31] Mittleman, M. H., Phys. Rev. **A42**, 5645 (1990)
[32] Atoms in Intense Laser Fields, ed. Gavrila, M. (Academic Press, New York, 1992)
[33] Time-Dependent Methods for Quantum Dynamics, ed. Kulander, K., Comp. Phys. Comm. **63** (1991)
[34] Volkov, D. M., Z. Phys. **94**, 250(1935)

ANALYSIS OF DISSOCIATIVE RECOMBINATION OF ELECTRONS WITH $ArXe^+$ USING $ArXe^*$ CALCULATIONS

A. P. Hickman, D. L. Huestis, and R. P. Saxon

Molecular Physics Laboratory
SRI International
Menlo Park, CA 94025, USA

INTRODUCTION

Electron scattering from a molecular ion is closely connected with the behavior of the highly excited electronic states of the corresponding neutral molecule. Both systems are governed by the same Hamiltonian. This chapter reviews recent calculations that illustrate the methodology that can be applied in such situations. The examples described are drawn from our work over the past few years developing theoretical techniques to treat inelastic collision processes important in laser systems. For example, in the atomic xenon laser, the following processes are important:

$$Xe^*(n\ell KJ) + He \rightarrow Xe^*(n'\ell'K'J') \tag{1}$$

$$Xe^*(n\ell KJ) + Ar \rightarrow Xe^*(n'\ell'K'J') \tag{2}$$

$$ArXe^+ + e \rightarrow Xe^*(n\ell KJ) + Ar \tag{3}$$

Processes (1) and (2) are fine structure changing collisions. The quantum numbers specify the atomic orbital (nl), the details of the angular momentum coupling between the valence electron and the ionic core (K), and the total angular momentum (J). We have recently calculated the excited state potential curves and matrix elements necessary for a quantum mechanical treatment of the dynamics of these collisions.[1,2] Process (3), dissociative recombination (DR), involves the same atomic particles (in a different arrangement) as processes (1) and (2). Using information from our recent calculations on the low-lying excited states that govern processes (1) and (2), we have been able to obtain the necessary potential curves and matrix elements to analyze DR.

Two features characteristic of heavy atoms such as argon and xenon make calculations more difficult than corresponding work on ligher atoms. The first is that spin orbit and relativistic contraction effects must be included in the calculations of the electronic wave functions. In our calculations, these are handled by techniques that replace the inner shell electrons with an effective potential, and then treat explicitly only the outermost electrons. In this fashion, argon and xenon can be treated as eight electron systems: the active electrons for

Computational Methods for Electron-Molecule Collisions
Edited by W.M. Huo and F.A. Gianturco, Plenum Press, New York, 1995

argon are $3s^23p^6$ and for xenon are $5s^25p^6$ or $5s^25p^5n\ell$. The second feature is that there are many more excited state energy levels to consider in the heavier molecules. Methods that are feasible for simpler systems become too cumbersome to apply. We have adopted several techniques of Multichannel Quantum Defect Theory (MQDT), which provides a framework for estimating a very large number of excited state potential curves in a systematic fashion.

This chapter describes the process of DR in general terms, indicate its role in the atomic xenon laser, describe our recent calculations of low-lying electronic potential curves, matrix elements, and scattering cross sections for processes (1) and (2), describe the extension of these results to obtain the potential curves for highly excited states, and finally arrive at an estimate for the rate of DR of electrons with $ArXe^+$

DESCRIPTION OF DR

The physical picture of DR, treated by Bardsley[3] in 1968, may be understood using the potential curves sketched in Fig. 1. This diagram shows $V^+(R)$, the potential for a molecular ion AB^+, and $V^*(R)$, the potential for a highly excited state (denoted AB^{**}) that dissociates to $A+B^*$. Symbolically, we write

$$e + AB^+ \leftrightarrow AB^{**} \rightarrow A + B^* \tag{4}$$

The electronic matrix element $V_{el}(R)$ (not shown) couples the initial scattering state ($e + AB^+$) with the AB^{**} state. The first step of DR is the capture of the incident electron into the state AB^{**}. Then the nuclei follow the dissociating potential curve V^*. The excess electronic energy of the resonant state is converted to kinetic energy of the dissociating products. These products must separate to the point where V^* crosses V^+ for the recombination to be complete. Competing with the stabilization of the resonant state by dissociation is electronic autoionization [indicated by the leftward pointing arrow in Eq.(4)]. The size of the DR cross section will be sensitive to the relative values of the autoionization rate and the speed of dissociation.

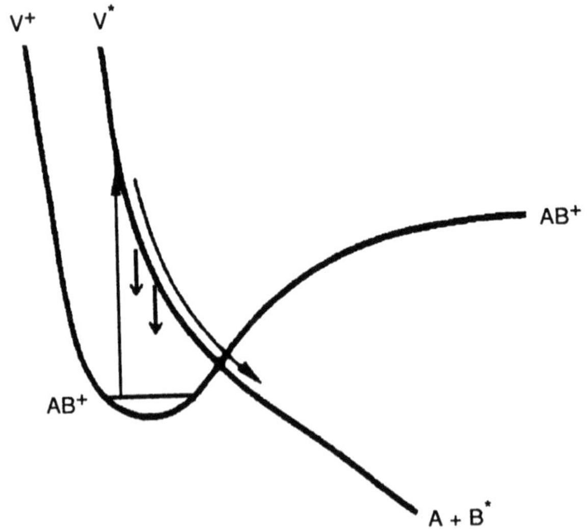

Figure 1. Schematic diagram of dissociative recombination.

The details of the theoretical analysis of DR are available in the literature.[3-6] Another mechanism, which we do not consider here, is called indirect DR and is a two step process involving another intermediate state in the process. The results of the analysis of the process shown in Fig. 1 show that the cross section is given by

$$\sigma_{DR} = \frac{\pi^2}{\varepsilon} \frac{\omega}{2} \frac{\hbar K}{M} \left| \frac{\frac{1}{C} \int F(R)\, V_{el}(R)\, \chi_o(R)\, dR}{1 - i\pi g_{oo}} \right|^2 \tag{5}$$

where ε is the incident electron energy, the statistical factor is the ratio of the multiplicity of the V^* state to that of V^+, K and M are the wave vector and reduced mass for the heavy particle motion, C is the Wronskian K/2M, F is the wave function for the heavy particle motion, V_{el} is the electronic coupling matrix element, χ_o is the initial vibrational wave function of AB^+, and g_{oo} is the matrix element of the Greens function G(R',R):

$$g_{oo} = \int dR' \int dR\, \chi_o(R')\, V_{el}(R')\, G(R_<, R_>)\, \chi_o(R)\, V_{el}(R) \tag{6}$$

The explicit form of G(R',R) is available in the literature.[6]

ROLE OF DR IN THE ATOMIC XENON LASER

The atomic xenon laser operates on several IR transitions from 5d to 6p levels of excited xenon. These transitions are indicated on the energy level diagram in Fig. 2. The operation of the laser provides clear evidence that the 5d levels are populated, but the details of the kinetics are not fully understood.[7,8] The general scheme may be sketched as follows. The laser gas mixture typically consists of a few percent xenon in a buffer of helium and argon. Xenon is ionized by electron beam pumping, or in some cases by so-called nuclear pumping[7]:

$$Xe \rightarrow Xe^+ + e \tag{7}$$

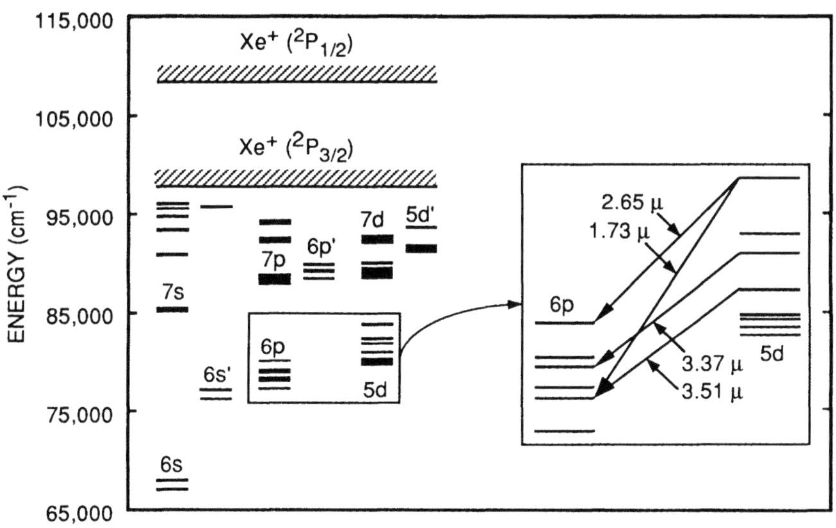

Figure 2. Energy level diagram for atomic xenon. Energies are given relative to the ground state, Xe($5p^6$). Both Xe^+ fine structure levels are shown. The inset shows selected 5d→6p transitions for the atomic xenon laser.

Various three body association reactions are thought to lead to the formation of $ArXe^+$:

$$Xe^+ + 2\,Ar \rightarrow ArXe^+ + Ar \tag{8}$$

DR can then produce various excited states of Xe:

$$ArXe^+ + e \rightarrow Xe^{**} + Ar \tag{9}$$

The distribution of excited states Xe^{**} is not known. It is possible that some of the 5d upper laser levels are produced directly by DR, but probably more likely that additional collisional and radiative processes convert more highly excited Xe^{**} to 5d levels.

Xe_2^+ may also be produced in the laser. It is thought[8] that DR of Xe_2^+ directly produces the lower laser levels $Xe^*(6p)$. This process, which would degrade the population inversion, is suppressed in practice by avoiding large concentrations of xenon.

POTENTIAL CURVES AND MATRIX ELEMENTS

Electronic structure calculations to determine excited-state potential curves for heavy atom systems are difficult because of the large number of inner shell electrons, which exhibit relativistic effects, and the large spin-orbit interactions. Theoretical treatment of the dynamics is also a challenging problem. A significant physical feature is the structured core of the xenon atom. The $Xe^+(5p^5)$ core has spin and orbital angular momenta that can participate in the collision. This complexity is absent in the more frequently studied case of collisions of excited alkali atoms with rare gases, because the core of the alkali atom is a closed shell.

Applying standard methodology to calculate potential curves and matrix elements, and then using the results in a coupled-channel calculation is not feasible in the present situation. Available electronic structure methods do not readily provide all the information necessary for a scattering calculation. At the present time, the most tractable approach to treat the electronic structure uses an effective potential to represent the ion core, and includes the spin-orbit matrix elements at the configuration-interaction (CI) level. Unfortunately, this method has not been extended to provide the radial coupling matrix elements necessary for a scattering calculation. Therefore, we have developed an analytic model that enables us to combine the information determined from *ab initio* electronic structure calculations with available spectroscopic information. The method allows us to determine realistic potentials and coupling matrix elements.

The details of our calculations of the potential curves have appeared elsewhere.[1] Here we will only summarize the key points of the method. The first step is to perform *ab initio* electronic structure calculations for Xe^*Ar. These calculations used an approach developed by Pitzer and co-workers.[9,10] In this approach, the effect of the inner shell electrons on each orbital outside the core is represented by a relativistic effective potential (REP). The REP for each orbital depends on the angular momentum quantum numbers ℓ and j and is determined from a numerical solution of the relativistic Dirac-Fock equation. It is convenient to replace the REPs for the two orbitals having $j=\ell+1/2$ and $j=\ell-1/2$ by their weighted average and difference. The averaged relativistic effective potential (AREP) depends only on ℓ and can be easily implemented with standard electronic structure codes. The difference between the REPs provides an approximate description of the true spin-orbit interaction. The spin-orbit operator is incorporated in the calculation at the CI level. The final result is that XeAr can be treated as a 16 electron problem.

Further refinement of the *ab initio* calculations was necessary before proceeding with the scattering calculations. Because the energy levels of Xe (including fine structure) are so closely spaced ($\Delta E \sim 100\text{-}1500\,cm^{-1}$), particular attention was paid to the accuracy of the asymptotic limits of the calculated potential curves. Although accurate by conventional standards, the *ab initio* calculations did not provide asymptotes of the exceptional precision needed for the scattering calculations. In addition, the *ab initio* calculations provided only the adiabatic potential curves, that is, the eigenvalues of the total Hamiltonian at each internuclear distance R. Also necessary for a scattering calculation are the matrix elements coupling the various potential curves. To insure accurate asymptotes, and to obtain the necessary coupling matrix elements we developed a method to refine and extend the *ab initio* calculations.

Our method is based on defining a model Hamiltonian. As illustrated in Fig. 3, the physical scattering system is composed of an incident rare gas atom and an excited xenon atom, which consists of an excited electron and an ion core. We partition the total Hamiltonian into components that describe the interactions between each pair of components of the full system. The essential physical approximations in our method are the same as those used successfully to describe fine structure changing collisions of excited ^2P alkali atoms with rare gases.[11,12] The only difference in the present case is that the spin and orbital angular momenta of the ion core are nonzero and must be included in the electronic wave function. This extra feature can be fully incorporated by treating the atomic portion of the model Hamiltonian (He-Xe^+) as a two electron system, that is, one valence electron and one core hole. The details of this method have been developed by Condon and Shortley.[13] The result is that our model Hamiltonian is based on the same physics used to treat simpler fine structure changing collisions; only the algebra is more complicated.

The model Hamiltonian depends on a small set of physically sensible parameters. Our strategy is to determine the parameters of the model Hamiltonian that fit the *ab initio* potentials, and then selectively modify certain parameters that can be more accurately determined from spectroscopy or other experimental data. The model Hamiltonian is fully determined by the final set of parameters, and it is possible to determine both the adiabatic potential curves and the coupling matrix elements needed for the scattering calculations.

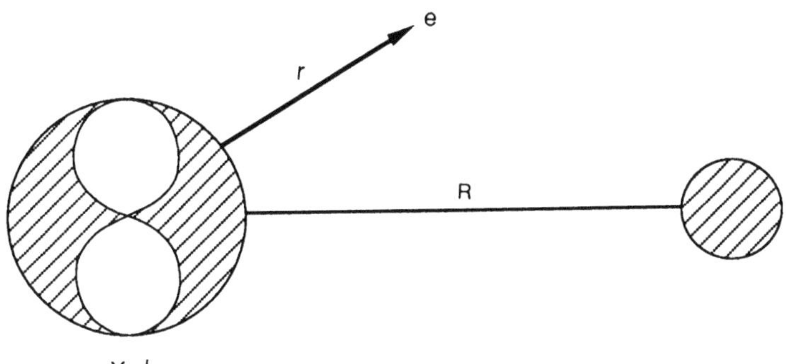

Figure 3. Schematic illustration of the model used to formulate the interaction of an excited xenon atom with a rare gas. The Xe^+ core has an unfilled 5p electron orbital.

SCATTERING CALCULATIONS FOR Xe^* + Ar

The potential curves and coupling matrix elements determined by the methods previously described provide the input to coupled channel scattering calculations. These calculations are described in detail elsewhere;[2] here we only note for completeness the general form of these equations:

$$\left[\frac{d^2}{dR^2} + \mathbf{U}(R)\right]\Psi(R) = 0 \quad (10)$$

$$\mathbf{U}(R) = (2m/\hbar^2)\,[\mathbf{E} - \mathbf{V}(R) - \mathbf{V}_c(R)] \quad (11)$$

$$\Psi(R) \sim \mathbf{J}(R) + \mathbf{N}(R)\mathfrak{R} \quad (12)$$

Solution of these equations determines cross sections for transitions among the various excited states of Xe. Eq.(7) determines the total wave function ψ in terms of the coupling matrix **U**, which contains electronic terms **V** and centrifugal terms $\mathbf{V_c}$. $\mathbf{V_c}$ also includes rotational coupling. The asymptotic form of the wave function ψ is determined by the reactance matrix $\mathfrak{R}$, which specifies the linear combination of regular and irregular scattering functions in the **J** and **N**. For the present calculations, we included all states corresponding to the 6s', 6p, and 5d levels of the isolated Xe atom. The total number of channels was 34 in each of two symmetries, and the log derivative algorithm[14] was used to solve the coupled equations.

Several experimental groups[15-22] have measured collisional mixing and quenching rate constants at T=300, and so for comparison with these data we calculated rate constants from the calculated cross sections. An overview of the results can be seen in Fig. 4, which presents total quenching rate constants. The quenching rate is defined as the sum of all the inelastic rates from the specified level to any other level. The most striking feature of the calculations and the experimental data shown in Fig. 4 is the extreme selectivity of the rate constants with respect to the initial level and the collision partner. Some levels exhibit a large quenching rate, and other nearby levels do not. There are cases where a level is strongly quenched by He but not by Ar, and where the opposite is true.

In summary, our potential curves and coupled channel scattering calculations have provided an explanation for a large amount of experimental data. Remarkable overall agreement with measured total quenching rates has been achieved; a few questions remain concerning the branching ratios. The extreme selectivity of the rates with respect to specific states and collision partners can be explained in terms of the details of the potential curves.

POTENTIAL CURVES FOR DR

The work just described provided a great deal of knowledge about the low-lying excited state potential curves of Xe^*Ar. Since the same Hamiltonian describes inelastic collisions of Xe^* with Ar and DR of electrons with $ArXe^+$, we can extend our work to develop methods for determining the highly excited potential curves that govern DR. This section sketches the preliminary work we have accomplished in this direction.

Some of the adiabatic potential curves already calculated are illustrated in Fig. 5. This figure emphasizes where these curves lie with respect to the molecular ion potential curves. It is clear that the curves already calculated are well below the ion potential curves. What we would like to do is extrapolate the low lying potential curves already available, in order to obtain curves all the way up to the ionization limit.

A standard method for extrapolating energy levels is quantum defect theory.[23] If there is only a single ion core level, at energy I, the procedure is very simple. An energy level at E = E_n can be used to determine the quantum defect μ, which is defined such that the energy

$$E_n = I - \frac{1}{2(n-\mu)^2} \quad (13)$$

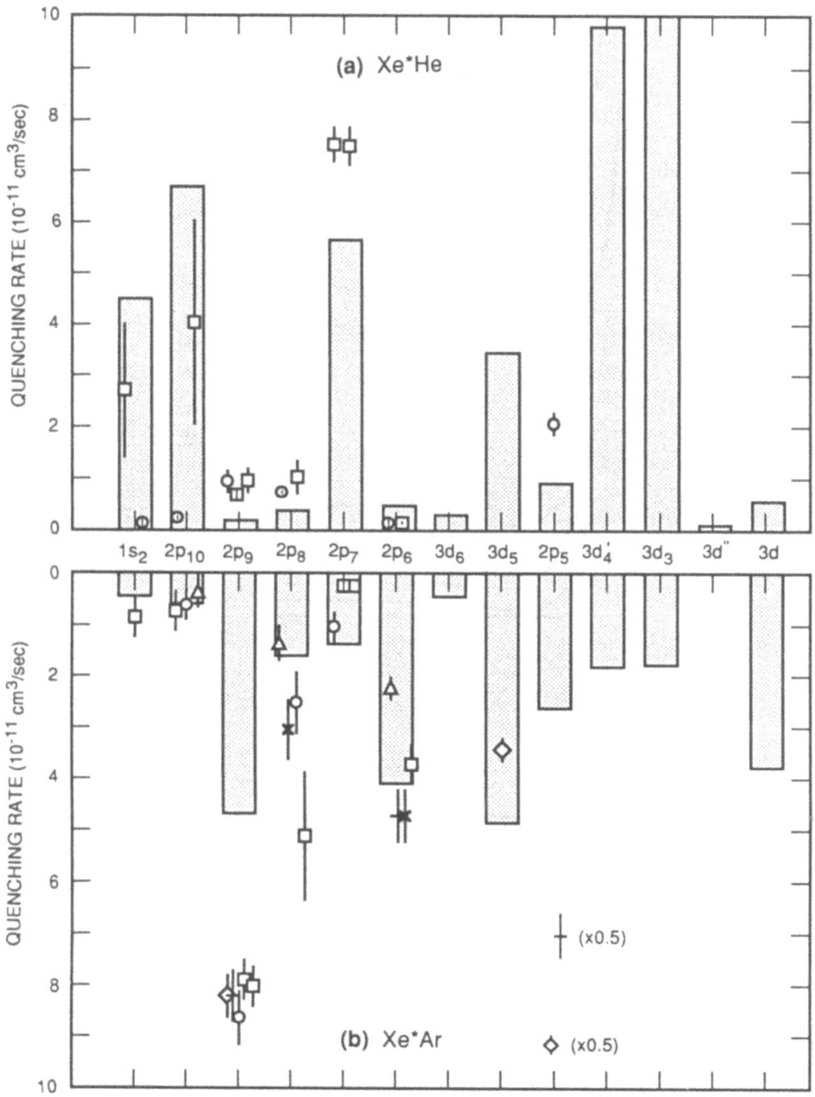

Figure 4. Total quenching rates of specific excited states of Xe by He (panel a) and by Ar (panel b). The shaded bars are the present calculations, and the points are various experiments: □, Alford; O, Xu and Setser; ◇, Bruce et al.; —, Ku and Setser; ∇, Inoue et al.; X, Horiguchi et al.

It is usually the case that the quantum defect μ depends weakly on the energy, so that the binding energies of levels for higher quantum numbers are found by replacing n with successively higher integers in Eq. (13). This approach is easily generalized to include R dependent potential curves by replacing μ with μ(R); I becomes the potential energy curve V(R) of the single ion state, and the calculation is performed separately for each value of R.

It is evident from Fig. 5 that this simple approach is inadequate for Xe^*Ar because several molecular ion levels are involved. One must generalize the approach using Multichannel Quantum Defect Theory (MQDT).[23] In the present ase, there are three ionic levels, which can be written I_1, I_2, and I_3, or $V_1(R)$, $V_2(R)$, $V_3(R)$. These potentials correspond to the X, A_1, and A_2 potential curves. The single quantum defect μ(R) must be

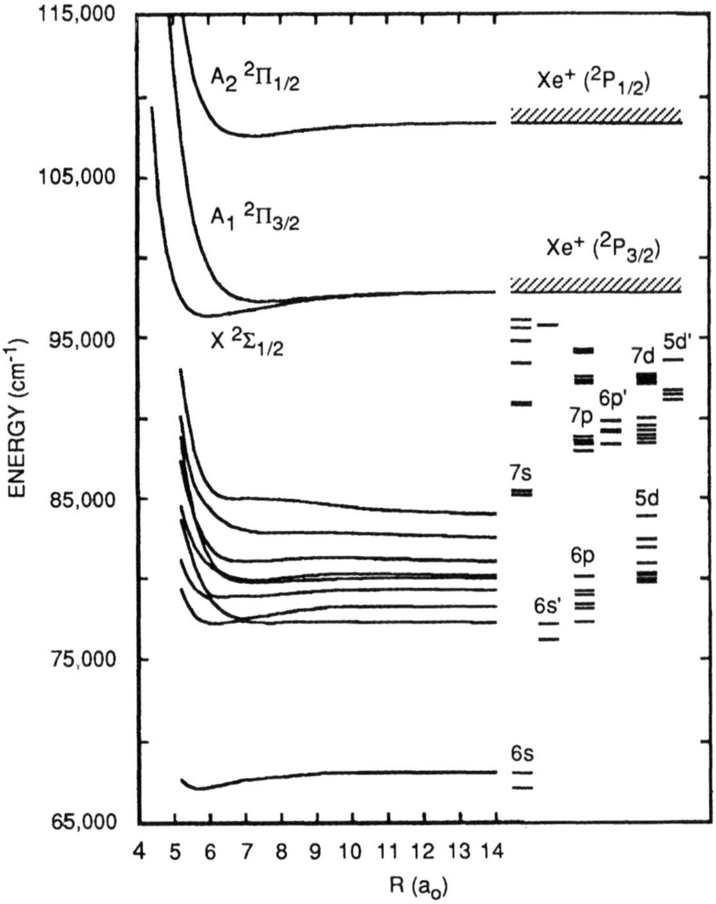

Figure 5. Adiabatic potential curves with $\Omega = 0^+$ calculated by the methods described in Ref. 1. Also shown are the three lowest potential curves of $ArXe^+$ and several excited states of the xenon atom.

replaced by a matrix **K**(R). The elements of **K**(R) correspond to channels that are labeled by the ionic energy levels i and possibly other electronic quantum numbers α. The energies E of bound states satisfy the determinental equation

$$\det|\tan\pi\nu(E)+\mathbf{K}(R)| = 0 \tag{14}$$

the solutions to this equation, ehich must be found separately at each R, detemine the energies of the excited electronic states of the system. In Eq. (14), ν(E) is a diagonal matrix ($\nu_{ij} = \nu_i\delta_{ij}$) that defined the energy differences between any E and each of the ionic energy levels, according to

$$E = I_i - \frac{1}{2\nu_i^2} \tag{15}$$

The general formula, Eq. (14), reduces to the single channel limit if one makes the association $K = \tan\pi\mu$.

In order to apply MQDT, we must identify the parent ion core states for our basis functions. For our original scattering calculations, we expressed all matrix elements in a basis of atomic eigenfunctions. This representation was appropriate for collisions of Ar with Xe^*, but is unsatisfactory for the analysis of DR. To treat collisions of electrons with $ArXe^+$, we must transform to a representation in which the $ArXe^+$ core quantum numbers are well defined.

The relationship between these two representations can be discussed by referring to the schematic picture in Fig. 3. The Xe^+ core has orbital and spin angular momenta L_c and S_c, and the electron has orbital and spin angular momenta ℓ and s. For our purposes, the argon is structureless. For DR, we want to couple the Xe^+ with Ar to form $ArXe^+$, which has angular momentum quantum numbers J_c and M_{Jc}, and then let $ArXe^+$ collide with an electron. The scheme is then

$$\begin{aligned} L_c + S_c &= J_c \\ \ell + s &= j \end{aligned} \tag{16}$$

and the basis functions may be denoted $|L_cS_cJ_cM_{J_c};\ell s j m_j>$. For inelastic scattering, we want to couple Xe^+ with the electron to form Xe^*, and then let the Xe^* collide with the Ar. The scheme that most accurately reflects the physical situation is

$$\begin{aligned} L_c + S_c &= J_c \\ J_c + \ell &= K \\ K + s &= J \end{aligned} \tag{17}$$

and the basis functions are $|L_cS_cJ_c\ell Ks;J\Omega>$, where $\Omega = M_J$. (Following conventional notation, we have used K in Eq. (17) and elsewhere to denote an approximate quantum number of Xe^*; this K should not be confused with the **K** matrix defined by MQDT). For the calculations we have reported,[1,2] we actually used a coupling scheme determined by numerical diagonalization of the appropriate Hamiltonian. The scheme thereby determined is similar to that of Eq. (17), but has a few notable differences that were pointed out. The basis functions actually used for the numerical diagonalization were the jj coupling functions, and it will be convenient to return to these for our further discussion. These basis functions are denoted $|L_cS_cJ_c;\ell s j m_j;JM_J>$, and they are based on the following coupling scheme:

$$\begin{aligned} L_c + S_c &= J_c \\ \ell + s &= j \\ J_c + j &= J \end{aligned} \tag{18}$$

The transformations between all of these basis sets may be written explicitly using Clebsch-Gordan coefficients. For the jj basis and the basis used for DR, the result is

$$|L_cS_cJ_cM_{J_c};\ell s j m_j> = \sum_J \sum_{M_J} < J_cM_{J_c}\, j m_j \mid JM_J > |L_cS_cJ_c;\ell s j m_j;JM_J> \tag{19}$$

We note that the electronic states with $\Omega=|M_J|=0$ in the atomic basis are divided into two groups with symmetries $\Omega=0^+$ and $\Omega=0^-$. It is possible to choose a set of molecular ion core eigenfunctions that preserves the same symmetry. Those states with $\Omega=0$ correspond to state $|L_cS_cJ_cM_{J_c};\ell s j m_j>$ such that $M_{J_c}+m_j=0$. Taking linear combinations of the states with M_{J_c},m_j and with $-M_{J_c},-m_j$ leads to states of the desired symmetry. (Since M_{J_c} and m_j are half integral, the plus and minus states are distinct.) In this way we can separately transform each set of atomic states with $\Omega=0^+,0^-,1,2,3,4$ to a corresponding set of states with a well defined J_c and $|M_{J_c}|$. These are exactly the quantum numbers that define the molecular ion potential curves.

In the $|L_cS_cJ_c;\ell s j m_j;JM_J>$ basis, the general form of the Hamiltonian (for any fixed R) is as follows:

(20)

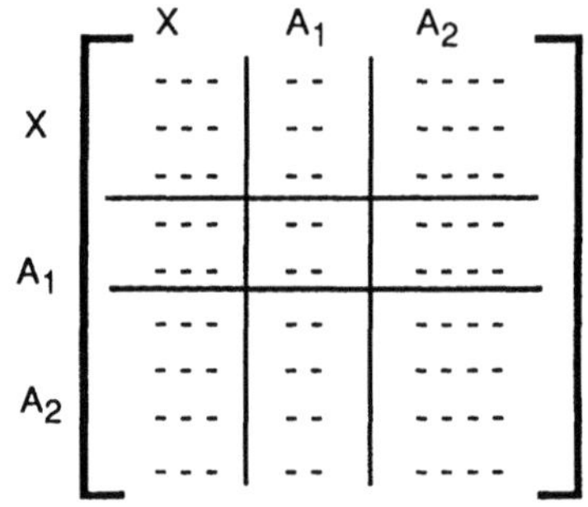

The blocks are labeled by X, A_1, and A_2, corresponding to the state of the molecular ion core. In the present case, the levels 6s, 6s', 6p, 6p', 5d, and 5d' were included, and the blocks have from 0 to 8 elements. The fact that there are off-diagonal elements connecting different blocks reflects the fact that, in general, the electronic eigenfunctions are linear combinations of different core states.

It is conceptually useful to apply a transformation that separately diagonalizes each diagonal block of the above Hamiltonian. Such a transformation still leaves matrix elements connecting one block with another. The Hamiltonian then looks like this:

(21)

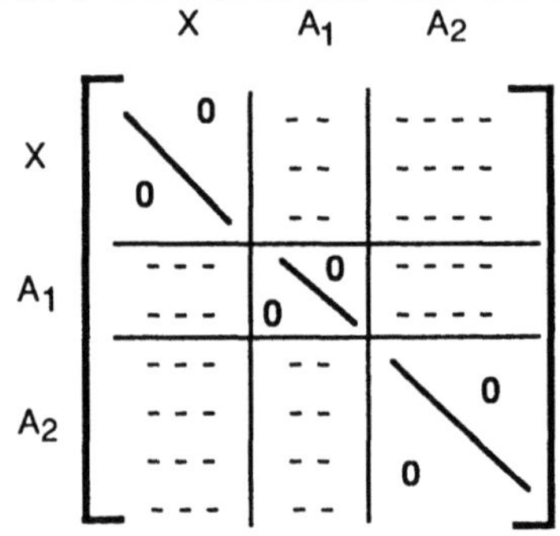

This procedure gives diabatic states that correspond to freezing the ion core. The diagonal elements are the diabatic energies, and the off-diagonal elements can be related to autoionization widths for these states.

The next step in a rigorous approach would be to determine the matrix **K**(R) such that the energies that satisfy the determinental equation agree with those determined by direct diagonalization of Eq. (18). The **K** matrix would be obtained by a fitting process. Such a determination of **K** is planned by the present workers, and some related work has already been performed by other investigators.[20] When **K** is available, the entire spectrum of excited state potential curves as well as autoionization widths can be determined from MQDT.

We have used two methods to make a preliminary determination of **K**. The first method is the simplest; one neglects the off-diagonal elements of Eq. (18). This leads to a solution in which **K** is diagonal, and each element can be determined by single channel quantum defect theory. Each of the diagonal energies in Eq. (18) corresponds to a single (R-dependent) quantum defect, and this single quantum defect generates an entire series of Rydberg states in the standard way.

Preliminary calculations of the potential curves based on this approximation are shown in Fig. 6. The curves shown correspond to the lowest electronic energies for each "frozen" molecular ion core level. Those curves with the X or the A_1 core roughly correspond to the diabatic counterpart of the adiabatic curves shown in Fig. 5. (The adiabatic curves may have an A_2 component as well, but it is typically small.) The curves with the A_2 core represent new

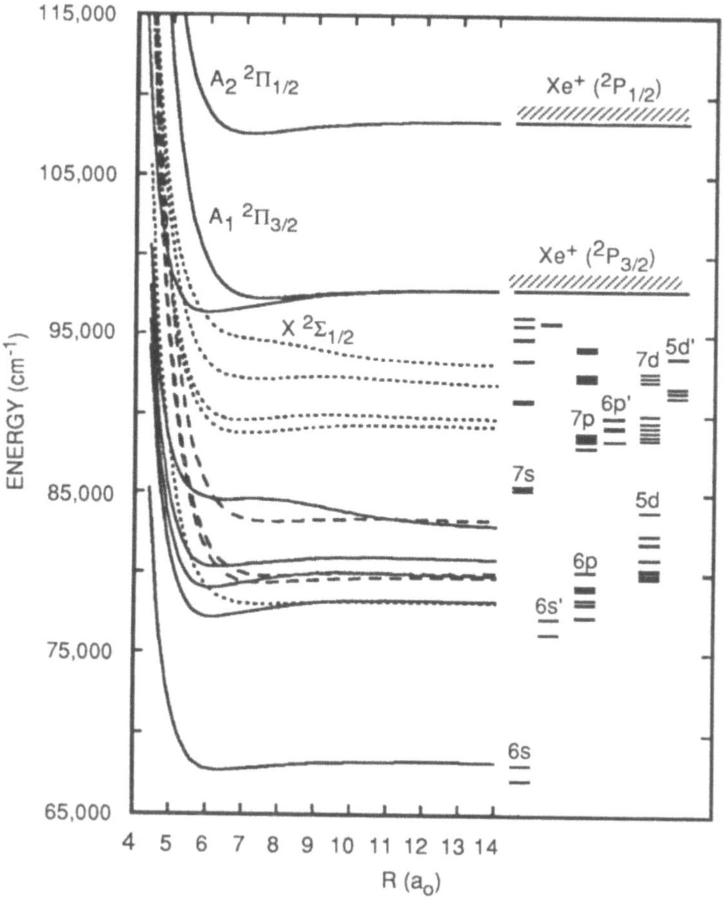

Figure 6. Preliminary diabatic potential curves $\Omega = 0^+$. The curves are the eigenvalues of a modified Hamiltonian that has no coupling between different molecular ion core states. The solid curves correspond to the X core, the dashed curves to the A_1 core, and the dotted curves to the A_2 core. Additional curves have been calculated but are not shown.

information, that was not available from the earlier calculations. As the diagram shows, the dotted curves cross the ground state (X) $ArXe^+$ potential; above this ionic curve the dotted curves represent autoionizing states, and are candidate states for DR. The dashed curves shown, which belong to the A_1 core, do not provide obvious pathways for DR. Higher Rydberg potentials in this series were calculated but are not shown to avoid a cluttered diagram. These higher energy states on the A_1 core also cross the X $ArXe^+$ potential curve and provide potential pathways for DR.

There clearly are a large number of dissociating autoionizing states of Xe^*Ar that may lead to DR. Autoionizing states with both the A_1 and A_2 cores are available. All of the diabatic states generated so far dissociate to fairly high lying states of atomic xenon. However, we cannot rule out processes that could directly populate the lower lying 5d levels, because we have not yet analyzed the possibilities for curve crossings on the outgoing potential curves. The number of such crossings is potentially enormous, particularly if rotational coupling is important. It will undoubtedly be easier to obtain estimates of the total cross sections for DR before attempting detailed calculations of the distribution of final states.

ESTIMATE OF RATE FOR DR OF $ArXe^+$

The calculations of K(R) described at the end of the previous section do not provide the off-diagonal coupling terms needed for the application to dissociative recombination. An improved method has been implemented that provides these terms. The matrix elements of our model Hamiltonian are expressed in the following form:

$$\mathbf{H}(R) = \mathbf{E}_{ion}(R) - \frac{1}{2\,[\mathbf{n}^*(R)]^2} \tag{22}$$

$$\mathbf{K}(R) = -\tan\pi\mathbf{n}^* \tag{23}$$

where $\mathbf{E}_{ion}$ is a diagonal matrix whose elements are the potential curves of the lowest states of $ArXe^+$. One can solve for the matrix $\mathbf{n}^*$ at each R and then calculate $\mathbf{K}(R)$, which is the fundamental matrix required for MQDT. We note that the above equations are formally similar to the simple results for quantum defect theory in one channel. In that case, n^* is the effective quantum number of the Rydberg level whose energy is giben by Eq. (13). Our equations are exact in this single-channel limit and also in the multichannel case in which the off-diagonal elements are zero. The equations are not exact in the general multichannel case because the matrices involved are not diagonal and need not commute. However, our testing using the asymptotic atomic energy levels has indicated that the above formulas are fairly accurate. Also, we foresee that in many applications one will treat the parameters of the model Hamiltonian $\mathbf{H}(R)$ as fitting parameters and vary them until the optimum value of $\mathbf{K}(R)$ is obtained. In this case, the above equations should be completely satisfactory.

The expression for $\mathbf{K}(R)$ can be combined with general expressions from MQDT to calculate (1) more accurate potential curves than those presented in Fig. 6, (2) the matrix elements V_{el} coupling these states with the continuum, and (3) cross sections for DR. The MQDT expressions are more general than Eq. (5), because they allow for the availability of several possible capture channels. However considerable further effort would be needed to account for the various curve crossings in the outgoing channels.

Work to address the full complexity of DR has not yet been completed. At present, we are able to make a rough estimate of the rate for DR based on the work that has been done so far. This estimate is described in this section. The coupling matrix elements are determined from the $\mathbf{K}(R)$ matrix as follows:

$$V_{el}(R)_{ij} = \pi n^{-3/2}\, K(R)_{ij} \tag{24}$$

where n is the effective quantum number of an $ArXe^{**}$ dissociating state, and the typical scaling for Rydberg wave functions is assumed. The subscripts refer to an appropriate pair of initial and final states. The energy width of a state i due to coupling to a continuum state j is $\Gamma_{ij}(R) = 2\pi|V_{el}(R)_{ij}|^2$.

We simplify the calculation by evaluating the coupling matrix only at the equilibrium distance $R=6a_0$ of the initial vibrational wave function of $ArXe^+$. Because of the narrow range of R in which this wave function is large, this approximation should be reasonable. In this case, n has only one of two possible values, depending on whether the A_1 or the A_2 excited state of $ArXe^+$ is the parent of the autoionizing state.

We further simplify the calculation by neglecting the competition between the various entrance channels and just treating separately all plausible incoming channels and adding the cross sections. This approach does not include saturation effects, in the sense that capture by any one state may reduce the amount of flux that can be captured by another state of the same symmetry. We note, however, that Eq. (2) fully includes survival effects: for large values of the coupling matrix elements, the electron is quickly captured but also quickly reemitted. Only those collisions are counted for which the resonance state survives long enough to dissociate.

We evaluated approximate potential curves for several possible dissociating $ArXe^{**}$ states. Some of these are indicated schematically in Figure 7. It is known that the cross section given by Eq. (2) is largest for those dissociating curves that cross the initial vibrational state near the bottom of the well. By examining the cross sections for various curves, we formulated a criterion to determine those curves expected to make the dominant contribution to DR. Those curves are those for which the classical turning point at the appropriate energy is within the range 5.75 a_o to 6.5 a_o; we counted 143 such curves by direct evaluation of all curves. Note that we count each distinct value of Ω as a separate state and then apply Eq. (2), using the statistical factor for singlet states.

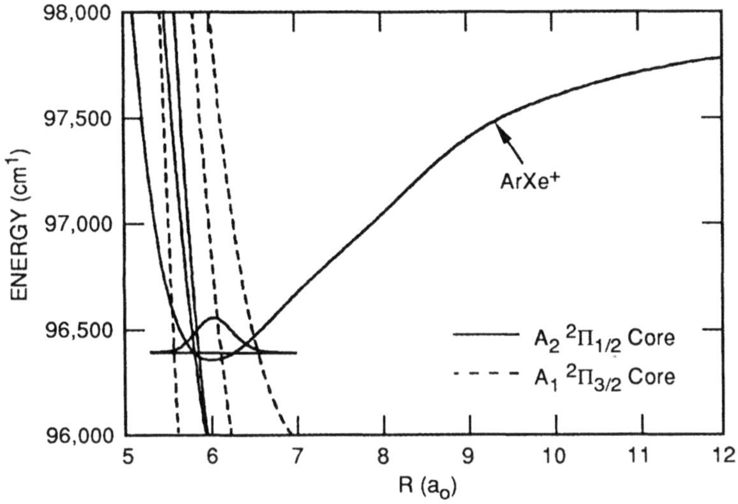

Figure 7. Potential curves for dissociative recombination of electrons with $ArXe^+$. The ground state vibrational wave function of $ArXe^+$ is shown, along with several representative $ArXe^*$ states. Dissociating states with both excited core levels are available.

We found that typical cross sections were in the range 10 to 30 A^2 for the energy ε_0 = kT for T=300 (k is Boltzmann's constant) and for dissociating curves that satisfied our criterion . Since the cross section has a $1/\varepsilon$ energy dependence, it can be shown that the rate constant α has a $T^{-0.5}$ temperature dependence:

$$\alpha(T) = \alpha(300)\,(300/T)^{-0.5} \tag{25}$$

where $\alpha(300)$, the rate at T=300 is given by

$$\alpha(300) = \sigma_{DR}(\varepsilon_0)\,\overline{v}(300) \tag{26}$$

where $\overline{v}$ is the average thermal velocity

$$\overline{v}(T) = \left(\frac{8\ kT}{\pi \mu_e}\right) \tag{27}$$

and μ_e is the electron mass. Since $\overline{v}(300) \sim 10^7$ cm/sec, a typical cross section of 20 A^2 for a single dissociating state corresponds to a rate constant $\alpha(300)$ of about 2×10^{-8} cm^3/sec. Multiplying this by the 143 states yields a product of 3×10^{-6} cm^3/sec. This figure undoubtedly overestimates the rate, for reasons discussed earlier; our final best estimate is that the T=300 rate constant should be of order 10^{-6} cm^3/sec.

CONCLUSIONS

Several conclusions can be drawn from the above estimates. We find that both the A_1 and the A_2 excited $ArXe^+$ states contribute to the DR process. Many of the dissociating resonance states belonging to each of these parent ion states can contribute to the dissociation. Our calculations clearly show that, for the expected size of the coupling matrix elements, and for the wavelengths of the translational and vibrational wave functions involved, the cross sections can be significant. Because such a large number of channels (~100) contribute to the cross sections, we may expect a fairly broad distribution of final Xe^* states. One question related to the kinetics of the atomic xenon laser was whether direct population of 5d levels was possible, or whether higher levels were populated first and then cascaded to the lower levels. The results we have so far suggest that there are many more viable channels correlating to the higher states.

The estimated magnitude for the rate is consistent with the available rates for other heavy diatomic ions. Measured rates[4-6] for Ar_2^+, Kr_2^+, and Xe_2^+ are 0.91 ± 0.1, 1.6 ± 0.2, and 2.3 ± 0.2 times 10^{-6} cm^3/sec, respectively. The rate determined here is also consistent with rates for $ArXe^+$ that have been determined empirically for use in laser models.[8,25]

REFERENCES

1. A. P. Hickman, R. P. Saxon, and D. L. Huestis, J. Chem. Phys. **96**, 2099 (1992).
2. A. P. Hickman, R. P. Saxon, and D. L. Huestis, J. Chem. Phys. **98**, 5419 (1993).
3. J. N. Bardsley, J. Phys. B. **1**, 349 (1968).
4. J. N. Bardsley and M. A. Biondi, Adv. Atom. Mol. Phys. **6**, 1 (1970).
5. A. Giusti, J. Phys. B. **13**, 3867 (1980).
6. A.P. Hickman, J. Phys. B. **20**, 2091 (1987).
7. W. J. Alford and G. N. Hays, J. Appl. Phys. **65**, 3760 (1989).
8. M. Ohwa, T. J. Moratz, and M. J. Kushner, J. Appl. Phys. **66**, 5131 (1989).
9. R. M. Pitzer and N. W. Winter, J. Phys. Chem. **92**, 3061 (1988).
10. A.H.H Chang and R. M. Pitzer, J. Am. Chem. Soc. **111**, 2500 (1989).
11. R.H.G. Reid, J. Phys. B **6**, 2018 (1973).
12. E. E. Nikitin, Adv. Chem. Phys. **28**, 317 (1975).
13. E. U. Condon and G. H. Shortley, *The Theory of Atomic Spectra* (Cambridge University Press, New York, 1967).
14. B. R. Johnson, J. Comp. Phys. **13**, 445 (1973).
15. J. K. Ku and D. W. Setser, J. Chem. Phys. **84**, 4304 (1986).

16. W. J. Alford, IEEE J. Quant. Elect. **26**, 1633 (1990).

17. M. R. Bruce, W. B. Layne, C. A. Whitehead, and J. W. Keto, J. Chem. Phys. **92**, 2917 (1990).

18. J. Xu and D. W. Setser, J. Chem. Phys. **92**, 4191 (1990).

19. J. Xu and D. W. Setser, J. Chem. Phys. **94**, 4243 (1991).

20. W. J. Alford, J. Chem. Phys. **96**, 4330 (1992).

21. H. Horiguchi, R.S.F. Chang, and D. W. Setser, J. Chem. Phys. **75**, 1207 (1981).

22. G. Inoue, J. K. Ku, and D. W. Setser, J. Chem. Phys. **81**, 5760 (1984).

23. M. J. Seaton, Rep. Prog. Phys. **46**, 167 (1983).

24. N. Y. Du and C. H. Greene, J. Chem. Phys. **90**, 6347 (1989).

25. J. Baker and C. K. Rhodes, J. Chem. Phys. **73**, 2626 (1980).

ELECTRON-SCATTERING FROM POLYATOMIC MOLECULES USING A SINGLE-CENTER-EXPANSION FORMULATION

F.A. Gianturco[1], D.G Thompson[2], and A. Jain[3]

[1]Dept. of Chemistry, The University of Rome, Città Universitaria 00185 Rome, Italy
[2]Dept. of Applied Mathematics and Theoretical Physics The Queen's University, Belfast, BT71NN, N. Ireland, U.K.
[3]Dept. of Physics, Florida A & M University, Tallahassee, FL 32307 USA

1. INTRODUCTION

Low-energy electron scattering from gaseous molecules of ever increasing complexity has already been studied for more than half a century[1-3] both through experimental analysis and with the development of several theoretical treatments. Various types of experiments, in fact, have been performed in order to measure the energy and angular dependence of the magnitude of the cross sections involved in the elastic, inelastic, ionizing and dissociative processes induced by collisions of an electron with a molecule. For nearly the same period of time, reasonably successful theories have been proposed to explain the behaviour of these cross sections.[2,3]

As a general, qualitative rule it has often been found desirable for the general understanding of the basic physics involved in the scattering phenomena to divide the process into resonant and direct contributions. Physically, such a division is often useful and meaningful since, at a resonance, there are always one or more partial waves which undergo constructive interference within the space of the target electronic density and therefore they could be considered as those which contribute the most to the process at hand. Depending on the magnitude of the effect of such an intramolecular interference structure with the target, the ensuing scattering process can therefore be treated as resonant or not and the formulation of its theoretical interpretation, or of its computational reproduction, can be carried out by analysing the behaviour of a reduced set of contributing partial waves.[1]

Computational Methods for Electron-Molecule Collisions
Edited by W.M. Huo and F.A. Gianturco, Plenum Press, New York, 1995

This means that, when treating complicated, non-linear molecular targets one may expect the highly anisotropic nature of the electron-molecule interactions (local and nonlocal) will couple more effectively a limited number of partial waves during a resonant process, thereby allowing one to carry out the calculations by treating more carefully just that part of the scattering while employing a more approximate treatment for all the other contributing angular momenta of the continuum electron. In all cases, however, the more simple language of potential scattering theory becomes a very powerful tool for discussing the basic physics at hand and therefore suggests that molecular collision with electrons could be usefully treated directly as processes where one dominant scattering centre provides most of the significant variables.

Thus, the use of Single-center Expansion (SCE) treatments finds one of its justifications with the desire to keep the descriptive language of low-energy electron-molecule collisions at as simple a level as possible, especially when it comes to discussing low-energy resonances a rather important and ubiquitous feature for a broad variety of molecular partners. Historically, such a picture has been justified as also being the natural extension of the description of electron-atom scattering processes, especially for the diatomic molecular targets which were the first ones to be considered, both theoretically and experimentally (e.g. H_2, N_2, CO, O_2, etc.) in the earlier stages of the field.[4] Another chapter of the present Volume will deal specifically with linear systems and diatomics, while we will concentrate instead on polyatomic non-linear partners and discuss below, in some detail, various aspects of their interaction with low-energy electrons and of the behaviour of the differential and integral cross sections, both partial and total.

Since we will talk here only about theoretical and computational results, it is fair to say at the outstart that most treatments which have employed the SCE approach to electron-molecule collisions have considered only scattering in the electronically elastic channels, i.e. have taken into consideration excitation processes which involved the additional nuclear motion but did not include specific excitations to higher electronic states by electron impact processes. Thus, the present discussion of the theoretical and computational methods involving polyatomic targets will be limited to electronically elastic excitation processes.

Even with such limitation, however, the interaction of a slow electron with a highly anisotropic molecular target presents a series of computational and conceptual hurdles which need to be clarified and overcome during any practical implementation of the SCE approach. It is well known, in fact, that the behaviour of low-energy electrons with molecules is controlled chiefly by three kinds of interactions: electrostatic, exchange and polarisation. Outside the region where the charge density is localized, a region usually referred to as the target core, the dominant interaction will be the last of the above three types and can be understood as an induced effect arising from distorsions of the molecular electronic wavefunction by the charged projectile. Rigorously speaking, it arises in quantum collision theory as virtual excitations of energetically closed target electronic states, including those in the continuum. Of the previous two, the exchange effect originates from the well-known need to make the scattering wavefunction antisymmetric with respect to the interchange of the (N+1) electronic coordinates for an N-electron target molecule, thereby giving rise to the nonlocal part of the full scattering potential.

All such contributions need to be included in any realistic treatment of the scattering and the following Sections will show more in detail how this can be done for nonlinear molecules. Furthermore, we will also show how the nuclear motion, an additional complication for molecular targets with respect to atomic scattering, can be treated in some approximate fashion in order to make the overall computational problem more tractable within the framework of a generalized Close-coupling (CC) expansion of the continuum wavefunction.

Thus, the various ways of handling the interaction forces, either exactly or approximately, are discussed in Section 2, while Section 3 presents more in detail the construction of the radial equations within a symmetry-adapted form of the overall expansion in the SCE context.

The specific methods employed to solve the radial equations are discussed in Section 4, where we also briefly discuss the method which has been chiefly used to handle nuclear motion in polyatomic targets. Finally Section 5 discusses some of the actual applications of the SCE method and compares the results of calculations with existing experiments. Section 6 reports briefly our conclusions while Section 7 list the main references to this subject which we found important to the present discussion.

2. FORMULATION OF THE INTERACTION FORCES

2.1. Definition of SE and SEP Approximations

We consider scattering of an electron by a rigid non-rotating molecule. The wavefunction for the total system, $\Psi(\mathbf{x}_1 \ldots \mathbf{x}_{N+1})$ satisfies

$$(H-E)\Psi = O \tag{1}$$

where

$$H = H_{mol}(1 \ldots N) + V_{\text{int}}(1 \ldots N+1) - \frac{1}{2}\nabla^2_{N+1} \tag{2}$$

$$V_{\text{int}} = -\sum_{i=1}^{N} \frac{1}{|\mathbf{r}_{N+1} - \mathbf{r}_i|} + \sum_{j=1}^{M} \frac{Z_j}{|\mathbf{r}_{N+1} - \mathbf{R}_j|} \tag{3}$$

The molecule has N electrons, position $\mathbf{r}_i$, and M nuclei, position $\mathbf{R}_i$ and charge Z_j, the $\mathbf{x}$ coordinate includes both space and spin.

The molecular wavefunctions Φ_m satisfy

$$(H_{mol}-E_m)\,\Phi_m\,(1 \ldots N) = O \tag{4}$$

The total wave function can be expanded in terms of the molecular eigenfunctions

$$\Psi = A \sum_m \Phi_m F_m \tag{5}$$

where A is an antisymmetrising operator and F is the scattered electron function.

The static exchange approximation (SE) is obtained by retaining only the ground state wave function in this expansion. We define a polarisation approximation (SEP) as the inclusion of any terms to correct for this truncation process.

2.2. The SE Equation for F(r)

We obtain the SE equation for F(r) from

$$\int \Phi^*(1..N)[H(1...N+1)-E]\Psi(1...N+1)=0 \tag{6}$$

we note that Ψ can be expressed as

$$\Psi(1...N+1)=\frac{1}{\sqrt{N+1}}\sum_{i=1}^{N+1}\Phi(i^{-1})F(i)(-1)^{i+1} \tag{7}$$

where i^{-1} means all coodinates 1,...N+1 except the ith.

Let us look now at the integral

$$\int \Phi^*(1..N)[H(1...N+1)-E]\Phi([N-1]^{-1})F(N-1)d\tau \tag{8}$$

Interchanging N and N-1, which are dummy variables, we obtain

$$\int \Phi^*(1,...N-2,N,N-1)[H(1...N,N-1,N+1)-E]\Phi(N^{-1})F(N)d\tau \tag{9}$$

But H is symmetric for interchange of coordinates and Φ is antisymmetric, so that we obtain

$$-\int \Phi^*(1...N)[H(1...N+1)-E]\Phi\left([N-1]^{-1}\right)F(N)d\tau \tag{10}$$

By repeating this analysis for other terms in the expansion of Ψ equation (6) which defined the SE approximation becomes

$$\int \Phi^*(1..N)[H(1...N+1)-E]\{\Phi([N+1]^{-1})F(N+1)-N\Phi(N^{-1})F(N)\}d\tau=0 \tag{11}$$

To reduce this further we assume that we are dealing with a closed shell molecule and that the molecular wave function is known in the Hartree-Fock approximation to be of the single determinant form

$$\Phi(1...N)=\frac{1}{\sqrt{N!}}\varepsilon_{\alpha\beta...\pi}u_\alpha(1)...u_\pi(N) \tag{12}$$

The u_a are single electron orbital functions, the u_α being an appropriate complete set of quantum numbers

$$\varepsilon_{\alpha\beta\ldots\pi} = \begin{cases} 1 \; \textit{if} \; \alpha,\beta\ldots\pi \; \textit{is an even permutation of} \; 1,2\ldots N \\ -1 \; \textit{if} \; \alpha,\beta\ldots\pi \; \textit{is an odd permutation of} \; 1,2,\ldots N \\ 0 \; \textit{otherwise} \end{cases}$$

The notation is to sum over repeated indices.
Φ satisfies

$$\int \Phi^*(H_{mol} - E_o)\Phi d\tau = 0 \tag{13}$$

but we will assume

$$(H_{mol} - E_o)\Phi = 0 \tag{14}$$

Since we are dealing with closed shell molecules we can also assume that

$$\int u_\alpha^* F \; d\mathbf{r} = 0 \tag{15}$$

for all orbitals. This follows because Ψ also has the structure of a single determinant.

If we use the properties

$$\varepsilon_{\alpha\beta\ldots\pi} \; \varepsilon_{\alpha\beta\ldots\pi} = N! \tag{16a}$$

$$\varepsilon_{\alpha\beta\ldots\pi} \; \varepsilon_{\alpha'\beta\ldots\pi} \; f_{\alpha\alpha'} = (N-1)! \sum_\alpha f_{\alpha\alpha} \tag{16b}$$

we can easily show that F satisfies

$$[-\frac{1}{2}\nabla^2 - \frac{1}{2}k^2 + V_s(\mathbf{r})]F(\mathbf{r}) = \sum_\alpha \int u_\alpha^*(\mathbf{s}) \frac{1}{|\mathbf{r}-\mathbf{s}|} F(\mathbf{s}) d\mathbf{s} \, u_\alpha(\mathbf{r}) \tag{17}$$

where

$$V_s(\mathbf{r}) = \sum_\alpha \int u_\alpha^*(\mathbf{s}) \frac{1}{|\mathbf{r}-\mathbf{s}|} u_\alpha(\mathbf{s}) d\mathbf{s} + \sum_{j=1}^{M} \frac{Z_j}{|\mathbf{r}_{N-1} - \mathbf{R}_j|} \tag{18}$$

and

$$\frac{1}{2}k^2 = E - E_o \tag{19}$$

We therefore see that eq.(17) treats the evaluation of the continuum functions F($\mathbf{r}$) within the local and nonlocal interaction with a single-determinant (SD) description of the target molecule bound orbitals, without any additional contribution coming from the distortion of such orbitals due to the perturbative effect of the impinging electron. This is usually called the Static-Exchange (SE) approximation. When no further simplification is included, i.e. when the integrals in eq.(17) are treated exactly using only single-particle MO's for the bound electrons, the search for the continuum functions of (17) is considered to be with the Exact Static-Exchange (ESE) approximation. We will show in the following different procedures which could be employed to solve eq.(17), both exactly and with more approximate, local forms for the exchange integrals.

2.3. The Iterative Exchange Method

This is one possible way of handling the evaluation of the specific exchange integral of eq.(17):

$$\sum_{\alpha} \int u_{\alpha}^{*}(s)|\mathbf{r}-\mathbf{s}|^{-1} F^{(P\mu)}(\mathbf{s})d\mathbf{s}u_{\alpha}(\mathbf{r}) \tag{20}$$

where we now have labelled the continuum function by the symmetry indices of its Irreducible Representation (IR) as we shall discuss below. The corresponding matrix formulation of the ensuing close-coupled (CC) equations can be given as:

$$L\ F_{i}^{(P\mu)} = W\ F_{i-1}^{(P\mu)} \tag{21}$$

where the $W\ F_{i-1}^{(p\mu)}$ is the exchange term one starts with for i=1, given by some initial estimate which allows one to obtain an approximate form of the continuum functions. The iterative procedure therefore consists in generating successive $F_{i}^{(p\mu)}$ matrices until a predetermined level of convergence is achieved in the structure of the final K matrix (see below) as a level of invariance for its matrix elements[5,6]. Such a procedure usually works rather slowly and, even with various accelerating devices suggested earlier by us, it usually requires several iterations (say, between 20 and 30) before a satisfactory convergence is reached. Furthermore, because of the oscillatory behaviour of the control quantity, i.e. of the K matrix elements, one may find that unstable solutions may be reached at intermediate level.

A marked improvement can be obtained if one further modifies the iterative procedure by taking advantage of the Schwinger variational method as applied earlier to both photoionisation[8] and electron-molecule collisions[9] (see another Chapter in this Volume). The use of this method has been implemented in two different ways in electron-polyatomic molecule codes. Both approaches are based on writing the scattering equations in terms of a Lippmann-Schwinger equation for the scattering by a potential, V_e, which is given by the difference between some initial, local form of the exchange potential, V_{lex}, and the exact exchange potential V_{eex}

$$\Psi_p = \phi_p + G_a V_e \Psi_p \tag{22}$$

where ϕ_p is a solution to the purely local potential problem

$$H_a \phi_p = E \phi_p \tag{23}$$

The local hamiltonian H_α is therefore $-\frac{1}{2}\nabla^2 + V_\alpha$, where V_α is the sum of local potentials

$$V_\alpha = V_{static} + V_{polarisation} + V_{lex} \tag{24}$$

The potential V_e is defined as

$$V_e = V_{eex} - V_{lex} \tag{25}$$

and the corresponding Green's function is defined as

$$(E - H_\alpha)G_\alpha = 1 \tag{26}$$

If one now defines the total $\boldsymbol{K}$ matrix as having the following elements:

$$K_{pq} = K_{pq}^{(\alpha)} + K_{pq}^{(e)} \tag{27}$$

where $\boldsymbol{K}^{(\alpha)}$ is the K-matrix due to scattering by V_α only, then the $\boldsymbol{K}^{(e)}$ can be obtained from the following relationship

$$K_{pq}^{e} = -2\langle \phi_p | V_e | \Psi_q \rangle \tag{28}$$

In one possible approach one could obtain the relevant K-matrix using a simple iterative method for the wavefunction and then a variational expression for the K-matrix itself.[10] The iterative procedure is defined as follows:

$$\Psi_p^{(i+1)} = \alpha(\phi_p + G_\alpha V_e \Psi_p^{(i)}) + (1-\alpha)\Psi_p^{(i)} \tag{29}$$

where α is a damping factor and $\Psi_p^{(1)} = \phi_p$. At the and of each iteration the K-matrix can be obtained by a linear Schwinger variational expression:

$$K_{pq}^{(e)} = -2\{\langle \phi_p | V_e | \Psi_q^{(i)} \rangle + \langle \Psi_p^{(i)} | V_e | \phi_q \rangle - \langle \Psi_p^{(i)} | V_e - V_e G_\alpha V_e | \Psi_q^{(i)} \rangle\} \tag{30}$$

One may now note that the continuum functions obtained from the standard propagation procedures can be written using eq.(29) without damping factor:

$$\overline{\Psi}_p^{(i+1)} = \phi_p + G_\alpha V_e \Psi_p^{(i)} \tag{31}$$

once the iterative procedure (21) is implemented for the exchange interaction. Thus, eq.(30) can be rewritten as:

$$K^{(e)} = -2\{\langle \phi \,| V \,| \Psi^{(i)} \rangle - \langle \Psi^{(i)} | V \,| \overline{\Psi}^{(i+1)} - \Psi^{(i)} \rangle\} \tag{32}$$

Using such a variational procedure the final convergence is attained much more rapidly and the K-matrix reaches stability after a few iterations only (typically, less than 10). A further improvement on the use of iterative procedures could be introduced by employing Pade's approximants in a way similar to that implemented for the photoionisation of linear molecules.[11] One could then start by defining a sequence of functions $\phi_p^{(i)}$ which are given by:

$$\phi_p^{(i)} = \langle G_a V_e \rangle^i \phi_p \tag{33}$$

such that $\phi_p^{(o)} = \phi_p$ is just the solution of a purely local problem with same modeling of the exchange interaction. The [N/N] Pade's approximant can be obtained from the following expression:

$$K_{pq}^{(e)}[N/N] = \sum_{i,j=O,N-1} \langle \phi_p | V_e | \phi_q^{(i)} \rangle D_{ij}^{-1} \langle \phi_p | V_e | \phi_q \rangle \tag{34}$$

where D_{ij}^{-1} is the (ij)th element of the D^{-1} matrix, which is the inverse of the following D-matrix:

$$D_{ij} = \langle \phi_p^{(i)} | V_e - V_e G_a V_e | \phi_q^{(i)} \rangle \tag{35}$$

Such a procedure was also found to converge rather quickly to a stable structure for the K-matrix.[10]

2.4. Local Exchange: The Free Electron Gas

To circumvent the difficulty of having to solve integro-differential radial scattering equations, the necessary exchange effects have been accurately modelled through the use of simpler, energy-dependent potential forms which can be added to the exact static potential discussed before. One of the most widely used for diatomic and polyatomic targets has been the Free-Electron-Gas (FEG) potential introduced by Hara[12] as an extention to scattering problems of the older Slater's average exchange potential for bound states.[13]

The actual form of this exchange potential derives from two approximations in the integral exchange terms of the exact-exchange scattering equations. First, the molecular electrons are treated as a free-electron gas, with a given charge density $\rho_\Gamma(\mathbf{r},\mathbf{R})$ determined for the Γth electronic state of target (initially its ground electronic state). Second, the distortion due to the interaction on the continuum function is neglected and the impinging projectile is treated as a plane wave. The resulting FEG potential is then given by the following form:

$$V_{\mathrm{Mex}}^{\mathrm{FEG}}(\mathbf{r},\mathbf{R}) = -\frac{2}{\pi} K_{\mathrm{F}}(\mathbf{r},\mathbf{R}) \left[\frac{1}{2} + \frac{1-\eta^2}{4\eta} \ln \left| \frac{1+\eta}{1-\eta} \right| \right] \tag{36}$$

where the wavevector up to the top of the Fermi surface is given by the usual free-electron relation

$$K_F(\mathbf{r},\mathbf{R}) = \left\{3\pi^2\rho_\Gamma(\mathbf{r},\mathbf{R})\right\}^{1/3} \tag{37}$$

and η holds the ratio between the actual wavevector for the scattering electron and the one at the top of the Fermi surface for the electron gas, $\eta = k/K_F$, where the local momentum is given as

$$k(\mathbf{r}) = \left\{2(E_{coll} + I_p) + K_F^2(\mathbf{r},\mathbf{R})\right\}^{1/2} \tag{38}$$

the collision energy E_{coll} is in turn given by the initial-channel energy $1/2\ k_o^2$. I_p is now the ionisation potential for the neutral target molecule.

It turns out that what is called the Hara FEG potential model for exchange interaction with additional orthogonality constraint, HFEGE, has been shown to be an accurate approximation for a wide variety of many-electron molecules[14,15,16], while it has been less successful for systems in which the target has fewer electrons.[17] Obviously, it becomes increasingly less accurate to consider the molecular electrons of a few-electron molecule as an homogeneous electron gas. However, a model exchange potential of the form (36) can be used even for light molecules provided the I_p potential is treated as an adjustable parameter. The resulting Tuned-Free-Electron-Gas-Exchange (TFEGE) potential has been shown to give fairly accurate elastic integral and differential cross sections for e^--H_2 scattering over three orders of magnitude of collision energy. The tuning procedure is usually done by reproducing some specific features in the experimental cross sections and then one employs the final potential over the whole range of energies to be evaluated.

2.5. The Local Exchange Semiclassical Approximation

Several studies have dealt with possibility of casting the integro-differential coupled equations given before (eq. (17)) into a simpler form which could allow one to treat the exchange interaction in a local way that also depended on the local, classical velocity of the impinging electron. These approximate schemes usually started by casting the equation for the continuum function into a local form:

$$\left\{-\frac{1}{2}\nabla^2 + V_{static}(\mathbf{r}) - \frac{1}{2}k^2\right\}u_o = -L(\mathbf{r},k^2)u_o(\mathbf{r}) \tag{39}$$

which can also be written as:

$$\left\{\nabla^2 + k_o^2(\mathbf{r})\right\}u_o(r) = 0 \tag{40}$$

where

$$k_o^2(\mathbf{r}) = k^2 - 2V_{static}(\mathbf{r}) - 2L(\mathbf{r},k^2) \tag{41}$$

The last term therefore represents, in a semiclassical picture, the local velocity of the incoming electron within the region of interaction with the molecular charge cloud. The quantity L is usually treated as a function of the asymptotic velocity rather than as an operator.

The contributions to the total exchange interaction contained in L can be written as a product of a slowly varying amplitude A($\underline{r}$) times the rapidly oscillating function of the impinging electron $u_o(\underline{r})$. This means that the application of the ∇^2 operator on each product contained in L leads to the following expression for each contributing amplitude A($\underline{r}$) in terms of $k_o^2(\underline{r})$

$$A(\mathbf{r}) = \frac{4\pi u_i^*(\mathbf{r})}{k_o^2(\mathbf{r})} + \frac{2}{k_o^2(\mathbf{r})}\left\{\frac{\nabla u_o(\mathbf{r})}{u_o(\mathbf{r})}\cdot\left[\nabla\left(\frac{4\pi u_i^*(\mathbf{r})}{k_o^2(\mathbf{r})}\right)\right]\right\} + \ldots \tag{42}$$

where the u_i's are now the undistorted, occupied molecular orbitals of the target molecule. By using now the leading term of the above equation one can simplify the sum over occupied orbitals within the L term. If one then substitutes this simplified sum into eq.(41), one obtains a quadratic equation for $L(\underline{r},k^2)$. The acceptable, physical solution may then be determined by using the condition that the correct $L(\underline{r},k^2)$ must vanish as $k^2 \to \infty$, which gives the following equation

$$V_{Mex}(\mathbf{r},k^2) = \frac{1}{2}(E_{coll} - V_{static}(\mathbf{r})) - \frac{1}{2}[(E_{coll} - V_{static}(\mathbf{r}))^2 + \alpha^2\,]^{1/2} \tag{43}$$

where E_{coll} is the asymptotic collision energy $= 1/2k^2$ and α^2 is given by

$$\alpha^2 = 8\pi\sum_{i=1}^{n} N_i|u_i(\mathbf{r})|^2 \tag{44}$$

where n is the index for the highest occupied molecular orbital and N_i is the occupation number for each of them. For doubly occupied spatial orbitals one can therefore write

$$\alpha^2 = 4\pi\rho(\mathbf{r},\mathbf{R}) \tag{45}$$

where ρ is now the total electron density of the molecular target and the 'frozen' geometry of its nuclei has been explicitely introduced by the representative coordinate of all the nuclear positions. The potential of eq.(43) thus intoduces an additional attractive term into the scattering equation which is called the semiclassical exchange (SCE) model potential.[20]

In conclusion, the essence of the above SCE modelling of the non local interaction is that the local momentum of the bound electrons can be disregarded with respect to that of the impinging electron. Such an assumption is possibly more likely to hold at rather large collision energies, where one can accept the picture in which the electron-electron collisions only slightly modify the continuum electron velocity. On the other hand, at lower velocities of the scattering electron the relative momenta are more comparable and therefore one requires a different treatment for operators like the following one

$$-\frac{4\pi}{|\nabla_{u_o}+\nabla_{u_i}|^2}u_i^*(\mathbf{r})u_o(\mathbf{r}) \tag{46}$$

This has been done for polyatomic targets a while ago, by using the FEG model mentioned before to treat the electron distribution of the target molecule, thereby obtaining the following expression for the expectation value of the average of the square of the local wavevector for the bound electons

$$\left\langle k^2(\mathbf{r})\right\rangle = \frac{3}{5}K_F^2(\mathbf{r}) \tag{47}$$

where

$$K_F^2(\mathbf{r}) = \left\{3\pi^2\rho(\mathbf{r})\right\}^{1/3} \tag{48}$$

One can therefore write that

$$\nabla_{u_i}^2 u_i^*(\mathbf{r}) = -\frac{3}{5}\left\{3\pi^2\rho(\mathbf{r})\right\}^{2/3} u_i^*(\mathbf{r}) \tag{49}$$

Once the above equation is employed in the previous eq.(46) then a new, physically acceptable solution for a local form of the SCE model can be obtained. It was called the Modified semiclassical Exchange (MSCE) approximation and has been used for treating exchange interaction in several polyatomic targets[21,22]

$$V_{MSCE}(\mathbf{r},k^2) = \frac{1}{2}\left\{E_{coll} - V_{static}(\mathbf{r}) + \frac{3}{10}\left(3\pi^2\rho(\mathbf{r})\right)^{2/3}\right\} - \frac{1}{2}\left\{\left[E_{coll} - V_{static}(\mathbf{r}) + \frac{3}{10}\left(3\pi^2\rho(\mathbf{r})\right)^{2/3}\right]^2 + 4\pi^2\rho\right\}^{1/2} \tag{50}$$

In such a model one actually employs the best available target electronic density and can extend the treatment of exchange in a semiclassical framework down to rather low collision energies for the impinging electron[23,24].

2.6. The Local Exchange Separable Approximation

The development of a discrete-basis-set representation of the bound and continuum orbitals in an electron-molecule scattering process has led earlier on to the use of separable approximations for handling the non local exchange kernel discussed in this Section[25]. Recently such a development has been extended to polyatomic targets[27] and will be discussed below in a schematic outline.

The non local expression for the familiar exchange kernel is given by

$$K(\mathbf{r},\mathbf{r}') = \sum_{i=1}^{n} u_i^*(\mathbf{r})\frac{1}{|\mathbf{r}-\mathbf{r}'|}u_i(\mathbf{r}') \tag{51}$$

and we want to express it as a different product written as

$$K(\mathbf{r},\mathbf{r}') = \sum_{\alpha,\beta} \phi_\alpha(\mathbf{r})W_{\alpha\beta}\phi_\beta(\mathbf{r}') \tag{52}$$

where the ϕ_α are still unknown quantities as are the $W_{\alpha\beta}$ coefficient. The separable approximation essentially consists in representing the r.h.s. of eq. (51) using a finite basis set of STO or GTO functions that span a wider functional space than that of the occupied molecular orbitals employed to describe the target molecule

$$\sum_{\alpha,\beta} \phi_\alpha(\mathbf{r})W_{\alpha\beta}\phi_\beta(\mathbf{r}') = \sum_{\alpha,\beta} |\alpha\rangle\langle\alpha|K|\beta\rangle\langle\beta| \tag{53}$$

where the $|\alpha\rangle$, $|\beta\rangle$ are the finite basis employed to represent the nonlocal kernel of eq.(51).

The matrix elements of such a kernel over the SCF wavefunctions of the molecule are given by:

$$\tilde{K}_{\gamma\tau} = \int d\mathbf{r}\int d\mathbf{r}'\; n_\gamma^*(\mathbf{r})K(\mathbf{r},\mathbf{r}')n_\tau(\mathbf{r}') \tag{54}$$

using the new finite basis one can also write

$$\tilde{K}_{\gamma\tau} = \sum_{\alpha,\beta} \int d\mathbf{r}\; n_\gamma^*(\mathbf{r})\phi_\alpha^*(\mathbf{r})W_{\alpha\beta}\int d\mathbf{r}'\phi_\beta(\mathbf{r}')n_\tau(\mathbf{r}') \tag{55}$$

or

$$\tilde{K}_{\gamma\tau} = \sum_{\alpha,\beta} S_{\gamma\alpha}W_{\alpha\beta}S_{\beta\tau} \tag{56}$$

which can give us the matrix elements of the unknown W-matrix as

$$W = S^{-1}\tilde{K}\; S^{-1} \tag{57}$$

It is also convenient to define a slightly different matrix[25]

$$\Omega = S^{-1/2}\tilde{K}\; S^{-1/2} = S^{-1/2}W\; S^{-1/2} \tag{58}$$

One now wants to obtain a separable representation of the exchange kernel over an orthogonal basis set

$$K(\mathbf{r},\mathbf{r}') = \sum_{\gamma} \tilde{\phi}_\gamma(\mathbf{r})\varepsilon_\gamma\tilde{\phi}_\gamma(\mathbf{r}') \tag{59}$$

where

$$\tilde{\phi}_{\gamma}(\mathbf{r}) = \sum_{\alpha} U_{\gamma\alpha}\phi_{\alpha}(\mathbf{r}') \tag{60}$$

which implies a diagonalisation of the Ω-matrix

$$T\ \Omega\ T^{+} = \varepsilon \tag{61}$$

or

$$T\left\{S^{-1/2}\tilde{K}\ S^{-1/2}\right\}T^{+} = \varepsilon \tag{62}$$

which allows one to obtain the matrix elements of eq.(56) in a new form

$$\tilde{K}_{\gamma\tau} = \sum_{\lambda}\ \sum_{\alpha,\beta} U_{\lambda\alpha}\phi_{\alpha}(\mathbf{r}')\varepsilon_{\lambda\beta}\phi_{\beta}(\mathbf{r}') \tag{63}$$

or

$$\tilde{K} = U^{+}\varepsilon\ U \tag{64}$$

which then allows one to write

$$U\ \tilde{K}\ U^{+} = \varepsilon \tag{65}$$

The above result means that the orthogonal new basis can be used to represent the exchange kernel and one could therefore employ for that basis a set of SCE functions of the corresponding symmetry for the target molecule[26]

$$\tilde{\phi}_{\alpha}(\mathbf{r}) = \chi_{\alpha}^{A_1}(r)X^{A_1}(\hat{r}) \tag{66}$$

where the symmetry-adapted angular functions are linear combinations of spherical harmonics and will be discussed in the following Section. One can therefore proceed with the diagonalisation of eq.(65) to obtain the coefficients of the orthogonal functions as obtained from the eigenvalues and eigenvectors of the Ω−matrix.

2.7. The Correlation-Polarisation Potentials

As mentioned before, another important type of interaction between the impinging electron and the target molecule comes from the description of the response function of the target to the incoming projectile.

Rigorously speaking, this effect arises in quantum collision theory from the virtual excitations into the closed electronic states of the (N) bound electrons and of the (N+1) continuum electrons. In practice, however, the infinity of such states precludes such an approach and therefore one is left either with a direct method that solves the scattering and the many-body problem in a combined fashion over a discrete representation of the most important states (e.g. the R-matrix approach, as discussed in another chapter of this Volume) or one can introduce and evaluate an optical potential. The latter is not uniquely defined since, due to its energy dependence, there is an infinite set of optical potentials yielding the same optical wavefunction. Its implementation within the Multichannel Schwinger Variational Method is being also discussed in another chaper of the present Volume.

One further, more model-oriented, approach requires the search for an effective, local potential form which describes the modified behaviour of polarisation forces within the overlap region between the scattered electron and the target undistorted charge density. What usually makes such an approach rather difficult in its formulation is not the representation of the distortion of the target electronic density as a function of a charge which is being held fixed some distance away from the origin of all charges, but rather the additional effects which come into play near and within the target 'core' of its charge distribution. At large distances from it, in fact, the velocity of the slow electron can be considered low enough that the bound molecular electons can respond adiabatically to the impinging projectile without a specific dependence on its local velocity. As the projectile nears the target, however, the attraction from the Coulomb core increases the local velocity of the projectile and therefore strongly modifies its motion via correlation processes similar to multiple scattering effects. The ensuing polarisation potential at short range should therefore be not only energy dependent but also nonlocal.

The model correlation-polarisation potentials, V_{cp}, that have been implemented within a SCE treatment of polyatomic molecules are therefore distinguishing between a long-range region of interaction (where perturbative polarisation effects are dominant, the potential acquires correctly a local analytic form and is both adiabatic on the projectile's velocity and independent of the sign of its charge) and a short-range region where non-adiabatic, non local effects play an important role and differences appear between the behaviour of electrons and positrons as projectiles. It is the description of the latter region which has been the subject of several studies in the last few years. One therefore writes the overall Correlation-Polarisation interaction as given by

$$V_{cp}(\mathbf{r}) = V_{pol}(\mathbf{r}) \quad for \quad \mathbf{r} > \mathbf{r}_c \tag{67a}$$

$$= V_{corr}(\mathbf{r}) \quad for \quad \mathbf{r} < \mathbf{r}_c \tag{67b}$$

where the connecting spatial factors $\mathbf{r}_c$ are usually obtained from the crossing radii of the lower coefficients in the expansion

$$V_{cp}(\mathbf{r}) = \sum_{p\mu lh} V_{cp}^{(p\mu lh)}(r) X_{lh}^{p\mu}(\hat{r}) \tag{68}$$

where the indices and angular coeficients are discussed in the following Section. In other words, one makes the rather ad hoc assumption that the chosen form for V_{corr} goes smoothly into its perturbative form as soon as the overlap between the target charge distribution and the impinging electron has become negligible.

The short-range correlation has been obtained from different prescriptions, each within an increasing level of complexity:
(i) Earlier in the implementation of such an approach, the asymptotic polarisation potentials from the dipole polarisability coefficients were simply multiplied by a cut-off function that contained an adjustable parameter, r_c. Thus, for a diatomic target:

$$V_{pol} = -\left[\frac{\alpha_o}{2r^4} + \frac{\alpha_2}{2r^4} P_2(\hat{r}\cdot\hat{R})\right] f_c(r\,;\,r_c) \tag{69}$$

and the parameter r_c was chosen to optimize the agreement between calculated and experimental values of some specific feature in the cross sections[28];
(ii) A further modification employed the FEG modelling of the correlation between the bound electrons, given as the SCF density for the target ground electronic state, and the impinging electron. The corresponding E_{corr} turns out to be an analytic function of such densities within the Kohn-Sham description of the single-particle orbitals[29] and therefore one obtains an analytic expression for both V_{pol} and V_{corr} in eq.s(67), thereby producing the corresponding crossing region to obtain the required r_c value. The latter is therefore not anymore an adjustable parameter but comes directly from the choice of both potential contributions in that equation[30,31];
(iii) A more recent implementation[32] has replaced the FEG picture for the evaluation of correlation forces with a Density Functional Theory (DFT) that describes, in local form, the dependence of correlation energy on the density and density gradients of the target electronic state[33]. The ensuing potential is called the LYP (Lee-Yang-Parr)[34] correlation potential and describes in more sophisticated terms the overall dependence of short-range polarisation forces on the target electronic density.
(iv) A final, more recent implementation of the cut-off function of earlier models employs a semiclassical model[35] which dampens the asymptotic value of the polarisation potential using directly the polarised-orbital picture employed earlier for atomic scattering[36]. Such an approach, therefore, involves no empirical parameters which have to be adjusted but relies instead on the knowledge of the electronic density for the target molecule.

3. THE SCE RADIAL EQUATIONS

3.1. Expansion of the Bound Orbitals

It is well-known that most polyatomic molecules possess a certain set of symmetry operations (rotations, reflection, inversion, etc.) under which their Hamiltonians is invariant. If the Hamiltonian H of the molecule, with eigenfunction $\psi_\alpha(\mathbf{r}_N)$, is totally symmetric with respect to a given group of symmetry operators ($\hat{R}_i$, i=1,2,...), then the $\hat{R}_i\psi_\alpha$ are also eigenfunctions of H with the same eigenvalue[37]. Let us now assume that ε_α is an l_α-fold degenerate eigenvalue and let the corresponding set of l_α orthogonal eigenfunctions be $\psi_{\alpha k}$ (k=1,.....l_α). By the above result, the operation with any symmetry operator $\hat{R}_i$ on anyone of the l_α functions produces yet another function (with the same eigenvalue) which can be expressed as a linear combination of this orthonormal set of degenerate functions, i.e. we can write:

$$\hat{\mathfrak{R}}\psi_{\alpha k} = \sum_{k'=1}^{l_\alpha} \tilde{\mathfrak{R}}_{kk'}\psi_{\alpha k'} \qquad (k = 1,\ldots l_\alpha) \tag{70}$$

where the $\overline{\mathfrak{R}}_{kk'}$ are matrix elements of the matrix $\tilde{\mathfrak{R}}$, which is a (l_α xl_α) matrix representation of the operator $\hat{\mathfrak{R}}$ and a collection of these matrices, one for each operation, is an l_α-dimensional representation of the point group for the chosen, fixed molecular geometry[38].

On the basis of the above discussion, one can therefore construct the symmetry-adapted wavefunctions (SAWF) which have the symmetry properties of each irreducible representation (IR) for the point group of the molecule in question. If the molecular point group has g elements $\hat{\mathfrak{R}}_i$, with c IR's, Γ_n (of dimension l_n each) and the character of the operation $\hat{\mathfrak{R}}_i$ for each Γ_n can be represented as $\chi_n(\hat{\mathfrak{R}}_i)$, then for each IR we can construct a corresponding projection operator

$$\hat{p}_n = \frac{l_n}{g}\sum_{i=1}^{g} \chi_n(\hat{\mathfrak{R}}_i)^* \hat{\mathfrak{R}}_i \tag{71}$$

such that each projected function $\psi_n^\alpha = \hat{p}_n\psi_\alpha$ belongs to the IR Γ_n and it is called a SAWF for the target molecule.

As we shall discuss below, the quality of the occupied, molecular robitals (MO's) employed to describe the electronic density of the target molecule is obviously an important ingredient for the evaluation of the final scattering cross sections. For instance, it has been discussed several times (see other chapters on the Kohn-Variational Method in the present Volume) that the intramolecular correlation effects are inextricably connected with the intermolecular correlation forces which describe the short-range interaction between the scattering and the bound electons. Thus, it seems reasonable to start with descriptions of the bound electrons which include as much as possible both types of correlation forces. Until now, however, most SCE calculations for polyatomic targets have relied on target electronic wavefunctions obtained at the near Hartree-Fock (HF) level of description from SCF calculations, using a truncated set of analytic functions[2]. For a closed-shell system, the total SCF wave function of the target is given by a single Slater determinant of one-electron functions $\Phi_i(\mathbf{r}_i;\mathbf{R})$ (which can be taken to be real) for each set of nuclear geometries $\mathbf{R}$ and for each of the bound electrons. Each of such functions, for the total of N electrons in the target, is in turn given by a truncated expansion over an optimized combination of analytic functions $\chi_j(\mathbf{r}_i)$:

$$\Phi_i(\mathbf{r}_i;\mathbf{R}) = \sum_{j=1}^{j_{\max}} C_{ij}(\mathbf{R})\chi_j^i(\mathbf{r}_i) \tag{72}$$

The expansion functions employed thus far to treat scattering problems have been:
(i) Slater-type orbitals (STO) centered around the heavy atom of the molecular target, usually nearly coincident with the center-of-mass (c.o.m.) of the molecule[28];
(ii) Gaussian-type orbitals (GTO) centered over the many atomic nuclei of the target, thereby producing a multicentered description of the relevant MO of eq.(72)[6].

In either case, each function Φ_i is taken to be real and is expanded over the molecular c.o.m. in terms of symmetry-adapted functions as discussed before:

$$\Phi_i(\mathbf{r}_i;\mathbf{R}) = \sum_s r_i^{-1} U^i_{l_s h_s}(r_i) X^{p_s\mu_s}_{l_s h_s}(\hat{r}_i) \tag{73}$$

where we have dropped on the r.h.s. of the equation the explicit indication of its dependence on the nuclear geometry. The X functions are linear combinations of spherical harmonics (see below) and belong to the μ_sth component of the p_sth IR, with h_j distinguishing between different basis with the same set of $|p_s\mu_s l_s\rangle$ indices. l_s describes here the particular value of the partial wave index of the spherical harmonics in X The latter coefficients are, in fact, defined as[39,40]

$$X^{p\mu}_{lh}(\hat{r}) = \sum_{m=o}^{+l} b^{p\mu}_{lhm} S^{p\mu}_{lm}(\hat{r}) \tag{74}$$

where the real spherical harmonics are defined below.

The radial coefficients of eq.(73) are therefore obtained either by direct use of the STO's chosen in the previous expansion (72) or by numerical quadratures of the multicenter MO's obtained with Gaussian functions[6].

In the latter instance, each GTO function is given by a set of contracted, primitive cartesian gaussian functions

$$\chi^i_{kj}(x_k) = N(a,b,c;\alpha)x^a \cdot y^b \cdot z^c \cdot \exp(-\alpha x^2) \tag{75}$$

where k labels the atomic center, i the specific MO and j labels the function in the expansion (72). The α is the non-linear parameter in the GTO and N the normalisation function:

$$N(a,b,c;\alpha) = \left(\frac{2}{\pi}\right)[(2a-1)!!(2b-1)!!(2c-1)!!](\alpha)^{\frac{a+b+c}{2}+3/4} \tag{76}$$

The corresponding radial coefficients of eq.(73) can therefore be obtained by angular quadratures over the corresponding multicenter MO's:

$$U^{i}_{l_s}\hbar_s(r_i;\mathbf{R}) = \sum_k \sum_j \sum_m \int_0^{\pi} sin\vartheta d\vartheta \int_0^{2\pi} b^s_{lm} S^s_{lm}(\vartheta,\varphi) C^i_{kj}(\mathrm{R}) \cdot$$

$$\cdot N(a,b,c;\alpha) x^a y^b z^c \exp(-\alpha x^2) d\varphi \tag{77}$$

where the first two terms on the r.h.s. describe explicitely the contributions in expansion (74), with s=|pμ>. The quadratures can be carried out via Gauss-Laguerre grids using a discrete, variable radial grid for each point of which the spherical grid in (ϑ,φ) was evaluated. Usually, one needs to generate about 500 radial values in the BF frame with a set of (100x100) points in the angular region for each of those distances[60].

3.2. Continuum Functions and Radial Equations

In order to reduce the SE equation for F(**r**) to radial form by expanding all function in single centre form, we have to be able to write:

$$F(\mathbf{r}) = r^{-1} \sum_{p\mu l h} f^{p\mu}_{lh}(r) X^{p\mu}_{lh}(\hat{\mathbf{r}}) \tag{78}$$

Strictly speaking, we should expand about the centre of mass at the molecule, but it is usual to expand about one of the heavy atoms in the molecule when, for example, one has MO's already available centered around such heavy atomic center.

The $X^{p\mu}_{lh}(\hat{\mathbf{r}})$ are linear combinations of spherical harmonics (which can be real for the closed shell molecules usually studied)

$$X^{p\mu}_{lh}(\hat{\mathbf{r}}) = \sum_{m=0}^{l} b^{p\mu}_{lhm} S^{mq}_{l} \tag{79a}$$

$$S^{mq}_{l} = \left[\frac{(2l+1)(l-m)!}{2\pi(l+m)!(1+\delta_{mo})}\right]^{1/2} P^m_l(\cos\vartheta) f^q(\phi) \tag{79b}$$

$$f^1(\phi) = \cos m\phi, \qquad f^{-1}(\phi) = sin m\phi \tag{80}$$

The real spherical harmonics can be expanded in terms of complex spherical harmonics

$$S^{mq}_{l} = [(q+1) + i(q-1)]\frac{1}{2}\frac{1}{\sqrt{2}}\left\{Y^m_l + (-1)^m q\ Y^m_l\right\}\frac{(-1)^m}{(1+\delta_{mo})^{1/2}} \tag{81}$$

The b coefficients are chosen so that the X functions transform as the μth component of the pth irreducible representation of the molecular point group, with h distinguishing between functions having the same l. The coefficients are described and tabulated in Altmann and Cracknell.[40]

Multiplying equation (17) by $X_{lh}^{p'\mu'}(\hat{\mathbf{r}})$ and integrating over all $\mathbf{r}$ we obtain a set of coupled equations which are diagonal in p and independent of μ.

$$\left\{\frac{d^2}{dr^2}+k^2-\frac{l_i(l_i+1)}{r^2}\right\}f^{(i)}(r)=2\sum_i V^{(i,j)}(r)f^{(j)}(r)+\sum_j\int_o^\infty K^{(i,j}(r,s)f^{(j)}(s)ds \quad (82)$$

The indices i and j represent a "channel" lh.

3.3. Single Centre Expansion of the Static Potential

Each bound one-electron function is also expanded about one center in terms of the X functions.

$$u_j(\mathbf{r})=r^{-1}\sum_{lh} u_{lh}(r)X_{lh}^{p\mu}(\hat{\mathbf{r}}) \quad (83)$$

We wish to obtain the static potential as a single center expansion. To this end, consider first the one electron density function

$$\rho(\mathbf{r}_i)=\int|\Phi(\mathbf{x}_1\ \mathbf{x}_2\ldots\mathbf{x}_N)|^2\,d\mathbf{x}_2\ldots d\mathbf{x}_N=2\sum_j |u_j(\mathbf{r}_i)|^2 \quad (84)$$

The factor 2 is due to the sum over spin, and the j sum is over each doubly occupied orbital. The quantity $\rho(\mathbf{r})$ can be expanded in terms of symmetry-adapted functions belonging to the A_1 irreducible representation

$$\rho(\mathbf{r})=\sum_{lh} r^{-1}\rho_{lh}(r)X_{lh}^{A_1}(\hat{r}) \quad (85)$$

we can obtain the potential from ρ as given by

$$V(\mathbf{r})=\int\rho(\mathbf{s})\frac{1}{|\mathbf{r}-\mathbf{s}|}ad\mathbf{s}=\sum_{j-i}^{m} Z_j/|\mathbf{r}-R_j|=\sum_{lh} V_{lh}(r)X_{lh}^{A_1}(\hat{\mathbf{r}}) \quad (86)$$

Note the potential is also expanded in terms of A_1 functions.

In practice it can be convenient to consider the $u_j(\mathbf{r})$ in terms of S functions, rather than the X functions of equation (79a)

$$u_j(\mathbf{r})=\sum_i r^{-1}V_{\alpha_i}^j(r)S_{l_i}^{m_i A_i}(\hat{\mathbf{r}}) \quad (87)$$

where α_i is shorthand notation for $(n_i l_i m_i)$. The V are analytic functions multiplied by a coefficient, c, determined by the usual variational procedures, or are numerical functions calculated from multicentre Gaussian codes. We see also that

$$\rho(\mathbf{r}) = \sum_{ijk} r^2 V^k_{\alpha_i} V^k_{\alpha_j} S_{l_i}^{m_{\beta_i}} S_{l_j}^{m_{\beta_j}} \tag{88}$$

which can be written

$$\rho(\mathbf{r}) = \sum_{LMP} r^{-2} \bar{\rho}_{LMP} S_L^{MP} \tag{89}$$

Thus to obtain $\bar{\rho}$ we must pick out the coefficient of S_L^{MP} from each of the products $S_l^{mp} S_\lambda^{\mu\pi}$ occurring in the sum. We only have to consider those products for which

$|l-\lambda| \leq L \leq l + \lambda$
$(l + \lambda + L)$ is even
$M = (m + \mu)$ or $|m-\mu|$
and $P = p\pi$

The required coefficients are then

$$\beta \left\{ \frac{(2l + i)(2\lambda + i)}{4\pi(2L + i)} \right\}^{1/2} C^{l\lambda L}_{ooo} \tag{90}$$

where β depends on m and μ

$(i)\ m \neq o,\ \mu \neq o,\ M = m + \mu;\ \beta = (p + \pi + 1 - P) C^{l\lambda L}_{m\mu M} / 2\sqrt{2}$

$(ii)\ m \neq o,\ \mu \neq o,\ M = m - \mu$

$(a)\ m > \mu \qquad \beta = (P + 1 + \pi - p) C^{l\lambda L}_{m-\mu M} (-1)^\mu / 2\sqrt{2}$

$(b)\ m < \mu \qquad \beta = (P + 1 + p - \pi) C^{l\lambda L}_{m-\mu M} (-1)^m / 2\sqrt{2}$

$(c)\ m = \mu \qquad \beta = \frac{1}{2}(P + 1) C^{l\lambda L}_{m-mo} (-1)^m$

(iii) *for all other cases.* $\qquad \beta = C^{l\lambda L}_{m\mu M}$

The electronic contribution to the potential is

$$V_{el}(\mathbf{r}) = \int \rho(\mathbf{s}) \frac{1}{|\mathbf{r} - \mathbf{s}|} ds = \sum_{LMP} V_{LMP}(r) S_L^{MP} \tag{91}$$

where

$$V_{LMP}(r) = \frac{4\pi}{2L+1}\int \bar{\rho}_{LMP}(s)\gamma_L(s,r)ds \tag{92}$$

and $\gamma_L = r_<^L / r_>^{L+1}$

The nuclear contribution to the potential can be expanded in a similar way. One divides then the nuclei into groups, according to the same R_i and Z_i. Of course if $R_i = 0$ we have a group contribution of Z_i/r, but in general for each group we must consider instead

$$Z_i \sum_{\lambda\mu} \gamma_\lambda(R_i,r)\frac{4\pi}{2\lambda+1}Y_\lambda^\mu(\hat{\mathbf{r}})\left\{\sum_i Y_\lambda^\mu(\hat{\mathbf{R}}_i)\right\} \tag{93}$$

The group sum $C_{\lambda\mu} = \sum_i Y_\lambda^\mu(\hat{\mathbf{R}}_i)$ is either a real or an imaginary number and we find that eq. (93) becomes

$$Z_i \sum_\lambda \sum_{\mu\geq 0} a_{\lambda\mu}(R_i,r)\frac{4\pi}{2\lambda+1}\gamma_\lambda(R_i,r)S_\lambda^{\mu\zeta} \tag{94}$$

where $\zeta = 1$ when $C_{\lambda\mu}$ is real, -1 when $C_{\lambda\mu}$ is imaginary and

$$\alpha_{\lambda o} = C_{\lambda o}$$

$$\alpha_{\lambda\mu} = (-1)^\mu\sqrt{2}\ C_{\lambda o} \quad if \quad \zeta = 1$$

$$\alpha_{\lambda\mu} = -i(-1)^\mu\sqrt{2}\ C_{\lambda\mu} \quad if \quad \zeta = -1$$

We see that the complete potential can be obtained as an expansion in real spherical harmonics. It only remains to use this to obtain the expansion given by equation (86) as an expansion in symmetry adapted functions.

There is no difficulty for the dihedral groups (eg H_2O-C_{2v} and NH_3-C_{3v}) because each $X_{hl}(\hat{\mathbf{r}})$ is a single real spherical harmonics. We can drop the h suffix and label the symmetry adapted functions $X_{lm} = S_l^{mq}$ with a single $b_{lm} = 1$ coefficient.

The problem is more complicated for the tetrahedral group (eg. CH_4). Consider some general function $\psi_l(\mathbf{r})$ (of particular symmetry) which can be expanded in terms of either X_{lm} or S_l^{mq}

$$\Psi_l = \sum_{h=1}^{H} \alpha_h(r)X_{lh}(\hat{\mathbf{r}})$$

$$= \sum_{m=0}^{L} \beta_m(r)S_l^{mq}(\hat{\mathbf{r}}) \tag{95a}$$

Expanding the X_{lh} in terms of the S_l^{mq} we see that

$$\beta_m = \sum_{h=1}^{H} \alpha_h b_{lhm} \qquad (95b)$$

We note that if H=1 we can determine α from a knowledge of just one β_m:

$$\beta_m / b_m \qquad (95c)$$

but in general we need to know a sequence H of different β_m, for each given l, in order to recover potentials in the form of equation (82).

3.4. The Symmetry-adapted Coefficients

As mentioned before the evaluation of the b_{lm} coefficients requires the use of the character tables of the relevant molecular point groups. They can be obtained also from published tables but only up to rather small values of l.

A more general procedure that has been introduced in the more recent coding of the SCE scattering procedure involves a numerical quadrature approach[10].

One needs to start by computing the following matrix elements

$$D^{(l)}(\mathfrak{R})_{\min} = \int Y^*_{lm'}(\vartheta,\phi)[\mathfrak{R}Y^*_{lm'}(\vartheta,\varphi)]\sin\vartheta d\vartheta d\phi \qquad (95d)$$

where $\mathfrak{R}$ is one of the symmetry operators of the point group to which the molecule belongs and is one which moves a point $\hat{r}(\vartheta,\phi)$ into $\hat{r}(\vartheta,\varphi)$ without changing r. The above integral can therefore be computed exactly using quadrature formulae over the relevant angles:

$$D^{(l)}(\mathfrak{R})_{\min} = \sum_{j,k} Y^*_{lm'}(\vartheta_j,\phi_k) Y^*_{lm'}(\vartheta',\phi')W_j W'_k \qquad (95e)$$

where the indeces {j} and {k} label the grid points in the variables ϑ and ϕ respectively, and the W's are the corresponding weights. The (ϑ'_j,ϕ'_k) are obtained from the inverse operation $\mathfrak{R}^{-1}$ which transforms (ϑ_j,ϕ_k) into (ϑ'_j,ϕ'_k) so that

$$Y_{lm}(\vartheta'_j,\phi'_k) = \mathfrak{R}Y_{lm}(\vartheta_j,\phi_k) \qquad (95f)$$

Once all the *D* matrix elements have been computed, one can further evaluate an irreducible matrix representation for each symmetry type within the given molecule.[38]

For representations of dimension 1 this comes directly from the character table. For higher dimensional representations the matrices need to be specifically computed. To compute an irreducible representation, first a function belonging to the representation must be found. This is done by using the projection operator $P^{(j)}$ which projects onto the *j*th representation, $\Gamma^{(j)}$.

Once this vector has been obtained, the other basis vectors needed to construct the irreducible representation are obtained by operating on the original basis vector using the symmetry operations. Once the number of orthonormal vectors obtained in this manner is the same as the dimensionality of the representation then the corresponding matrix representation can be computed. In order to be able to use the abelian subgroup, a unitary transformation is applied to the basis vectors to that they individually form representations of one of the symmetry types of the abelian subgroup. Thus, when the symmetry functions are formed they will simultaneously belong to a representation of the full group and to the abelian subgroup.

Once irreducible matrix representations for the symmetry types have been constructed the symmetry functions for a given l are obtained by the application of the projection operator $P_{kk}^{(j)}$ to obtain the first element of a representation and then the operators $P_{\lambda k}^{(j)}$ is diagonalized to obtain the set of functions belonging to row k of the representation j. The operator $P_{kk}^{(j)}$, as defined by Tinkham,[38] yields zero unless it operates on a function belonging to the k row of $\Gamma^{(j)}$. And if $P_{\lambda k}^{(j)}$ operates on a function from the k row of $\Gamma^{(j)}$. We shall see below applications of the above procedure to the T_d point-group of the tetrahedral molecules that are currently of great experimental interest (e.g. CH_4, SiH_4, CF_4 etc.).

4. SOLUTION OF THE SE AND SEP RADIAL EQUATIONS

We need to solve

$$\left(\frac{d^2}{dr^2} + k^2 - \frac{l_i(l_i+1)}{r^2}\right) f_{(r)}^{(i)} = \sum_j V^{(i,j)} f_{(r)}^{(j)} + \sum_j \int K^{(i,j)}(r,s) f^{(j)}(s)\,ds \tag{96}$$

A full discussion is given in a recent book edited by Burke at al.[41] We describe here some methods we have used for polyatomic molecules.

4.1. An Iterative Method

This is a very simple method to program. One starts by estimating a solution, say $f_o^{(j)}$, and uses it to calculate the exchange integral on the r.h.s. of eq.(96). One then solves the ensuing coupled inhomogeneous equations for $f_1^{(i)}$, with i=1,2,....N until some prearranged convergence on the final K-matrix elements is achieved. We have discussed various aspects of it in the previous subsection 2.3 and details of the procedure were given in the various publications quoted in that section. As mentioned there, the rate of convergence can be slow if the initial estimate is not good. Improved estimates can also be made by noting that the scattered wavefunctions are usually nearly independent of k for small r values.

Since f(k') ≈ f(k) for small r we can use f(k) as the initial approximation for f(k'). Convergence can be obtained in about five iterations (cf. the preliminary results of ref. 42 for H_2O). Further accelerating procedures have been considered and discussed in the previous Section, when we analysed the properties of the iterative exchange method, the latter being one of the essential ingredients for applying the above scheme.

4.2 A Linear Algebraic Method

A program has been written to solve the exchange equations directly using some of the ideas of Seaton for electron-atom scattering.

Let R be such that the exchange terms are negligible for r>R and consider a grid of points r_i, i=0,1,...N+1, $r_0=0$, $r_{N+1}=R$, $h_{i+1}= r_{i+1}-r_i$, not necessarily equidistant. We calculate solutions f_i at the grid points.

The differential equation

$$\frac{d^2}{dr^2}f = gf + k \tag{97}$$

is replaced by the generalized Numerov algorithm

$$(\alpha_n^{(o)} - \beta_n^{(o)}g_n)f_n + (\alpha_n^{(1)} - \beta_n^{(1)}g_{n+1})f_{n+1} + (\alpha_n^{(2)} - \beta_n^{(2)}g_{n+2})f_{n+2}$$
$$= \beta_i^{(o)}k_n + \beta_n^{(1)}k_{n+1} + \beta_n^{(2)}k_{n+2} \tag{98}$$

where

$$\alpha_n^{(0)} = h_{n+1}$$
$$\alpha_n^{(1)} = -(h_n + h_{n+1})$$
$$\alpha_n^{(2)} = h_n$$
$$\beta_n^{(0)} = \frac{1}{12}h_{n+1}\{h_n^2 + h_nh_{n+1} - h_{n+1}^2\}$$
$$\beta_n^{(1)} = \frac{1}{12}(h_n + h_{n+1})\{-h_n^2 + h_nh_{n+1} + h_{n+1}^2\}$$
$$\beta_n^{(2)} = \frac{1}{12}h_n\{-h_n^2 + h_nh_{n+1} + h_{n+1}^2\}$$

The integral $\int$ K(r,s) f(s) ds is approximated by a quadrature formula

$$\int K(r,s)f(s)ds \approx \sum_j \gamma_{ij}K(r_i, s_j)f_i \tag{99}$$

By enforcing boundary conditions at r=0 and R, and collecting terms, the problem is reduced to the solution of a set of linear equations

$$\mathbf{A}\ \mathbf{f} = \mathbf{b} \tag{100}$$

where $\mathbf{f} = \left[\mathbf{f}^{(1)}, \mathbf{f}^{(2)}, \ldots \mathbf{f}^{(M)}\right]^T$

(M being the numbers of channels)

and

$$\mathbf{f}^{(i)} = \left[f_i^{(i)}, \ldots f_N^{(i)}\right]^T \tag{101}$$

The most computationally expensive part of this procedure is calculating the exchange contribution to A. Fortunately it is energy independent and has only to be calculated once for each scattering state.

4.3. S and K Matrices, and Scattering Amplitudes

Let us recall that we need to solve M coupled (integro) differential equations for each state (pμ), and let us rewrite these solutions as $f_j^{p\mu}$ (j= 1,2...M; j=lh). For M coupled equations there will be a set of M independent solutions which satisfy f(0)=0 and we need to introduce an additional index i=1,...M, to write the solutions as $f_{ij}^{p\mu}$. The $\underline{S}$ matrix is defined by looking for solutions with the asymptotic form

$$f_{ij}^{p\mu} \underset{r \to \infty}{\approx} e^{-i(kr-\frac{1}{2}l_j\pi)}\delta_{ij} - S_{ij}^{p\mu}\, e^{i(kr-\frac{1}{2}l_j\pi)} \tag{102}$$

we can form a general solution $F_j^{p\mu}$ which is a linear combination of these independent solutions

$$F_j^{p\mu} = \sum_{i=1}^{M} a_i^{p\mu} f_{ij}^{p\mu} \tag{103}$$

from which we can construct a total function for the scattered electron

$$F(\mathbf{r}) = \sum_{jp\mu} r_i^{-1} F_j^{p\mu}(r) X_{l_j h_j}^{p\mu}(\hat{r}) \tag{104}$$

We choose the $a_i^{p\mu}$ so that

$$F \sim e^{i\mathbf{k}.\mathbf{r}} + f(\vartheta,\phi)\, e^{ikr}/r \tag{105}$$

where $f(\vartheta,\phi)$ is the fixed nuclei scattering amplitude.

To calculate the a_i we note first that

$$e^{i\mathbf{k}\cdot\mathbf{r}} = \sum_l (2l+1) i^l P_l(\hat{\mathbf{k}}\cdot\hat{\mathbf{r}}) \frac{1}{kr} j_l(kr) \tag{106}$$

where

$$P_l(\hat{\mathbf{k}}\cdot\hat{\mathbf{r}}) = \sum_{h p\mu} \frac{4\pi}{2l+1} X_{lh}^{p\mu}(\hat{\mathbf{k}}) X_{lh}^{p\mu}(\hat{\mathbf{r}}) \tag{107}$$

and $j_l(kr)$ is spherical Bessel function with asymptotic form

$$j_l(kr) \underset{T\to\infty}{\sim} \sin(kr - \frac{1}{2}l\pi)$$

so that

$$e^{i\mathbf{k}\cdot\mathbf{r}} \underset{r\to\infty}{\sim} \frac{4\pi}{2ikr} \sum_{l,h,p\mu} X^{p\mu}_{lh}(\hat{\mathbf{r}}) X^{p\mu}_{lh}(\hat{\mathbf{k}}) \left\{ e^{ikr} - (-1)^l e^{-ikr} \right\} \tag{108}$$

Comparing the coefficient of e^{-ikr}/r in the asymptotic form of equations (105) and (106) we can show that

$$a^{p\mu}_i = \frac{4\pi}{2ik} X^{p\mu}_{l_i h_i}(\hat{\mathbf{k}})(-1)^{l_i} \tag{109}$$

The scattering amplitude is then obtained by comparing coefficients of e^{-ikr}/r in the two equations giving

$$f(\hat{\mathbf{k}}\cdot\hat{\mathbf{r}}) = \sum_{ijp\mu} \frac{4\pi}{2ik} X^{p\mu}_{l_i h_i}(\mathbf{k}) X^{p\mu}_{l_j h_j}(\mathbf{r}) i^{l_i - l_j} \left\{ S^{p\mu}_{ij} - \delta_{ij} \right\} \tag{110}$$

As computations are usually done with real functions we introduce a K matrix which is defined from the asymptotic form

$$f^{p\mu}_{ij} \underset{r\to\infty}{\approx} \sin(kr - \frac{1}{2}l_j\pi)\delta_{ij} + K^{p\mu}_{ij} \cos(kr - \frac{1}{2}l_j\pi) \tag{111}$$

The **S** and **K** matrices are related by

$$\mathbf{S} = (\mathbf{I} - i\mathbf{K})^{-1}(\mathbf{I} + i\mathbf{K}) \tag{112}$$

The scattering calculation is carried out in two parts. In the internal region where $o \le r \le R$, we include all terms in the integro differential equation and generate M independent solutions f_{ij}. In the external region, $r>R$, the exchange terms are negligible and the direct local potential can be expanded as an inverse series in r.

We can also generate independent solutions in the external region which have the familiar asymptotic form

$$F_{ij}(r) \underset{r\to\infty}{\approx} \sin(kr - \frac{1}{2}l_j\pi)\delta_{ij} \tag{113}$$

$$G_{ij}(r) \approx \cos(kr - \frac{1}{2}l_j\pi)\delta_{ij} \tag{114}$$

The solutions in the two regions are then joined at R by the condition

$$\sum_{i=1}^{M} A_{ik} f_{ij} = F_{kj} + K_{kj} G_{kj} \tag{115}$$

and also enforcing continuity of the continuum function derivatives. There are then $2M^2$ equations for the unknown A_{ik} and K_{ik}. (Note that we can also obtain $2M^2$ equation by joining the two solutions at two points R_1 and R_2).

4.4. The Total Cross Sections

The integral, elastic cross section is often called the total cross section, σ, and describes the scattering from randomly oriented molecules. It is obtained by integrating $|f(\hat{k},\hat{r})|^2$ over all orientations $\hat{r}$ of the scattered electron and by averaging over all the orientations $\hat{k}$ of the impinging beam:

$$\sigma = \frac{1}{4\pi} \int |f(\hat{k},\hat{r})|^2 d\hat{r}\; d\hat{k} \tag{116}$$

which is easily shown to be equivalent to

$$\sigma = \frac{\pi}{k^2} \sum_{i,j,p,\mu} |S_{ij}^{p\mu} - \delta_{ij}|^2 \tag{117}$$

It is also instructive to express σ in terms of eigenphases which are in turn defined via the further diagonalization of each S-matrix from each contributing molecular symmetry, $S^{p\mu}$. We could obtain a unitary transformation U such that $\Lambda^{p\mu}$ defined by

$$\Lambda^{p\mu} = U\; S^{p\mu}\; U^{+} \tag{118}$$

is a diagonal matrix. If we now define the diagonal elements of $\Lambda^{p\mu}$ to be given by $\exp(2i\eta_l^{p\mu})$ where the $\eta_l^{p\mu}$ are the quantities known as the eigenphases for each contributing symmetry $|p\mu\rangle$, then one can easily show that

$$\sigma = \frac{4\pi}{k^2} \sum_{lp\mu} \sin^2 \eta_l^{p\mu} \tag{119}$$

In the simpler case of the rotational symmetry of the spherical atom, then the (2l+1) eigenphases for each contributing l are equal and therefore the integral cross section reduces to the familiar formula

$$\sigma = \frac{4\pi}{k^2} \sum_{l} (2l+1) \sin^2 \eta_l \tag{120}$$

4.5. The Differential Cross Sections

To obtain an expression for the differential cross sections one needs first to introduce a coordinate system (XYZ) which is fixed in space, the LAB system, and another coordinate system which is fixed to the molecular figure axis (xyz) and is called the MOL system or BODY system. The latter is obtained from (XYZ) by a rotation through the appropriate Euler angles ($\alpha\beta\gamma$).

The previously discussed expression for the scattering amplitudes contains angular factors, $Y_l^m(\hat{\mathbf{r}})$, where $\hat{\mathbf{r}}$ is defined with respect to the MOL system. The corresponding expression in the LAB frame is obviously given by

$$Y_l^m(\hat{\mathbf{r}}) = \sum_{\mu} Y_l^m(\hat{\mathbf{r}}) D^l_{\mu m}(\alpha,\beta,\gamma) \tag{121}$$

where $\hat{\mathbf{r}}$ is defined with respect to (xyz) and $\hat{\mathbf{r}}$ with respect to (XYZ) and the D^a_{bc} are the usual Wigner Coefficients.

The necessary symmetry adapted functions can be written in terms of real or complex spherical harmonics

$$X^{p\mu}_{lh} = \sum_{m\geq 0,q} b^{p\mu}_{lhm} S_l^{mq} \tag{122a}$$

$$= \sum_{m} \bar{b}^{p\mu}_{lhm} Y_l^m \tag{122b}$$

where the b coefficients are given by

$$\bar{b}^{p\mu}_{lhm} = (-)^m \exp[i\pi(q-1)/4] b_{lhm}/\sqrt{2} \qquad \text{for } m>0 \tag{123a}$$

$$= q \ \exp[i\pi(q-1)/4] b_{lh|m|} \qquad \text{for } m<0 \tag{123b}$$

The corresponding symmetry-adapted functions defined before are given by

$$X^{p\mu}_{lh}(\hat{\mathbf{r}}) = \sum_{m\nu} \bar{b}^{p\mu}_{lhm} Y_l^{\nu}(\hat{\mathbf{r}}') D^l_{\nu m}(\alpha\beta\gamma) \tag{124a}$$

and

$$X^{p\mu}_{lh}(\hat{\mathbf{r}}) = \sum_{m} \bar{b}^{p\mu}_{lhm} \left(\frac{2l+1}{4\pi}\right)^{1/2} D^l_{\mathrm{o}m}(\alpha\beta\gamma) \tag{124b}$$

where we have simplified the previous function by taking $\hat{\mathbf{k}}'$ along z, i.e. by having $\hat{\mathbf{r}}' = \mathbf{0}$.

The differential cross section for scattering by randomly oriented molecules is obtained by averaging $|f(\hat{\mathbf{k}}'\cdot\hat{\mathbf{r}}';\alpha\beta\gamma)|^2$ over all $(\alpha\beta\gamma)$ values

$$\frac{d\sigma}{d\Omega}(\hat{\mathbf{k}}'\cdot\hat{\mathbf{r}}') = \frac{1}{8\pi^2}\int |f(\hat{\mathbf{k}}'\cdot\hat{\mathbf{r}}';\alpha\beta\gamma)|^2 d\alpha\ sin\beta\ d\beta d\gamma \tag{125}$$

One also can make use of the following properties:[43]

$$D^{li}_{\nu m_i} D^{lj}_{om_j} = \sum_j (2J+1)\begin{pmatrix} l_i & l_j & J \\ \upsilon & o & -\nu \end{pmatrix}\begin{pmatrix} l_i & l_j & J \\ m_i & m_j & M_J \end{pmatrix} D^{J*}_{-\nu,M_J}$$

$$\int D^{a*}_{bc} D^{A*}_{BC} d\alpha\ sin\beta\ d\beta d\gamma = \frac{8\pi^2}{(2a+1)}\delta_{aA}\delta_{bB}\delta_{cC} \tag{126}$$

$$Y^{\nu}_{l_i} Y^{\nu*}_{\bar{l}_i} = (-1)^{\nu}\sum_L \left[(2l_i+1)(2\bar{l}_i+1)\right]^{1/2}\frac{1}{4\pi}\begin{pmatrix} l_i & \bar{l}_i & L \\ o & o & o \end{pmatrix}\begin{pmatrix} l_i & \bar{l}_i & L \\ \nu & -\nu & o \end{pmatrix} P_L \tag{127}$$

$$\sum_{\nu}\begin{pmatrix} l_i & l_j & J \\ \nu & o & -\nu \end{pmatrix}\begin{pmatrix} \bar{l}_i & \bar{l}_j & J \\ \nu & o & -\nu \end{pmatrix}\begin{pmatrix} l_i & \bar{l}_i & L \\ \nu & -\nu & o \end{pmatrix}(-1)^{\nu} = (-1)^{L+J} W(l_i l_j \bar{l}_i \bar{l}_j; JL)\begin{pmatrix} l_j & \bar{l}_j & L \\ o & o & o \end{pmatrix}$$

where W(abcd;ef) is a Racah coefficient and $\begin{pmatrix} a & b & c \\ \alpha & \beta & \gamma \end{pmatrix}$ is a Wigner 3-j symbol. One can therefore show that

$$\frac{d\sigma}{d\Omega} = \sum_L A_L P_L(\cos\vartheta) \tag{128}$$

where

$$A_L = \frac{(-1)^L(2L+1)}{4k^2}\sum_{JM_J i_j \bar{i}_j} l\,(2J+1)(-1)^J i^{l_i - l_j}(-i)^{\bar{l}_i - \bar{l}_j}$$

$$\left[(2l_i+1)(2l_j+1)(2\bar{l}_i+1)(2\bar{l}_j+1)\right]^{1/2} W(l_i l_j \bar{l}_i \bar{l}_j; JL) \tag{129}$$

$$\begin{pmatrix} l_i & \bar{l}_i & L \\ o & o & o \end{pmatrix}\begin{pmatrix} l_j & \bar{l}_j & L \\ o & o & o \end{pmatrix} M^{JM_J}_{ij} M^{JM_J *}_{\bar{i}\bar{j}}$$

where

$$M_{ij}^{JM_J} = \sum_{m_i m_j p\mu} \bar{b}_{l_i h_i m_i}^{p\mu} T_{ij}^{p\mu} \bar{b}_{l_j h_j m_j}^{p\mu} \begin{pmatrix} l_i & l_j & J \\ m_i & m_j & M_J \end{pmatrix} \tag{130}$$

and

$$T = S - I \tag{131}$$

If one now integrates over all solid angles one obtains the required total cross section

$$\sigma = 4\pi A_o \tag{132}$$

which, of course, reduces to the expression for σ already given in the previous section.

A further quantity of interest is the momentum transfer cross section

$$\sigma_m = \frac{1}{8\pi^2} \int (1 - \cos\vartheta) \frac{d\sigma}{d\Omega} \sin\vartheta d\vartheta d\varphi = 4\pi\left(A_o - \frac{1}{3}A_1\right) \tag{133}$$

4.6. Transitions Involving Nuclear Motion

So far we have ignored the participation of the rotational and vibrational molecular degrees of freedom in the scattering process. However, in spite of the added complexity that such additional couplings introduce in the treatment of the collision, one can still extract some useful information by using an approximate treatment of the dynamics that goes under the name of the adiabatic approximation.[44]

Briefly, the physics involved requires us to consider the effect of the kinetic energy operator of nuclear motion as negligible with respect to that of the corresponding operator on the electronic motion. As a consequence, the molecules are seen as having 'infinite' mass and to stand still during the scattering event, when the energy spacings of the molecular levels is treated as negligible with respect to the local kinetic energy of the impinging electron. Such an approximation obviously fails for very slow electrons and for very long interaction times. It is also less likely to hold for light molecules and for polar targets.

The well-known formulation by Chase[44] writes down the scattering amplitude $f_{ij}(\hat{\mathbf{k}}\cdot\hat{\mathbf{r}})$ for the transitions from a rotational state |i> to a rotational state |j> as approximately given by

$$f_{ij}(\hat{\mathbf{k}}\cdot\hat{\mathbf{r}}) = \int \Phi_j^*(\Omega) f(\hat{\mathbf{k}}\cdot\hat{\mathbf{r}};\Omega)\Phi_i(\Omega)d\Omega \tag{134}$$

where Φ_i, Φ_j are the initial and final rotational states and the $f(\hat{\underline{k}}\cdot\hat{\underline{r}},\Omega)$ is the scattering amplitude defined in the previous section 4.3. Full details about the specific expressions for asymmetric top, symmetric top and spherical top molecules have been given before and will not be discussed here.

For example, in the case of a symmetric-top target molecule the rotational wavefunction is given by the well-known formula

$$\Phi_i = \Psi_{JKM}(\alpha\beta\gamma) = \left(\frac{2J+1}{8\pi^2}\right)^{1/2} D^{J*}_{KM}(\alpha\beta\gamma) \tag{135}$$

where M gives the projection of the total angular momentum J along the LAB axis Z, while K gives its projection along the molecular figure axis, MOL frame, z. In this case both K and M are good quantum numbers.

One therefore write down the explicit scattering amplitude as calculated from eq.(134), f(JKM→J'K'M') and obtain from it the differential cross sections summed over all final degenerate states M' and averaged over all initial degenerate states M

$$\frac{d\sigma}{d\Omega}(JK \rightarrow J\,K') = \frac{k'}{k}\sum_{M,M'}\frac{1}{(2J+1)}|f(JKM \rightarrow J\,K'\,M')|^2 \tag{136}$$

where $k'^2 = k^2 + 2\,(E_{JK} - E_{J'K'})$ and the formula above is appropriate for molecules like NH_3, just to mention one example.

In the case of asymmetric top molecules (e.g. the H_2O target) $\underline{K}$ is no longer a good quantum number and the rotational wavefunction needs to be expressed as a linear combination of symmetric top functions

$$\Phi_{J\tau M} = \sum_{K=-J}^{J} a^J_{\tau K}\Psi_{JKM} \tag{137}$$

The scattering amplitude can therefore be obtained once more from the (134) formula and the corresponding cross section, summed over final states M' and averaged over initial states M, is given by

$$\frac{d\sigma}{d\Omega}(J\tau \rightarrow J\,\tau') = \frac{k'}{k}\sum_{M,M'}\frac{2}{2J+1}|f(J\tau M \rightarrow J\,\tau'\,M')|^2 \tag{138}$$

where $k'^2 = k^2 + 2\,(E_{J\tau} - E_{J'\tau'})$.

The corresponding rotational wavefunction for a spherical top molecule is the same as that for the symmetric top. There is a further degeneracy in that the $(2J + 1)^2$ states with the same J, but with different K and M, have the same energy E_J, where

$$E_J = B_e J(J+1) \tag{139}$$

where B_e is the molecular rotational constant (in units of energy). The corresponding differential cross section is therefore given by

$$\frac{d\sigma}{d\Omega}(J \rightarrow J') = \frac{k'}{k} \sum_{MM'} \frac{1}{(2J+1)^2} |f(JKM \rightarrow J'K'M')|^2 \tag{140}$$

Such an expression is applicable to a CH_4 molecule as a scattering target.

As long as the simplifying conditions discussed before are satisfied, the adiabatic approximation can be applied to vibrational motion also and one obtains the inelastic scattering amplitude for all the contributing symmetries as[45]

$$f_{ij}(\hat{\mathbf{k}} \cdot \hat{\mathbf{r}}) = \int \Phi_j^*(\Omega) X_j^*(Q) f(\hat{\mathbf{k}} \cdot \hat{\mathbf{r}}, \Omega, Q) \Phi_j(\Omega) X_i(Q) d\Omega \, dQ \tag{141}$$

where Q is the normal coordinate associated with the chosen nuclear motion and is given by a linear combination of internal coordinates.

One should note, however, that the additional quadrature of above may not be able to take advantage of the symmetry factorisation of the scattering equations discussed before for nonlinear molecules. In case the vibrational motion being considered is, in fact, involving a normal mode which is not totally symmetric, then the fuller symmetry of the (N+1) electron + (M) nuclei problem should be considered. In that case one finds that that all the partial waves are coupled together and much bigger problem than for fixed nuclei needs to be treated. This is, perhaps, one of the reasons why so little has been done on the vibrational excitations of polyatomic, non linear molecules.

5. EXAMPLES OF SPECIFIC CALCULATIONS

Since the earliest applications of the SCE approach to polyatomic targets (in 1972[46]) several molecular systems have been studied using directly target wavefunctions which were given by expansion over STO's basis functions centered at the c.o.m. of the molecular partner[2]. Usually, such treatments have tried to progressively improve on the sophistication of the potential forms used to treat static, polarisation and exchange forces. The main workhorse of such calculations has been the CH_4 system, for several reasons:

(i) it is a nonlinear molecule with still a limited number of electrons and only one 'heavy' atom located at the center of mass of the system;

(ii) the highly symmetric nuclear structure allows well for a factorisation of the sets of coupled equations for each contributing Irreducible Representation;

(iii) the many, fairly accurate, experimental data on integral and differential cross sections indicate two distinct features of the measurements: A broad maximum of the integral elastic cross section around 8 eV of collision energy and a marked Ramsauer-Townsend RT) minimum well below 1 eV collision energy.

The various attempts at computing the integral and differential cross sections have indeed succeeded, over the years, in reproducing both features via SCE calculations by successive improvements on the quality of the interaction forces employed in the calculations. Thus, the earlier work treated exchange as a model, free-electron gas approximation (FEGE) and correlation effects were included via a parametric cut-off function of the long-range polarisation. Further calculations introduced semiclassical exchange and density functional polarisation forces, until exact exchange and model polarisation were employed later on, with additional calculation. The agreement with the experimental data was found to be quantitatively better as the treatment of the interaction forces improved.

Just to give an example of the quality of the results which can be obtained using an SCE treatment of the scattering process, we show in Figure 1 a comparison between computed and measured integral elastic cross sections for the CH_4 target. In this case, the solid line corresponds to calculations[47] which started from a multicenter SCF wavefunction of GTO basis, while the thin line shows the same calculations using the earlier SCE-SCF wavefunction of STO orbitals centered on the Carbon atom[21]. In both bases the static and the exchange interaction were treated exactly (ESE) while the correlation-polarisation potential was described via a DFT approach. The experimental points are given by open squares[48], filled squares[49], filled rhombuses[50] and and open rhombuses[51].

One clearly sees that the agreement is very good and that the computations agree with measurements down to very low collision energies. They also reproduce rather well the RT minimum. A discussion of similar results obtained by using a multicenter approach will be presented in another chapter of this Volume.

Another polyatomic system which has been studied rather extensively over the years, due to the important role which it plays in the study of the decomposition in plasmas and of the preparation methods for the manifacturing of electronic devices, is the tetrahedral molecule of Silane, SiH_4. The experimental work has been quite intensive and was carried out by several groups, which focussed again on two distinct features of the integral cross sections: the RT minimum near 0.35 eV of collision energy and a broad shape resonance around 3 eV. However, fairly large discrepancies still exist among experimental findings when they report either the 3 eV shape resonance or the position of the RT minimum. A detailed analysis of the experimental situation and of the existing disagreement among experimentalists is presented in refs. [10].

An example of the quality achieved by SCE scattering calculations which use ESE treatment and model polarisation forces is shown in Figure 2 for the region around the RT minimum. Various calculations are presented there, together with some of the experimental points.

It is interesting to see that the solid line, which gives the SCE results using exact exchange and model polarisation, follows the experimental points[52] better than other model calculations of before (dashed curve[53] and filled dots[54])

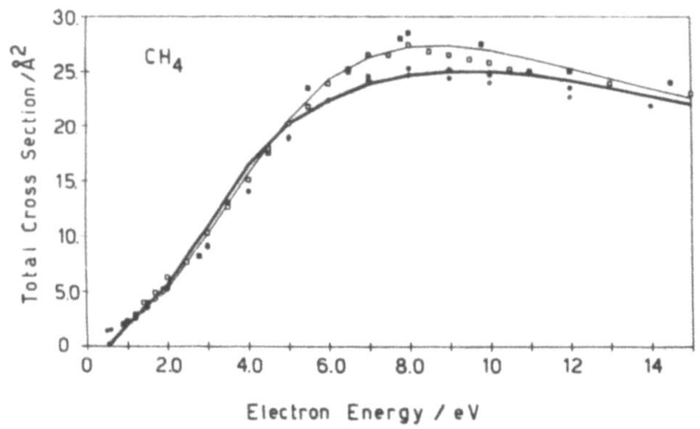

Fig. 1 - Computed (solid and thin lines) elastic integral cross sections and experimental data for the same quantities. Filed squares: ref. 49; open squares: ref. 4; filled rhombuses: ref. 50; open rhombuses: ref. 51.

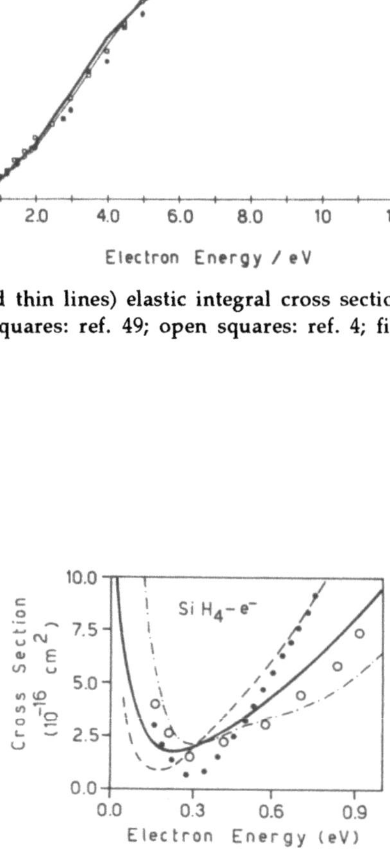

Fig. 2 - Computed and measured integral cross sections for Silane around the RT minimum region. solid line: SCE calculations with exact exchange. Dashed line: SCE calculations with model exchange. Dot-dashed: complex Kohn filled dots: model calculations. Open circles: experiments.

and with comparable quality with the multicenter calculations that used the complex Kohn variational method[55].

That current sophistication in SCE calculations for polyatomic targets is now reaching very good accord with experiments and with more sophisticated theoretical models and is further shown by the comparison between various cross section data in the region of the shape resonance for the Silane target, reported in figure 3.

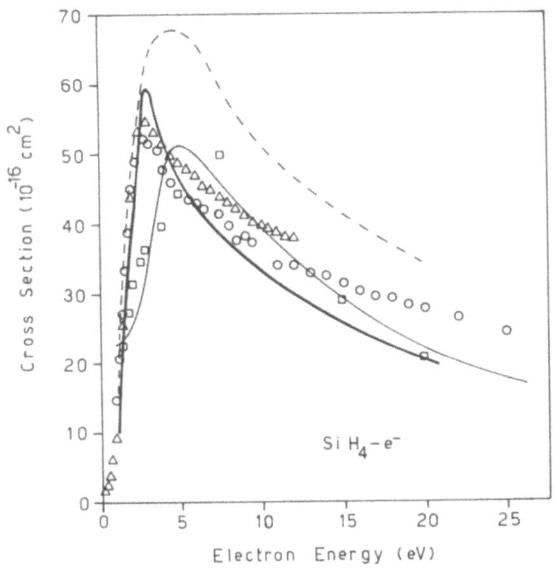

Fig. 3 - Comparison of measured and computed integral cross sections for SiH_4 in the region of the shape resonance. Solid line: SCE calculations[10], thin line: calculations from ref [9]. Dashed line: model calculations from ref. [56]. The experiments are given by: open circles[57]; open squares[58]; open triangles[52].

It is interesting to note that the SCE calculations from a multicenter target wavefunction[10], using exact exchange and model polarisation, follow very closely the experimental points, better than the previous calculations[9] (thin line) which used the complex Kohn variational method. Earlier model calculations (dashed line)[56] which used SCE wavefunctions but employed a model exchange and polarisation show only qualitative accord with measurements and markedly overestimate the cross section around the resonance position. Essentially all the SCE calculations which have been carried out thus far employed model forms of polarisation potentials which treated the short-range part of the polarisation via some form of local density functional theory (DFT), as discussed earlier on in Section 2 of this chapter. Once both the static and exchange forces are treated as exactly as possible, comparison with experiments or the analysis of specific features like angular distributions or resonance peaks can give a good indication on the quality of the treatment of the Correlation-Polarisation forces. An example of this is provided by the SCE treatment of low-energy electron scattering from acetylene (C_2H_2, $^1\Sigma_g$) where a marked shape resonance was found experimentally around 2.6 eV. Recent SCE calculations were carried out using a multicenter wavefunction of near-HF quality and treating the exchange interaction via the separable exchange approximation[27]. The polarisation potential was once more included via a local form of DFT, as discussed earlier. The total cross sections turn out to be very close to the experiments, the best ever obtained from 1 to 30 eV. Around the resonance position the calculations overestimate experiments but correctly locate the resonance. Figure 4 shows the various contributions coming from the most important Irreducible Representations of the molecular point group.

One sees clearly from the results in the Figure that the shape resonance is of Π_g symmetry and dominated by an l=2 partial wave. All other symmetries, apart from the Σ_g one, contribute negligibly around the resonance position. The calculations also showed that the RT minimum was indeed a feature dominated by the Σ_g symmetry for the continuum electron.

The use of multicenter target wavefunctions to describe the molecular partners in the scattering of slow electrons from polyatomic molecules has opened up broader possibilities in term of the systems which can be realistically treated. One can, in fact, move away from the constraint of having to deal mostly with AH_n system if SCE wavefunctions have to be used to begin calculations. Two examples of more complex systems which have been of great experimental and theoretical interest over many years are given by Figures 5 and 6, where molecules like CF_4 and SF_6 are briefly presented.

In the case of CF_4, the interest in having elastic cross sections available for that molecule stems from the many areas of physical chemistry in which such a system presents important applications, from atmospheric chemistry to chemical vapour deposition kinetics, from the fluorocarbon reactions to the capability of CF_4 to be used as an electron quencher in lieu of SF_6, another molecule which we shall see below. The calculations were recently carried out using a very extended multicenter expansion for the target molecule[59]. The exchange and static interactions were then treated exactly while the polarisation was included using a DFT correlation-polarisation potential.

Because of the presence of heavy atoms away from the molecular c.o.m., one naturally expects that convergence will be harder to achieve. We see, in fact, that the earlier results where the continuum electron was expanded up to $l_{max}=15$ produce a shape resonance in fair agreement with experiments. On the other hand, as more partial waves are added in the expansion the resonance moves closer to measurements, while the computed cross section changes little

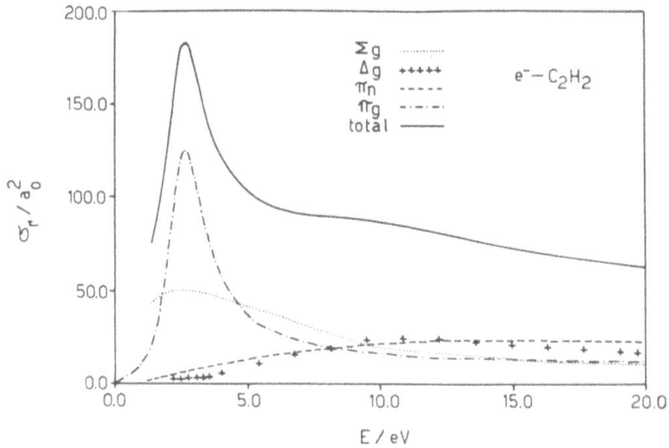

Fig. 4 - Computed components to the total integral cross section (solid line) for C_2H_2. ESE and model polarisation calculations[27].

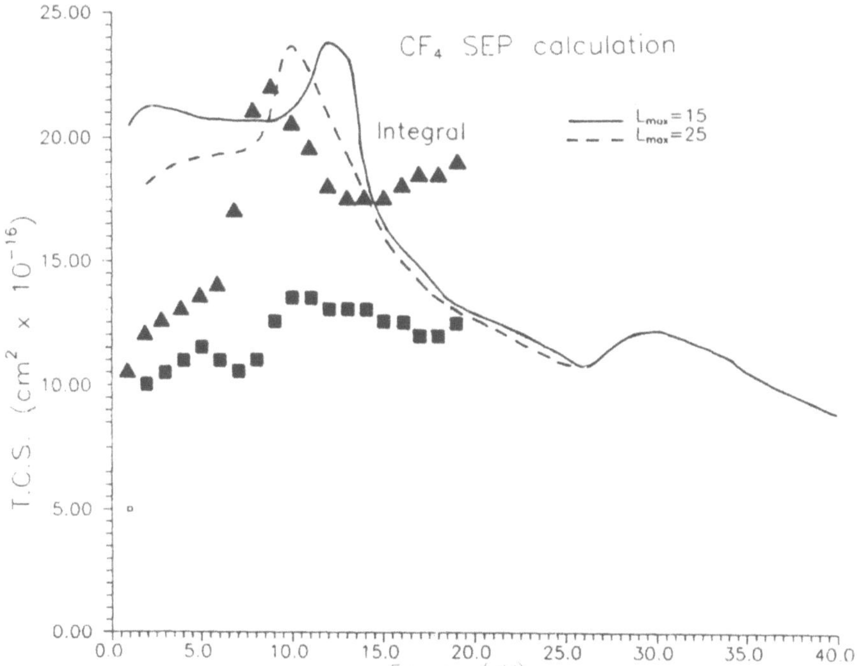

Fig. 5 - Computed and measured integral cross sections for CF_4 SEP calculations using 25 partialwaves for the continuum electron (dashed line)[39] and 15 partialwaves (solid lines)[60]. The experiments are given by filled squares[61] and filled triangles[62].

in the higher energy region[60]. A test of angular distributions suggest that convergence is essentially attained with the l_{max}= 25 calculations, as it was also tested at a few collision energies around the resonance and found to be unchanged in going up to l_{max}= 30.

An even bigger system is the well known SF_6 molecule, an extensively studied molecular target with very high point group symmetry (O_h) and of great importance as an electron 'scavenger' and a gaseous quencher of discharges in molecular mixtures. The first ever accurate calculations on this system are shown in figure 6, where exact exchange and static potentials were used and the DFT polarisation model was employed.

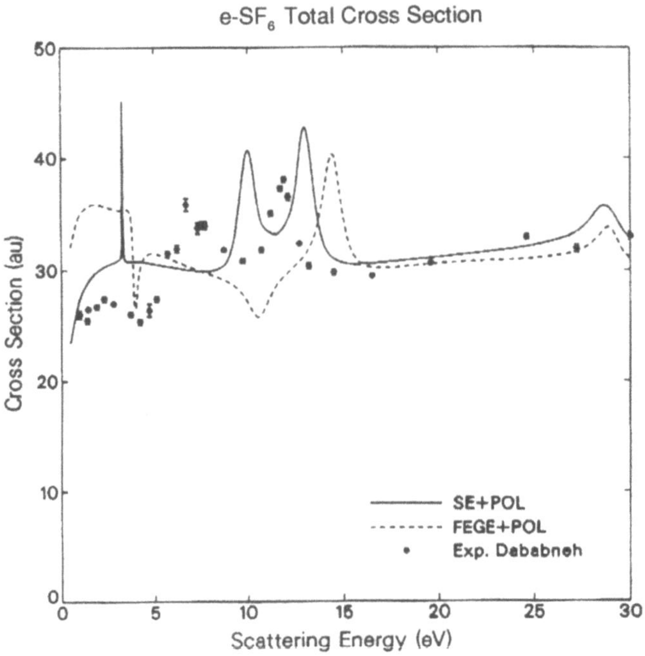

Fig. 6 - Computed and measured integral elastic cross sections for SF_6 targets. The solid line shows ESE calculations with model polarisation, while the dashed line shows calculations with also model exchange[63]. The experiments are given by filled squares[64].

It is interesting to see that the use of an accurate exchange interaction for this system certainly makes a big difference in the final results. The FEGE model exchange, usually found to be a reasonable approximation, modifies completely the shape of the cross section, expecially at lower energies.

The calculations with exact exchange suggest the presence of a very narrow resonance around 2.5 eV, of a_{1g} symmetry, which may be responsible for the system efficiency in trapping electrons at very low energies. Furthermore, the two broader resonances between 10 and 15 eV agree only qualitatively with the experimental positions. The continuum function was expanded here up to $l_{max}=30$. Obviously more work is needed in refining the convergence studies and in selecting better target wavefunctions. On the other hand it is reassuring to see that the SCE approach is now capable of describing relatively well such complex targets and can be used to analyse several features of electron scattering from increasingly larger polyatomic molecules.

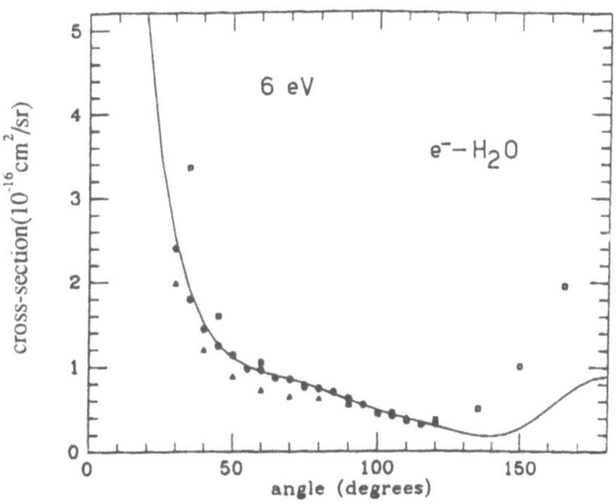

Fig. 7 - Computed and measured DCS (rotationally summed) for H_2O targets at 6 eV. Solid line shows SCE calculations at the ESEP level[65]. The experiments are given by open squares[66], filled triangles[67] and filled circle[68].

The scattering of electrons from polar polyatomics presents a well-known divergence of the cross sections computed in a Body-fixed frame of reference, a feature which has been analysed elsewhere and which can be remedied with some specific corrections that we will not be discussing here, as they have been given various times before[23,24]. It is, on the other hand, interesting to see how some of such systems behave in terms of their differential cross sections (DCS) at low collision energies. Molecules like H_2O, NH_3 and H_2S have been analysed before within the SCE approach and the agreement with experiments was found to be rather good. Examples of the behaviour of the DCS, rotationally summed and rotationally inelastic are given in Figures 7 and 8, respectively.

It is interesting to see that the SCE calculations agree well with experiments at most angles but disagree with measurements in the large angle region. However, because of steric effects from the detector, this is the region which shows the largest relative errors and is often obtained from experiments only indirectly. Earlier SCE calculations using model exchange indicated that momentum transfer cross sections require in this system that the computed DCS be larger at larger angles than suggested above.

Another helpful test is given by the calculation of direct rotationally inelastic cross sections, which are usually rather large for polar systems. The SCE results for H_2O are given, within exact SE calculations and the DFT polarisation model mentioned before, in Figure 8 for the (00→10) rotational transition at a collision energy of 6 eV. The corresponding DCS are shown in the Figure.

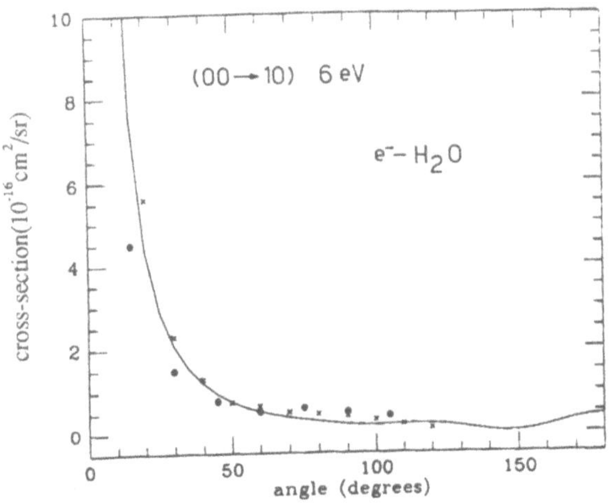

Fig. 8 - Computed and measured DCS for the rotational transition (00→10) in H_2O, at 6 eV. The ESE calculation with model polarisation are shown by the solid line[65]. The crosses are earlier SCE calculations using a model Exchange[23]. The filled circles are the experiments[69].

In this example, two different calculations are compared with experiments, both using the SCE approach. The solid line shows the exact SE results with DFT model polarisation[65]. The crosses use the same wavefunction and DFT and static potential but differ in that they employed a semiclassical model for exchange[23]: both results agree with each other and are also very close to experiments[69].

Another example of a polar molecule for which the SCE approach should work well is given by the H_2S molecule. SCE calculations have been carried out for it, using model potentials, and were found in rather surprisingly good accord with the existing measurements In Figure 9 we compare exact SE calculations, within the SCE approach, but using a model polarisation potential as before. The new experimental DCS are also shown.

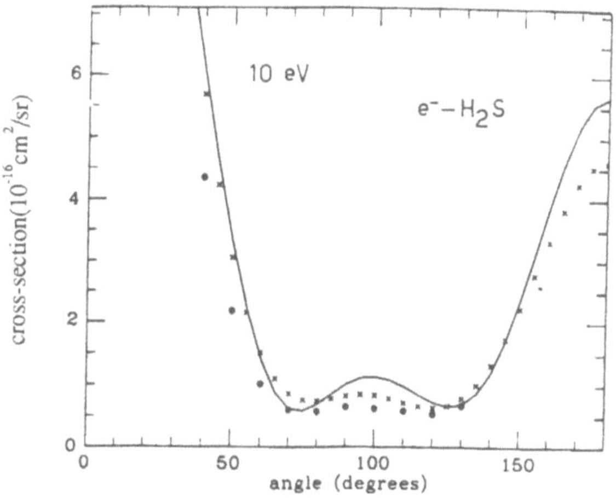

Fig. 9 - Computed and measured DCS for H_2S at 10 eV. The solid line shows the SCE calculations at the ESEP level[65], while the crosses are the same calculations at the SE level[65]. The experiments are given by filled circles[70].

The exact treatment of exchange produces here reasonable agreement with experiments, although differences are seen as the scattering angle goes into the backward direction. Furthermore, the little effect of the polarisation potential at these collision energies is not entirely surprising since in polar molecules the static interaction dominates over a much larger range of relative distances. On the other hand, the fact that the polarisation potential makes the results to be less in accord with experiments is obviously surprising and must be related both to the quality of the wavefunction employed and to the overall convergence of the scattering calculations.

6. CONCLUSIONS

We have discussed in the present chapter a specific procedure, the single-centre expansion approach, for the study of different scattering models which can be employed to treat electron (and positron) scattering from polyatomic targets. Most of the discussion has dealt with electron as projectiles, although a fair amount of work exists already for positron colliding with polyatomic targets. They will not be discussed here, however, since it is generally understood that most of the considerations made when dealing with electrons can be fairly simply extended to positron scattering, provided one can disregard the effect of positronium formation at the energies considered.

We have shown that the symmetry-adapted partial wave expansion exhibits for molecules a series of rather appealing features which help us to understand more transparently the microscopic processes that take place during the scattering event. In particular:

(i) we can break down the scattering, for fixed nuclei processes, into separate calculations for each of the contributing irreducible representations;

(ii) the anisotropic coupling which originates from the various forms of interaction forces can be analysed in terms of contributing partial waves for each I.R.,

(iii) the selection rules which apply to rotational and vibrational excitation processes can be related to the specific symmetries of the mediating interaction during the scattering;

(iv) fairly standard integration algorithms can be applied to generate phaseshifts and $\underline{K}$-matrices since the coupled equations are in the end solved along the scattering variable of the selected frame of reference.

The mayor drawback comes, of course, from the inevitably slow convergence of expanding the Coulomb cusps and of the charge distributions located away from the scattering center onto a different reference point. This means that several partial waves are needed in the process and that the size of the coupled equations increase rapidly for the heavier atoms and for the larger and less symmetric molecular targets. On the other hand, because of the largely technical nature of this major difficulty, we have already shown that present-day computational tools are able to handle rather realistically systems like CF_4 and SF_6 and one can therefore expect that the rapid progress in the area of multitasking and of parallel processing should render the above effort increasingly more accessible for the treatment of even larger systems.

7. REFERENCES

1. N.F. Lane, *Rev. Mod. Phys.*, 52, 29 (1980)
2. F.A. Gianturco and A.K. Jain, *Rep. Progr. Phys.*, 143, 347 (1986)
3. M.A. Morrison, *Adv. At. Mol. Phys.*, 24, 51 (1988)
4. Y. Itikawa, *Phys. Rev.*, 46, 11 (1978)
5. P. Mc Naughten and D.G. Thompson, *J. Phys. B*, 21, L703 (1991)
6. F.A. Gianturco, V. Di Martino and A. Jain, *Nuovo Cim.*, D14, 411 (1992)
7. A. Jain, F.A. Gianturco and D.G. Thompson, *J. Phys. B*, 24, L255 (1991)
8. R.R. Lucchese, *J. Chem. Phys.*, 92, 4203 (1990)
9. C. Winstead and V. Mc Koy, *Phys. Rev.*, A42, 5357 (1990)
10. F.A. Gianturco, R.R. Lucchese, N. Sanna and A. Talamo, in "Electron scattering from molecules, clusters and sufraces", H. Ehrhardt and L. Morgan Eds., Plenum Press, New York, 194
11. R.R. Lucchese and V. Mc Koy, *Phys. Rev.* A, A28, 1382 (1983)
12. S. Hara, J. Phys. *Soc. Japan*, 22, 710 (1967)
13 M.E. Riley and D.G. Truhlar, *J. Chem. Phys.*, 63, 2182 (1975)
14. S. Salvini and D.G. Thompson, *J. Phys.* B 14, 3797 (1981)
15. A. Jain and D.G. Thompson, *J. Phys.* B 17, 443 (1983)
16. A. Jain and D.G. Thompson, *J. Phys.* B 16, 3077 (1983)
17. M.A. Morrison and L.A. Collins, *Phys. Rev.*, A17, 918 (1978)

18. M.A. Morrison and L.A. Collins, *Phys. Rev.*, A23, 127 (1981)
19. J.B. Furness and I.E. Mc Carthy, *J. Phys.* B 6, 2280 (1973)
20. D.G. Truhlar and M.A. Brandt, *J. Chem. Phys.*, 65, 3092 (1976)
21. F.A. Gianturco and A. Scialla, *J. Phys.* B 20, 3171 (1987)
22. F.A. Gianturco and A. Scialla, *J. Chem. Phys.*, 87, 6468 (1987)
23 F.A. Gianturco, *J. Phys.* B 24, 3837 (1991)
24. F.A. Gianturco, *J. Phys.* B 24, 4627 (1991)
25 T.N. Rescigno and A.E. Orel, *Phys. Rev.* A 24, 1267 (1981)
26 T.N. Rescigno and A.E. Orel, *Phys. Rev.* A 25, 2402 (1982)
27. F.A. Gianturco and T. Stoecklin, *J. Phys.* B, submitted (1994)
28. F.A. Gianturco and D.G. Thompson, *J. Phys.* B, L383, (1976)
29. J. O'Connell and N.F. Lane, *Phys. Rev.* A27, 1893 (1983)
30. N.T. Padial and D.W. Norcross, *Phys. Rev.* A29, 1742 (1984)
31. F.A. Gianturco, A.Jain and L.C. Pantano, *J. Phys.* B 19, 571 (1986)
32. F.A. Gianturco and J.A. Rodriguez-Ruiz, *J. Phys.* A47, 1975 (1993)
33. J.P. Perdew and A. Zunger, Phys. Rev. B 23, 5048 (1981)
34. C. Lee, W. Yang and R.G. Parr, *Phys. Rev.* B 37, 785 (1988)
35. F.A. Gianturco, D. De Fazio, J.A. Rodriguez, K.T. Tang and J.P. Toennies, *J. Phys.* B 27, 303 (1994)
36. A. Temkin and J.C. Lamkin, *Phys. Rev.* 121, 788 (1961)
37. E.P. Wigner "Group Theory", Academic Press, London (1959)
38. M. Tinkham, "Group Theory and Quantum Mechanics", Mc Graw Hill, New York (1964)
39. S.L. Altman, *Proc. Cambr. Phyl. Soc.* 53, 343 (1957)
40. S.L. Altman andA.P. Cracknell, *Rev. Mod. Phys.* 37, 19 (1965)
41. P.G. Burke and K.A Berrington, "Atomic and Molecular Processes", IOP Publishing, Bristol (1993)
42. A. Jain, F.A. Gianturco, K.L. Baluja and V. Di Martino, *Chem. Phys. Lett,.* 1 83, 34 (1991)
43. A. Jain and D.G. Thompson, *Comp. Phys. Com.*, 32, 367 (1984)
44. D.M. Chase, Phys. Rev., 104, 838 (1956)
45. D. Thirumalai, K. Onda and and D.G. Thruhlar, *J. Chem. Phys.*, 74, 6792 (1981)
46. P.G. Burke, N. Chandra and F.A. Gianturco, *J. Phys.* B5, 2212 (1972)
47. F.A. Gianturco, N. Sanna, A. Talamo, unpublished results (1994)
48. A. Zecca, *J. Phys.* B24, 2747 (1991)
49. M.S. Dababneh, *Phys. Rev.* A38, 1207 (1988)
50. B. Lohmann and S. buckmann, *J. Phys.* B19, 2565 (1986)
51. J. Ferch, B. Granitza, W. Raith, *J. Phys.* B 18, L445 (1985)
52. H.-X. Wan, J.H. Moore and J.A. Tossel, *J. Chem. Phys.*, 91, 1340 (1989)
53. A.K. Jain, A.N. Tripathi, A. Jain, *J. Phys.* B 20, L389 (1987)
54. Jammin Yan, *J. Phys.* B 22, 2589 (1989)
55. W. Sun,C.W. Mc Curdy and B.H. Lengsfield III, *Phys. Rev.* A45, 6323 (1992)
56. F.A. Gianturco, A. Jain, L.C. Pantano, *Phys. Rev.* A 36, 4367 (1987)
57. O. Sueoka and S. Mori, private communication (1987)
58. H. Tanaka, L. Boesten, M. Sato, M. Kimura, M.A. Dillon, D. Spence, *J. Phys.* B 23, 577 (1990)
59. F.A. Gianturco, N. Sanna and R.R. Lucchese, *J. Chem. Phys.* 100, 1532 (1994)
60. F.A. Gianturco, F. Raganelli, R.R. Lucchese, andN. Sanna, in preparation (1994)
61. R.K. Jones, *J. Chem. Phys.*, 84, 813 (1986)
62. A. Mann and F. Linder, *J. Phys.* 25, 533 (1992)
63. F.A. Gianturco, R.R. Lucchese, unpublished results
64. M.S. Dababneh, Y.-F. Hisieh, W.E. Kauppila, C.K. Kwan, S.J. Smith, T.S. Stein and M.N. Uddin, *Phys. Rev.* A 38, 1207 (1988)
65 R.A. Greer and D.G. Thompson, *J. Phys.* B, in press (1994)

66. T.A. Shyn and S.Y. Cho, *Phys. Rev.* A36, 5138 (1987)
67. A. Danyo and H. Nishimura, *J. Phys. Soc. Jpn.* , 54, 1224 (1985)
68. W.H. Johnstone and W.R. Newell, *J. Phys.* B 24, 3633 (1991)
69. K. Jung, Th. Antoni, R. Müller, K.H. Kochem and H. Ehrhardt, *J. Phys.* B 15, 3535 (1982)
70. R.J. Gulley, M.J. Brunger and S.J. Buckman, XVII ICPEAC, Conference Abstracts, Aarhus, Denmark, 1993, pg. 255.

A STUDY OF THE PORTING ON SIMD AND MIMD MACHINES OF A SINGLE CENTRE EXPANSION CODE TO TREAT ELECTRON SCATTERING FROM POLYATOMIC MOLECULES

F. A. Gianturco[1], N. Sanna[2], and R. Sarno[3]

[1] Department of Chemistry, The University of Rome, Cittá Universitaria, 00185, Rome, Italy
[2] CASPUR, The University of Rome, Cittá Universitaria, 00185 Rome, Italy
[3] INFN, Sezione di Roma I, The University of Rome, Cittá Universitaria, 00185, Rome, Italy

1. INTRODUCTION

We report an application of a new computational treatment of the scattering processes which occur in the collision of electrons with polyatomic non-linear targets. We focus our attention on one of the most time-consuming parts of our set of codes, the Single Center Expansion of the bound wavefunction, thus developing a parallel code to solve the huge computational task needed to describe larger molecular targets, that are out of reach for any computational serial code. The methods and models adopted for the porting of the code on parallel machines are shown and a comparison of performances for each of them on SIMD and MIMD architectures will also be reported.

2. THE SINGLE CENTRE EXPANSION (SCE) METHOD

A general discussion on the computational aspects of the present model has been given before[1] and specific results have also been analysed elsewhere.[2, 3] We will therefore only outline here the procedure for obtaining the bound state wavefunction to be included later into the scattering equations.[4] In the present computational approach one needs to interface an initial, general-purpose quantum chemistry code that is employed to generate the Single Determinant description, (near to the Hartree-Fock (HF) limit) of the target electronic wavefunction, with a numerical procedure that can give us all the necessary quantities as being referred to the molecular c.o.m.[1]

In most of the numerical methods employed to solve the scattering equations one then converts the CC equations into a set of coupled radial equations by making first a single centre expansion (SCE) of the bound and continuum functions and then by integrating over all the angular coordinates. One therefore writes down the bound and continuum orbitals as expansions around the centre of mass, from which the body-fixed (BF) frame of reference originates:

$$\phi_k(x_i) = \sum_{hl} r^{-1} u_{hl}^k(r) X_{hl}^{pk\mu k}(\theta,\phi) \quad , \tag{1a}$$

$$F_{p\mu}(x_i) = \sum_{hl} r^{-1} f_{hl}^{p\mu}(r) X_{hl}^{p\mu}(\theta,\phi) \quad . \tag{1b}$$

Here k labels a specific, multicentre molecular orbital (MO), which contributes to the density of the bound electrons in the nonlinear target, while the indices $|p\mu>$ for the continuum functions label one of the relevant IRs and one of its components, respectively. The index h labels a specific basis, at a given angular momentum l, for the p-th IR that one is considering.[5]

One important point is the construction of the symmetry-adapted, generalized harmonics X as linear combinations of spherical harmonics $Y_l^m(\theta,\phi)$ which, for given l, form a basis of the $(2l+1)$ dimensional IR of the full rotation group. If such generalized harmonics satisfy the usual orthonormality relations, then the coefficients over the Y's correspond to a unitary transformation between the X's and the spherical harmonics:

$$X_{hl}^{p\mu}(\theta,\phi) = \sum_{m} b_{hlm}^{p\mu} Y_l^m(\theta,\phi) \tag{2}$$

and the coefficients b's can be obtained from a knowledge of the character tables for each of the IR, appearing in the relevant molecular point group.[6]

In order to perform expansions (1), one needs, therefore, to start from the multicentre wavefunction which describes the target molecule and then generate by quadrature each $u_{hl}^k(r)$ coefficient; they were obtained for the first time by numerical quadrature of the multicenter GTO's given as Cartesian Gaussian functions:[2]

$$g_v^{kj}(\alpha_j,\overline{r}_k) = N_j(a,b,c;\alpha)\, x^a \cdot y^b \cdot z^c\, e^{-\alpha \overline{r}_k^2} \tag{3}$$

labelled by the k-th atomic center, of which $g_v^{kj}(\alpha_j,\overline{r}_k)$ is the j-th function, and by the global index v of the GTO $(a,b,c;\alpha)$ parameters included in the normalisation constant N:

$$N_j(a,b,c;\alpha) = \left(\frac{2}{\pi}\right)\left[(2a-1)!!(2b-1)!!(2c-1)!!\right](\alpha)^{\frac{a+b+c}{2}+\frac{3}{4}} \tag{4}$$

The corresponding radial coefficients of eq. (1a) are thus given by the angular quadrature:

$$u_{hl}^k(r;R) = \sum_k\sum_j\sum_v\sum_{lm}\int_0^{\pi} \sin(\theta)\, d\theta \int_0^{2\pi} b_{lm}^i S_{lm}^i C_{kj}^i(R) \cdot d_v^{kj} \cdot g_v^{kj}(\alpha_j,\overline{r}_k)\, d\phi \tag{5}$$

where the first two terms on the r.h.s. describe explicitly the generalized real, symmetry-adapted harmonics X_{hl} with $|p\mu> = |i>$, C_{kj}^i the GTO coefficient of the v-th GTO at a given molecular geometry R and the d_v^{kj} the contraction coefficients of the primitive Gaussians $g_v^{kj}(\alpha_j,\overline{r}_k)$. The quadrature were carried out via Gauss-Legendre grids using a discrete, variable radial grid, for each point of which the spherical grid in the (θ,ϕ) points was evaluated.

3. COMPUTATIONAL DETAILS

3.1. Code description and parallel strategies adopted

The original code to calculate the SCE wavefunction of the target molecule is written in FORTRAN 77 (with some F90 extensions) and is about 1000 lines long. It goes as usual, through an input phase where all the initial parameters are read. The integration grid is then formed and the weights and abscissas of a Gauss-Legendre quadrature calculated, for both θ and ϕ angles. At this stage, the SPHINTEG subroutine is called to perform the angular integration over θ and ϕ to produce the $\Phi_k(x_i)$ of Eq. (1a) expanded over the r grid points. The angular integration is performed with several nested loops as the following pseudo-code listing can shown:

```
do for each MO orbital
   do for each r values
      do for each θ value
         do for each φ value
            CALCULATE Ψ^MO_GTO(r,θ,φ) = C^i_kj(R) · d^kj_v · g^kj_v(α_j, r̄_k)
            do for each lm couple
               CALCULATE u^k_hl(r;R)
            end do
         end do
      end do
   end do
end do
```

A first look at the loop structure soon reveals an obvious choice to make the code parallel. Since the calculations for each MO orbital of the bound state wavefunction are completely independent, one can think to spawn N processes on different nodes each of them with the task of performing all the inner loop calculations. This is what is usually called *embarassing parallelism* where the physics of the problem is intrinsically parallel. While this method is not so interesting from an algorithmic point of view, we nonetheless found useful to implement it in the first preliminary version of the parallel code in order to evaluate some features of the hardware it had to run on (data distribution, network performances and topology, etc.). The second method used to parallelize the original serial code was the well known one called *domain decomposition.* The $\theta \times \phi$ domain has been equally distributed on all the nodes which now have to run all the loops but on a $\delta\theta \times \delta\phi$ subset of the total $\theta \times \phi$ space. This method has in this particular case, an intrinsic load balancing feature due to the equally spaced $\delta\theta \times \delta\phi$ subsets, still maintaining the low coupling among the parallel processes since no boundary conditions exists on θ and ϕ for the solution of the angular integration. We implemented this method on both the SIMD and MIMD version of the code, with no significant differences on the programmed algorithm, using different languages and parallel tools.

3.2. SIMD hardware and software description

The SIMD machine we use for the porting of our SCE code was the QUADRICS Qxx series from Alenia Spazio S.p.A. which is the commercial version of the INFN

APE100 supercomputer. Briefly we will sketch below the hardware and software used, while a more detailed reference can be found elsewhere.[7, 8]

The QUADRICS supercomputer is constituted by several *crates* that can be assembled (at present time up to 16) to build a larger parallel machine. Each crate contains 16 FPU board devoted to floating point operations and 1 ZCPU board which performs all the communications with the host (a SUN workstation) and addresses, among other things, branching control and integer manipulation. Each board has 8 FPU with one single precision MWord (4MB) of RAM each running at 25 Mhz with dual fetch capability for a total of 400 MFLOPS peak performance per board (6.4 GFLOPS per crate). We used a Q1 (1 FPU + 1 ZCPU boards) for developing purpose and up to a Q16 (16 FPU + 1 ZCPU boards) to run test calculations as reported in the next section.

The standard language of the QUADRICS machine is TAO, a fortran-like language built over a parser called ZZ. During the developing phase we were able to efficiently implement the whole program. Thanks to the high level of programming given by the TAO compiler, also very hard-coded part of the original serial version (like the Associated Legendre polynomials generation) were successfully addressed.

3.3. MIMD hardware and software description

The hardware we use for the porting of the MIMD version of our SCE code is listed below:

- A cluster of 8 DEC Alpha AXP-3000/500 (150 Mhz clock speed) with 128 MB of RAM (or more), each running OSF/1.3A both on Ethernet and FDDI networks, the latter using a Gigaswitch, a Cross Bar Switching device from DEC which allows full peer-to-peer FDDI bandwidth (up to 200 Mb/s on full duplex).[9]
- An IBM SP1 (Envoy) with 16 POWER-1 RISC nodes with 128 MB of RAM, with proprietary high bandwidth switching device (FCS - Fiber Channel Switch) with up to 240 Mb/s burst internode communication bandwidth.[10]

In this case we adopted two well known parallel tools like PVM (V. 3.2.6)[12] and TCGMSG (V. 4.04)[11] within the SPMD (Single Program Multiple Data) programming model. In the case of the IBM-SP1 test, we also used the enhanced version of PVM developed at the IBM-ECSEC research centre, PVM6000, which is the most performing version of this tool (up to 10 MB/s burst bandwidth on SOCC network) but available on IBM-RS risc hardware only.

4. RESULTS AND DISCUSSION

4.1. The test case

The CH_4 molecule has been chosen as test case and the input data have been taken from an *ab-initio* SCF calculation with the Gaussian-92[13] program, using DZP[14] GTO basis set (40 primitive functions) with R_{CH} =2.063 a.u. and the tetrahedral angle $\widehat{HCH}$=109.5°.

The Single Centre Expansion was performed on a grid of 600000 points (n_r=75, n_θ=80, n_ϕ=100) and up to l_{max}=12 for all the occupied orbitals, a_1, a_2, t_{2x}, t_{2y}, t_{2z}.

While this is not the larger test we could arrange, it seemed to be the best one due to the presence of a huge amount of data available on the methane molecule and so simplifying the comparison of the obtained results between the serial and both versions of parallel codes.

4.2. SIMD version

Within the *domain decomposition* model we adopted, we chose to distribute all the angular domains over one board (that is, 8 FPUs) thus giving to each processor a 20x50 $\theta \times \phi$ (1000 points) subset of the whole $\theta \times \phi$ space (8000 points). This distribution was replicated on all the available boards, each with a different value for r. In this way, once the inner loops on MO, θ, ϕ and lm have been calculated, one can generate as many r points as the number of boards available namely 1, 4, 16, r values for a Q1, Q4 and a Q16 machine respectively.

At the startup, the code is loaded on all the nodes, then the first node (0,0,1) start to read input data from host disk and distribute them to all the others, following the distribution scheme described so far. This stage takes less than 1 sec on a Q16 machine (the input data are ca. 20 KB) and almost all the time is spent in copying data (in parallel) between neighboring nodes. Thanks to the particular architecture of the Quadrics machine, this step is carried out very efficiently and no additional overload has been observed. At this point the input data are accessible to all the nodes and the calculation of the MO, θ, ϕ, and lm inner loops could begin producing n (n = no. of boards) values of $u^k_{hl}(r;R)$ in the unit of time needed to calculate one $u^k_{hl}(r;R)$ value, that is 5.0 sec.

The results we obtained are shown in Table 1 while in Figure 1 and 2 we report respectively the elapsed time vs. the number of processors and the corresponding speedup as the number of processors increases.

Table 1. SCE SIMD version results

# Procs	Elaps. time (sec)	Speedup
8	374.4	1.0
16	146.2	4.0
128	23.4	16.0

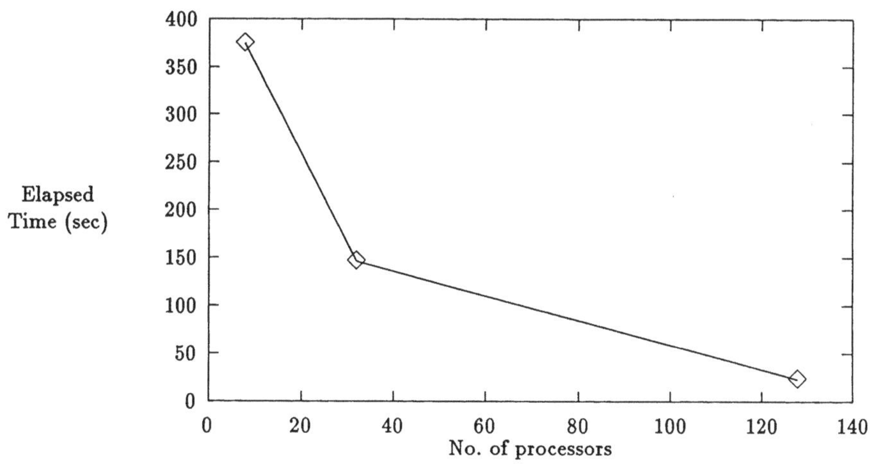

Figure 1. SCE SIMD version. Elapsed time in second vs. no. of processors.

As one can clearly sees, the speedup is perfectly linear as expected. This is one of the best features one can have when using a SIMD machine like the Quadrics and when, as in our case, no synchronisation deadlocks are present in the code. Furthermore,

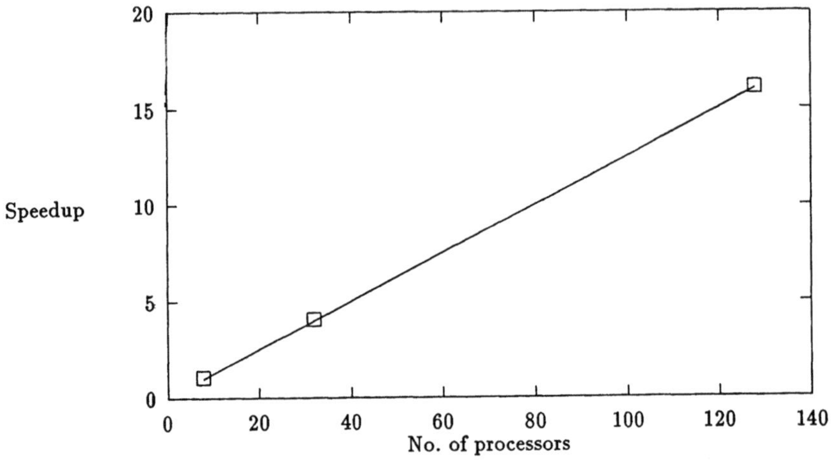

Figure 2. SCE SIMD version. Speedup vs. no. of processors.

once one has the code written for a one board machine (but keeping in mind the whole machine architecture) it is straightforward to move on to the full one. One has to just change few parameters into an include file and recompile (as in our test case) or alternatively write a dynamic data allocation code to comply with the actual number of boards (nodes) present. This is, in our opinion, a very powerful feature which permits us to have access to the full hardware CPU power with a minimal effort: it is well-known that linear *scalability* (in the sense of computational time) is not easy to obtain and very often is this the main difficulty to overcome on other massively parallel machines. In our case, we were able to obtain a very good scalability also in the data allocation sense; in fact in our test on methane molecule, less than 25% of the available memory was used keeping later room for larger computer simulations on bigger molecules like CF_4, SF_6, etc.

Another point worth noting is the check on the numerical results depending on the single precision arithmetic implemented on the Quadrics machine. We carefully compared the results on using the calculated SCE bound wavefunction in the subsequent scattering code calculation and we observed a small deviation on calculated e-CH_4 Total Cross Section (σ_t) of 1% or less. In the first implementation of the parallel code we had quite a larger deviation (about 5%) which was in turn almost eliminated when using the *binary tree summation technique*[15] in the SPHINTEG integration routine.

A last comment on this version of the SCE program is about the code *efficiency*. The Quadrics machine has a peak rate of 400 MFLOPS per board and, as is well-known, this represents only a theoretical limit since in general no user application would perform at that rate. In the SIMD version of the SCE code we were able to obtain between 40 and 80 MFLOPS rate with an *efficiency* up to 20 %. This is not, of course, the best result one could obtain on this machine but in our case, with the code still under development, seems to be highly appreciable the fact that, with some refinements, we could obtain a factor of two or more on calculated speedups. In particular, one has to take into account how the FPU pipeline is loaded, thus trying to fill it at its maximum with techniques like *loop unrolling* or moving unused code inside loops. In this version of the SCE code we paid attention to producing highly reliable numerical results, while in the future versions we will full exploit the hardware capabilities.

4.3. MIMD version

In the case of the MIMD version of the SCE parallel code, we implemented both models explained so far: *embarassing parallelism* and *domain decomposition.* The former has been implemented mainly to have a rough comparison with previous preliminary parallel versions, while the latter has been adopted as the final model and hence investigated in full details.

In order to better understand relative performances and results, we decided to report here the PVM data only, due to the fact that this was the tool present on both machines we used (DEC-Alpha cluster and IBM-SP1). TCGMSG results are available on request for the Alpha cluster environment.[9] Using the PVM V3.2.6 parallel tool, we used the SPMD programming model through the master-slave mechanism: that is, the master had only to accomplish the initial I/O, spawn the **n** equal slaves with their own data and finally collect from them the results.

In the case of *embarassing parallelism* concept, each molecular orbital (MO) was spawned to each node together with all the needed data to perform the Gauss-Legendre quadrature. In Table 2 we report the results obtained on the Alpha cluster and SP1 machine, together with older data from a preliminary parallel version running on a IBM 3090-600J, 6-CPU mainframe.

Table 2. SCE MIMD version results. *Embarassing Parallelism* model

	DEC-Alpha-cluster		IBM-SP1		IBM-3090-600J	
# nodes	E.T. (sec)	Speedup	E.T. (sec)	Speedup	E.T. (sec)	Speedup
1	197.1	1.0	220.0	1.0	330.0	1.0
5	44.9	4.4	52.8	4.2	78.0	4.2

Although a direct comparison of the obtained results is not directly feasible [1/], a rough idea of the improvement given by the new RISC chips architecture is evident. A first look soon reveals that, while the relative speedup is almost constant, the RISC machines perform up to a factor of 1.75 better than the old IBM-3090 mainframe. At any rate, in the case we tested no use of the IBM-3090 vector facility feature has been done, so the compared results are among *scalar* results which is in turn the only way to get a homogeneous view although in the case of the IBM-3090 machine, the available hardware capability has not been fully exploited.

Between the same parallel and internal chip architecture, the DEC-Alpha seems to give better serial performances, about 10% better with respect to the IBM-POWER1 as expected,[16] probably due to the fact that in our code the integer manipulation is not negligible, a situation where the *superpiped* architecture of the DEC-Alpha RISC chip seems to give the best performance results.

The obtained results using the *domain decomposition* model are shown in Table 3 and reported in Figure 3 and 4.

The implementation of this model on the MIMD architectures reported, was carried out using one master process to perform all the I/O operations while spawning n equal slaves on all the available nodes. Each slave had a subset of $\theta \times \phi$ domain (e.g. 10x100=1000 points in the case of 8 nodes) and all the data to perform the MO, r, θ, ϕ and lm loops calculation. Once the slaves have finished their jobs the (partial) calculated $u^k_{\alpha\beta,hl}(r;R)$ is sent back to the master which sums all the partial contributions to

[1/] The IBM-3090-600J is a Shared Memory Multiprocessor architecture machine while the Alpha cluster with Gigaswitch and IBM-SP1 are a Distributed Memory MIMD machines.

Table 3. SCE MIMD version results. *Domain Decomposition* model

	DEC-Alpha-cluster		IBM-SP1	
# nodes	E.T. (sec)	Speedup	E.T. (sec)	Speedup
1	195.1	1.0	216.8	1.0
4	50.1	3.9	57.0	3.8
8	26.1	7.6	29.3	7.4
10	N.A.	N.A.	23.8	9.1
12	N.A.	N.A.	20.2	10.7
15	N.A.	N.A.	17.0	12.7

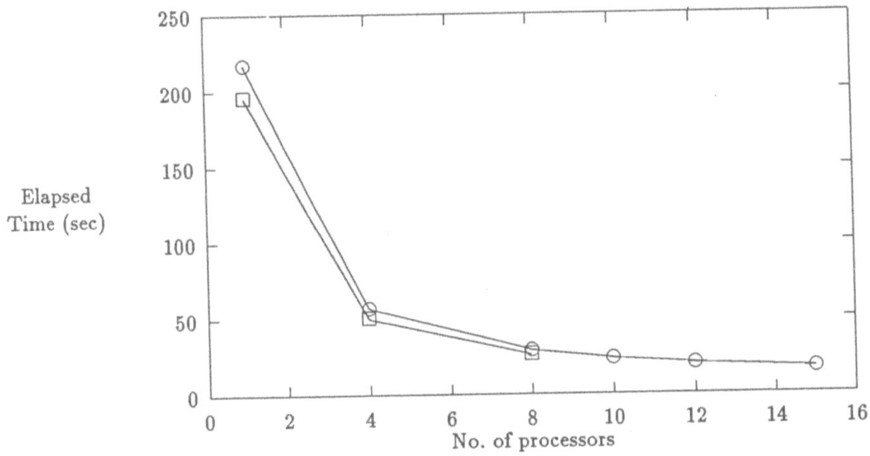

Figure 3. SCE MIMD version. Elapsed time (in seconds) vs. no. of processors. Open squares: DEC-Alpha cluster+Gigaswitch. Open circles: IBM-SP1.

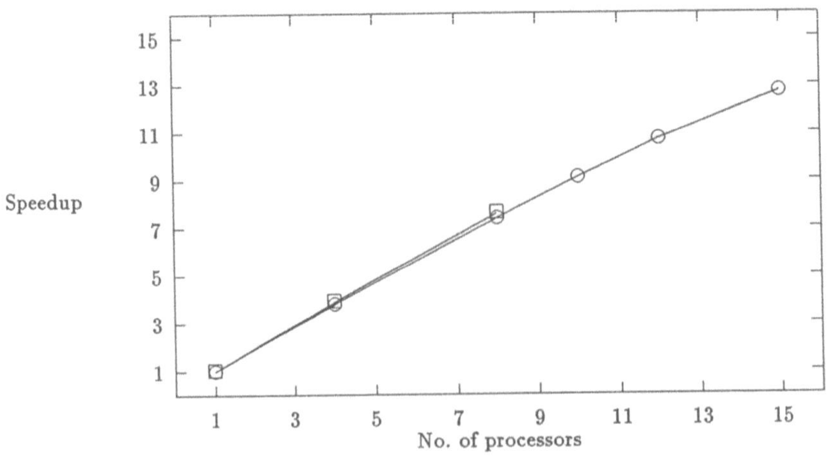

Figure 4. SCE MIMD version. Speedup vs. no. of processors. Open squares: DEC-Alpha cluster+Gigaswitch. Open circles: IBM-SP1.

produce the final $u_{hl}^k(r;R)$. In this way one can easily obtain a load balancing feature because all the data subsets sent from the slaves are equal while in the *embarassing parallelism* model, the amount of data sent and received by each slaves was dependent on the molecular orbital MO expansion over the blm's. A first look to the results soon reveals that up to 8 processors the performances ratio between the IBM and DEC RISC chips is maintained and they are almost equivalent from a parallel point of view. In fact, our application is typically CPU bounded and these results confirm this assumption, since the communication time is not significant. One point worth noting is that all the communications on the Alpha cluster were carried out on the FDDI network while on the SP1 the slaves spawning was done on Ethernet thus causing a little delay in the measured elapsed time. This is not a serious problem and depends on the application size; in fact, as the size of the problem increases the delay disappears as expected, and it could be relevant only on overloaded Ethernet networks. This aspect of the SP1 architecture influenced a bit the results we obtained on SP1 with more then 8 processors, due to the fact that the Ethernet network at ECSEC was heavely in use during our measurements and we feel that higher speedup ratios could be addressed in a more quiet network environment. Nonetheless, the IBM-SP1 gave the best absolute performance and permitted us to exploit the scalability of our parallel code up to 15 nodes with a factor of 1.8 respect to the 8 nodes results. This result give us the idea that the present implementation of the MIMD SCE version could be easily ported and used on more massively parallel architectures so giving the needed speedup to attack larger electron-molecule calculations.

5. FUTURE DEVELOPMENTS

The complete set of codes used in our group to perform electron-molecule scattering calculation, is briefly outlined in the flow-chart of Figure 5. The present work has been carried out on one code (SCE) which is the starting point of the scattering calculation procedure: the calculation of the bound state wavefunction. The results we obtained here in parallelizing the Gauss-Legendre quadrature inside the SCE code, open the opportunity of efficiently perform it in parallel throughout the set of codes which could give a dramatic speedup in the overall demand of computer CPU time needed to perform a complete e-molecule scattering calculation.

In fact, in the single-center expansion all functions are written as

$$f(r,\theta,\phi) = \sum_{l,m} f_{lm}(r)\, Y_{lm}(\theta,\phi) \quad . \tag{6}$$

The most computationally intensive step in the scattering calculation is forming a product of two functions in the computation of exchange integrals. The product of two functions is evaluated by first transforming the angular momentum representation of f into a coordinate representation using

$$f_{\alpha\beta}(r) = \sum_{l,m} f_{lm}(r)\, U_{lm,\alpha\beta} \quad , \tag{7}$$

where $U_{lm,\alpha\beta} = Y_{lm}(\theta_\alpha, \phi_\beta)$ and $f_{\alpha\beta}(r) = f(r,\theta_\alpha,\phi_\beta)$. In the coordinate representation, a product of two functions is just a point-by-point product of the form

$$(fg)_{\alpha\beta}(r) = f_{\alpha\beta}(r)\, g_{\alpha\beta}(r) \quad . \tag{8}$$

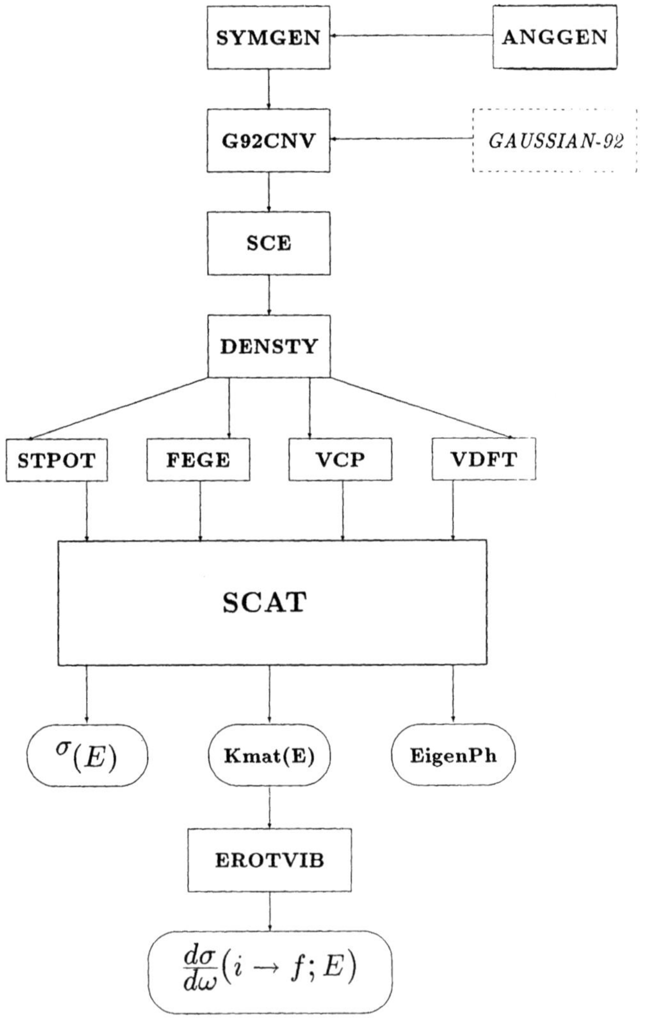

Figure 5. Flow-chart of the whole set of codes to perform electron-molecule scattering calculations.

The angular momentum representation can then be recovered by transforming the coordinate representation using

$$f_{lm}(r) = \sum_{\alpha,\beta} f_{\alpha\beta}(r)\, V_{\alpha\beta,lm} \tag{9}$$

where

$$V_{\alpha\beta,lm} = Y_{lm}(\theta_\alpha, \phi_\beta)\, W_\alpha\, W'_\beta \;. \tag{10}$$

The weights and points for the θ integration are those given by the Gauss-Legendre integration formulae so that for an n point integration the point θ_α is the αth zero of the Legendre polynomial $P_n(cos\theta)$ and the weight W_α is given by

$$W_\alpha = \frac{-2}{(n+1)\, P'_n(cos\theta_\alpha)\, P_{n+1}(cos\theta_\alpha)} \;. \tag{11}$$

The weights and points for the ϕ integration are those given by the Gauss-Chebyshev integration formulae applied once to the $0 \rightarrow \pi$ region and once to the $\pi \rightarrow 2\pi$ region so that for a $2n$ point integration in the region $0 \rightarrow 2\pi$ the weight W'_β is given by

$$W'_\beta = \frac{\pi}{n} \tag{12}$$

and the point ϕ_β is given by

$$\phi_\beta = \frac{(2\beta - 1)\pi}{2n} \;. \tag{13}$$

The back tansformation in the angular momentum representation given by (10) require the evaluation of the integral

$$V_{\alpha\beta,lm} = \int Y_{lm}(\theta_\alpha, \phi_\beta)\, sin(\theta_\alpha)\, d\theta_\alpha d\phi_\beta \tag{14}$$

where the formulation is identical to (5) apart from the fact that in the case of ϕ domain of integration the Gauss-Chebyshev quadrature formula is used. Since the analogy of the solution of (10) with the present work is clearly evident, we would port in parallel the general Gauss-Legendre(Chebyshev) subroutines used throughout our codes using the schemes and methods described so far, hoping to obtain the speedup needed to attack larger e-molecule scattering calculation in terms of both number of partial waves ($l_{max} = 50$) and bound state wavefunction description (above 100 primitive GTO basis functions).

References

[1] F.A. Gianturco, V. Di Martino and A. Jain, Electron Scattering from nonlinear molecules using multicentre wavefunctions expansion, Nuovo Cim. D, **14**, 411 (1992).

[2] F.A. Gianturco, N. Sanna and A. Talamo, Electron scattering from polyatomic molecules with a multicentre target description, submitted to Phys. Rev. A, (1994).

[3] F.A. Gianturco, R.R Lucchese and N. Sanna, Calculation of low-energy elastic cross sections for electron-CF_4 scattering, to appear in J. Chem. Phys., (1994).

[4] F.A. Gianturco, R.R. Lucchese, N. Sanna and A. Talamo, A Generalized Single Centre Approach For Treating Electron Scattering From Polyatomics Molecules, in: "Electron Scattering Processes in Molecules Clusters and Surfaces", H. Erhardt and L. Morgan Eds, Plenum Publ. Co., New York (1994).

[5] S.L. Altmann, The Cubic group, Proc. Camb. Phil. Soc., **53**, 343 (1957).
[6] Tinkham "Group Theory and Quantum Mechanics", McGraw-Hill, New York (1964).
[7] A. Bartoloni et al., The Software Of The APE100 Processor, Int. J. Mod. Phys. C, **4**, 955 (1993).
[8] A. Bartoloni et al., The Hardware Implementation Of The APE100 Architecture, Int. J. Mod. Phys. C, **4**, 969 (1993).
[9] For a detailed configuration see CASPUR, Internal Bulletin No. 2, or send a request via e-mail to Sanna@caspur.it.
[10] For a detailed configuration send a request via e-mail to Ing. Renzo Di Antonio, dantonio@romesc.vnet.ibm.com, IBM-ECSEC, Rome (ITALY).
[11] R. J. Harrison, Int. J. Quant. Chem., **40**, 847(1991).
[12] A. Beguelin, J. Dongarra, A. Geist, R. Manchek and V. Sunderam, Oak Ridge Natl. Laboratory Tech Report **TM-11826:1** (1991).
[13] M. J. Frisch, M. Head-Gordon, G. W. Trucks, J. B. Foresman, H. B. Schlegel, K. Raghavachari, M. Robb, J. S. Binkley, C. Gonzalez, D. J. De Frees, D. J. Fox, R. A. Whiteside, R. Seeger, C. M. Melius, J. Baker, R. L. Martin, L. R. Kahn, J. J. P. Stewart, S. Topiol and J. A. Pople, **Gaussian 92**, Gaussian Inc., Pittsburgh, PA, (1992).
[14] R. Poire, R. Kari and I. G. Csizmadia, Handbook of Gaussian Basis Sets, Elsevier, Amsterdam (1985).
[15] What Every Computer Scientist Should Know About Floating-Point Arithmetic, issue of Computing Survey, March 1991.
[16] J. J. Dongarra, Performance of Various Computers Using Standard Linear Equations Software Oak Ridge Natl. Laboratory Tech Report, (1994).

HOW TO CALCULATE ROTATIONAL AND VIBRATIONAL CROSS SECTIONS FOR LOW-ENERGY ELECTRON SCATTERING FROM DIATOMIC MOLECULES USING CLOSE-COUPLING TECHNIQUES

Michael A. Morrison and Weiguo Sun[1]

Department of Physics and Astronomy
University of Oklahoma
Norman, Oklahoma 73019

[1]Permanent Address:
Department of Chemistry
The Sichuan Union University
Chengdu, Sichuan 610065
People's Republic of China

INTRODUCTION

This chapter is not a review; electron-molecule dynamics is already replete with fine reviews, many of which appear in books devoted entirely to this topic.[1-6] These reviews discuss the applied importance of this field,[7-10] survey the status of electron-molecule collision data,[11-40] and address specialized topics such as resonance scattering,[15] vibrational excitation,[16] near-threshold scattering,[17,18] particular theoretical approaches such as the R-matrix method,[19] numerical methods for solving the Schrödinger equation,[20-22] and scattering from polar[23,24] and polyatomic[25,26] targets. Neither is this chapter primarily pedagogical; readers can find elsewhere a wealth of pedagogically useful tutorial introductions and reviews that narrate the major developments in the field's long rich history and survey recent advances that have made it the focus of intense activity during the last 20 years.[27-30] Rather, this chapter is a "ready reference" of the key equations for the application of one very widely used theoretical strategy—the eigenfunction-expansion or "close-coupling" method—to one very important class of problems: quantum scattering (at incident energies less than about 10 eV) from a closed-shell diatomic molecule accompanied, perhaps, by rotational and/or vibrational (but not electronic) excitation of the target. As exemplary of an extremely powerful method for reducing multi-variable integro-differential equations to more tractable sets of fewer-variable equations, this class of problems illustrates strategies used in many other theoretical contexts, both within and outside of electron-molecule collisions.

Table 1. Electron-molecule scattering theories: an acronym guide.

LAB-CAM	laboratory-frame close-coupling	coupled ang. mom. $\hat{\mathbf{J}} = \hat{\mathbf{j}} + \hat{\boldsymbol{\ell}}$
LAB-RCAM	laboratory-frame close-coupling	recoupled ang. mom. $\hat{\boldsymbol{\ell}}_t = \hat{\mathbf{j}} - \hat{\mathbf{j}}_0 = \hat{\boldsymbol{\ell}} - \hat{\boldsymbol{\ell}}_0$
RR	rigid-rotor	internuclear separation fixed at $R = R_{eq}$
BF-FNO	body-frame fixed-nuclear-orientation	$\hat{R}$ fixed for duration of collision
BF-FN	body-frame fixed nuclei	$\hat{R}$ and R fixed for duration of collision
BF-FONDA	first-order non-degenerate adiabatic	BODY T-matrix calculated off-shell
ANR	adiabatic-nuclear rotation	based on BF-FNO scattering matrices
ANV	adiabatic-nuclear vibration	based on BF-FN scattering matrices

Four features of the non-relativistic Schrödinger theory of electron-molecule collisions deserve special note. First, the quite different physics of the electronic and "nuclear" (i.e., rotational and vibrational) degrees of freedom demands careful attention; here we consider the very common situation where the Born-Oppenheimer approximation[31,32] accurately represents the ground-state electronic target wave function. Second, the electron-molecule interaction potential is not spherically symmetric, so the angular dependence of the projectile's scattering function cannot simply be separated out of the problem as can be done for spherical potentials. Third, this potential is non-local—a consequence, as we shall see, of the anti-symmetrization requirement on the system wave function—so the equations resulting from eigenfunction expansion are *integro-differential* in character. And fourth, differences between the rotational and vibrational dynamics of the target admit a variety of special approximations that may dramatically simplify low-energy scattering calculations;[27–30] the numerical difficulties attending this class of calculations have focused attention on the validity conditions for these approximations and on what options are available if they fail. This plethora of approximations is partly responsible for the sometimes bewildering alphabet soup of acronyms that confronts newcomers and experienced practitioners alike; to aid the intrepid we offer Table 1 as a guide.

In addition to collecting key equations using consistent notation and conventions we provide practical suggestions for calculating high-precision integral and differential cross sections for elastic scattering, momentum transfer, and rotational/vibrational excitation of small-molecule targets. As most problems arise in the computationally problematic very-low-energy region, this chapter emphasizes collisions at few eV and sub-eV energies.

We have excluded, however, computational details concerning the static, exchange, and correlation/polarization components of the electron-molecule interaction potential. The need for such discussion is largely obviated by careful presentations of theory, models, and codes in recent papers and reviews.[33] Finally, we have not assayed the exhaustive reference listing that would be required to adequately credit the many papers in which first appeared most of the fundamental equations quoted in this chapter. Space limitations preclude so extensive a listing, which in any case would be inappropriate to the goals of the present discussion; moreover such detailed reference lists are available in several recent reviews.[18,22,23,26,29] Instead, we refer readers first to reviews pertinent to the topic at hand and second to a few selected papers where they can find more detailed derivations or discussion concerning key theoretical issues, giving preference to papers which themselves contain fairly complete reference lists. Our chapter, then, should be considered home base for sojourns into the rich and rewarding original literature on low-energy electron-molecule scattering via eigenfunction expansion methods.

THEORETICAL CONCERNS

The obvious coordinate system in which to formulate collision theory is a space-fixed laboratory (LAB) frame. When working with molecules, however, alternatives prove desirable; in the Born-Oppenheimer theory of diatomic molecular structure, for example, the electronic wave function is most easily calculated in a body-fixed (BODY) reference frame whose z axis is aligned with the internuclear axis $\mathbf{R}$. Both frames serve in electron-molecule collision theory. Ultimately, of course, one must return to the lab, where experiments are done, and many early methods were formulated entirely in the LAB frame. But a BODY formulation may be simpler conceptually and computationally—especially if the physical conditions admit an adiabatic treatment of the rotational or vibrational dynamics. We shall define and denote these coordinates as

$$\text{(BODY)} \qquad \mathbf{r} \equiv (r, \theta, \phi) \qquad \hat{\mathbf{e}}_z = \hat{R} \tag{1a}$$

$$\text{(LAB)} \qquad \mathbf{r}' \equiv (r, \theta', \phi') \qquad \hat{\mathbf{e}}_{z'} = \hat{k}'_0. \tag{1b}$$

The BODY coordinates in Eq. (1a) are those of the theory of molecular electronic structure, the LAB coordinates in Eq. (1b) are conventional in descriptions of cross section measurements.[1]

The relationship between LAB and BODY coordinate systems follows from a conventional coordinate rotation[34–38] executed via an Euler angle rotation operator $\hat{\mathcal{R}}(\alpha, \beta, \gamma)$. According to Eqs. (1), the angles of this operator are just the spherical angles θ_R and ϕ_R of the internuclear axis in the LAB frame:[2]

$$\hat{\mathcal{R}}(\alpha = \theta_R, \beta = \phi_R, \gamma = 0): \text{ LAB } (\mathbf{r}') \longrightarrow \text{BODY } (\mathbf{r}). \tag{2}$$

This operator plays a vital role in relating orbital angular momentum eigenfunctions in the BODY and LAB frames: in a single-center formulation for a diatomic molecule (more generally, for any molecule belonging to the point groups $D_{\infty h}$ or $C_{\infty v}$), these are just the usual spherical harmonics. These functions are distinguished, of course, by the axis with respect to which the projection of the (single-electron) orbital angular momentum $\boldsymbol{l}$ is quantized. The LAB harmonics $Y_\ell^{m_\ell}(\hat{r}')$ are eigenfunctions of $\hat{\ell}^2$ and $\hat{\ell}_{z'}$ while the BODY harmonics $Y_\ell^\Lambda(\hat{r})$ are eigenfunctions of $\hat{\ell}^2$ and $\hat{\ell}_z$, with m_ℓ as the LAB magnetic quantum number and Λ its BODY counterpart.

---

[1] The alignment of the BODY z axis with the internuclear axis $\mathbf{R}$ is essential to the fixed-nuclear-orientation approximation, which as discussed below inheres in body-frame electron molecule calculations. Our alignment of the LAB z' axis with the incident wave vector $\mathbf{k}'_0$ is merely convenient; it eliminates sums over LAB magnetic orbital angular momentum quantum numbers from equations that have more than enough sums to begin with. Since the radial coordinates are identical in the two frames, we don't tack a prime on this coordinate, r.

[2] The third Euler angle γ is arbitrary (for scattering from a linear molecule); following convention, we choose $\gamma = 0$. Then the molecular x axis is in the $z\,z'$ plane, and for $\beta = \phi_R = \pi/2$, the internuclear axis is in the LAB $x'\,y'$ plane.

In particular, the LAB and BODY harmonics are related by matrix elements of the rotation operator (2) between LAB eigenkets; these are just elements of the unitary, unimodular Wigner rotation matrices,[3]

$$\mathcal{D}^{\ell}_{m_\ell,m_\ell'}(\hat{R}) = \langle \ell m_\ell \mid \hat{\mathcal{R}} \mid \ell m_\ell' \rangle. \tag{3}$$

Since $\hat{\mathcal{R}}$ transforms (LAB) eigenkets of $\hat{\ell}_{z'}$ into (BODY) eigenkets of $\hat{\ell}_z$, the essential relationship between LAB and BODY spherical harmonics is

$$Y_\ell^\Lambda(\hat{r}) = \sum_{m_\ell=-\ell}^{+\ell} Y_\ell^{m_\ell}(\hat{r}')\, \mathcal{D}^{\ell}_{m_\ell,\Lambda}(\hat{R}) \tag{4a}$$

$$Y_\ell^{m_\ell}(\hat{r}') = \sum_{\Lambda=-\ell}^{+\ell} Y_\ell^\Lambda(\hat{r})\, \mathcal{D}^{\ell *}_{m_\ell,\Lambda}(\hat{R}), \tag{4b}$$

where the arguments on the D matrices are $(\theta_R, \phi_R, 0)$.

The Target Molecule. In close-coupling formulations one almost immediately eliminates the *electronic* wave functions of the target molecule, either replacing them with model potentials or burying them in coupling matrix elements.[18,29] But as we shall see, the accuracy of these functions—both their numerical precision and whatever approximations inhere in their generation—may subtly and unexpectedly influence electron-molecule cross sections. For *electronically elastic* collisions, the Born-Oppenheimer approximationapplies, and we can treat the electronic and "nuclear" (rotational and vibrational) degrees of freedom separately. Thus we approximate each target function by a product of an electronic function $\Phi_0^{(e)}(\tau_e; R)$ and a ro-vibrational function $\chi_\nu(\mathbf{R})$. The electronic function is labeled by a subscript collectively denoting its electronic numbers α, where we use the subscript zero to denote the ground state. This function further depends on the space and spin coordinates τ_e of the N_e bound molecular electrons, with the spatial variables written in a BODY frame whose origin sits on the center-of-mass of the molecule. The nuclear function is labelled by ro-vibrational quantum numbers $\nu \equiv (v, j, m_j)$ and depends on the spatial coordinates of the internuclear axis $\mathbf{R}$. In particular, for an electronic state α_0 whose total electronic orbital angular momentum has projection zero along $\hat{R}$ (e.g., a Σ state, such as the ground state $\alpha = X^1\Sigma_g^+$ typical of homonuclear diatomics), the ro-vibrational factor in the Born-Oppenheimer wave function is the product of a reduced vibrational wave function $\varphi_v^{(v)}(R)$ and a function that represents the rotational motion. While for an arbitrary electronic state the latter is a symmetric-top eigenfunction (in the BODY frame),[37] for a Σ molecular state it simplifies to a spherical harmonics, and the nuclear wave function is just

$$\chi_\nu(\mathbf{R}) = \frac{1}{R}\varphi_v^{(v)}(R)\, Y_j^{m_j}(\hat{R}). \tag{5}$$

The corresponding eigenvalues $\epsilon_\nu = \epsilon_{vj}$ of this function with respect to the ro-vibrational Hamiltonian

$$\hat{\mathcal{H}}_m^{(n)} = \hat{T}^{(n)} + V_{nn} + \mathcal{E}_0^{(e)}(R) \tag{6}$$

---

[3] For rotation matrices, coupling coefficients, and the other accouterments of the quantum theory of angular-momentum, we adopt the conventions of Ref. 36; Table 4.2 of that reference relates these conventions to those of the myriad other treatments of angular-momentum.

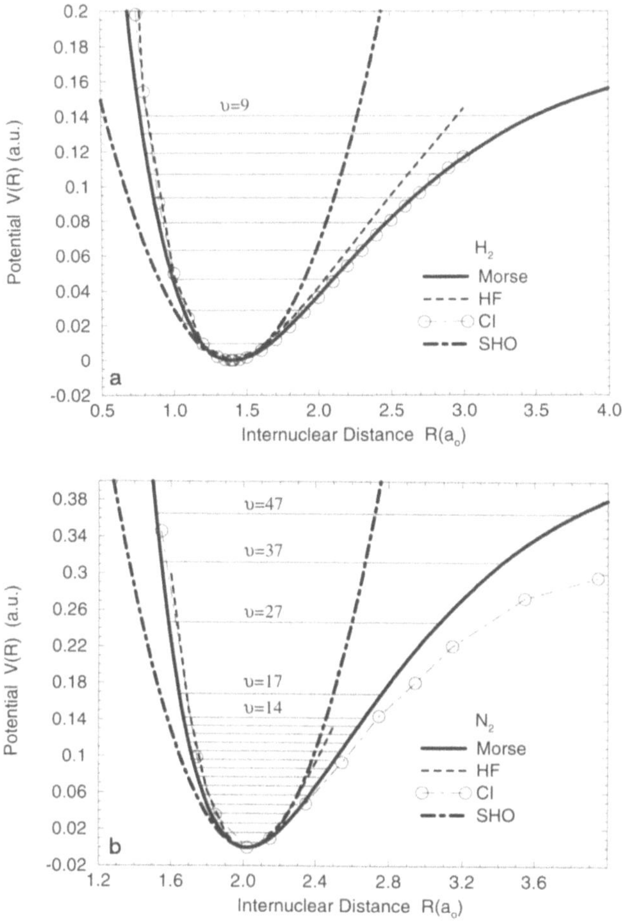

Figure 1. Potential energy curves for the nuclear motion (electronic energy plus internuclear repulsion) and low-lying vibrational energy levels for (a) H_2 and (b) N_2. Curves are based on *ab initio* theoretical calculations using the Hartree-Fock (HF) and configuration interaction (CI) methods and on simple-harmonic-oscillator (SHO) and Morse potential models. In order to clarify differences in the shapes of these potentials, we have shifted the HF and CI curves so they share a common equilibrium separation and value with the Morse and SHO potentials. Parameters in the Morse and SHO potentials were determined to reproduce the experimental energies of the states shown. In (b), we omit some states for $v > 14$ for clarity.

approximate the ro-vibrational target energies (in the ground electronic state). Here $\hat{T}^{(n)}$ is the operator for the kinetic en^rgy of rotation and vibration and V_{nn} is the nuclear-nuclear Coulomb repulsion operator. Consistent with Born-Oppenheimer theory, the energy $\mathcal{E}_0^{(e)}(R)$ of the ground electronic state, the eigenvalue of the electronic Hamiltonian for this state, appears in (6) in the guise of a potential energy for the nuclear motion. We shall soon return to the critical importance of these eigenvalues for low-energy electron-molecule theory, where they determine the thresholds for inelastic collisions. In Fig. 1 we illustrate the energy structure of typical diatomic targets.

The Electron-Molecule Schrödinger Equation. For low-energy scattering from light molecules, non-relativistic quantum mechanics pertains. In the time-independent incarnation of this theory, the continuum stationary-state wave functions of the system are eigenfunctions of the Hamiltonian

$$\hat{\mathcal{H}} = \hat{T}_e + \hat{\mathcal{H}}_m^{(e)} + \hat{T}^{(n)} + V_{nn} + V_{Coul}. \tag{7}$$

Here $\hat{T}_e$ is the projectile kinetic energy operator, $\hat{\mathcal{H}}_m^{(e)}$ is the electronic Hamiltonian of the target, and V_{Coul} is the electrostatic interaction potential, the sum of all two-particle bound-free Coulomb interactions. For a molecule with N_α nuclei at positions $\mathbf{R}'_\alpha$ in the LAB frame, this potential in atomic units (a.u.)[4] is the sum of electron-nucleus and electron-electron terms,

$$V_{Coul} = V_{en} + V_{ee} = -\sum_{\alpha=1}^{N_\alpha} \frac{Z_\alpha}{|\mathbf{r}' - \mathbf{R}'_\alpha|} + \sum_{i=1}^{N_e} \frac{1}{|\mathbf{r}' - \mathbf{r}'_i|}, \tag{8}$$

where $\mathbf{r}'_i$ denotes the (LAB) coordinates of the $i^{\rm th}$ bound electron and $\mathbf{r}'$ that of the projectile.

The first step in nearly all studies of *electronically elastic* electron-molecule scattering is to get rid of the molecular electronic coordinates. We don't need them to describe actual electronic excitation, and we can approximate their (extremely important) role in describing *virtual* electronic excitation—the phenomena of bound-free correlation and induced polarization of the target—via one of a cornucopia of model (or, more rigorously, optical) potentials. To this end we reduce the eigenfunctions of Eq. (7) by projecting out the ground electronic target state.[5] The usual algebra[18] yields a Schrödinger equation for the "reduced" system wave function

---

[4] In collision theory, the most useful atomic units are those for length [the first Bohr radius $1\,a_0 = 0.529\,177 \times 10^{-10}\,$m], energy [1 Hartree $E_h = 27.211\,396\,$eV], mass [the electron mass $m_e = 9.109\,398 \times 10^{-31}\,$kg], and, of course, cross section [$1\,a_0^2 = 0.28002\,\text{Å}^2$]. Also widely used, often with much attendant confusion, is the energy unit known as the Rydberg: this conversion factor, which absorbs an unsightly factor of 1/2 from energy terms in the Schrödinger equation, is just the Rydberg constant in the approximation of infinite nuclear mass, $R_\infty = 13.605\,698\,$eV. Finally, in converting between experimental and atomic units, the following are useful:

$$\begin{aligned} 1\,\text{eV} &= 1.602\,177 \times 10^{-19}\,\text{J} \\ 1\,\text{amu} &= 1822.89\,\text{au} \\ 1\,\text{cm}^{-1} &= 4.5563352672 \times 10^{-6}\,\text{au} \\ 1\,\text{Å} &= 1.88972599\,\text{au} \end{aligned}$$

For lots more atomic units and conversion factors, see Ref. 49.

[5] That the spin coordinates of the bound and continuum electrons do not explicitly appear in these equations is a happy consequence of our restriction to closed-shell targets. But to derive these equations one must explicitly take into account electron spin. In particular, the bound-free antisymmetrizer yields, after elimination of spin variables, the exchange kernel in the non-local exchange potential $\hat{\mathcal{V}}(\mathbf{r}, \mathbf{r}')$ in the equation for the reduced system wave function. A general pedagogical derivation that clarifies these matters appears in Ref. 29.

$\Psi_{\mathbf{k}_0',\nu_0}(\mathbf{r}', \mathbf{R})$, which we write in LAB coordinates and label by the incident wave vector $\mathbf{k}_0$ and initial ro-vibrational quantum numbers $\nu_0 = (v_0, j_0 m_{j_0})$. Since we have eliminated the electronic kinetic and potential energies of the target, this equation contains only the nuclear Hamiltonian (6) and an *averaged* interaction potential V_{int}:

$$\left[\hat{T}_e + \hat{\mathcal{H}}_m^{(n)} + V_{\text{int}} - E\right] \Psi_{\mathbf{k}_0',\nu_0}(\mathbf{r}', \mathbf{R}) = 0. \tag{9}$$

The total energy E in the entrance (initial) and exit (final) channels respectively is

$$E = \frac{1}{2}k_0^2 + \epsilon_0 = \frac{1}{2}k_v^2 + \epsilon_\nu. \tag{10}$$

The averaged interaction potential in Eq. (9) contains three terms:

$$V_{\text{int}}(\mathbf{r}', \mathbf{R}) = V_{\text{st}}(\mathbf{r}', \mathbf{R}) + \hat{\mathcal{V}} + V_{\text{c/p}}(\mathbf{r}', \mathbf{R}). \tag{11}$$

The *static potential* V_{st} is the average of the Coulomb interaction potential (8) over the ground-state electronic wave function. The *exchange potential* $\hat{\mathcal{V}}$ (whose non-local nature we emphasize by using a script symbol) results from the required antisymmetry of the electron-molecule wave function [the eigenfunctions of (7) before elimination of the molecular electrons] with respect to interchange of the space and spin coordinates of the projectile and any bound electron. (Of course, this wave function must also be antisymmetric under bound-bound electron interchange, but we assume that whoever generated the Born-Oppenheimer target electronic wavefunctions properly incorporated this requirement.) The final interaction term is a *correlation/polarization potential* that incorporates physical effects lost when we jettisoned excited target electronic states in deriving the reduced Schrödinger equation (9). Depending on exigencies of the problem we seek to solve (and practicalities of the computer on which it must be solved), we can approximate these effects in a variety of ways—e.g., rather rigorously using a non-local optical potential[39,40] or less so using a local model potential.[41–44] Extensive discussions of these and other options appears in Ref. 29; for pedagogical background see Ref. 30.

We can easily write down the static and exchange potentials. For a closed-shell molecule with N_{occ} doubly-occupied spatial orbitals (in BODY coordinates) $\{\xi_i^{\Lambda_i}(\mathbf{r}_j; R), i = 1, \ldots, N_e\}$, where Λ_i is the magnetic quantum number of the molecular electrons with respect to the internuclear axis, the static interaction assumes the form of an integral over a dummy variable $\mathbf{r}''$,

$$V_{\text{st}}(\mathbf{r}', \mathbf{R}) = V_{\text{en}} + 2\sum_{i=1}^{N_{\text{occ}}} \int \xi_i^{\Lambda_i *}(\mathbf{r}''; R) \frac{1}{|\mathbf{r}' - \mathbf{r}''|} \xi_i^{\Lambda_i}(\mathbf{r}''; R)\, d\mathbf{r}''. \tag{12}$$

The non-local exchange potential contributes an integral to the reduced Schrödinger equation— which, rather awkwardly for those who would solve it, thence becomes integro-differential. The action of this potential on the system wave function introduces one of the banes of electron-molecule calculations, the exchange kernel:[6]

$$\hat{\mathcal{V}}(\mathbf{r}', \mathbf{R})\Psi_{\mathbf{k}_0',\nu_0}(\mathbf{r}', \mathbf{R}) = \int \mathcal{K}(\mathbf{r}', \mathbf{r}'')\Psi_{\mathbf{k}_0',\nu_0}(\mathbf{r}'', \mathbf{R})\, d\mathbf{r}'', \tag{13}$$

[6] In this discussion of exchange we have ignored (i.e., set to zero) orthogonality terms involving the continuum function and bound molecular orbitals of the same symmetry. Doing so is formally valid for scattering from a *closed-shell* that, as in the Hartree-Fock approximation, is represented by a single-configuration wave function. For more complicated representations or for open-shell targets, these terms must be considered anew. For information on orthogonality see Ref. 45, for information and further references on the role of these terms in electron-molecule scattering see Ref. 46 and the review Ref. 29.

where the exchange kernel for a closed-shell target is more-or-less simply

$$\mathcal{K}(\mathbf{r}',\mathbf{r}'') = -\sum_{i=1}^{N_{\text{occ}}} \xi_i^{\Lambda_i}(\mathbf{r}';R)\frac{1}{|\mathbf{r}'-\mathbf{r}''|}\xi_i^{\Lambda_i\,*}(\mathbf{r}'';R). \tag{14}$$

Although today the rigorous inclusion of exchange effects is computationally (if not intuitively) simpler than so treating correlation/polarization effects, the computational problems induced by the non-locality of the exchange interaction remain formidable, particularly for systems with more than a dozen or so electrons. [For discussions of two numerical methods for solving the "exact-exchange" equations of electron-molecule scattering, see Ref. 47 and Ref. 48.] Hence many theorists have invested effort in the quest for accurate local model exchange potential.[26,29,30]

Boundary Conditions. In principle, having generated an interaction potential, one "merely" solves the Schrödinger equation (9), extracts the scattering amplitude $f_{\nu,\nu_0}(\hat{r}') = f(\mathbf{k}'_\nu, \nu \leftarrow \mathbf{k}'_0, \nu'_0)$ from the asymptotic ($r \to \infty$) form of the scattering function $\Psi_{\mathbf{k}'_0,\nu_0}(\mathbf{r}',\mathbf{R})$, calculates from it the desired cross sections, and goes home. As signified by the subscripts dangling from this function (the incident wave vector of the projectile $\mathbf{k}'_0$ and the initial ro-vibrational target state ν_0), it obeys the usual plane-wave boundary conditions (BC), which in the LAB frame (with $\hat{k}'_\nu = \hat{r}'$) are[7]

$$\Psi_{\mathbf{k}'_0,\nu_0}(\mathbf{r}',\mathbf{R}) \xrightarrow[r\to\infty]{} (2\pi)^{-3/2}\left[e^{i\mathbf{k}'_0\cdot\mathbf{r}'}\chi_{\nu_0}(\mathbf{R}) + \sum_{\nu}^{\text{(open)}} \frac{e^{ik_\nu r}}{r} f_{\nu,\nu_0}(\hat{r}')\,\chi_\nu(\mathbf{R})\right], \tag{15}$$

where terms appear in the sum only for open channels—i.e., energetically accessible final states ν.

The relationship between this amplitude and its wave function is further clarified by the equation

$$f_{\nu,\nu_0}(\hat{r}') = -(2\pi)^{1/2}\frac{m_e}{\hbar^2}\int e^{-i\mathbf{k}'_\nu\cdot\mathbf{r}'}\,\chi_\nu^*(\mathbf{R})\,V_{\text{int}}(\mathbf{r}',\mathbf{R})\,\Psi_{\mathbf{k}'_0,\nu_0}(\mathbf{r}',\mathbf{R})\,d\mathbf{r}'\,d\mathbf{R}, \tag{16}$$

where we have momentarily abandoned atomic units (m_e is the electron mass). This may be more familiar in its alternate guise as an integral equation for the matrix element of the transition operator $\hat{T}$ between asymptotic free plane-wave states. The two are related by a proportionality constant as[8]

$$f_{\nu,\nu_0}(\hat{r}') = \frac{4\pi m_e}{\hbar^2}\langle \mathbf{k}'_\nu, \nu \mid \hat{T} \mid \mathbf{k}'_0, \nu_0\rangle. \tag{17}$$

In practice, one almost never computes cross sections directly from this scattering amplitude because it's so hard to calculate $\Psi_{\mathbf{k}'_0,\nu_0}(\mathbf{r}',\mathbf{R})$. We have inadvertently made

[7] Implicit in this equation is the choice that of Dirac delta function normalization for the asymptotic plane-wave states. This choice affects many of the subsequent equations in this chapter; alternatives are considered in the Appendix to this chapter.

[8] This equation introduces the choice of proportionality constant between the scattering amplitude and the momentum-space T matrix; when partial-wave channels are introduced (see below), this choice directly influences the relationship between the S and T matrices. So this constant, although arbitrary, has consequences for many of the subsequent equations in this chapter. The Appendix shows how to implement choices other than ours, $\mathsf{S} = 1 + 2i\mathsf{T}$.

solving the scattering problem even more difficult than it has to be; by demanding that this wave function satisfy plane-wave boundary conditions, we have ignored the constants of the motion of the collision. If instead we seek solutions that satisfy boundary conditions in which the asymptotic states are eigenstates of the constants of the motion, then we can simplify greatly the computational chores that lie ahead. In particular, we shall exploit the degeneracy of $\Psi_{\mathbf{k}_0',\nu_0}(\mathbf{r}',\mathbf{R})$ to construct linear combinations for total energy E that obey such boundary conditions, doing so via a succession of physically motivated unitary transformations.

Rotational invariance of the electron-molecule Hamiltonian (7) implies that this Hamiltonian commutes the total angular momentum operator, which for the problem at hand is the sum of operators for the projectile's orbital and target's rotational angular momenta: $\hat{\mathbf{J}} \equiv \hat{\imath} + \hat{\mathbf{j}}$. So for any total energy E we can find solutions to the reduced Schrödinger equation that are eigenfunctions of $\hat{J}^2$ and the projection $\hat{J}_{z'}$ on the LAB z axis. Unlike the T matrix in Eq. (17), the transition matrix *in a basis of asymptotic free eigenstates of these operators* is diagonal in the corresponding quantum numbers J and M_J—a property that greatly facilities its calculation!

The problem is that $\Psi_{\mathbf{k}_0',\nu_0}(\mathbf{r}',\mathbf{R})$ satisfies the plane-wave boundary conditions (15). Since plane-wave functions are eigenfunctions of the linear, not the angular, momentum, these boundary conditions are incompatible with a description in terms of asymptotic eigenstates of the orbital angular momentum. This problem, as we shall see, is easily solved by a familiar transformation. We can then implement a second transformation to couple the (projectile) orbital and (target) rotational angular momenta and generate the desired asymptotic eigenstates of $\hat{J}^2$ and $\hat{J}_{z'}$. In the first transformation, we consider just the projectile. To go from a description of the collision in terms of asymptotic plane waves to one that identifies a well-defined orbital angular momentum for the entrance and exit channels, we use the unitary transformation between the two kinds of free waves. The only tricky aspect of this transformation is (as usual) normalization. We shall adopt the following conventions for our plane and spherical free-wave functions:[9]

$$\langle \mathbf{r}' \mid \mathbf{k}' \rangle = (2\pi)^{-3/2} e^{i\mathbf{k}'\cdot\mathbf{r}'} \tag{18a}$$

$$\langle \mathbf{r}' \mid E\ell m_\ell \rangle = \left(\frac{2m_e}{\pi k}\right)^{1/2} \frac{1}{r}\,\hat{j}_\ell(kr)\,Y_\ell^{m_\ell}(\hat{r}'), \tag{18b}$$

where $\hat{j}_\ell(kr)$ is the Ricatti-Bessel function $\hat{j}_\ell(kr) = (kr)\,j_\ell(kr)$. For plane and angular-momentum free kets normalized as in Eq. (18), the elements of the unitary transformation between these states is

$$c_{\ell m_\ell}(\mathbf{k}') = \langle E\ell m_\ell \mid \mathbf{k}' \rangle = \frac{i^\ell}{\sqrt{m_e k}}\, Y_\ell^{m_\ell *}(\hat{k}'), \tag{19a}$$

where $\mathbf{k}' = k\hat{r}'$ and $E = \hbar^2 k^2/(2m_e)$. Because we defined the LAB frame so that in the entrance channel $\hat{k}_0' = \hat{\mathbf{e}}_{z'}$, this transformation coefficient simplifies to

$$c_{\ell_0 m_{\ell_0}}(\mathbf{k}_0') = c_{\ell_0} = \frac{i^{\ell_0}}{\sqrt{4\pi m_e k_0}} \sqrt{2\ell_0+1}\,\delta_{0,m_{\ell_0}}. \tag{19b}$$

We can now couple the orbital and rotational angular momenta, generating eigenstates with well-defined total angular momentum quantum numbers J and

[9] We have energy normalized the angular-momentum free states $\mid E\ell m_\ell\rangle$. See the Appendix for modifications required to accomodate k-normalization.

$M_J = m_\ell + m_j$, via the familiar Clebsch-Gordan series.[34,35] Together, these two transformations relate our original LAB system wave function, which obeys plane-wave boundary conditions (15), to the new eigenfunctions of the coupled angular momentum (CAM) operators, to wit:

$$\Psi_{\mathbf{k}'_0,\nu_0}(\mathbf{r}',\mathbf{R}) = \sum_{\ell_0=0}^{\ell_{\max}} \sum_{m_{\ell_0}=-\ell_0}^{\ell_0} \sum_{J=J_{\min}}^{J_{\max}} c_{\ell_0 m_{\ell_0}}(\mathbf{k}'_0)\, C(j_0\,\ell_0\,J; m_{j_0}\,m_{\ell_0}\,M_J)\, \Psi'^{J}_{E_0,v_0 j_0 \ell_0}(\mathbf{r}',\mathbf{R}). \tag{20}$$

[With the continuum normalization of Eq. (18), this transformation preserves normalization of the system wave function.] Note that the Clebsch-Gordan coefficient imposes a triangle relation $\Delta(j_0\,\ell_0\,J)$ that restricts the sum over J to range from $J_{\min} = |j_0 - \ell_0|$ to $J_{\max} = j_0 + \ell_0$. No sum over m_{ℓ_0} appears in Eq. (19) because we cleverly chose the LAB z axis, $\hat{\mathbf{e}}_{z'} = \hat{k}'_0$ so that $m_{\ell_0} = 0$.

Into the BODY frame. This completes construction of the theoretically optimal system wave function . . . if we're content to work solely in the LAB frame. But as noted above, under conditions that admit an adiabatic treatment of the rotational motion of the molecule, a BODY formulation offers additional computational simplifications, reducing the complexity of channel coupling and the number of coupled channels required for convergence.[52–55] Such conditions pertain except near rotational thresholds[55,56] and in regions (where special strategies can be employed).[18] And so the preponderance of electron-molecule calculations are based on the fixed-nuclear-orientation (FNO) approximation,[27,52,53] according to which the rotation of the internuclear axis is ignored for the duration of the collision. This approximation amounts to neglecting the rotational kinetic energy in the nuclear Hamiltonian (6) for the solution of the Schrödinger equation.

Table 2. Dynamical approximations underlying various formulations of electron-molecule scattering.

formulation	treat rotations adiabatically?	treat vibrations adiabatically?	separate continuum function?	target-state degeneracy?[1]
LAB-CAM	no	no	no	no
ANR (BF-FNO)	yes	no	yes	$\epsilon_{vj'} = \epsilon_{vj}$
ANV (BF-FN)	yes	yes	yes	$\epsilon_{v'j'} = \epsilon_{vj}$
BF-FONDA	yes	no	yes	no

[1]In ANR calculations based on BF-FNO K matrices, only rotational states within the same vibrational manifold are treated as degenerate.

The practical power of BODY formulations arises from a new (approximate) constant of the motion. In the FNO approximation, the projection Λ of the projectile's orbital angular momentum along the internuclear axis (the BODY z axis) is conserved in the collision. The small price we must pay to exploit this fact is construction of yet another system wave function—one in terms of BODY coordinates whose boundary conditions identify asymptotic free eigenstates of the vibrational Hamiltonian (quantum number v), the projectile orbital angular momentum (ℓ), and the projection of this quantity along the internuclear axis (Λ). To relate this new BF-FNO system wave function $\Psi^{\Lambda}_{E_0,v_0\ell_0}(\mathbf{r},\mathbf{R})$ to the LAB-CAM wave function we just constructed requires yet another rather-more-subtle unitary transformation, the rotational frame transformation we shall discuss below. Note that the super- and subscripts festooning this wave function specify the incident kinetic energy E_0 and the initial quantum numbers appropriate to the BF-FNO description, $(v_0, \ell_0; \Lambda)$.

Table 3. Fundamentals of widely-used formulations of electron-molecule collision theory: wave functions, asymptotic constants of the motion (COM), and channel definitions.

formulation	wave function	asymptotic COM	channels
LAB (plane-wave BC)	$\Psi_{\mathbf{k}_0', \nu_0}(\mathbf{r}', \mathbf{R})$	$\hat{\mathbf{p}}_e$, $\hat{\mathcal{H}}_m^{(\text{n})}$	$\mathbf{k}', \nu$
LAB-CAM	$\Psi^J_{E_0, v_0 j_0 \ell_0}(\mathbf{r}', \mathbf{R})$	$\hat{\mathcal{H}}_m^{(\text{v})}$, $\hat{\mathcal{H}}_m^{(\text{r})}$, $\hat{\ell}^2$, $\hat{j}^2$, $\hat{j}_{z'}$	$(v, j, \ell; JM_J)$
LAB-RCAM[1]	—	$\hat{\mathcal{H}}_m^{(\text{v})}$, $\hat{\mathcal{H}}_m^{(\text{r})}$, $\hat{\ell}^2$, $\hat{\ell}_t^2$, $\hat{\ell}_{tz'}$	$(v, j, \ell; \ell_t m_t)$
BF-FNO	$\Psi^\Lambda_{E_0, v_0 \ell_0}(\mathbf{r}, R)$	$\hat{\mathcal{H}}_m^{(\text{v})}$, $\hat{\ell}^2$, $\hat{\ell}_z$	$(v, \ell; \Lambda)$
BF-FN (RR)[2]	$\Psi^\Lambda_{E_0 \ell_0}(\mathbf{r}; R)$	$\hat{\ell}^2$, $\hat{\ell}_z$	$(\ell; \Lambda)$

[1]The recoupled angular momentum operator $\hat{\boldsymbol{l}}_t$ is defined in the text.
[2]In the rigid-rotor (RR) approximation, the internuclear separation R is a parameter.

To aid the reader bemused by this smorgasbord of alternative wave functions, we offer the summaries of formulations in Table 3 and of scattering theories in Table 2. The next step in either the LAB-CAM or BF-FNO formulations is to write down the much desired alternatives to the plane-wave boundary condition (15). To this end we now introduce eigenfunction expansions.

Coupled Equations

Converting the three-dimensional integro-differential reduced Schrödinger equation (9) into something we can solve proceeds via the time-honored undergraduate-quantum-mechanics method of expanding the unknown wave function in sets of known eigenfunctions that are complete in as many variables as possible.[57] For each of the formulations in Table 3, we can eliminate from the unknown function all the angular variables ($\hat{r}'$ and $\hat{R}$ in the LAB-CAM theory and $\hat{r}$ in the BF-FNO theory) and the internuclear separation R. In each case we get *sets of coupled integrodifferential radial (single-variable) equations* which are easier (though not easy) to solve.

Let's begin with the (algebraically) simplest theory, the BF-FNO formulation. We can convert the BF-FNO scattering equation to coupled radial equations in two steps. We first expand $\Psi^\Lambda_{E_0, v_0 \ell_0}(\mathbf{r}, R)$ in the complete set of target vibrational states $\{\varphi_v^{(\text{v})}(R)\}$, thereby introducing the *coupled-state scattering functions* $\xi^\Lambda_{v v_0}(\mathbf{r})$ which play the mathematical role of expansion coefficients. Second we expand these functions in angular-momentum eigenfunctions appropriate to the BF-FNO theory, i.e., eigenfunctions of $\hat{\ell}_z$ with eigenvalue Λ. This step introduces the reduced radial functions $u^{\Lambda}_{v\ell,v_0\ell_0}(r)$ for which we ultimately solve. The whole thing looks like[10]

$$\Psi^{\Lambda}_{E_0,v_0\ell_0}(\mathbf{r},R) = \sum_{v=0}^{v_{\max}} \xi^{\Lambda}_{vv_0}(\mathbf{r})\,\varphi^{(\mathrm{v})}_v(R) = \left(\frac{2m_e}{\pi k_0}\right)^{1/2} \frac{1}{r}\sum_{v=0}^{v_{\max}}\sum_{\ell=0}^{\ell_{\max}} u^{\Lambda}_{v\ell,v_0\ell_0}(r)\,\varphi^{(\mathrm{v})}_v(R)\,Y^{\Lambda}_{\ell}(\hat{r}). \tag{21}$$

In principle these sums over vibrational states and spherical harmonics should include the infinity of functions in these complete sets; in practice, of course, we truncate these sums at the minimum number of such functions required for global convergence to the desired precision of whatever scattering quantity we seek—being careful to take account of the variation of the number of required terms with energy, scattering angle, etc. [More detailed discussions of convergence criteria appear in Ref. 20.]

The total number of BF-FNO *channels*—and hence the number of simultaneous equations we must solve and the dimensionality of the solution and scattering matrices—is the product of the number of vibrational states N_v times the number of partial waves N_ℓ. This number is reduced somewhat by the (approximate) conservation of $\hat{\ell}_z$. This consequence of the FNO approximation means that channels with different values of Λ "don't talk to each other." So the coupled equations fall into independent sets distinguished by the modulus of the projection quantum number Λ, as $|\Lambda| = 0, 1, 2, \ldots$. If the target belongs to the point group $D_{\infty h}$, a further delightful bifurcation of these sets of equations occurs according to the parity (gerade or ungerade as ℓ is even or odd).[11]

The LAB-CAM theory, by contrast, isn't amenable to so conceptually clean an eigenfunction expansion. Formulating the problem in terms of eigenstates of $\hat{J}^2$ and $\hat{J}_z$ nicely exploits the collision constants but leads to CAM basis functions that involve both projectile and target coordinates.[58] These functions, which are complete in $\hat{r}'$ and $\hat{R}$, assume the form

$$\mathcal{Y}^{JM}_{j\ell}(\hat{r}',\hat{R}) \equiv \sum_{m_j=-j}^{+j}\sum_{m_\ell=-\ell}^{+\ell} C(j\,\ell\,J;m_j\,m_\ell\,M_J)\;Y^{m_j}_j(\hat{R})\,Y^{m_\ell}_\ell(\hat{r}'). \tag{22}$$

The corresponding expansion, which also involves the vibrational eigenfunctions (complete in R), is

$$\Psi^{J}_{E_0,v_0j_0\ell_0}(\mathbf{r}',\mathbf{R}) = \left(\frac{2m_e}{\pi k_0}\right)^{1/2}\frac{1}{r}\sum_{v=0}^{v_{\max}}\sum_{\ell=0}^{\ell_{\max}}\sum_{j=0}^{j_{\max}} u^{J}_{vj\ell,v_0j_0\ell_0}(r)\,\varphi^{(\mathrm{v})}_v(R)\,\mathcal{Y}^{JM}_{j\ell}(\hat{r}',\hat{R}). \tag{23}$$

---

[10] The normalization constant $[2m_e/(\pi k_0)]^{1/2}$ conforms to energy normalization of the orbital angular momentum free states $|E\ell m_\ell\rangle$ and is required, since the radial function $u^{\Lambda}_{v\ell,v_0\ell_0}(r)$ reduces to $\hat{j}_{\ell_0}(k_0r)\delta_{\ell,\ell_0}$ in the absence of a potential. The radial scattering functions, then, are normalized with respect to k just like the Riccati-Bessel functions,

$$\int_0^\infty \hat{j}_\ell(kr)\,\hat{j}_\ell(k'r)\,dr = \frac{\pi}{2}\,\delta(k-k') = \frac{\pi k}{2m_e}\,\delta(E-E').$$

Although this prefactor disappears from the (homogeneous) scattering equations for $u^{\Lambda}_{v\ell,v_0\ell_0}(r)$, it persists in some theoretical formulations—if, for example, the scattering equations include an inhomogeneous term. Note that some authors, e.g., Taylor in Ref. 51, include a phase factor i^{ℓ_0} in their definition of the radial function; we consign this phase to the transformation (19*a*).

[11] The parity of the LAB-CAM channel follows from that of the constituent spherical harmonics as $(-1)^{j+\ell}$. (For $J = 0$ only even parity states exist). In the BF-FNO formulation, the parity of each channel is simply $(-1)^\ell$. Furthermore, reflection symmetry in the scattering plane (which in the BODY frame contains $\hat{R}$) causes the scattering function to be independent of the sign of Λ. So although in principle $-\ell \le \Lambda \le +\ell$, in practice we need only consider $|\Lambda| \ge 0$.

This expansion is not quite as formidable as it looks; the basis states that appear in it are constrained not only by truncation (at $v_{\max}$, $\ell_{\max}$, and $j_{\max}$) but also by triangle rules spawned by the LAB-CAM coupling potentials we shall peruse below.

Table 4. Basis sets for eigenfunction expansion of electron-molecule wave functions.

formulation	wave number k_p	coordinates	basis functions	parity	K matrix
LAB-CAM	k_{vj}	$\mathbf{r}', \mathbf{R}$	$\varphi^{(v)}(R)\mathcal{Y}_{j\ell}^{JM}(\hat{r}', \hat{R})$	$(-1)^{j+\ell}$	$K^{J}_{vj\ell, v_0 j_0 \ell_0}$
BF-FNO	k_v	$\mathbf{r}, R$	$\varphi^{(v)}(R)Y_\ell^\Lambda(\hat{r})$	$(-1)^\ell$	$K^{\Lambda}_{v\ell, v_0 \ell_0}$
BF-FN (RR)	k_b	$\mathbf{r}; R$	$Y_\ell^\Lambda(\hat{r})$	$(-1)^\ell$	$K^{\Lambda}_{\ell, \ell_0}(R)$

In Table 4 we summarize the expansion bases $\{\, \Phi_p^q \,\}$ for our various formulations. We have introduced a generic channel label of the form $(p; q)$. The quantum numbers denoted collectively by q, which are segregated by semi-colons in each channel label, correspond to constants of the motion; those denoted by p label asymptotic free states. The special status of the quantum numbers q is that the scattering matrix is diagonal with respect to them, so, as noted above, the (large) set of coupled radial equations we must solve separates into (not quite so large) sets identified by q.

The last line of Table 4 introduces the rigid-rotor (RR) approximation, according to which we completely ignore the vibrational motion of the target and "freeze" the internuclear separation at its equilibrium value $R = R_{\text{eq}}$ [See also Table 3]. Because it completely eliminates one variable from the scattering problem (and hence one complete set from the LAB-CAM or BF-FNO bases), the RR approximation greatly simplifies the calculation of elastic and rotational excitation cross sections. Not surprisingly, this approximation has been a mainstay of theoretical studies of vibrationally elastic collisions since the early days of this field.[23,28–30]

Today, however, the needs of various technological applications for these cross sections[7–10] and the increased accuracy of experimental measurements[11] have given rise to circumstances in which the error introduced by the RR approximation is unacceptable. In Fig. 2 we illustrate these errors for two different vibrationally-elastic cross sections. Most of this error can be eliminated using a computationally simple *vibrational averaging* procedure according to which one replaces the equilibrium potential energy by its average over the ground vibrational state in the RR scattering equations.[59] This tactic, which requires only a single additional quadrature (over R) neglects effects due to coupling to excited vibrational states but includes the zero-point motion of the target, usually the most important influence on elastic and rotational excitation cross sections. Throughout the remainder of this chapter we shall not draw explicit attention to the RR option; when physically justified it can be implemented by obvious alterations in the BF-FNO or LAB-CAM equations.

In any formulation, the integrodifferential equations that result from eigenfunction expansion, which don't care what basis set we use, have the generic form

$$\left[\frac{d^2}{dr^2} - \frac{\ell(\ell+1)}{r^2} + k_p^2\right] u^q_{p,p_0}(r) = 2\sum_{p'} \left[V^q_{p,p'}(r)\, u^q_{p',p_0}(r) + \hat{\mathcal{V}}^q_{p,p'}\, u^q_{p',p_0}(r)\right]. \tag{24}$$

Two kinds of coupling matrix elements appear on the right hand side. Matrix elements $V^q_{p,p'}(r)$ couple channels $(p; q)$ through *local* terms in the potential: those due to the static and intermediate- and long-range polarization interaction. The elements $\hat{\mathcal{V}}^q_{p,p'}$, however, couple through non-local terms: those due to exchange [Eq. (13)] and, if

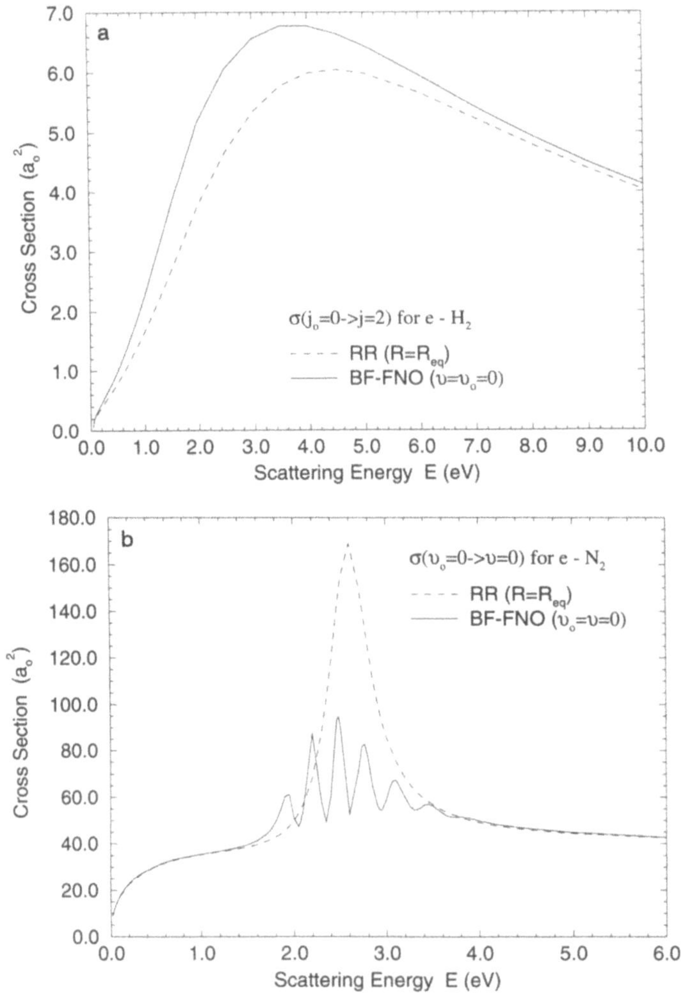

Figure 2. The effect of vibrational dynamics on two *vibrationally elastic* cross sections. (a) Rotational e–H_2 cross sections $\sigma^{(r)}_{0\to 2}$ from two treatments of the vibrational dynamics. The dashed curve shows results based on K-matrices calculated in the rigid-rotor (RR) approximation with the internuclear separation fixed at $R = R_{\rm eq} = 1.40\,a_0$; the solid curve is the vibrationally elastic ($v_0 = 0 \to v = 0$) rotational cross section from converged BF-FNO calculations, in which the vibrational dynamics are fully included. (b) Elastic e–N_2 cross sections from RR calculations (dashed curve) with $R = R_{\rm eq} = 2.02\,a_0$ and results from a converged 15-state BF-FNO calculation (solid curve). The elaborate structure from 1.5 to 4.0 eV in (b) arises from an intermediate-duration shape resonance at 2.39 eV; the RR approximation does not allow for the vibrational dynamics to influence the projectile and so does not predict this structure. [The resonance energies and widths from the RR calculations are 2.539 eV and 0.470 eV, respectively.]

present, short-range bound-free correlation terms in, say, an optical potential.[40] In the case of exchange, the effect of the non-local matrix elements on the scattering function is given in terms of the exchange kernel (14) as

$$\hat{\mathcal{V}}^q_{p,p'}\, u^q_{p,p_0}(r) = \int \mathcal{K}^q_{p,p'}(r,r')\, u^q_{p',p_0}(r')\, dr'. \tag{25}$$

The kernel matrix elements do, of course, depend on the particular expansion basis $\{\, \Phi^q_p \,\}$ one has adopted. If BF-FNO is the theory of choice, then

$$\mathcal{K}^q_{p,p'}(r',r'') = r'\,r'' \int\!\!\int\!\!\int \varphi^{(\mathrm{v})*}_{v}(R)\, Y^{\Lambda*}_{\ell}(\hat{r}')\, \mathcal{K}(r',r'')\, Y^{\Lambda}_{\ell'}(\hat{r}'')\, \varphi^{(\mathrm{v})}_{v'}(R)\, d\hat{r}'\, d\hat{r}''\, dR. \tag{26}$$

Whatever the formulation, the sum over the channel index p includes a sum over the projectile angular momentum quantum number ℓ. Usually one must include more terms in the sum over local matrix elements than in the sum over non-local (exchange) elements. Depending on the numerical algorithm one adopts, the addition of non-local exchange terms may greatly increase the computational demands of the scattering calculation, so it's worth determining these upper limits prior to production.[47]

Table 5. Radial free functions $\hat{f}_\ell(k_\nu r) = k_\nu r\, f_\ell(k_\nu r)$ used in the definitions of the S, T, and K matrices. The last three columns designate which functions appear in the partial-wave free-particle Green's function for the corresponding integral equation. Conventions as in Ref. 51.

Function	defining equation	asymptotic behavior	K	S	T
$\hat{\jmath}_\ell(k_\nu r)$ (regular)	$\sqrt{\frac{\pi}{2}}(k_\nu r)^{1/2} J_{\ell+\frac{1}{2}}(k_\nu r)$	$\sin\left(k_\nu r - \frac{1}{2}\ell\pi\right)$	√		√
$\hat{n}_\ell(k_\nu r)$ (irregular)	$(-1)^\ell \sqrt{\frac{\pi}{2}}(k_\nu r)^{1/2} J_{-\ell-\frac{1}{2}}(k_\nu r)$	$\cos\left(k_\nu r - \frac{1}{2}\ell\pi\right)$	√		
$\hat{h}^-_\ell(k_\nu r)$ (incoming)	$\hat{n}_\ell(k_\nu r) - i\,\hat{\jmath}_\ell(k_\nu r)$	$e^{-i(k_\nu r - \frac{1}{2}\ell\pi)}$		√	
$\hat{h}^+_\ell(k_\nu r)$ (outgoing)	$\hat{n}_\ell(k_\nu r) + i\,\hat{\jmath}_\ell(k_\nu r)$	$e^{+i(k_\nu r - \frac{1}{2}\ell\pi)}$		√	√

Boundary Conditions At Last. The boundary conditions we impose on the radial functions that solve these equations determine the scattering matrix. Depending on how we encode their solution, we might wind up with any of three such matrices, S, T, or K. Which matrix we get depends on which of the free radial functions in Table 5 we use in the boundary conditions. Thus the S matrix is obtained from solutions of Eq. (24) that incorporate the Ricatti-Hankel functions:[51]

$$u^q_{p,p_0}(r) \xrightarrow[r\to\infty]{} \frac{i}{2}\left[\delta_{p,p_0}\, \hat{h}^-_{\ell_0}(k_0 r) - \sqrt{\frac{k_0}{k_p}}\, S^q_{p,p_0}\, \hat{h}^+_\ell(k_p r)\right]. \qquad \text{(open channels)} \tag{27}$$

Alternatively, we could identify the transition matrix directly by re-writing (27) as,

$$u^q_{p,p_0}(r) \xrightarrow[r\to\infty]{} \delta_{p,p_0}\, \hat{\jmath}_{\ell_0}(k_0 r) + \sqrt{\frac{k_0}{k_p}}\, T^q_{p,p_0}\, \hat{h}^+_\ell(k_p r), \qquad \text{(open channels)} \tag{28}$$

with $\mathsf{S}^q = 1 + 2i\mathsf{T}^q$. If we prefer to work with real functions, we can use Ricatti-Bessel and -Neumann functions in the boundary conditions to get the real, symmetric reactance (K) matrix:[12]

$$u^q_{p,p_0}(r) \xrightarrow[r\to\infty]{} \delta_{p,p_0}\,\hat{\jmath}_{\ell_0}(k_0 r) + \sqrt{\frac{k_0}{k_p}}\,K^q_{p,p_0}\,\hat{n}_\ell(k_p r). \qquad \text{(open channels)} \tag{29}$$

All these forms pertain to open channels only. But except in a BF-FN calculation, where all channels are degenerate, we may have to cope with closed channels, for which no scattered flux reaches the asymptotic region. We ensure this by imposing on the closed-channel elements of the radial wave function matrix the condition

$$u^q_{p,p_0}(r) \xrightarrow[r\to\infty]{} e^{-|k_p|r}, \qquad \text{(closed channels)} \tag{30}$$

where, of course, the exit-channel wave number k_p is imaginary for a closed channel.[13] Ultimately, of course, all theoretical paths must return to the LAB scattering amplitude. This step, if no other, is quite easy: we simply undo the transformations that took us from plane-wave boundary conditions to those of our chosen formulation.

Nuts and Bolts: Convergence Matters. A crucial practical concern in scattering studies based on eigenfunction expansions is convergence with respect to the basis appropriate to our chosen formulation. The fundamental expansion commands us to sum over all quantum numbers collectively denoted by p in our channel label $(p;q)$. In a BF-FNO expansion a là Eq. (21), for example, this mandates sums over the vibrational and projectile angular momentum quantum numbers v and ℓ. We thus introduce two numerical parameters: the number N_v of vibrational target states and the number N_ℓ of (BODY) spherical harmonics. Each must be determined via careful convergence studies.

Such studies are so important because these parameters establish the dimensionality of the radial wave function [$u^\Lambda_{v\ell,v_0\ell_0}(r)$ in the BF-FNO theory], the resulting scattering matrix [e.g., $T^\Lambda_{v\ell,v_0\ell_0}$], and other matrices involved in the solution of the coupled equations (see, for example, the integral equations algorithm discussed below). Within some algorithms, we can reduce (by truncation) the dimensionality of these matrices once the radial wave function propagation has left the region of strong coupling,[63] but even so it's best to determine at the outset how many channels we really need to attain convergence. To yield to the temptation to just dump a lot of basis functions into the calculation (e.g, setting N_v and N_ℓ to values we hope will exceed the actual demands of the calculation) is to ask for trouble from a smorgasbord of numerical glitches. The litany of things we've seen go wrong when we've used unphysical basis functions include linear dependence in the radial function owing to

[12] There is, of course, a price to be paid for the luxury of avoiding complex functions. The K matrix, which can be related to the arctangent of a phase angle, exhibits sharp structures at certain angles [See Ref. 60]. Furthermore in the presence of a shape or Feschbach resonance, key elements of the K matrix may vary wildly over a short energy range. In many cases, then, the (complex) T matrix is a smoother, more tractable function than its real counterpart.

[13] The normalization of the continuum function implied by Eqs. (27) and (28) are by no means standard. There is no standard. With the normalization choices in Eqs. (18) and the relationship $\mathsf{S}^q = 1 + 2i\mathsf{T}^q$ between the S and T matrices, however, Eq. (27) follows inexorably; from it and the definition of the Ricatti Hankel function, so does Eq. (28). By contrast, Eq. (29) requires re-normalization of the radial function, since the relationship between the S and K matrices, if directly inserted into Eq. (27), yields the boundary conditions of Eq. (29) multiplied by $(1 - i\mathsf{K}^q)^{-1}$. In practice, this causes no difficulty: all we do with K^q is turn it into T^q.

high-order partial waves, numerical instabilities arising from closed-channels that couple insignificantly to the open channels of interest, severe inaccuracies in high-order free-wave functions at small radial distances, and more.[20,21]

Convergence strategies can be found in most theoretical papers published during the last 20 years. To summarize, we consider it essential to first choose the particular scattering quantity we want to converge (rather than to just try "to converge the calculation") and a specific numerical precision, then to seek global convergence for a (preferably small) set of test cases. Testing several cases is essential because convergence properties are often acutely sensitive to several factors: to the strength and asphericity of the interaction potential (e.g., scattering from H_2 versus N_2), to the energy range of interest (e.g., below or above an eV, resonant or non-resonant), to the excitation of interest (e.g., elastic scattering versus $0 \rightarrow 1$ vibrational excitation), and to the values of the constant of the motion in the scattering equations being solved (e.g., the electron-molecule symmetry in a BF-FNO calculation, the total angular momentum J in a LAB-CAM calculation).

To illustrate one such sensitivity we show in Fig. 3 the convergence properties of resonant and non-resonant e–N_2 elastic cross sections with respect to number of vibrational states using in the BF-FNO expansion (21). Near the 2.39 eV shape resonance, the resonance cross section in Fig. 3b is especially sensitive to N_v because of subtle interactions between the nuclear dynamics and the transient N_2^- complex formed during the resonant encounter.[64]

A rather more subtle aspect of convergence in vibrational states is illustrated in Fig. 4 with BF-FNO cross sections[65] at two energies for the $0 \rightarrow 1$ vibrational excitation of H_2. At issue here is the region of configuration space in which vibrational coupling influences the cross section. At these energies, $N_v = 5$ is sufficient to converge this cross section to 1%. By contrast, including only the initial and final states $\varphi_0^{(v)}(R)$ and $\varphi_1^{(v)}(R)$ in the BF-FNO expansion (21), which yields the two-state results in the figure, is inadequate. That is, the converged cross section $\sigma_{0\rightarrow 1}^{(v)}$ manifests effects of coupling to "intermediate" vibrational states—states that, whether open or closed, influence the cross section of interest only through coupling to other states. [The thresholds of the intermediate states in this calculation, $v = 2$, 3, and 4, are given roughly by v times the $0 \rightarrow 1$ threshold energy 0.5 eV.]

To see where vibrational coupling of the initial and final states and coupling of these to intermediate states influence the cross section, we performed a series of *gedanken-calculations* in which we artificially "switched off" all vibrational coupling to intermediate states and between the initial and final states outside the "truncation radius" r_t, which is the horizontal axis in Fig. 4. [This is easily accomplished by setting to zero the relevant vibrational coupling potentials $w_{v,v'}^{\lambda}(r)$ (discussed below) for $r \geq r_t$.] The results demonstrate the inherently short-range character of vibrational coupling: i.e., this feature of the dynamics is much more important near the target than far from it. Similar studies (not shown) in which we switched off coupling *only to intermediate states* at r_t, leaving the initial and final states coupled, showed that coupling to intermediate states is important only at radii less than $1\,a_0$; by contrast, coupling of initial and final states influences the cross section from the origin to radii larger than this value. Although such studies are, obviously, based on experimentally unrealizable situations, their results may yield important insights for defining and refining approximations both to the collision dynamics, which often treats near- and far-regions differently, and to the interaction potential, which is dominated by physically distinct effects in these two regions.

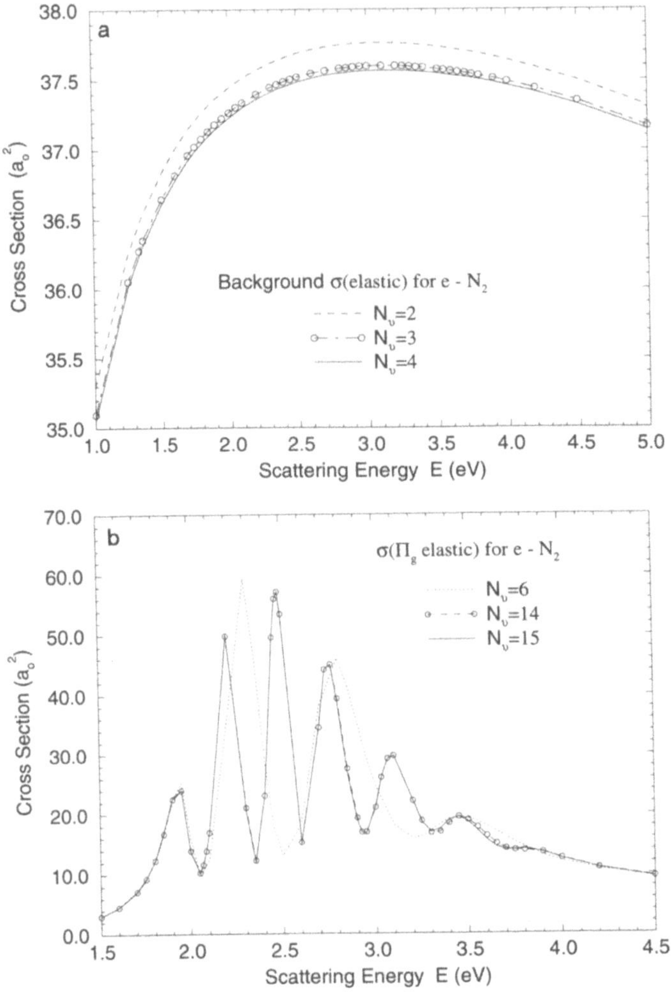

Figure 3. Convergence of the *elastic* e–N_2 cross section in the number of vibrational states N_v included in a BF-FNO eigenfunction expansion of the system wave function. (a) Convergence of the non-resonant (background) contribution, which includes contributions from Σ_g, Σ_u, and Π_u electron-molecule symmetries. (b) Convergence of the resonant (Π_g) contribution in the vicinity of the low-energy shape resonance at 2.4 eV.

S, T, and K Matrices and Relationships Between Them

While the most easily generated scattering matrix is usually K, one writes scattering amplitudes and cross sections in terms of T. Thus, for example, equations for the ro-vibrational differential and integral cross sections are written in terms of the LAB-CAM T matrix.[58] So we must convert. The S, K, and T matrices are related by the Heitler damping equation, which enforces conservation of probability.[62] So if your K matrix is Hermitian, then the corresponding S matrix will be properly unitary. From the Heitler equation follow simple relations between these scattering matrices.

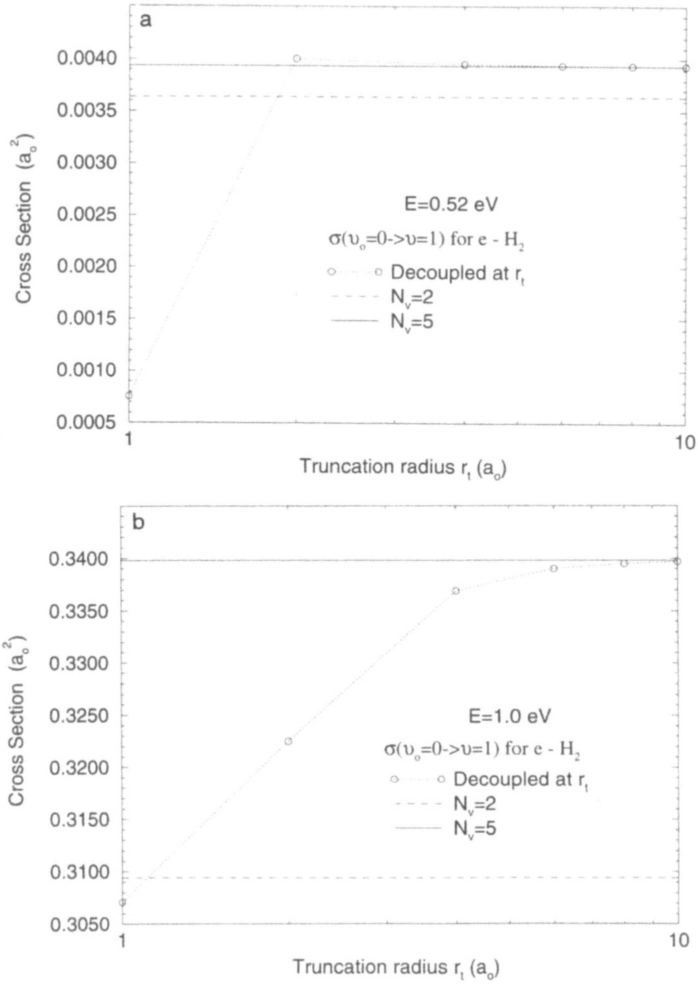

Figure 4. Convergence of the e–H_2 vibrational excitation cross section $\sigma_{0\to1}^{(v)}$ in the number of vibrational states N_v included in a BF-FNO eigenfunction expansion of the system wave function at (a) 0.52 eV and (b) 1.0 eV. The horizontal lines show the fully converged results (solid) and those obtained when only the initial and final states are included in the vibrational expansion (dashed). The dotted lines show the variation of this cross section with the region of configuration space in which vibrational coupling is allowed in this (hypothetical) scattering event.

With the convention[14]

$$\mathsf{S} = 1 + 2i\,\mathsf{T}, \tag{31}$$

[14] A rich variety of relationships between S and T appear in the scattering literature, and this choice influences many subsequent equations. Especial care is required since many authors don't explicitly specify their conventions. See the Appendix for alternatives.

the S and K matrices are related *by definintion* as

$$\begin{aligned} \mathsf{S} &= (1 + i\mathsf{K})\,(1 - i\mathsf{K})^{-1} \\ &= \left[(1 - \mathsf{K}^2) + 2i\mathsf{K}\right](1 + \mathsf{K}^2)^{-1}. \end{aligned} \tag{32}$$

From this we obtain the crucial relationship between K and T *appropriate to Eq. (31)*,

$$\mathsf{T} = \mathsf{K}\,(1 - i\mathsf{K})^{-1}. \tag{33}$$

Rotational and Vibrational Frame Transformations. In the BF-FNO formulation an additional step intrudes between the solution of the scattering equations and calculation of cross sections: we must transform the BF-FNO scattering matrix into the LAB frame in the CAM representation. We accomplish this by applying a rotational frame transformation (RFT).[67,68] If we chose the BF-FN approximation and also hold the internuclear separation R fixed, then we must further transform from the BF-FN representation, in which R is a parameter, to one in which channels are labelled by the vibrational quantum number v. This is accomplished by a vibrational frame transformation (VFT).[68] The physical assumptions underlying these frame transformations have been discussed and reviewed extensively and will not be reiterated here.[18,29] The RFT is a change of basis that effects both a rotation from the BODY to the LAB frame and a coupling of the target rotational and projectile orbital angular momentum a là the Clebsch-Gordan series (22). We wind up where we want to be: back in the LAB-CAM representation:

$$\underbrace{\text{eigenfunctions of } \hat{\ell}^2 \text{ and } \hat{\ell}_z}_{\text{BF-FNO}} \quad \xrightarrow[RFT]{} \quad \underbrace{\text{eigenfunctions of } \hat{J}^{\,2} \text{ and } \hat{J}_{z'}}_{\text{LAB-CAM}} \tag{34}$$

It's not hard to work out the form of the RFT matrix: we need only evaluate the matrix element of the BODY $\rightarrow$ LAB rotation operator between LAB-CAM and BF-FNO basis functions.[18] Our tools are completeness of the BODY spherical harmonics, the definition (22) of the LAB-CAM basis functions, the orthogonality of the spherical harmonics, and the standard relation for the integral of three Wigner rotation matrices.[36] After a little algebra, we get the deceptively simple result

$$A_{j\ell}^{J\Lambda} = \sqrt{\frac{2j+1}{2J+1}}\; C(j\,\ell\,J; 0\,\Lambda\,\Lambda)\,. \tag{35}$$

Knowing the properties of the Clebsch-Gordan coefficient, we're not surprised to find that this transformation is real and unitary:

$$\sum_j A_{j\ell}^{J\Lambda}\, A_{j\ell'}^{J\Lambda'} = \delta_{\ell,\ell'}\,\delta_{\Lambda,\Lambda'} \tag{36a}$$

$$\sum_\Lambda A_{j\ell}^{J\Lambda}\, A_{j'\ell'}^{J\Lambda} = \delta_{\ell,\ell'}\,\delta_{j,j'}. \tag{36b}$$

The summations over j and Λ in these relations run over all values allowed by the Clebsch-Gordan coefficients in the RFT (35). Because of its unitarity, the RFT commutes with the $\mathsf{K} \leftrightarrow \mathsf{T}$ transformation (33), so it doesn't matter in which order one performs them.

Applying the RFT in the entrance and exit channels, we obtain the fundamental relationship between the BF-FNO and LAB-CAM T matrices:[15]

$$T^{J}_{vj\ell,v_0 j_0 \ell_0} \approx \sum_{\Lambda=-\ell}^{+\ell} A^{J\Lambda}_{j\ell}\, T^{\Lambda}_{v\ell,v_0\ell_0}\, A^{J\Lambda}_{j_0\ell_0}. \tag{37}$$

Formally the sum over Λ includes both positive and negative values. But as noted above, symmetry properties of the FNO Hamiltonian render the BF-FNO T matrix independent of the sign of Λ. So we can simply include all matrix elements with $\Lambda > 0$ twice. The RFT, then, generates an *approximate* LAB-CAM T matrix, calculated under the assumption that the rotational motion of the molecule can be treated adiabatically; we can use the resulting T-matrix in equations for the desired ro-vibrational (LAB) cross sections. This insertion leads to the widely used adiabatic-nuclear-rotation (ANR) method.[52–54] Embedded in this method is the further approximation that the entrance- and exit-channel energies are equal, i.e., that the energy lost by the electron in exciting the target can be neglected compared to the initial and final rotational energies—so in effect the target states are degenerate. The assumptions of adiabatic rotational motion and target-state degeneracy are often valid; the most common exception occurs in at energies near a rotational threshold; this exception is important in scattering from H_2, whose rotational thresholds are anomalously large.[55]

A molecule's lowest vibrational threshold may be 10–1000 times its rotational threshold. So the assumption of *vibrational* target-state degeneracy is invalid over a comparably wider range of scattering energies than its rotational counterpart.[70] Nevertheless, at high enough incident energies (and for non-resonant scattering), one can use this assumption together with an adiabatic treatment of vibrations to approximate vibrational cross sections using the adiabatic-nuclear-vibration (ANV) method.[71] This approach, which is computationally far easier than a vibrationally close-coupled calculation, begins with a fixed-geometry BF-FN scattering calculations at a range of internuclear separations. From examination of the range of R values embraced by the initial and final vibrational states of interest and the variation of the overlap of the corresponding vibrational wave functions, we can choose a sensible grid for these calculations. From the resulting scattering matrices $\mathsf{T}^{\Lambda}(R)$, we can then approximate the BF-FNO scattering matrix via a vibrational frame transformation (VFT). This is a canonical transformation from the BF-FN representation, in which R is a parameter, to one in which asymptotic channels are labeled by vibrational quantum numbers:

$$T^{\Lambda}_{v\ell,v_0\ell_0} \approx \langle \varphi^{(\mathrm{v})}_{v} \mid T^{\Lambda}_{\ell,\ell_0}(R) \mid \varphi^{(\mathrm{v})}_{0} \rangle. \tag{38}$$

We then just insert the resulting approximate BF-FNO T matrix into the RFT (37) to obtain an approximate LAB-CAM ro-vibrational transition matrix for subsequent calculation of cross sections.[16]

The final step is determination of the LAB scattering amplitude $f_{\nu,\nu_0}(\hat{r}')$ in terms of which the differential and integral cross sections for the scattering process

[15] In Ref. 68 Chang and Fano introduced the RFT as a way to transform the radial scattering function matrix from the BF-FNO formulation (which is physically most appropriate near the target) to the LAB-CAM formulation (most appropriate far from the target). Because of practical difficulties involved in implementing this transformation at a non-asymptotic radius, as demonstrated in Ref. 69, the RFT and its vibrational counterpart the VFT are nearly always applied asymptotically, to the BF-FNO transition matrix.

[16] Under conditions where one must rigorously conserve energy in the entrance and exit channels (e.g., near a rotational or vibrational threshold) and hence cannot use the ANR or ANV methods.

$\nu_0 = (v_0, j_0, m_{j_0}) \to \nu = (v, j, m_j)$ are defined. This involves *uncoupling* the rotational and orbital angular momentum, then *undoing* the transformation (15) from plane-wave to angular-momentum boundary conditions, to wit:

$$f_{\nu,\nu_0}(\hat{r}') = \frac{4\pi m_e}{\hbar^2} \sum_{\gamma} c^*_{\ell m_\ell}(\mathbf{k}'_\nu)\, C(j\,\ell\,J; m_j\, m_\ell\, M_J) \times T^J_{vj\ell,v_0 j_0 \ell_0}\, C(j_0\,\ell_0\,J; m_{j_0}\, m_{\ell_0}\, M_J)\, c_{\ell_0,m_{\ell_0}}(\mathbf{k}'_0). \tag{39a}$$

Using the plane-wave to angular-momentum transformation coefficient, this becomes[17]

$$f_{\nu,\nu_0}(\hat{r}') = \frac{4\pi}{\sqrt{k_\nu k_0}} \sum_{\gamma} i^{\ell_0 - \ell}\, Y_\ell^{m_\ell}(\hat{k}'_\nu) \times C(j\,\ell\,J; m_j\, m_\ell\, M_J)\, T^J_{vj\ell,v_0 j_0 \ell_0}\, C(j_0\,\ell_0\,J; m_{j_0}\, m_{\ell_0}\, M_J)\, Y_{\ell_0}^{m_{\ell_0}\,*}(\hat{k}'_0). \tag{39b}$$

Note that in the LAB frame $\hat{k}'_\nu = \hat{r}'$, and since $\hat{k}'_0 = \hat{\mathbf{e}}_{z'}$, the transformation coefficient for the entrance-channel $c_{\ell_0,m_{\ell_0}}(\mathbf{k}'_0)$ simplifies as in Eq. (19). The summations in Eqs. (39) runs over all the quantum numbers γ required to undo these transformations, i.e.

$$\sum_{\gamma} = \sum_{\ell=0}^{\ell_{\max}} \sum_{m_\ell=-\ell}^{+\ell} \sum_{\ell_0=0}^{\ell_{\max}} \sum_{J=J_{\min}}^{J_{\max}} \sum_{M_J=-J}^{+J}, \tag{40}$$

where for given partial-wave quantum numbers, the sums over J and M_J are limited by the triangle rule enforced by the Clebsch-Gordan coefficients.

Integral Scattering Equations and their Solution

Because of the prominence in quantum collision theories of coupled radial differential equations like Eqs. (24), theorists and numerical analysts have developed a host of methods for their solution.[21,74] One that proves particularly suitable for low-energy electron-molecule scattering is based on standard techniques for transforming a differential equation and its boundary conditions into a single integral equation.[20,60,75] In the present case, we incorporate into the integral equation the boundary conditions (27) or (29) using the partial-wave Green's function

$$G_p(r, r') \equiv -\frac{1}{k_\nu} G^1_\ell(k_\nu r_<)\, G^2_\ell(k_\nu r_>), \tag{41}$$

where $r_<$ and $r_>$ are the minimum and maximum of r and r'. Then the set of integral equations equivalent to the differential equations (24) assume the form

$$u^q_{p,p_0}(r) = \delta_{p,p_0} G^1_\ell(k_\nu r) + 2 \sum_{p'} \int_0^\infty G_p(r, r')\, V^q_{p,p'}(r')\, u^q_{p',p_0}(r')\, dr'. \tag{42}$$

---

it is still possible to implement an adiabatic treatment of the rotational or vibrational motion and avoid the considerable demands of a full close-coupling calculation -- by calculating the BF-FNO K matrix off the energy-momentum shell, an idea first suggested in Ref. 72. Off-shell calculations of rotational and vibrational e–H_2 cross sections (in LAB-CAM and BF-FNO theories, respectively) have been carried out using a particular implementation of this idea, the first-order non-adiabatic (FONDA) method described in Ref. 73.

[17] The prefactor in this equation is independent of how one normalizes the asymptotic plane-wave and orbital-angular-momentum free states [see Eqs. (18)] but not of the relationship between the S and T matrix. See the Appendix for alternatives to our $S^q = 1 + 2iT^q$.

Note that the potential in the coupling matrix elements $V^q_{p,p'}(r')$ may be non-local. The single-variable Green's functions $G^1_\ell(k_\nu r)$ and $G^2_\ell(k_\nu r)$ in Eq. (41) control the behavior of $u^q_{p,p_0}(r)$ at the origin and in the asymptotic region respectively. The factor of 2 appears because here as elsewhere our potential energy is in atomic units (Hartrees, not Rydbergs).

The generic form (42) pertains to any of the formulations in Table 3 (in the BF-FN approximation the reduced radial wave function depends parametrically on R). To facilitate their further manipulation, we shall write these equations in matrix form. To this end we define diagonal Green's functions matrices whose dimensionality equals the number of channels N_p. We further define $N_p \times N_p$ matrices $\mathsf{u}(r)$ for the radial function and $\mathsf{V}(r)$ for the coupling potential and write (42) as

$$\mathsf{u}(r) = \mathsf{G}^1(r) + 2\int_0^\infty \mathsf{G}(r, r')\,\mathsf{V}(r')\,\mathsf{u}(r')\,dr'. \tag{43}$$

The elements of $\mathsf{G}^1(r)$ depend on the channel index p: for example, in the LAB-CAM theory with $(p;q) = (vj\ell; JM_J)$ and for K or T-matrix boundary conditions, its elements are [see Table 5]

$$[\mathsf{G}^1(r)]_{p,p'} = G^1_\ell(k_{vj}r) = \hat{\jmath}_\ell(k_{vj}r). \tag{44}$$

The matrix $\mathsf{G}^2(r)$ determines the asymptotic behavior of the solution of Eq. (43). As discussed above, two choices are commonplace: K-matrix and T-matrix boundary conditions. One of the beauties of the integral equations formulation is the easy of switching between these choices. If we define

$$[\mathsf{G}^2(r)]_{p,p'} = G^2_\ell(k_{vj}r) = \hat{n}_\ell(k_{vj}r) + \gamma\,\hat{\jmath}_\ell(k_{vj}r), \tag{45}$$

then we get K-matrix boundary conditions (29) for $\gamma = 0$ and T-matrix boundary conditions (28) for $\gamma = i$.

To put Eq. (43) into a form recognizable to mathematicians, we split its integral in two: one runs from 0 to r, the other from r to ∞. We can now use the definition of the partial-wave Green's function Eq. (41) to unmask the resulting sum of integrals as a *Volterra equation of the second kind*:

$$\mathsf{u}(r) = \mathsf{g}(r) + 2\mathsf{k}^{-1}\int_0^r [\mathsf{G}^1(r)\mathsf{G}^2(r') - \mathsf{G}^2(r)\mathsf{G}^1(r')]\,\mathsf{V}(r')\,\mathsf{u}(r')\,dr', \tag{46}$$

where k^{-1} is a diagonal matrix with elements k_ν^{-1} and

$$\mathsf{g}(r) \equiv \mathsf{G}^1(r) - 2\mathsf{k}^{-1}\mathsf{G}^1(r)\int_0^\infty \mathsf{G}^2(r')\,\mathsf{V}(r')\,\mathsf{u}(r')\,dr'. \tag{47}$$

Written thusly, the integral equation (43) admits a solution of the form

$$\mathsf{u}(r) = \mathsf{u}^{(0)}(r)\,(1 + \mathsf{C}), \tag{48}$$

where the "homogeneous solution" $\mathsf{u}^{(0)}(r)$ satisfies the more easily solvable equation

$$\mathsf{u}^{(0)}(r) = \mathsf{G}^1(r) + 2\mathsf{k}^{-1}\int_0^r [\mathsf{G}^1(r)\mathsf{G}^2(r') - \mathsf{G}^2(r)\mathsf{G}^1(r')]\,\mathsf{V}(r')\,\mathsf{u}^{(0)}(r')\,dr'. \tag{49}$$

As detailed elsewhere,[20] we now merely impose a conventional quadrature on this equation and straightforwardly propagate its solution from $r = 0$ (or some suitably small value) to $r_{\max}$ (some large "asymptotic" value) where we extract the scattering matrix. In so doing, it's convenient to introduce the intermediate matrix quantities

$$\mathsf{I}^1(r) = 2\mathsf{k}^{-1} \int_0^r \mathsf{G}^1(r')\, \mathsf{V}(r')\, \mathsf{u}^{(0)}(r')\, dr' \tag{50a}$$

$$\mathsf{I}^2(r) = 1 + 2\mathsf{k}^{-1} \int_0^r \mathsf{G}^2(r')\, \mathsf{V}(r')\, \mathsf{u}^{(0)}(r')\, dr'. \tag{50b}$$

This renders the homogeneous solution (49) into the transparent form

$$\mathsf{u}^{(0)}(r) = \mathsf{G}^1(r)\, \mathsf{I}^2(r) - \mathsf{G}^2(r)\, \mathsf{I}^1(r). \tag{51}$$

Although our ultimate goal is the scattering matrix, we now detour briefly to reconstruct the physical solution in terms of Eqs. (50). With the constant matrix

$$\mathsf{C} = \left[1 - \mathsf{I}^2(\infty)\right] \left[\mathsf{I}^2(\infty)\right]^{-1}, \tag{52}$$

(where in practice we evaluate the I integrals at $r_{\max}$, not ∞), we obtain from (46) an elegant result for the physical solution and its asymptotic behavior:[18]

$$\mathsf{u}(r) = \mathsf{u}^{(0)}(r) \left[\mathsf{I}^2(\infty)\right]^{-1} \xrightarrow[r\to\infty]{} \mathsf{G}^1(r) - \mathsf{G}^2(r)\, \mathsf{I}^1(\infty) \left[\mathsf{I}^2(\infty)\right]^{-1}. \tag{53}$$

Generation of the scattering matrix proceeds immediately from the asymptotic I-matrices. For example, if we adopt real boundary conditions [and the corresponding choices of $\mathsf{G}^1(r)$ and $\mathsf{G}^2(r)$ as specified in Table 5,] then we obtain the K matrix from the I matrices as

$$\mathsf{K}^q = -\mathsf{k}^{1/2}\, \mathsf{I}^1(\infty) \left[\mathsf{I}^2(\infty)\right]^{-1} \mathsf{k}^{-1/2}. \tag{54}$$

Similar equations obtain for the S and T matrices, with the appropriate choices of Green's functions. Substitution of the definitions of the I matrices and re-introduction of the physical solution $u^q_{p,p_0}(r)$ unmasks Eq. (54), to no one's surprise, as the usual integral equation for the K matrix (with our choice of continuum normalization),[19]

$$K^q_{p,p_0} = -\frac{2}{\sqrt{k_\nu k_0}} \sum_{p'} \int_0^\infty \hat{\jmath}_\ell(k_\nu r')\, V^q_{p,p'}(r')\, u^q_{p',p_0}(r')\, dr'. \tag{55}$$

---

[18] However elegant, this result is of little practical use. Unlike direct solution of the coupled differential equations via, say, the Numerov propagator used in Ref. 76, the integral-equations technique is an awkward way to generate the physical solution. In nearly all low-energy applications, numerical difficulties occasioned by the need for large-order partial waves at small radii in the wave function expansion requires intermediate matrix transformations (e.g., the stabilization procedures discussed in Ref. 63 and Ref. 77 enforce linear independence of columns of the solution matrix). In and of themselves, these difficulties are not crippling—strategies for solving them are discussed in Ref. 20. But to generate the physical solution one must save all intermediate matrix transformations and, after propagation is completed, undo them, as in the iterative static-exchange method of Ref. 47—a requirement that may exceed the storage capacity of one's computer. These problems do not attend generation of the K matrix.

[19] The channel indices on K^q_{p,p_0} are a little deceptive, because they imply an association of the second index with the entrance channel, which we're denoting by zero subscripts. At the end of all these transformations and calculations, we seek cross sections for some transition specified by initial-state quantum numbers ν_0 (contained in channel index p_0) and final-state quantum numbers ν (contained in p), so it is vital to know whether we can correctly associate the row and column indices on a given scattering matrix with the channels of the physical transformation of interest. Rigorously and in general, such an association is correct *only* for the S and T matrices. Its elements are either proportional to scattering amplitudes or, in more complicated problems like electron-molecule scattering, bear a mathematical relation to the scattering amplitude (through the unitary transformations discussed above) that preserves this association [see Eq. (39)]. But *only as the final-state electron energy approaches zero, the*

This corresponds to the boundary condition (29), which looks like

$$\mathsf{u}(r) \xrightarrow[r\to\infty]{} \mathsf{G}^1(r) + \mathsf{G}^2(r)\,\mathsf{k}^{-1/2}\,\mathsf{K}\,\mathsf{k}^{1/2}. \tag{56}$$

Readers can find in Ref. 20 details concerning calculation of the homogeneous solution by numerical quadrature as well as suggestions for how to ensure numerical accuracy at low energies, where large-order Green's functions are required.

Integral Equations Strategies in Separable Methods. The integral equations method is especially powerful in methods based on a separable approximation to the electron-molecule wave function. In the ANV and FONDA methods, for example, we can use Eqs. (38) and (37) to derive a convenient expression for the LAB-CAM K-matrix in terms of radial integrals over the BF-FN wave function. [Doing so lets us take advantage of the enormous library of well-tested computer packages for performing BF-FN scattering calculations—see, for example, Ref. 33.] In particular, we apply vibrational and rotational frame transformations to the integrals

$$I^{\Lambda}_{\ell'\lambda}(\ell,\ell_0;k,R) \equiv \int_0^{\infty} \hat{\jmath}_\ell(kr)\, v_\lambda(r,R)\, u^{\Lambda}_{\ell',\ell_0}(r;R)\, dr \tag{57}$$

to obtain an approximate LAB-CAM K matrix as

$$K^{J}_{vj\ell,v_0j_0\ell_0} = -\frac{2}{\sqrt{k_\nu k_0}} \sum_{\ell'=0}^{\ell'_{\max}} \sum_{\lambda=0}^{\lambda_{\max}} \sum_{\Lambda=-\ell}^{+\ell} A^{J\Lambda}_{j\ell}\Big[g_\lambda(\ell\ell';\Lambda)\,\langle\varphi^{(\mathrm{v})}_v \mid I^{\Lambda}_{\ell'\lambda}(\ell,\ell_0;k,R) \mid \varphi^{(\mathrm{v})}_0\rangle\Big] A^{J\Lambda}_{j_0\ell_0}. \tag{58}$$

This reactance matrix is approximate in that it treats the rotational and vibrational motion of the target adiabatically. One can use this form either in a conventional ANV calculation, making the further approximation that the exit-channel and entrance-channel wave numbers are equal $k_{vj} = k_0 = k_b$ and hence generating (57) (and implicitly the BF-FN K matrix) on the energy shell, or if this approximation is invalid, correctly conserving energy between the entrance and exit channels and thus generating the integrals (57) off shell.[72] In the FONDA method, for example,[73] the BF-FN radial function in Eq. (57) is evaluated at the entrance-channel energy $k_0^2/2$ and the free spherical function $\hat{\jmath}_\ell(k_\nu r)$ at the exit-channel energy $k_{vj}^2/2$. In the ANV method, by contrast, the "body energy" $k_b^2/2$ is the energy at which the spherical free wave in Eq. (57) is evaluated.[20]

---

limit where the First Born Approximation applies, is the K matrix actually proportional to the T matrix. So only in this limit can one apply this association to the K matrix. Near threshold the real part of the T matrix vanishes and T = K. But at higher energies, the K ↔ T transformation (33) mixes rows and columns of the K matrix (the rows corresponding to various channels p, the columns to various linearly independent solution vectors of the scattering equations). This fact is of especial importance in interpreting approximate K matrices calculated by direct integration of the product of an R-dependent K matrix and designated initial- and final-state wave function (in the ANV or FONDA theories of Ref. 71 and Ref. 73, respectively). Its practical importance depends on the extent to which those additional elements of K that are mixed into the particular element of T of interest actually affect the desired cross section. This is a numerical question best explored empirically.

[20] Because the AN approximation ignores the difference between the entrance- and exit-channel energies, it leaves unspecified the energy at which its core BF-FN scattering calculations—which yield, for example, the K matrices in Eq. (38)—are to be performed. There appears, therefore, an ambiguous "body energy" $k_b^2/2$. Most investigators take this to be the incident energy $k_0^2/2$, but others have proposed using for it the final-state electron energy $k_\nu^2/2$ or the geometric mean of these two; these options are discussed in Ref. 18 and Ref. 29. This ambiguity is resolved in off-shell treatments, which use the correct energies in the entrance and exit channels.

The Interaction Potential and its Matrix Elements

As noted above, the electron-molecule interaction potential (11), which appears in the reduced Schrödinger equation, contains three kinds of terms: electrostatic (Coulomb), exchange (from the antisymmetrization requirement), and correlation/polarization. Each of these terms depends, in general, on projectile and target radial and angular coordinates, and the last two terms are, in whole or part, non-local. The first two decay rapidly to zero outside the charge cloud; the latter continues to infinity, assuming at large r a simple asymptotic form involving permanent and induced moments. Each of these characteristics affects the scattering function, for this potential appears in coupling matrix elements whose form they influence.

Long-range Behavior. In single-center formulations we can facilitate the evaluation of angular integrals in the coupling matrix elements by expanding V_{int} in Legendre polynomials. Thus for the static potential we have[78,79]

$$V_{\text{st}}(\mathbf{r}', \mathbf{R}) = \sum_{\lambda=0}^{\lambda_{\max}} v_\lambda^{(\text{st})}(r, R)\, P_\lambda(\cos\Theta), \tag{59}$$

where Θ is the angle between the LAB vectors $\mathbf{r}'$ and $\mathbf{R}$. [In the BODY frame, with $\hat{\mathbf{e}}_z = \hat{R}$, the corresponding expansion of $V_{\text{st}}(\mathbf{r}, R)$ involves θ, the polar angle of $\mathbf{r}$.]

Table 6. Atomic-unit conversions useful for treating long-range potentials of electron-molecule scattering.

physical quantity	atomic unit	conversion to CGS units
dipole moment	ea_0	2.5418×10^{-18} esu-cm
quadrupole moment	ea_0^2	1.345×10^{-26} esu-cm^2
polarizability	a_0^3	1.4818×10^{-25} cm^3

[1] In converting permanent moments the following are useful:
$1e = 4.803250 \times 10^{-10}$ esu
$1a_0 = 0.52917716 \times 10^{-8}$ cm
1 debye $= 1 \times 10^{-18}$ esu-cm

Only the $\lambda = 0$ Legendre projection $v_\lambda^{(\text{st})}(r, R)$ defined by Eq. (59) decays exponentially to zero as r increase; all others go to zero like an inverse power of r, as

$$v_\lambda^{(\text{st})}(r, R) \xrightarrow[r\to\infty]{} -\frac{c_\lambda(R)}{r^{\lambda+1}}. \qquad (\lambda \geq 1) \tag{60}$$

(The long-range form of each static Legendre projection is negative because the projectile is negatively charged.) We use atomic units for the permanent moments $c_\lambda(R)$ and for V_{st} and $v_\lambda(r)$ (Hartrees) and so present in Table 6 useful relevant conversion factors. The point-group of the molecule controls which moments are non-zero: a heteronuclear molecule manifests permanent moments of all order, so the lowest-order moment is the dipole $[c_1(R) = d(R)]$. For a homonuclear molecule, however, inversion symmetry proscribes odd harmonics in the expansion Eq. (59), so the lowest-order permanent moment is quadrupole $[c_2(R) = q(R)]$. And so forth.

In practice, given a static potential calculated from some ground-state electronic molecular wave function, we determine the moments $c_\lambda(R)$ from the corresponding

Legendre projection at a radial value r large enough that $v_\lambda^{(\mathrm{st})}(r, R)$ has assumed its asymptotic form to the desired number of significant figures. For larger r, then, we evaluate coupling matrix elements in the scattering equations using the long-range multipole expansion defined by Eqs. (59) and (60) with the extracted moments.

A key issue in the calculation of the static matrix elements in terms of the Legendre projections $v_\lambda^{(\mathrm{st})}(r, R)$ is the maximum order λ required for convergence. The issue is *not* how to converge the static potential (59) itself; to do so might require a excessive number of terms—each term contains a cusp (at $r = R/2$), and the sum of all these cusps is the off-center singularity in V_{en} at the nuclei. By focusing instead on the number of terms required to converge the particular scattering quantity of interest (e.g., a differential cross section or an eigenphase sum) to some well-defined convergence criterion we can minimize the required number of static projections.[80] Moreover, with increasing order λ the chore of calculating these projections simplifies dramatically because the Legendre projection $v_\lambda^{(\mathrm{st})}(r, R)$ is increasingly dominated by the electron-nuclear term in the Coulomb potential Eq. (8):

$$v_\lambda^{(\mathrm{st})}(r, R) = v_\lambda^{(\mathrm{en})}(r, R) + v_\lambda^{(\mathrm{ee})}(r, R) \xrightarrow[\lambda \to \infty]{} v_\lambda^{(\mathrm{en})}(r, R). \tag{61}$$

For example, the electron-nuclear static projection for a homonuclear diatomic with charge Z on each nucleus is simply

$$v_\lambda^{(\mathrm{en})}(r, R) = -2Z \frac{\rho_<^\lambda}{\rho_>^{\lambda+1}} = -2Z \begin{cases} r^\lambda \left(\dfrac{2}{R}\right)^{\lambda+1} & r < \frac{1}{2}R \\[2ex] \dfrac{1}{r^{\lambda+1}} \left(\dfrac{R}{2}\right)^{\lambda} & r > \frac{1}{2}R \end{cases}, \tag{62}$$

where $\rho_<$ and $\rho_>$ are the minimum and maximum of r and $R/2$. Generalization of this result to other types of systems is straightforward: thus for electron scattering from CO_2, which is linear in its ground state with a carbon atom ($Z = 6$) at the origin, we simply add $-6/r$ to the $\lambda = 0$ electron-nuclear projection (62).

Vibrational Averaging of the Static Potential. The electron-nuclear cusps in the static potential—purely a consequence of the rigid-rotor (RR) approximation—force us to include in eigenfunction expansions of the electron-molecule wave function such as Eq. (21) partial waves of unphysically high order ℓ, as required by the strength of the static coupling matrix elements. These cusps therefore artificially increase the difficulty of close-coupling scattering calculations. One can obviate this difficulty and eliminate most of the error introduced by the RR approximation using a vibrational averaging procedure.[59] This simple alternative can be carried out using codes written to implement RR formulations such as the BF-FN theory, in which R is "frozen" at its equilibrium value R_{eq}. One merely replaces Legendre projection $v_\lambda^{(\mathrm{st})}(r, R_{\mathrm{eq}})$ in the static coupling matrix elements with its average over the ground vibrational state, viz.,

$$v_\lambda^{(\mathrm{st})}(r, R_{\mathrm{eq}}) \quad \longrightarrow \quad \int_0^\infty \varphi_0^{(\mathrm{v})\,*}(R)\, v_\lambda^{(\mathrm{st})}(r, R)\, \varphi_0^{(\mathrm{v})}(R)\, dR. \tag{63}$$

In solving the BF-FN (or LAB-CAM RR) scattering equations using these vibrationally averaged matrix elements, one typically requires fewer channels to attain convergence (because the averaging has "smoothed out" the cusps) and obtains more accurate cross sections (because this procedure incorporates some of the vibrational dynamics). The price is small: generating each Legendre projection on a grid of internuclear separations and performing the quadrature implied by (63) via a low-order Gauss-Hermite scheme. Depending on the system and the energy, vibrational averaging may improve the accuracy of elastic and/or rotational cross sections by as much as 10%.

The Long-range Polarization Potential. The term in the interaction potential we have designated $V_{c/p}$ represents two related physical effects. At short-range (roughly, inside the target charge cloud), this potential represents bound-free correlation effects. These are non-local many-body effects the inclusion of which remains a major hurdle to low-energy electron-molecule scattering theory.[39,81,82] Outside the charge cloud these effects sensibly diminish, and at intermediate and large radial distances, polarization effects take over. These are easily understood via a semi-classical model where they arise from distortion of the target electron charge cloud by the velocity- and position-dependent electric field set up by the charged projectile. (In principle, then, the resulting polarization potential should depend on the scattering energy; for scattering from light molecules this complication can be neglected.) In a rigorous quantum mechanical formulation, polarization effects arise from *virtual* excitations of energetically inaccessible ("closed") excited electronic target states.[83] In any case, these intermediate- and long-range effects give rise to local terms the calculation of which is much simpler than their non-local short-range counterparts.[84]

Outside the charge cloud the velocity dependence of the polarization potential is negligible (for reasonably light targets), so we can confidently treat the motion of the projectile as adiabatic, the target electrons fully relaxing in its presence.[41] This assumption facilitates a variational calculation of the polarization potential as the change (lowering) in the total energy of the electron-molecule system due to the fixed electron. To do so for a molecular-orbital representation of the target electronic state, we first generate polarized orbitals (for a particular projectile location), then calculate the difference between the expectation values of the system Hamiltonian (7) with respect to the polarized and undistorted ground-state molecular wave functions. This difference is $V_{c/p}(\mathbf{r})$ for that projectile location.[21] Repeating this process for various values of $\mathbf{r}$, we generate the polarization potential on a radial and angular grid. Far from the target (i.e., at large r), the resulting adiabatic polarization potential assumes a simple, well-known analytic form.

In the intermediate-r region the variationally determined adiabatic potential can be treated on a par with the static potential in the construction of coupling matrix elements, e.g., with an expansion analogous to Eq. (59). But with increasing r the bound-free Coulomb interactions diminish, and the distortion of the target orbitals becomes weak enough to be described via perturbation theory. In this limit the polarization potential is just the second-order perturbation-theory correction to the system energy.[85,86] At still larger radial distances, this second-order energy relaxes to an analytic expansion in *induced* multipoles, the spherical and non-spherical components of the target polarizability tensor. *For fixed internuclear separation* R, this form (in atomic units) is

$$V_{c/p}(r,R) \xrightarrow[r\to\infty]{} -\frac{\alpha_0(R)}{2r^4} - \frac{\alpha_2(R)}{2r^4} P_2(\cos\Theta). \tag{64}$$

[21] Such variational calculations require special care because the two energies being subtracted may be the same in several decimal places. Specifically, as r increases, one must ensure greater and greater accuracy in these energies in order to extract polarizabilities accurate to even two or three decimal places.

The fixed-R spherical and non-spherical components of the target polarizability tensor, $\alpha_0(R)$ and $\alpha_2(R)$ respectively, are defined in terms of the parallel and perpendicular polarizabilities as

$$\begin{aligned}\alpha_0(R) &= \frac{1}{3}\Big[\alpha_{\|}(R) + 2\alpha_{\perp}(R)\Big] \\ \alpha_2(R) &= \frac{2}{3}\Big[\alpha_{\|}(R) - \alpha_{\perp}(R)\Big]\end{aligned}. \tag{65}$$

The theoretical calculation of molecular polarizabilities (and higher-order induced moments) is an art in itself. In many situations, however, we eschew structure-theory methods for generating these target properties, preferring to ensure a consistent representation of the target in the scattering calculation. To this end we extract the induced moments from the $\lambda = 0$ and 2 Legendre projections of the adiabatic polarization potential $V_{\mathrm{c/p}}(\mathbf{r}, R)$; we base the latter on the near-Hartree-Fock electronic target wave function augmented by diffuse functions chosen to allow for induced distortion.[41] The "extraction radius" must, of course, be sufficiently large that these projections have settled down to their analytic forms to the desired accuracy.

Coupling Matrix Elements I: Vibrational Coupling. In the eigenfunction-expansion formulations we have discussed, the asymptotic channels [see Table 3] are coupled in the scattering equations (24) by matrix elements of the interaction potential V_{int} of Eq. (11) with respect to the basis functions $\{\Phi_p^q\}$ appropriate to that formulation [see Table 4]. For studies of vibrational excitation, the calculation of these matrix elements proceeds in two steps:

1. Construction of the *vibrational coupling potentials* by integrating the product of the interaction potential and pairs of vibrational wave functions over R.
2. Construction of the *angular coupling potentials* $V_{p,p'}^q(r)$ by integrating the result over the angular variables relevant to the formulation: e.g., in the LAB-CAM theory over $\hat{r}'$ and $\hat{R}$; in a BODY formulation, over $\hat{r}$.

For local terms in V_{int}, both steps are quite simple. For example, for the static interaction the simple quadrature over R required to evaluate the vibrational coupling potentials is similar to that of the vibrational averaging procedure in Eq. (63):

$$w_{v,v'}^{\lambda}(r) \equiv \langle \varphi_v^{(\mathrm{v})} \mid v_\lambda^{(\mathrm{st})} \mid \varphi_{v'}^{(\mathrm{v})} \rangle = \int_0^\infty \varphi_v^{(\mathrm{v})*}(R)\, v_\lambda^{(\mathrm{st})}(r, R)\, \varphi_{v'}^{(\mathrm{v})}(R)\, dR. \tag{66}$$

Evaluating these matrix elements becomes even simpler at large r, for here the potential is analytic. Combining the long-range polarization potential (64) with terms arising from the static potential, the interaction potential becomes

$$V_{\mathrm{int}} \xrightarrow[r\to\infty]{} -\frac{\alpha_0(R)}{2r^4} - \frac{d(R)}{r^2}P_1(\cos\theta) - \left[\frac{q(R)}{r^3} + \frac{\alpha_2(R)}{2r^4}\right] P_2(\cos\theta), \tag{67}$$

where, of course, the dipole r^{-2} term is present only for heteronuclear targets ($C_{\infty v}$ symmetry). We evaluate the vibrational coupling potential by simply replacing the fixed-R moments in this expression by their vibrational integrals: e.g., the long-range form of the $\lambda = 2$ vibrational coupling potential (in atomic units) is

$$w_{v,v'}^{2}(r) \xrightarrow[r\to\infty]{} -\frac{\langle \varphi_v^{(\mathrm{v})} \mid q(R) \mid \varphi_{v'}^{(\mathrm{v})} \rangle}{r^3} - \frac{\langle \varphi_v^{(\mathrm{v})} \mid \alpha_2(R) \mid \varphi_{v'}^{(\mathrm{v})} \rangle}{2r^4}. \tag{68}$$

As in the vibrational averaging procedure, we can evaluate these integrals by exploiting the close relationship between the vibrational wave functions and Hermite polynomials: we use Gauss-Hermite quadrature. Since the vibrational wave functions in the integrand vary smoothly with R and die off rapidly with increasing $|R - R_{\text{eq}}|$ (where R_{eq} is the minimum of the Born-Oppenheimer electronic potential), the required number of quadrature points and the range of integration is usually small.

As noted above, a key contribution to the fixed-R static Legendre projections $v_\lambda^{(\text{st})}(r, R)$ comes from the electron-nuclear potential. Not only are these functions analytic [and simple—see Eq. (62)], but they die to zero very rapidly with increasing $|r - R/2|$. So, depending on how we calculate our vibrational wave functions, we can evaluate the electron-nuclear vibrational coupling potentials either analytically or (trivially) numerically. If, for example, we obtain $\varphi_v^{(\text{v})}(R)$ by numerically solving the nuclear Schrödinger equation of the target, then we require only a simple quadrature. If, however, we approximate the vibrational wave function by the simple harmonic oscillator (SHO) functions $\varphi_v^{(SHO)}(R)$ defined below, then we can reduce these matrix elements even further. Exploiting the cusp in $v_\lambda^{(\text{en})}(r, R)$ at $R = 2r$, we change variables to $x \equiv R - R_{\text{eq}}$ in the electron-nuclear contribution and obtain

$$\begin{aligned} w_{v,v'}^{\lambda}(r) = &-2^{\lambda+2} Z\, r^{\lambda} \int_{-R_{\text{eq}}}^{2r-R_{\text{eq}}} \varphi_v^{(SHO)}(x)\,(x+R_{\text{eq}})^{-(\lambda+1)}\,\varphi_{v'}^{(SHO)}(x)\,dx \\ &- \frac{Z}{2^{\lambda-1} r^{\lambda+1}} \int_{2r-R_{\text{eq}}}^{\infty} \varphi_v^{(SHO)}(x)\,(x+R_{\text{eq}})^{\lambda}\,\varphi_{v'}^{(SHO)}(x)\,dx. \end{aligned} \tag{69}$$

We can estimate the range of the integration over x from the uncertainty $(\Delta x)_v$ inherent in the vth SHO wave function: using the symbols μ for the reduced mass of the nuclei and ω_e for the equilibrium natural frequency of the oscillator, we obtain for this uncertainty $(\Delta x)_v = \sqrt{(v+1/2)(\hbar/\mu\omega_e)}$.

Coupling Matrix Elements II: Local Terms. The second step in evaluating the coupling matrix elements $V_{p,p'}^{q}(r)$ in LAB-CAM or BF-FNO scattering equations involves angular integrals which, once performed through the magic of angular momentum algebra, need never be repeated. [To obtain the matrix elements for the RR counterparts of these theories, we simply replace the vibrational coupling potentials $w_{v,v'}^{\lambda}(r)$ in the equations that follow by their fixed-R counterparts $v_\lambda(r, R_{\text{eq}})$.] Because the BF-FNO basis functions depend only on angular variables of the projectile $\hat{r}$, the form of the BF-FNO coupling potential is especially simple:

$$V_{v\ell,v'\ell'}^{\Lambda}(r) = \sum_{\lambda=0}^{\lambda_{\text{max}}} g_\lambda(\ell\ell';\Lambda)\, w_{v,v'}^{\lambda}(r). \qquad \text{(BF-FNO)} \tag{70}$$

Here λ_{max} is the maximum order Legendre projection we must include in the expansion of the potential (59) *in order to converge the desired scattering quantity to the desired precision.* The BF-FNO angular coupling coefficient is

$$g_\lambda(\ell\ell';\Lambda) = \left[\frac{2\ell'+1}{2\ell+1}\right]^{1/2} C(\ell'\,\lambda\,\ell;\Lambda\,0)\; C(\ell'\,\lambda\,\ell;0\,0)\,. \tag{71}$$

The LAB-CAM matrix elements involve integration over angular variables of the projectile and the target nuclei and so, not surprisingly, their angular-coupling coefficients are more complicated. But the structure of the LAB-CAM matrix element is identical to that of the BF-FNO elements (70):

$$V^{J}_{vj\ell,v'j'\ell'}(r) = \sum_{\lambda=0}^{\lambda_{\max}} f_{\lambda}(j\ell, j'\ell'; J) w^{\lambda}_{v,v'}(r). \qquad \text{(LAB-CAM)} \tag{72}$$

The LAB-CAM angular coupling coefficients, often referred to as Percival-Seaton coefficients, are[58,87]

$$\begin{aligned} f_{\lambda}(j\ell, j'\ell'; J) = & \frac{(-1)^{j+j'-J}}{2\lambda+1} \left[(2j+1)(2\ell+1)(2j'+1)(2\ell'+1)\right]^{1/2} \\ & \times C(\ell\,\ell'\,\lambda; 0\,0)\; C(j\,j'\,\lambda; 0\,0)\; W(j\ell j'\ell'; J\,\lambda) \end{aligned} \tag{73}$$

where the last factor is the familiar Racah coefficient.[35]

The relationship of the matrix elements (and angular coupling coefficients) in these two formulations follows from the rotational frame transformation (35). Since this transformation effects a change of basis from the BF-FNO to the LAB-CAM descriptions, it's no surprise that

$$f_{\lambda}(j\ell, j'\ell'; J) = \sum_{\Lambda=-\ell}^{+\ell} A^{J\Lambda}_{j\ell}\, g_{\lambda}(\ell\ell'; \Lambda)\, A^{J\Lambda}_{j'\ell'} \tag{74}$$

$$V^{J}_{vj\ell,v'j'\ell'}(r) = \sum_{\Lambda=-\ell}^{+\ell} A^{J\Lambda}_{j\ell}\, V^{\Lambda}_{v\ell,v'\ell'}(r)\, A^{J\Lambda}_{j'\ell'}\,. \tag{75}$$

The Clebsch-Gordan and Racah coefficients in the LAB-CAM angular coupling coefficient induce a number of important restrictions on the sum over λ in Eq. (72). More importantly for practical applications, they control which channels $p' = (v'j'\ell')$ participate in the coupled equations. These restrictions include the two triangle relations $\Delta(j\,j'\,\lambda)$ and $\Delta(\ell\,\ell'\,\lambda)$ (these determine the values of ℓ that couple) and the restrictions that $j+j'+\lambda$ and $\ell+\ell'+\lambda$ must be even (these influence which Legendre projections participate in a particular matrix element). Finally, the LAB-CAM and BF-FNO angular coupling coefficients enforce the parity conditions[22]

$$\eta = (-1)^{j+\ell} = (-1)^{j'+\ell'} \qquad \text{(LAB-CAM)} \tag{76}$$

$$\eta = (-1)^{\ell} = (-1)^{\ell'}. \qquad \text{(BF-FNO)} \tag{77}$$

In principle the upper limits of the sums over λ in the LAB-CAM and BF-FNO matrix elements are constrained by Clebsch-Gordan coefficients to $\lambda_{\max} = 2\ell_{\max}$. This limit, in turn, controls the maximum order of partial waves in the eigenfunction expansion and hence the dimensionality of the radial wave function matrix—consideration vital for both memory and CPU-time. In many circumstances, however, the higher-order partial waves don't influence the scattering matrix (e.g., in some cases a satisfactory upper limit is $\lambda_{\max} = \ell_{\max}$). Great care is necessary if one

---

[22] We do not explicitly include the parity in the channel quantum numbers q because this information is redundant: knowing whether j and/or ℓ (only ℓ in the BF-FNO theory) is even or odd determines the parity. Nonetheless, only channels of the same parity are coupled. It is also useful to note that the "diagonal" coefficient $f_{\lambda}(j\ell, j\ell; J)$ is zero either (a) if $\lambda = 0$ or is odd, or (b) if $\lambda > 0$ but $\ell = 0$. Handy tables of Percival-Seaton coefficients for $\lambda = 2$ appear in Ref. 88.

chooses to explore this option, as convergence in partial waves may be exceedingly sensitive to the scattering energy—and is sometimes excruciatingly slow.

Coupling Matrix Elements III: The Exchange Potential. The non-local nature of the exchange term $\hat{\mathcal{V}}$ in the interaction potential (11) gives the radial scattering equations their integrodifferential character and influences the structure of the coupling matrix elements in these equations. The resulting numerical difficulties, which escalate with increasing number of electrons and non-sphericity of the interaction, are sufficiently formidable that many electron-molecule studies approximate this term with a local model potential.[29,30] The most widely used of these models approximates the exchange kernel with a free-electron-gas approximation for the target and a free-wave approximation for the continuum function.[46] Such approximations place the exchange potential on a mathematical par with the static and (local) polarization terms; all can be treated by Legendre expansion as described above. If, however, one must treat exchange effects as properly non-local, then special considerations enter the evaluation of the coupling matrix elements.

At the heart of the (generic) exchange matrix elements (25) is the kernel $\mathcal{K}^q_{p,p'}(r', r'')$. As illustrated for the BF-FNO theory in Eq. (26), this quantity is, in turn, the matrix element of the operator $\mathcal{K}(\mathbf{r}', \mathbf{r}'')$ which we defined (for a closed-shell target) in Eq. (14). The presence in this operator of the occupied molecular-orbitals (MOs) of the target means that to evaluate the exchange matrix elements (in a single-center treatment) we must expand these orbitals in the appropriate angular basis: the spherical harmonics. For example, in BODY coordinates, the (real) radial MO expansion coefficients for the ith (spatial) MO with electronic angular momentum projection Λ_i along $\hat{R}$ is

$$\xi_i^{\Lambda_i}(\mathbf{r}; R) = \frac{1}{r} \sum_{\ell'=0}^{\ell'_{\max}} \zeta_{i,\ell'}^{\Lambda_i}(r; R)\, Y_{\ell'}^{\Lambda_i}(\hat{r}). \tag{78}$$

In practice, we determine the summation limit $\ell'_{\max}$ empirically using known properties of the MO.

Inserting this expansion into the BF-FNO exchange matrix element leads to the ghastly form

$$\begin{aligned}\mathcal{K}^{\Lambda}_{v\ell, v'\ell'}(r, \bar{r}) = \sum_{i=1}^{N_{\text{occ}}} \sum_{\ell''=0}^{\ell''_{\max}} \sum_{\ell'''=0}^{\ell'''_{\max}} \sum_{\lambda=0}^{\lambda_{\max}} \langle \varphi_v^{(\text{v})} \mid \zeta_{i,\ell''}^{\Lambda_i}(\bar{r})\, \zeta_{i,\ell'''}^{\Lambda_i}(r) \mid \varphi_{v'}^{(\text{v})} \rangle \\ \times\, h_\lambda(\ell\, \ell'\, \ell''\, \ell'''; \Lambda\, \Lambda_i)\, \frac{r_<^\lambda}{r_>^{\lambda+1}},\end{aligned} \tag{79}$$

where $r_<$ and $r_>$ are the minimum and maximum of r and $\bar{r}$. The exchange coupling coefficient is

$$\begin{aligned}h_\lambda(\ell\, \ell'\, \ell''\, \ell'''; \Lambda\, \Lambda_i) = \left[\frac{(2\ell+1)(2\ell'+1)}{(2\ell''+1)(2\ell'''+1)}\right]^{1/2} C(\ell\, \lambda\, \ell''; -\Lambda, \Lambda - \Lambda_i)\; C(\ell\, \lambda\, \ell''; 0, 0) \\ \times\, C(\ell\, \lambda\, \ell'''; \Lambda, \Lambda_i - \Lambda)\; C(\ell\, \lambda\, \ell'''; 0, 0)\, .\end{aligned} \tag{80}$$

The various triangle rules spawned by the host of Clebsch-Gordan coefficients in this coefficient constrain the sums in the kernel matrix elements Eq. (79), but not much.

Vibrational Wave Functions and Their Energies

Of obvious centrality to the calculation of vibrational excitation cross sections are the vibrational wave functions of the target. As defined in Eq. (5), these functions $\varphi_v^{(\mathrm{v})}(R)$ comprise the radial part of the eigenfunctions of the nuclear Hamiltonian (6) that results from the Born-Oppenheimer separation of the molecular wave function. The corresponding eigenvalues $\epsilon_\nu = \epsilon_{vj}$, the bound stationary-state energies of the molecule in this approximation, appear in the total energy of the electron-molecule system through the energy conservation relation (10). These eigenvalues are conventionally matched to spectral data via the well-known expansion[89]

$$\epsilon_{vj} = \omega_e(v + \tfrac{1}{2}) - \omega_e x_e(v + \tfrac{1}{2})^2 + B_v j(j + 1) - D_v j^2(j + 1)^2. \tag{81}$$

The molecular constants—the natural frequency $\omega_e = \sqrt{k_e/\mu} = 2\pi\nu_e$ for force constant k_e and reduced nuclear mass $\mu \equiv (m_a m_b)/(m_a + m_b)$; anharmonicity constant $\omega_e x_e$, and rotational constants for the vth vibrational state $B_v = B_e - \alpha_e(v + 1/2)$ and D_v—can be found in numerous tabulations.[90,91]

The nuclear potential that produces these eigenvalues, the sum of the electronic energy and internuclear repulsion potential

$$V(R) \equiv \mathcal{E}_0^{(\mathrm{e})}(R) + V_{\mathrm{nn}}(R), \tag{82}$$

strikingly resembles a simple-harmonic-oscillator (SHO) potential—at least for low-lying vibrational states. This similarly, which we illustrate in Fig. 1, might tempt one to approximate the vibrational wave functions by SHO eigenfunctions.[57] Although doing so greatly simplifies evaluation of the many matrix elements that parade through these theories, it is risky. Even for electron scattering from N_2, whose deep, nearly symmetric nuclear potential seems to argue for an SHO approximation, low-energy vibrational cross sections are extremely sensitive to these eigenfunctions,[92] and cross sections obtained using the SHO approximation may differ significantly from those obtained using Morse eigenfunctions or by solving the nuclear Schrödinger equation. For e–H_2 scattering, where the potential well is weaker and more asymmetrical, SHO functions produce vibrational coupling potentials $w_{v,v'}^{\lambda}(r)$ of the wrong magnitude and, more seriously, of the wrong relative strength. In other words, the distribution of flux into various vibrational channels in the SHO approximation is wrong; not surprisingly, so are the resulting vibrational cross sections.

For most diatomic systems, one can neglect the dependence of the vibrational wave functions on the rotational quantum number j. Then the vibrational eigenvalue equation obtained by projecting the rotational state out of the nuclear Schrödinger equation simplifies considerably. It's convenient to write this equation not in terms of the eigenvalues ϵ_v themselves, which are negative and measured from the top of the nuclear potential well, but rather in terms of the positive vibrational energies $\bar{\epsilon}_v$, which are measured from the bottom of the well via the definition

$$\bar{\epsilon}_v \equiv D_e + \epsilon_v > 0, \tag{83}$$

where (by convention) $D_e > 0$. Using these quantities and the nuclear potential (82), we can write the vibrational eigenvalue equation as

$$\left[-\frac{\hbar^2}{2\mu}\frac{d^2}{dR^2} + V(R) + D_e - \bar{\epsilon}_v\right]\varphi_v^{(\mathrm{v})}(R) = 0. \tag{84}$$

Analytic Approximate Solutions. The simplest (though least accurate) analytic approximation to the solutions of Eq. (84) are the SHO eigenfunctions

$$\varphi_v^{(SHO)}(x) = N_v^{(SHO)}\, H_v(\alpha_S x)\, e^{-\alpha_S^2 x^2/2}. \tag{85}$$

Here the length scale factor α_S, written in terms of $\beta = \sqrt{\mu\omega_e/\hbar}$, is

$$\alpha_S \equiv R_{\text{eq}}\,\beta = R_{\text{eq}}\sqrt{\frac{\mu\omega_e}{\hbar}}, \tag{86}$$

and the normalization constant is

$$N_v^{(SHO)} = \left(\frac{\alpha_S^2}{\pi}\right)^{1/4} \frac{1}{\sqrt{2^v\, v!}}. \tag{87}$$

These functions satisfy the equation obtained by replacing the nuclear potential $V(R)$ in Eq. (84) by the familiar SHO potential, viz.,

$$\left[-\frac{\hbar^2}{2\mu}\frac{d^2}{dR^2} + \tfrac{1}{2}\mu\omega_e^2(R - R_{\text{eq}})^2 - \bar{\epsilon}_v\right] \varphi_v^{(\text{v})}(R) = 0. \tag{88}$$

The corresponding eigenvalues are, of course,

$$\bar{\epsilon}_v^{(SHO)} = (v + \tfrac{1}{2})\,\hbar\omega_e. \qquad (v = 0, 1, \ldots) \tag{89}$$

Considerably more accurate are the eigenfunctions obtained when the nuclear potential is approximated by the Morse potential. Lacking the symmetry of the SHO potential, the Morse potential more closely approximates the shape of the true nuclear potential $V(R)$. In terms of the dimensionless displacement variable $x \equiv (R - R_{\text{eq}})/R_{\text{eq}}$, this potential has the form

$$V^{(M)}(R) \equiv D_e\,(1 - e^{-\alpha_M x})^2 = D_e\,(1 + e^{-2\alpha_M x} - 2e^{-\alpha_M x}). \tag{90}$$

One can define the length scale parameter α_M in a variety of ways. In theoretical near-threshold studies aimed at comparison to experimental data, for example, it is essential that the vibrational energies be accurate; in such cases we can determine α_M so the Morse eigenvalues agree (to some desired precision) with experimental spectral data for as many vibrational states as possible. [The price we pay for this accuracy is that the resulting vibrational functions don't actually solve the vibrational Schrödinger equation (84) for the theoretical Born-Oppenheimer electronic energy $\mathcal{E}_0^{(\text{e})}(R)$.]

The relationship of the Morse and SHO potentials follows from the identification of the latter as the second-order term in a Taylor series expansion of the nuclear potential $V(R)$. If we expand the potential (90) about R_{eq} (i.e., about $x = 0$) as

$$V^{(M)}(x) \approx D_e\,(-1 + \alpha_M^2 x^2 + \cdots), \tag{91}$$

and identify

$$\alpha_M = \sqrt{\frac{\mu\omega_e^2 R_{\text{eq}}^2}{2D_e}}, \tag{92}$$

we obtain

$$V^{(M)}(x) \approx -D_e + \tfrac{1}{2}\mu\omega_e^2 R_{\text{eq}}^2 x^2 = -D_e + \tfrac{1}{2}\mu\omega_e^2 (R - R_{\text{eq}})^2. \tag{93}$$

Substituting this (truncated) expansion into the Schrödinger equation (84), we regain the SHO eigenvalue equation (88).

Nuts and Bolts: Morse Eigenfunctions. The Morse eigenvalue equation is also amenable to analytic solution, though not as easily as the SHO equation. We first define the intermediate variable

$$y \equiv \eta\, e^{-\alpha_M x}, \tag{94}$$

with α_M as in Eq. (92). The quantity η is yet another combination of molecular constants, $\eta \equiv 4D_e/(\omega_e \hbar)$. With the further definition $\kappa \equiv \eta - 2v - 1$, the Morse eigenfunctions are

$$\varphi_v^{(M)}(y) = N_v^{(M)}\, y^{\kappa/2}\, e^{-y/2}\, L_v^{(\kappa)}(y). \tag{95}$$

With the normalization constant

$$N_v^{(M)} = \left[\frac{\alpha_M}{R_{\text{eq}}} \frac{\kappa}{\Gamma(v+\kappa+1)}\, v!\right]^{1/2}, \tag{96}$$

the Morse eigenfunctions (95) satisfy the normalization integral[23]

$$\int_0^\infty |\varphi_v^{(M)}(R)|^2\, dR = \frac{R_{\text{eq}}}{\alpha_M} \int_0^\infty \frac{1}{y} |\varphi_v^{(M)}(y)|^2\, dy = 1. \tag{97}$$

The corresponding eigenvalues are

$$\bar{\epsilon}_v^{(M)} = D_e \left[1 - \left(1 - \frac{2(v+\frac{1}{2})}{\eta}\right)^2\right], \qquad (v = 0, 1, \ldots) \tag{98}$$

which conform to the first two terms in the experimental energies (81).

Although the form of the Morse eigenfunctions (95) is not particularly complicated, their calculation and use require some care—especially in the generation of the associated Laguerre polynomials. First, the order κ is not an integer and for moderately high-lying states may be quite large (e.g., for the vth vibrational state of N_2, we have $\kappa = 166.6 - 2v$). Under these conditions, many conventional schemes for generating Laguerre polynomials break down, become numerically inaccurate, or are simply inappropriate (e.g., recursion relations) [see Ref. 93 and Sec. 23.12 of Ref. 94]. In nearly all cases, however, one can correctly and efficiently generate these polynomials via the series

$$L_v^{(\kappa)}(y) = \sum_{n=0}^{v} \frac{(-1)^n\, (\kappa+v)!}{(v-n)!\,(\kappa+n)!\,n!}\, y^n, \tag{99}$$

a useful alternative form of which is

$$L_v^{(\kappa)}(y) = \frac{1}{v!}\left\{\sum_{m=0}^{v-1} \binom{v}{m} (-y)^m \prod_{i=m+1}^{v} (k+i) + (-y)^v\right\}. \tag{100}$$

[23] Strictly speaking this integral runs from $y = 0$ to an upper limit $y_0 = \eta e^{\alpha_M}$. But this upper limit is so large that the integrand is effectively zero for $y > y_0$ (e.g., for N_2, we have $y_0 = 3120.0 >> 1$). Hence we can extend the upper limit to ∞ with impunity.

If all else fails we can always solve the Laguerre differential equation numerically.

A second problem arises because the Gamma function in the normalization constant (96) also bears a large real non-integer argument $\eta - v$, which may render evaluation via the usual series expansion inaccurate or intolerably CPU-intensive. To dodge this difficulty we can reduce the magnitude of the argument by writing

$$\Gamma(\eta - v) = \Gamma(n + d) = \Gamma(d + 1)\left\{\prod_{i=1}^{n-1}(d + i)\right\}, \tag{101}$$

where the integer n and the quantity $d > 0$ are related by $d = \eta - v - n$. [One must be judicious in choosing n so that the argument of $\Gamma(d + 1)$ is sufficiently small to allow evaluation via series expansion and the number of terms in the product does not become excessive.]

Finally we note a potentially rich source of confusion in the widespread use of two different ways of writing the Laguerre differential equation and defining the associated polynomials [see p. 849 of Ref. 95]. We have used the following relationship for these polynomials in terms of confluent hypergeometric function:

$$L_v^{(\kappa)}(y) = \binom{v+\kappa}{\kappa}\, {}_1F_1(-v, \kappa + 1; y). \tag{102}$$

The function thus defined satisfies the Laguerre differential equation in the form

$$\left[y\frac{d^2}{dy^2} + (\kappa + 1 - y)\frac{d}{dy} + v\right] L_v^{(\kappa)}(y) = 0. \tag{103}$$

This is the convention used, for example, by `Mathematica` (W91).

Solution of the Vibrational Schrödinger Equation by Numerical Quadrature. If high accuracy of the theoretical threshold energies is not a high priority, then one may prefer to ensure internal consistency in the scattering calculation by determining vibrational wave functions as eigenfunctions for the theoretical electronic energy $\mathcal{E}_0^{(e)}(R)$. This choice guarantees that the same representation of the target electronic state generates the vibrational wave functions and the interaction potential. Since $\mathcal{E}_0^{(e)}(R)$ is known only for a discrete mesh of internuclear separations, one must solve the vibrational Schrödinger equation Eq. (84) numerically.

To efficiently solve this equation, we first fit the nuclear potential $\mathcal{E}_0^{(e)}(R) + V_{nn}(R)$ to one of the many well-known analytic parameterized functions[97,98] such as the Simmons-Parr-Finlan-Dunham (SPDF) form.[99,100] Written in terms of the dimensionless length variable $x = (R - R_{eq})/R_{eq}$ and N_p fitting parameters b_i, this form is

$$V^{(SPDF)}(x) \equiv b_0\, x^2 \left(1 + \sum_{i=1}^{i_{max}} b_i\, x^i\right) - V(R). \tag{104}$$

In practice we determine the number and values of the fitting parameters b_i (and hence the order of the polynomial) empirically.

We can easily solve the vibrational Schrödinger equation with the nuclear potential replaced by Eq. (104) via the linear variational method with a basis of SHO eigenfunctions (85),

$$\varphi_v^{(v)}(x) = \sum_{i=0}^{N_b} c_{iv}\, \varphi_i^{(SHO)}(R). \tag{105}$$

In setting up this calculation it's important to keep in mind our ultimate goal: using the functions (105) to calculate the vibrational coupling potentials $w^{\lambda}_{v,v'}(r)$ of Eq. (66). That is, we will be integrating a smoothly varying integrand over a region of R delimited by the overlap of two vibrational wave functions, $\varphi^{(\mathrm{v})}_{v}(R)\varphi^{(\mathrm{v})}_{v'}(R)$. We can again exploit the relationship between these functions and Hermite polynomials by executing the linear variational solution on a grid of internuclear separations R chosen to correspond to the points of Gauss-Hermite quadrature.[24]

Sensitivity: A Final Demonstration. Throughout this section we've emphasized the acute sensitivity of vibrational cross sections to the choice of vibrational wave functions used to calculate the coupling matrix elements. We therefore conclude this subsection with a couple of examples that reinforce this point. For H_2 the SHO approximation to the vibrational wave functions [Fig. 5a] look deceptively like the more accurate Morse wave functions. But Fig. 5b reveals that the $0 \to 1$ vibrational cross section calculated with SHO functions is quite inaccurate.

This sensitivity to vibrational wave functions is heightened for scattering near a resonance. To illustrate, we show in Fig. 6b the resonant (Π_g) e–N_2 cross section in the energy region of the low-lying shape resonance. The six-vibrational-state cross sections in this figure calculated using Morse and SHO eigenfunctions demonstrate the sensitivity to the vibrational target function (note especially the difference in the lowest two peaks in these two curves). To provide a benchmark, we also show converged cross sections using $N_v = 15$ Morse functions.

Integrated and Differential Cross Sections

In principle the final step in an electron-molecule scattering calculation—generation of the differential cross sections (DCSs)—should be trivial. Having solved the Schrödinger equation for the transition matrix, we convert it to the LAB scattering amplitude via Eq. (39) and calculate

$$\left.\frac{d\sigma}{d\Omega}\right|_{v_0 j_0 m_{j_0} \to v j m_j} \equiv \frac{k_{vj}}{k_0} \left| f_{vjm_j, v_0 j_0 m_{j_0}}(\hat{r}') \right|^2, \tag{106}$$

where we note that in the LAB frame, $\hat{r}' = \hat{k}'_{vj}$. Since most current experiments don't discriminate rotational magnetic sublevels,[11] we average over initial and sum over final magnetic rotational quantum numbers to obtain

$$\left.\frac{d\sigma}{d\Omega}\right|_{v_0 j_0 \to vj} = \frac{1}{2j_0 + 1} \frac{k_{vj}}{k_0} \sum_{m_{j_0}=-j_0}^{j_0} \sum_{m_j=-j}^{+j} \left| f_{vjm_j, v_0 j_0 m_{j_0}}(\hat{r}') \right|^2. \tag{107}$$

The integrated cross section (ICS) for this excitation, then, is just the integral over the scattering angle of this DCS. In practice, however, calculating these cross sections is nowhere near this trivial.

[24] A practical problem arises in scattering calculations that involve more than a few vibrational states. As v increases, so do the extent of $\varphi^{(\mathrm{v})}_{v}(R)$ and the number of oscillations this function executes before dying away to negligibility. The evaluation of matrix elements involving such functions may, therefore, require a dense, rather large integration mesh. We certainly don't want to solve the molecular structure problem [as required to compute $\mathcal{E}^{(\mathrm{c})}_0(R)$ for any R] on such a dense mesh! The obvious gambit—interpolation using a conventional scheme such as the cubic-spline method—resolves this conundrum except for large and small values of R. If one must use extrapolation procedures to allow for points in the integration mesh outside those values of R at which the electronic energy has been calculated, great care is in order, for most such procedures turn out to be unstable for nuclear potentials.

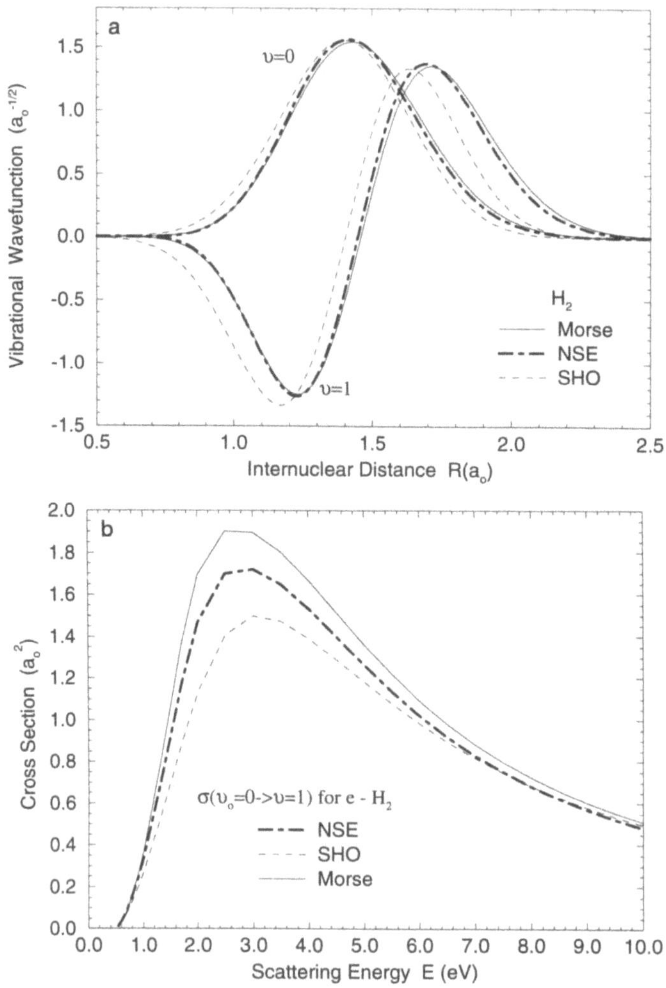

Figure 5. Sensitivity of the e–H_2 cross section for the $0 \to 1$ vibrational excitation to the vibrational target functions. In (a) we compare H_2 reduced vibrational functions from two analytic approximations—the Morse eigenfunctions (95) and SHO eigenfunctions Eq. (85) with parameters chosen to match the vibrational spectrum—and the solution of the vibrational Schrödinger equation (84) based on a near-Hartree-Fock electronic energy with $R_{eq} = 1.4011\, a_0$.] In (b) we show the corresponding cross sections as calculated in converged (4-state) BF-FNO scattering calculations that differ only in the vibrational wave function.

Readers who have stuck with us this far can probably guess the sources of difficulty. First, the number of summations in the expression for the DCS in terms of T matrix elements proliferates wildly because of the transformations we must undo to get from the LAB-CAM T matrix back to the LAB scattering amplitude that goes in Eq. (106). If our starting point is the BF-FNO T matrix, we must also perform a rotational frame transformation (RFT)—still more summations. Second, the number of terms generated by these sums may, depending on the system and scattering energy, be quite large. Although the number of total angular momentum states (J) that

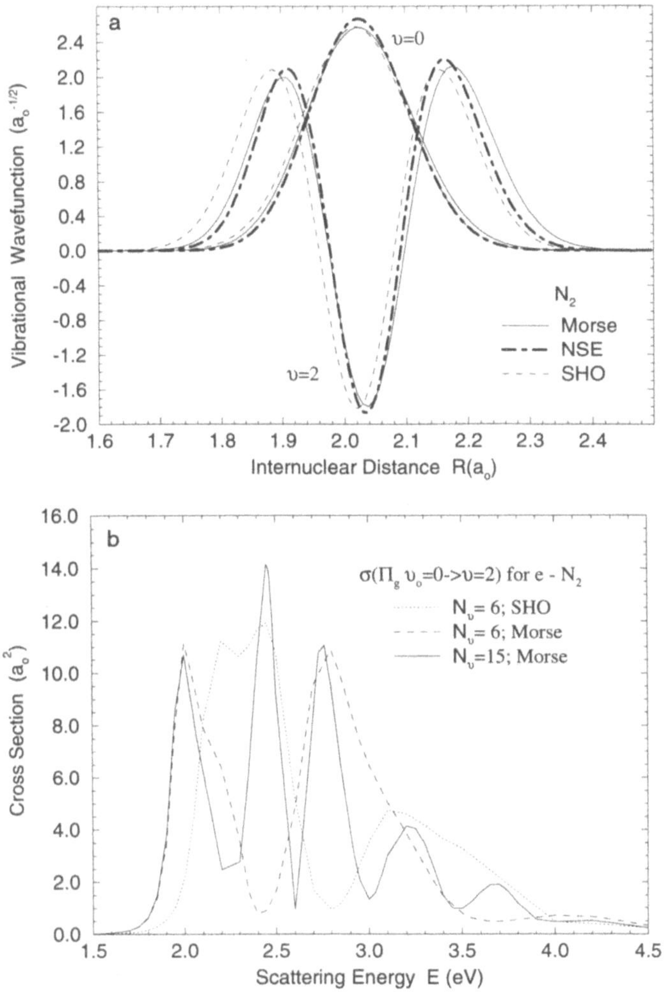

Figure 6. Sensitivity of the e–N_2 cross section for the $0 \to 2$ vibrational excitation near the 2.39 eV shape resonance to the vibrational wave function. In (a) we compare vibrational wavefunctions calculated as in Fig. 5a. In (b) we show the Π_g resonant cross sections as calculated with SHO and Morse wave functions.

contribute to electron-molecule cross sections is nowhere near that of heavy-particle cross sections, the number of partial waves (and in some cases, rotational states) required to converge a given DCS may be huge. Fortunately, methods exist for coping with both difficulties.

LAB-CAM Integrated Cross Sections. Integration over scattering angles wipes out many of these summations, so expressions for the ICS are simpler than those for the DCS. In both LAB and BODY formulations it is useful to "deconstruct" the ICS according to the appropriate constant of the motion: J in the LAB-CAM theory and Λ in a BODY theory. For example, we can write the LAB-CAM ICS in terms of cross sections partial in J as

$$\sigma_{v_0 j_0 \to v j} = \sum_{J=0}^{\infty} \sigma^{J}_{v_0 j_0 \to v j}, \tag{108}$$

where the partial cross sections in terms of LAB-CAM T-matrix elements are

$$\sigma^{J}_{v_0 j_0 \to v j} \equiv \frac{\pi}{k_0^2 (2j_0 + 1)} \sum_{\ell_0=|J-j_0|}^{J+j_0} \sum_{\ell=|J-j|}^{J+j} (2J+1) \left|T^{J}_{vj\ell, v_0 j_0 \ell_0}\right|^2. \tag{109}$$

Typically only a few of these partial cross sections need be included to converge the ICS, because of the very small magnitude characteristic of T-matrix elements for large partial-wave orders ℓ (which, due to the aforementioned Clebsch-Gordan triangle rules, tend to accompany large J). If high-J contributions are needed for convergence to high precision, they can accurately and efficiently be calculated using the Born Approximation [useful FBA equations appear in appendixes to Ref. 55, Ref. 101, Ref. 102, Ref. 63, and Ref. 103]. The decomposition (108), along with the various triangle and parity rules imposed by the LAB-CAM angular coupling coefficients (73), is a powerful source of insight into the often intricate issue of how the various constituents of the interaction potential in various regions of space control the cross section. [One can usually make similar deductions in the context of a BF-FNO calculation by first using the RFT (37) to transform to the LAB-CAM representation.] To illustrate, let us consider the relationship between the order λ of the Legendre projections of the potential [see Eq. (59)] and the total angular momentum quantum number J in Eq. (109).

The allowed values of J are determined by the Clebsch-Gordan coefficients in the definition of the LAB-CAM basis functions (22), which couples the target rotational and projectile orbital angular momenta. In particular, these coefficients spawn two key triangle rules: $\triangle(\ell_0\, j_0\, J)$ and $\triangle(\ell\, j\, J)$. To see how to use these rules, let's consider the (rotationally) elastic ($j_0 = 0 \to j = 0$) cross section.[55] Examination of the partial elastic cross sections (109) shows the dominant contribution to be that of $J = 0$. The most important contribution to this partial cross section is that of the diagonal $s \to s$ ($\ell_0 = 0 \to \ell = 0$) T-matrix element. This element, in turn, is wholly controlled by the $\lambda = 0$ (spherical) Legendre projection of the interaction potential. The off-diagonal elements such as $s \to d$, which contribute little to the elastic cross section, introduce effects due to $\lambda = 2$ projections.

These off-diagonal elements are crucial to rotational excitation. For example, at low energies the ICS for the $j_0 = 0 \to j = 2$ excitation is dominated by the $J = 1$ partial cross section, which in turn arises almost entirely from the $p \to p$ T-matrix element. This matrix element, as noted, brings into play both $\lambda = 0$ and 2 projections of the potential. This insight explains, for example, why the first Born approximation accurately predicts rotational cross sections at energies very near threshold.[104,105] It further explains the threshold behavior for these cross sections, a key diagnostic tool:[18] as the incident energy decreases to threshold for this excitation, the corresponding exit-channel energy $k_2^2/2 = E - \epsilon_2$ goes to zero. Since scattering in the exit channel is predominantly s wave, the rotational cross section must goes to zero at threshold linearly with k_j. With increasing energy, p waves become dominant, and the cross section exhibits k_j^3 behavior. Such threshold laws are useful for extrapolating calculated (or measured) cross sections downward in energy and for checking results of scattering calculations in the difficult, problematic very-low-energy region.

We shall conclude these remarks on ICS with the *elastic* momentum-transfer cross section. At low energies this cross section is of singular importance because of its central role in the analysis of transport quantities measured in swarm experiments, a primary resource for accurate determination of very-low-energy collision data.[106] (Swarm analysis actually determines the "grand-total momentum transfer cross section" that arises from elastic and all inelastic scattering. At very low energies, however, this quantity is entirely or primarily due to elastic scattering.) This cross section is the weighted integral over the (LAB) unit sphere of the elastic DCS

$$\sigma_{el}^{(m)} = \int \left.\frac{d\sigma}{d\Omega}\right|_{el} (1 - \cos\theta') \sin\theta' d\theta' d\phi'. \tag{110}$$

It can handily be expressed in terms of the LAB-CAM T matrix as

$$\sigma_{el}^{(m)} = \frac{\pi}{k_0^2} \sum_{J=0}^{\infty} \Big[(2J+1) |T_{00J,00J}^{J}|^2 - (J+1)\big(T_{00J,00J}^{J*} T_{00J+1,00J+1}^{J+1} + T_{00J,00J}^{J} T_{00J+1,00J+1}^{J+1*}\big)\Big] . \tag{111}$$

Integral Cross Sections from the BF-FNO T Matrix. Although the BF-FNO transition matrix is an additional step removed from LAB-frame cross sections, generation of ICS for elastic scattering or vibrational excitation from this matrix turns out to be quite simple.[25] We can transform the T matrix from the BODY to the LAB frame using the rotation matrices (4*b*) in the entrance and exit channels, then average the result over all molecular orientations $\hat{R}$ in the LAB frame to obtain

$$\sigma_{v_0 \to v} = \sum_{\Lambda=-\ell}^{+\ell} \sigma_{v_0 \to v}^{\Lambda}. \tag{112}$$

Here the BF-FNO cross section partial in Λ is

$$\sigma_{v_0 \to v}^{\Lambda} \equiv \frac{\pi}{k_0^2} \sum_{\ell=0}^{\ell_{\max}} \sum_{\ell_0=0}^{\ell_{\max}} |T_{v\ell,v_0\ell_0}^{\Lambda}|^2. \tag{113}$$

In principle, the sum over Λ in Eq. (112) includes an infinite number of terms. In reality, the magnitudes of the partial cross sections $\sigma_{v_0 \to v}^{\Lambda}$ decrease very rapidly with increasing Λ, so we need include only a small finite number of terms to converge $\sigma_{v_0 \to v}$. As usual, the summations in Eq. (113) are constrained by a triangle rule, here $\Delta(\ell\, \ell_0\, \Lambda)$. For homonuclear molecules the T matrix is block diagonal in the parity as well as in Λ, so these sums run over only even or odd ℓ, subject to the usual constraint $\ell \geq |\Lambda|$ and the requirement that all terms with $\Lambda > 0$ must be counted twice.

The partial BF-FNO cross sections defined in Eq. (112) are powerful tools for physical interpretation. They are designated by their electron-molecule symmetry—a convention that mirrors the separation of the coupled scattering equations into independent sets according to the irreducible representations of the molecular point group of the target. The first four symmetry designations are

$$\Sigma_g \quad \Lambda = 0 \quad \ell = 0, 2, 4, \ldots \tag{114a}$$
$$\Sigma_u \quad \Lambda = 0 \quad \ell = 1, 3, 5, \ldots \tag{114b}$$
$$\Pi_u \quad \Lambda = 1 \quad \ell = 1, 3, 5, \ldots \tag{114c}$$
$$\Pi_g \quad \Lambda = 1 \quad \ell = 2, 4, 6, \ldots \tag{114d}$$

[25] If one has used the ANV or FONDA methods instead of full vibrational close-coupling to generate approximate vibrational transition matrices, the equations of this subsection apply unchanged. If one has made the rigid-rotor (RR) approximation and so has in hand BF-FN T matrices for $R = R_{eq}$, the equations still apply; just erase the vibrational subscripts from them.

For heteronuclear systems, similar nomenclature applies; but since these systems lack inversion symmetry, their partial cross sections are not labeled with gerade or ungerade subscripts. The decomposition Eq. (112) is therefore the analog for electron-molecule scattering of the familiar partial-wave decomposition of a cross section for scattering from a spherical potential into s-wave, p-wave, d-wave, . . . constituents.[50] Within the FNO approximation this separation into electron-molecule symmetries is exact. Having noted this analogy, we must issue a caution: the problem at hand is *not* scattering from a spherical potential. So we must not inadvertently apply to this problem intuition we have derived from the far more familiar spherical-potential context, in which one can correctly identify independent partial-wave cross sections that correspond to orbital angular momentum eigenstates of the projectile. Such intuition may be completely erroneous for electron-molecule scattering (especially in the analysis of differential cross sections). Particularly for inelastic scattering, the coupling of orbital angular momenta and of target states complicates severely the simple potential-scattering picture, often rendering it useless or worse.

It is also important to appreciate the approximation lurking within Eqs. (112)–(113). Formally, the quantity $\sigma_{v_0\to v}$ there defined equals *the sum over all open channels of rotational cross sections calculated within the adiabatic-nuclear-rotation (ANR) approximation* for the given vibrational excitation $v_0 \to v$:

$$\sigma_{v_0\to v} = \sum_j \sigma^{\text{ANR}}_{v_0 j_0 \to v j}. \tag{115}$$

We can easily calculate the individual ANR rotational cross sections by applying the rotational frame transformation (35) to the BF-FNO T matrix and stuffing the result into Eq. (109). To the extent that the ANR theory is valid, the right-hand-side of Eq. (115) is independent of the initial rotational quantum number j_0—a useful diagnostic fact. Note, however, that this sum rule does not hold exactly for rotational cross sections calculated using variants the ANR approach designed to improve their behavior near threshold, as we shall discuss below.

As Eq. (113) shows, we can further decompose the partial BF-FNO cross sections $\sigma^{\Lambda}_{v_0\to v}$ in terms of particular pairs of orbital-angular-momentum quantum numbers, ℓ_0 for the entrance channel and ℓ for the exit channel. This facilitates an interpretive strategy akin to that for the LAB partial cross sections $\sigma^{J}_{v_0 j_0 \to v j}$ of Eq. (108). In fact, more information inheres in the BF-FNO partial cross sections than in the full ICS $\sigma_{v_0\to v}$, because, in effect, each electron-molecule symmetry "samples" different radial and angular regions of the interaction potential. So we may be able to use the partial cross sections $\sigma^{\Lambda}_{v_0\to v}$ as "theoretical detectors" (with perfect energy resolution!) to explore the potential.

Such analysis is possible, for instance, when a single symmetry overwhelmingly dominates the scattering in some energy range. This happens for the $0 \to 1$ vibrational excitation of H_2, which at below about 1 eV is dominated by the Σ_g partial cross sections and from 1 to several eV by the Σ_u cross section. So here we can, for example, meaningfully compare partial cross sections from calculations that treat a constituent of the potential via various models.

Even better, at very low energies the BF-FNO elastic e–H_2 cross section is overwhelmingly dominated by s-waves in the entrance and exit channels, all other elements of the radial wave function having exponentially decayed to irrelevance because of the strong centrifugal barriers for $\ell > 0$ [see Eq. (24)]. Here we can to an excellent approximation evaluate the elastic cross section from a single Σ_g T-matrix element as

$$\sigma_{\rm el} \approx \frac{\pi}{k_0^2} \left| T^0_{00,00} \right|^2. \qquad (\text{near } k_0 = 0) \tag{116}$$

Such happy simplifications, alas, are the exception rather than the rule.

Differential Cross Sections from the LAB-CAM T Matrix. Differential cross sections (DCS) are wonderful tools for understanding collisions and for comparing to experimental results, but they are a pain to calculate. If we insert the LAB amplitude $f_{\nu,\nu_0}(\hat{r}')$ of Eq. (39) into the definitions (106) and (107) for the ro-vibrational DCS, we obtain the innocuous looking result

$$\left.\frac{d\sigma}{d\Omega}\right|_{v_0 j_0 \to vj} = \frac{1}{4k_0^2} \sum_{L=0}^{L_{\rm max}} B_L(v_0 j_0 \to vj)\, P_L(\cos\theta'). \tag{117}$$

The nature of the difficulty becomes apparent when we look closely at the angle-independent coefficients in this form,[58]

$$\begin{aligned} B_L(v_0 j_0 \to vj) = \frac{(-1)^{j-j_0}}{2j_0+1} & \sum_{J_1=0}^{\infty} \sum_{J_2=0}^{\infty} \sum_{\ell_1=|J_1-j_0|}^{J_1+j_0} \sum_{\ell_2=|J_2-j_0|}^{J_2+j_0} \sum_{\ell_1'=|J_1-j|}^{J_1+j} \sum_{\ell_2'=|J_2-j|}^{J_2+j} \\ & \times Z(\ell_1 J_1, \ell_2 J_2; j_0\, L)\, Z(\ell_1' J_1, \ell_2' J_2; j\, L)\, T^{J_1 *}_{vj\ell_1', v_0 j_0 \ell_1}\, T^{J_2}_{vj\ell_2', v_0 j_0 \ell_2}, \end{aligned} \tag{118}$$

where algebraic coefficients in terms of Clebsch-Gordan and Racah coefficients are

$$\begin{aligned} Z(ab, cd; ef) \equiv & (-1)^{(f-a+c)/2} \left[(2a+1)(2b+1)(2c+1)(2d+1)\right]^{1/2} \\ & \times C(a\,c\,f; 0\,0\,0)\, W(abcd; ef). \end{aligned} \tag{119}$$

The only good news is that the many angular momentum coupling coefficients parading through this expression constrain the sums over partial wave orders in Eq. (118) to

$$\max\{\, |J - j_0|, |J - j|\,\} \le \ell, \ell_0 \le \min\{\, J + j_0, J + j\}. \tag{120}$$

In addition, Eq. (118) calls for a two-fold infinite summation over the total angular momentum quantum number J. At low energies, the contribution from each J may be quite small, and a huge number of terms may be needed to converge the DCS. (Similar problems afflict calculation of the DCS from a BF-FNO T matrix, where the offending sums are over Λ rather than J.) In practice, therefore, it is almost always preferable to introduce yet another basis transformation. This time we shall recouple the rotational and orbital angular momenta so as to permit judicious use of the First Born Approximation to complete various sums in the resulting DCS.[55] Doing so doesn't simplify the DCS equations much, but it does save CPU time and decrease numerical error in their evaluation.

Angular Momentum Recoupling. The end product of the angular-momentum recoupling transformation,[107] is a LAB transition matrix in a basis associated not with the total angular momentum $\hat{\mathbf{J}}$ but rather with the angular momentum $\hat{\iota}_t$ transferred from the orbital to the rotational degrees of freedom during the collision. In this basis $\hat{\ell}_t^{\,2}$ and $\hat{\ell}_{t_{z'}}$ are constants of the motion, so the ℓ_t-reduced T matrix connects asymptotic states with the same quantum numbers ℓ_t and m_t:

$$T^{\ell_t}_{vj\ell, v_0 j_0 \ell_0} \equiv \sum_{J=J_{\rm min}}^{J_{\rm max}} (-1)^J (2J+1) \begin{Bmatrix} j_0 & j & \ell_t \\ \ell & \ell_0 & J \end{Bmatrix} T^J_{vj\ell, v_0 j_0 \ell_0}. \tag{121}$$

For convenience we have used Wigner $6-j$ rather than Racah oefficients,

$$\begin{Bmatrix} j_1 & j_2 & j_{12} \\ j_3 & j & j_{23} \end{Bmatrix} \equiv (-1)^{j_1+j_2+j_3+j}\, W(j_1 j_2 j j_3;\, j_{12} j_{23}). \tag{122}$$

Now, since $\hat{\mathbf{J}}$ is conserved in the collision, we can write it in terms of constituent angular-momentum operators for the entrance and exit channels as

$$\hat{\mathbf{J}} = \hat{\ell}_0 + \hat{\mathbf{j}}_0 = \hat{\ell} + \hat{\mathbf{j}}\,. \tag{123}$$

Similarly, we write $\hat{\ell}_t$, which is also a constant of the motion, as

$$\hat{\ell}_t = \hat{\ell}_0 - \hat{\ell} = \hat{\mathbf{j}} - \hat{\mathbf{j}}_0\,. \tag{124}$$

One advantage to angular-momentum recoupling is that the sums over ℓ_t that result when we recast the DCS in terms of the transition matrix defined in (121), unlike those over J in the conventional LAB-CAM expression (118), are limited; their finite range is

$$\max\{\, |\ell - \ell_0|, |j - j_0|\,\} \le \ell_t \le \min\{\, \ell + \ell_0, j + j_0\,\}. \tag{125}$$

A second advantage appears when we express the LAB-CAM expansion coefficient in Eq. (117) in terms of the RCAM T matrix as

$$\begin{aligned} B_L(v_0 j_0 \to v j) = (-1)^L (2L+1) \sum_{\ell_t=0}^{\ell_{t\max}} \sum_{\ell_1=0}^{\infty} \sum_{\ell_2=0}^{\infty} \sum_{\ell'_1=|L-\ell_1|}^{L+\ell_1} \sum_{\ell'_2=|L-\ell_2|}^{L+\ell_2} b_{\ell_t}(\ell_1\ell_2, \ell'_1\ell'_2; L) \\ \times T^{\ell_t}_{vj\ell_1, v_0 j_0 \ell_2} T^{\ell_t *}_{vj\ell'_1, v_0 j_0 \ell'_2}. \end{aligned} \tag{126}$$

At first glance, the complexity of the new RCAM coefficient inspires little confidence:

$$\begin{aligned} b_{\ell_t}(\ell_1\ell_2, \ell'_1\ell'_2; L) \equiv (-1)^{\ell_t} i^{\ell_1-\ell_2-\ell'_1+\ell'_2} (2\ell_t+1) \left[(2\ell_1+1)(2\ell_2+1)(2\ell'_1+1)\right]^{1/2} \\ \times \left[(2\ell'_2+1)\right]^{1/2} \begin{pmatrix} \ell_1 & \ell_2 & L \\ 0 & 0 & 0 \end{pmatrix} \begin{pmatrix} \ell'_1 & \ell'_2 & L \\ 0 & 0 & 0 \end{pmatrix} \begin{Bmatrix} \ell_1 & \ell'_1 & L \\ \ell'_2 & \ell_2 & \ell_t \end{Bmatrix}. \end{aligned} \tag{127}$$

Again for convenience, we have here used Wigner 3j symbols

$$\begin{pmatrix} j_1 & j_2 & j_3 \\ m_1 & m_2 & m_3 \end{pmatrix} \equiv (-1)^{j_1-j_2-m_3} (2j_3+1)^{-1/2}\, C(j_1\, j_2\, j_3; m_1\, m_2\, -m_3)\,. \tag{128}$$

But in fact, recoupling has seized the advantage. Instead of confronting the two *infinite* sums over J in the LAB-CAM DCS, we face only the single *finite* sum in Eq. (121)—a far more desirable situation, since each J-term requires us to solve a set of coupled LAB-CAM radial equations. But (by the law of conservation of nastiness) we should expect two infinite sums to appear elsewhere in the evaluation of the DCS, and sure enough they are lurking in Eq. (126)—the sums over partial-wave orders ℓ_1 and ℓ_2. But these sums are somewhat more tolerable than those over J in Eq. (118), because we can easily determine a maximum partial-wave order ℓ_B beyond which the necessary T-matrix elements are given trivially and to high accuracy by the First Born Approximation. In any case, all these summations are constrained by still more triangle rules:

$$\triangle(\ell_1\,\ell'_1\,\lambda), \quad \triangle(\ell_2\,\ell'_2\,\lambda), \quad \triangle(\ell'_1\,\ell'_2\,\ell_t), \quad \triangle(\ell_1\,\ell_2\,\ell_t), \quad \triangle(j_0\,j\,\ell_t). \tag{129}$$

Differential Cross Sections from the BF-FNO T Matrix. The equations presented thus far for the ro-vibrational DCS in terms of the LAB-CAM T matrix are all applicable in the BF-FNO formulation since, as we remarked in discussing Eq. (115), the rotational frame transformation (RFT) handily converts the BF-FNO T matrix into an approximate LAB-CAM matrix; this can then be processed without further ado through Eqs. (117)—(119) or, more efficiently, through Eqs. (121)—(127) to generate approximate (ANR) cross sections.[61] (Alternatively we could incorporate the RFT directly into equations for the DCS in terms of the BF-FNO T matrix.)[26]

In comparing theoretical and experimental results we are usually less interested in the individual ro-vibrational DCS than in their sum, the DCS for a vibrational excitation $v_0 \to v$. In this case we can bypass the procedure sketched above and instead proceed directly from the BF-FNO reactance matrix to the vibrational DCS. The result—the differential counterpart of Eq. (115) for the ICS, is again approximate; formally it equals the sum of ANR ro-vibrational DCS for all open rotational channels.

We start with the relationship between the T matrix and the scattering amplitude *in the BF-FNO formulation*. This amplitude, which incorporates the FNO approximation, depends on the angle between the incident and outgoing wave vectors expressed *in the BODY frame* as[27]

$$f_{v,v_0}(\hat{k}_v, \hat{k}_0) = \frac{4\pi}{\sqrt{k_v k_0}} \sum_{\ell=0}^{\ell_{\max}} \sum_{\ell_0=0}^{\ell_{\max}} \sum_{\Lambda=-\ell}^{+\ell} i^{\ell_0-\ell}\, Y_\ell^\Lambda(\hat{k}_v)\, T^\Lambda_{v\ell,v_0\ell_0} Y^{\Lambda\,*}_{\ell_0}(\hat{k}_0). \tag{130}$$

We now rotate this quantity into the LAB frame using Eq. (4*b*) then average the result over all molecular orientations. For a diatomic target, this becomes an average over the unit sphere, viz.,[28]

$$\left.\frac{d\sigma}{d\Omega}\right|_{v_0\to v} \equiv \frac{1}{4\pi}\frac{k_v}{k_0}\int \left|f_{v,v_0}(\hat{r}')\right|^2 d\hat{R}. \tag{131}$$

We can develop a convenient expression for the FNO vibrational DCS by exploiting the combination properties of the Wigner rotation matrices[36] to reduce the integral over $d\hat{R}$ in this expression to an orthogonality relation for these matrices. By further combining spherical harmonics with respect to $\hat{r}'$ in the result, we can produce still more orthogonality relations. These let us subdue several of the remaining

---

[26] As in the LAB-CAM formulation, one can transform the BF-FNO T matrix to an recoupled-angular-momentum basis:

$$T^{\ell_t}_{v\ell,v_0\ell_0} \equiv \sum_\Lambda (-1)^\Lambda (2\Lambda+1)\, C\big(\ell\,\ell_0\,\ell_t;\Lambda,-\Lambda,0\big)\, T^\Lambda_{v\ell,v_0\ell_0}.$$

Here the sum over Λ runs in integral steps from $|\ell-\ell_0|$ to $\ell+\ell_0$. The advantages of this gambit, especially for scattering from polar molecules, are considerable [Ref. 101].

[27] This choice conforms to a relationship between the BODY S and T matrices of our usual form, $\mathsf{S}^\Lambda = 1 + 2i\mathsf{T}^\Lambda$. Alternatives are discussed in the Appendix.

[28] In general, we require three angles to specify the orientation of a molecule in the LAB frame, the Euler angles α, β, and γ that relate the BODY and LAB axes. So in general this average entails three angles, the pre-factor is $1/(8\pi^2)$, and the volume element is $d\alpha\,\sin\beta d\beta\,d\gamma$. But for a linear molecule the third Euler angle is arbitrary (we choose $\gamma = 0$). Such is not, of course, acceptable for scattering from a non-linear target.

summations to obtain, at long last,

$$\left.\frac{d\sigma}{d\Omega}\right|_{v_0\to v} = \frac{1}{4k_0^2}\sum_{L=0}^{L_{\max}} B_L(v_0 \to v)\, P_L(\cos\theta'). \tag{132}$$

As in the LAB-CAM DCS (117), we have here written the FNO DCS as a sum of Legendre polynomials in the scattering angle and buried the T matrix in an angle-independent coefficient,

$$B_L(v_0 \to v) = \sum_{\Lambda\bar{\Lambda}} \sum_{\ell\bar{\ell}} \sum_{\ell_0\bar{\ell}_0} d_L(\ell\ell_0, \bar{\ell}\bar{\ell}_0; \Lambda\bar{\Lambda})\, T^{\Lambda}_{v\ell,v_0\ell_0}\, T^{\bar{\Lambda}*}_{v\bar{\ell},v_0\bar{\ell}_0}. \tag{133}$$

All the angular-momentum coupling is manifest in the quantity

$$\begin{aligned} d_L(\ell\ell_0, \bar{\ell}\bar{\ell}_0; \Lambda\bar{\Lambda}) = i^{\ell_0-\ell+\bar{\ell}_0-\bar{\ell}} \frac{1}{2L+1} \left[(2\ell+1)(2\bar{\ell}+1)(2\ell_0+1)(2\bar{\ell}_0+1)\right]^{1/2} \\ C(\ell\,\bar{\ell}\,L;0,0)\; C(\ell\,\bar{\ell}\,L;\Lambda,-\Lambda)\; C(\ell_0\,\bar{\ell}_0\,L;0,0)\; C(\ell_0\,\bar{\ell}_0\,L;-\Lambda,\bar{\Lambda}). \end{aligned} \tag{134}$$

Finally, we can derive convenient expressions for the integral and momentum-transfer cross sections in terms of the coefficients (133) by integrating Eq. (132) over scattering angles $\hat{r}'$, viz.,

$$\sigma_{v_0\to v} = \frac{\pi}{k_0^2} B_0(v_0 \to v) \tag{135}$$

$$\sigma^m_{v_0\to v} = \frac{\pi}{k_0^2}\left[B_0(v_0 \to v) - \frac{1}{3}B_1(v_0 \to v)\right]. \tag{136}$$

In implementing the results of this section we must cope with now-familiar sums over partial-wave orders and, in this case, projection quantum numbers Λ and $\bar{\Lambda}$. Again, triangle rules resulting from Clebsch-Gordan coefficients ride to the rescue; in $d_0(\ell\ell_0, \bar{\ell}\bar{\ell}_0; \Lambda\bar{\Lambda})$ such rules preclude contributions to the total cross sections for $\Delta\ell \equiv |\ell - \bar{\ell}| > 0$. [Note also the restrictions on the momentum transfer cross section (136).] Finally, the values of L that contribute to the DCS are constrained for each pair of BF-FNO channels (indexed by i and j) by

$$\max\{\, |\ell_i - \ell_j|, |\ell_i' - \ell_j'|\,\} \le L \le \min\{\, \ell_i + \ell_j, \ell_i' + \ell_j'\,\}. \tag{137}$$

The maximum value of L that contributes to the DCS is determined by the maximum partial wave order included in Eq. (21), as $L_{\max} = 2\,\ell_{\max}$.

SPECIAL TECHNIQUES AND DIAGNOSTIC TOOLS

Problems and pitfalls in solving the electron-molecule Schrödinger equation mount as the scattering energy decreases. At energies below about 1 eV (and especially for near-threshold scattering), a variety of special methods become important to minimizing numerical error in the S matrix. Moreover, because of the numerically problematic character of low-energy calculations, the need for diagnostic tools for checking these matrices is especially acute. Space limitations preclude discussion of all such tools and methods [more can be found in Ref. 20]; instead we shall look briefly at some that have not received full elaboration elsewhere.

Calculating Near-threshold Cross Sections from BF-FNO T Matrices

If the physical conditions and requirements permit, it is far preferable to work in the BF-FNO formulation rather than the computationally more arduous LAB-CAM theory. Of course, if the energy is low enough, even the solution of the BF-FNO scattering equations can be numerically challenging (or, depending on one's algorithm, unstable). In such circumstances, alternate strategies sometimes serve. In one such instance[108] we used a recent implementation of modified effective range theory[109] along with BF-FNO K matrices from solutions of the scattering equations at a reasonably low energies to extrapolate these matrices to lower energies, where such solutions may be impossible or undesirable.

If inelastic cross sections are desired, further problems come into play. As discussed above, the use of a frame transformation to generate approximate adiabatic-nuclei (AN) cross sections is both efficient and illuminating. The most problematic regions for applying this strategy are those governed by shape resonances and those near excitation thresholds. In the latter case the problem isn't the adiabatic approximation inherent in fixed-nuclei approximations; it's the assumption that the electron loses no energy in the (inelastic) collision. Invalidity of this assumption causes conventional AN cross sections, for example, to go to a non-zero constant rather than to zero at threshold.[18]

Happily, this breakdown does not preclude use of methods based on an approximate (Born-Oppenheimer) separation of the electron-molecule wave function. The first-order non-degenerate adiabatic (FONDA) approximation mentioned above is one such method.[73] Although it requires calculating off-shell scattering matrices, the BF-FNO implementation of this method is a quite straightforward scheme for calculating and interpreting low-energy rotational and vibrational cross sections. Since details of this theory are available elsewhere,[65] we shall focus here on computationally simpler tactics one can implement within the context of a conventional (on-shell) AN study to improve markedly the accuracy of near-threshold cross sections, choosing rotational excitation as our example.

Rather than modify the scattering equations to enforce energy conservation (as do off-shell methods), these strategies seek to fix the S matrix (or resulting cross sections) after the fact. The problem they address is that the ANR approximation to the LAB-CAM transition matrix, obtained by applying a rotational frame transform (RFT) to the BF-FNO T matrix, does not obey the correct threshold laws. As discussed in Ref. 18, these laws require that each element of the LAB-CAM T matrix and, correspondingly, the inelastic cross section, go to zero at threshold as

$$T^{J}_{vj\ell,v_0j_0\ell_0} \xrightarrow[k_{vj}\to 0]{} k_{vj}^{\ell+1/2} \tag{138}$$

$$\sigma_{v_0j_0\to vj} \xrightarrow[k_{vj}\to 0]{} k_{vj}^{2\ell+1}. \tag{139}$$

That a theoretical scattering matrix obeys Eq. (138) may be especially important in determinations of low-energy DCS, whose shape often reflects subtle interference between elements of the T matrix.

Correcting the ANR Cross Section. The simplest way to correct the threshold value of the ANR cross section is to force it to zero as the exit-channel wave number approaches zero[54] by just multiplying the ANR cross section by the dimensionless kinematic ratio k_{vj}/k_0:

$$\sigma^{ANR}_{v_0j_0\to vj} \quad \longrightarrow \quad \tilde{\sigma}^{ANR}_{j_0\to j} \equiv \frac{k_{vj}}{k_0}\sigma^{ANR}_{v_0j_0\to vj}. \tag{140}$$

The resulting "flux-corrected" ANR cross sections $\tilde{\sigma}_{j_0 \to j}^{ANR}$ do, of course, go to zero. But they do not necessarily obey the threshold law (139) because not every important elements of the T matrix goes to zero at all, let alone according to Eq. (138).

The Scaled-Adiabatic-Nuclear-Rotation Method. We can ensure correct behavior of key elements of the scattering matrix by a simple correction based on the fact that near threshold, this matrix reduces to its First-Born approximate. Specifically, in either the LAB-CAM or BF-FNO theories, the elements of the reactance matrix are accurately given near threshold (except for special cases where the Born integrals are undefined) by the First-Born-Approximation (FBA) with a potential given by the long-range form (67):

$$K^J_{vj\ell, v_0 j_0 \ell_0} \xrightarrow[k_{vj} \to 0]{} {}^{\text{FBA}}K^J_{vj\ell, v_0 j_0 \ell_0} \tag{141}$$

$$K^\Lambda_{v\ell, v_0 \ell_0} \xrightarrow[k_v \to 0]{} {}^{\text{FBA}}K^\Lambda_{v\ell, v_0 \ell_0}. \tag{142}$$

The FBA approximate to the LAB-CAM T matrix, which we can obtain from the FBA K matrix using the transformation (33), does obey the threshold law Eq. (138). This is the basis of the scaled adiabatic-nuclear rotation (SANR) method[56] in which we "scale" each element of the ANR K matrix (obtained, as usual, by applying a RFT to our BF-FNO K matrix) by a ratio of corresponding FBA elements.[29]

Let's consider each element of this ratio in turn. Our goal is a quantity that will correct the threshold behavior of the ANR approximation to a particular LAB-CAM K matrix. For a given pair of LAB-CAM channels, then, the numerator of this ratio should be simply the First-Born approximate to this LAB-CAM reactance matrix element—which as noted obeys the desired threshold law. We can calculate this quantity, ${}^{\text{FBA}}K^J_{vj\ell, v_0 j_0 \ell_0}$, from the FBA integrals.[55] For theoretical consistency, the denominator of the SANR ratio must be a corresponding FBA K-matrix element that incorporates an adiabatic treatment of rotations. We can obtain this quantity by applying the RFT to the first Born approximate to the BF-FNO K matrix element.

Our starting point is the integral equation Eq. (55) for the BF-FNO K matrix in the FBA:

$${}^{\text{FBA}}K^\Lambda_{v\ell, v_0 \ell_0} = -\frac{2}{\sqrt{k_v k_0}} \int_0^\infty \hat{\jmath}_\ell(k_v r)\, V^\Lambda_{v\ell, v_0 \ell_0}(r)\, \hat{\jmath}_{\ell_0}(k_0 r)\, dr. \tag{143}$$

The potential matrix elements between the initial and final states are given by (70) with the vibrational coupling elements $w^\lambda_{v,v'}(r)$ of (66). Let's now write this K matrix in terms of radial FBA integrals

$${}^{\text{FBA}}R_\lambda(v\ell, v_0\ell_0) \equiv \frac{1}{k_v k_0} \int_0^\infty \hat{\jmath}_\ell(k_v r)\, w^\lambda_{v,v_0}(r)\, \hat{\jmath}_{\ell_0}(k_0 r)\, dr \tag{144}$$

and the BF-FN angular coupling coefficients Eq. (71), viz,

$${}^{\text{FBA}}K^\Lambda_{v\ell, v_0 \ell_0} = -2\,(k_v k_0)^{1/2} \sum_{\lambda=0}^{\lambda_{\max}} g_\lambda(\ell\ell_0; \Lambda)\, {}^{\text{FBA}}R_\lambda(v\ell, v_0\ell_0). \tag{145}$$

[29] This gambit can be applied even if rotational excitation is accompanied by vibrational excitation, but only with caution. *One must use the SANR fix-up only for near-threshold pure rotational excitation. Moreover, it should not be applied to rotationally elastic K-matrix elements; in the BF-FNO formulation the vibrational dynamics are treated properly, so these elements already obey the correct threshold laws.* Finally, note that if vibrational channels are open, one must take careful account of the non-degeneracy of asymptotic free channels in BF-FNO theory.

Our prescription for the numerator of the SANR ratio calls for transforming this matrix element into the LAB-CAM representation by applying the RFT (35). In the event, this simply transforms the angular coupling coefficients as in Eq. (74), leaving

$$ {}^{\mathrm{FBA}}K^{J}_{vj\ell,v_0j_0\ell_0} = -2\,(k_v k_0)^{1/2} \sum_{\lambda=0}^{\lambda_{\max}} f_\lambda(j\ell, j_0\ell_0; J)\, {}^{\mathrm{FBA}}R_\lambda(v\ell, v_0\ell_0). \tag{146} $$

We get the denominator by applying the RFT to the First Born approximate to the BF-FNO K matrix.

Last but hardly least, what is the quantity to be corrected? It is the LAB-CAM approximate obtained by applying the RFT to a K matrix calculated in a converged BF-FNO scattering calculation (using, of course, the full interaction potential, not just its long-range form):

$$ K^{J}_{vj\ell,v_0j_0\ell_0} \approx \sum_{\Lambda} A^{J\Lambda}_{j\ell}\, K^{\Lambda}_{v\ell,v_0\ell_0}\, A^{J\Lambda}_{j_0\ell_0}. \tag{147} $$

Applying the SANR ratio to this matrix element we obtain the key equation of this correction scheme:

$$ K^{J}_{vj\ell,v_0j_0\ell_0} \approx \sum_{\Lambda} A^{J\Lambda}_{j\ell}\, K^{\Lambda}_{v\ell,v_0\ell_0}\, A^{J\Lambda}_{j_0\ell_0} \quad\times\quad \frac{{}^{\mathrm{FBA}}K^{J}_{vj\ell,v_0j_0\ell_0}}{\sum_\Lambda A^{J\Lambda}_{j\ell}\, {}^{\mathrm{FBA}}K^{\Lambda}_{v\ell,v_0\ell_0}\, A^{J\Lambda}_{j_0\ell_0}}. \tag{148} $$

In effect, the entire SANR procedure takes place in the LAB-CAM formulation. The denominator of the SANR ratio "cancels" the offending threshold behavior of the frame-transformed BF-FNO matrix elements while the numerator "inserts" the correct behavior (138).

Although Eq. (148) may appear awkward, its implementation is quite simple, requiring only evaluation of two radial (FBA) integrals at each energy. Tests for e–H_2 scattering have demonstrated its ability to significantly improve the accuracy of near-threshold integrated and (especially) differential cross sections.[56] But the SANR prescription must be applied only at energies reasonably near threshold. At higher energies, this fix-up is inappropriate and will introduce rather than eliminate error, for here the energy dependencies of the BF-FNO and LAB-CAM K matrices are not given by their first Born approximates. Fortunately, an energy large enough for the SANR to foul up the cross sections is probably large enough for conventional ANR theory, perhaps corrected by the kinematic ratio as in (140), to be accurate. For e–H_2 scattering, the SANR correction is appropriate at energies below about 2.0 eV, while conventional ANR becomes reasonably accurate by about 1.0 eV. For e–N_2 scattering, this method is applicable (and necessary) below about 0.2 eV.

Diagnostic Tools

Possibilities for error in electron-molecule scattering calculations abound. So even the most mundane diagnostic tool is valuable: e.g., detailed balance,

$$ k_0^2\,(2j_0+1)\,\sigma_{v_0j_0\to vj} = k_{vj}^2\,(2j+1)\,\sigma_{vj\to v_0j_0}, \tag{149} $$

is a powerful test of scattering codes in both the LAB-CAM formulation (for ro-vibrational cross sections) and the BF-FNO formulation (for vibrational cross sections).

The Consequences of Unitarity. A rather more subtle set of diagnostic tools derive from the unitarity requirement on the S matrix. This requirement imposes relationships that must be obeyed among elements of the T matrix; it also relates the imaginary elements of this matrix to the integrated cross section (the optical theorem).[60,62] Moreover, integrated elastic, excitation, and total cross sections must obey the unitarity bounds.

Approximate dynamical schemes that violate unitarity, although not necessarily invalid, must be used with care. If, for example, the initial state is seriously depleted, then violation of unitarity may introduce serious error in inelastic cross sections. [One can, of course, guarantee unitarity of the S matrix by calculating it from a Hermitian K matrix as in Eq. (32).][30]

Checking the K Matrix. Another cluster of diagnostic tools appropriate and combine strategies hinted at in earlier subsections. For example, we can compare rotationally frame transformed BF-FNO K matrices to results from LAB-CAM scattering calculations using the same potential. If the scattering energy is large enough (and not near a shape resonance), the two K matrices should agree within the numerical precision of the calculation.

An even easier way to check close-coupling K matrices from either LAB-CAM or BF-FNO calculations is to call upon the First Born Approximation (FBA). At very low energies, FBA expressions for the elements of these matrices should agree with those from more rigorous scattering calculations. Even at rather larger energies, high-order (e.g., large ℓ) elements of these matrices should agree with their FBA approximates, because a strong centrifugal barrier controls the corresponding elements of the radial scattering function.

Even if conditions preclude quantitative verification using the FBA, we can adapt this reasoning to anticipate trends in the elements K^{q}_{p,p_0} with variation of the channel quantum numbers p and p_0. For example, if for a given excitation $v_0 \rightarrow v$, the elements of $K^{\Lambda}_{v\ell,v_0\ell_0}$ don't decrease with increasing partial-wave-order ℓ and ℓ_0, there better be a good reason (e.g., a resonance in the neighborhood). Especially valuable in checking and interpreting such trends are the "deconstructions" of the ICS in the LAB-CAM equations (108) and (109) and in the BF-FNO equations (112) and (113). We can use these decompositions to investigate contributions to the total cross sections (and, if a single symmetry dominates, to the ro-vibrational cross section) from particular J manifolds (LAB-CAM), symmetries (BF-FNO), or, in some cases, partial-waves ℓ, ℓ_0.

The ANR Ratio Test. Some tools also enable us to avoid onerous calculations. For example, at energies where the ANR approximation is valid, one can use the simple the RFT (35) to relate ANR cross sections for two *different* rotational excitations:

$$\sigma_{j_0' \rightarrow j'} = \sigma_{j_0 \rightarrow j} \frac{k_{j'}}{k_j} \left[\frac{C(j_0'\, 2\, j'; 0\, 0)}{C(j_0\, 2\, j; 0\, 0)} \right]^2 , \qquad (150)$$

[30] Some scattering codes artificially *impose* Hermiticity of the K matrix by calculating its upper triangle (only) and then equating corresponding elements in the two triangles. We prefer to use Hermiticity as a check on numerical accuracy of the radial wave function; breakdown of symmetry in the K matrix is one of the most useful indicators of linear dependence in the columns of the solution matrix. This condition, the bane of low-energy close-coupling calculations, depends subtly on machine precision, scattering energy, number of channels (partial waves), and strength of the coupling potential. If present but not too extreme, linear dependence can be rectified via a stabilization procedure, as described in Ref. 20 and in Ref. 77; this tactic, however, requires additional matrix operations that slow down calculation of the radial scattering function and may themselves introduce error.

where we have suppressed the vibrational quantum numbers.[31]

This trivial relationship is quite powerful. As a diagnostic tool, we can use it, for example, to verify LAB-CAM scattering calculations well above threshold.[55] It can also help us avoid work. If, for instance, we have solved the LAB-CAM scattering equations to obtain some rotational cross section (and/or our experimentalist colleagues have measured this cross section), then we can use Eq. (150) to generate other cross sections for the same Δj. We can thus assist the transport theory of swarm experiments, where the energy distribution of electrons in the swarm may admit many rotational excitations as important energy-loss mechanisms.[110]

The Eigenphase Shifts and Their Sum. There are no phase shifts in electron-molecule scattering. Although we sometimes talk the language of partial waves [e.g., in conjunction with the angular-momentum expansion (21) of the wave function or in identifying "contributions" such as $s \to d$ to a particular cross section], in fact the essential non-spherical character of the interaction precludes exact separation of the radial scattering equations (24) into independent sets each with a well-defined partial-wave order. We can, however, usefully define, calculate, and interpret *analogs* of phase shifts, the eigenphase shifts for a particular electron-molecule symmetry Λ (in a BF-FNO treatment) or J state (in a LAB-CAM treatment).

The phase shifts δ_ℓ of spherical potential scattering are the arctangents of the corresponding elements of the (diagonal) reactance matrix. The electron-molecule K matrix isn't diagonal because of partial-wave coupling. But unitarity and time-reversal invariance of the S matrix guarantee the existence of a unitary transformation that will diagonalize the K matrix:[60]

$$\langle p; q \mid \hat{K} \mid p'; q\rangle = \sum_{i=1}^{N_p} \langle p; q \mid i\rangle \, \tan\delta_i^q \, \langle i \mid p'; q\rangle. \tag{151}$$

The eigenphases δ_i^q are labelled by the particular symmetry Λ for the BF-FNO K matrix or a particular total angular momentum J for the LAB-CAM K matrix. In either case, the sum in Eq. (151) runs over the total number of channels p:

$$N_p = \begin{cases} N_v \times N_j \times N_\ell & \text{LAB-CAM} \\ N_v \times N_\ell & \text{BF-FNO} \\ N_\ell & \text{BF-FN} \end{cases} . \tag{152}$$

As its name implies, the eigenphase sum is the sum of the eigenphases:

$$\delta_{\text{sum}}^q \equiv \sum_{i=1}^{N_p} \delta_i^q. \tag{153}$$

These quantities are especially useful in the analysis of shape resonances.[111,112]

Indeed, when eigenphase sums can meaningfully be constructed, they characterize scattering (in a particular symmetry Λ or J state) in a way that is analogous to the partial-wave phase shifts of a spherical potential. At energies below the first inelastic threshold, in fact, we can write the elastic ICS entirely in terms of eigenphases. This is not true however for the differential cross section; the expression for

---

[31] This form is valid rigorously only for ANR rotational cross sections that have been "flux corrected" via Eq. (140). (If uncorrected ANR cross sections are used, the ratio of final-state wave numbers disappears.) In this form, however, this relationship is approximately valid for cross sections from various theoretical (and even experimental) sources.

this quantity in terms of eigenphases also involves mixture coefficients obtained from the diagonalization of the reactance matrix. If, as in some resonant collisions, the scattering process is dominated by a single symmetry and a single partial wave (e.g., e–N_2 scattering near the 2.39 eV shape resonance is identified with d-wave scattering in the Π_g symmetry), then we can express the resonant cross section in terms of the corresponding eigenphase.

Most applications of eigenphase analysis have been performed within the BF-FN formulation (i.e., in the rigid-rotor approximation). In this theory all channels are degenerate [see Table 2], so no ambiguity attaches to associating a particular eigenphase sum with a particular scattering event. But when the channels that define the K matrix are non-degenerate, as in the LAB-CAM or BF-FNO theories, a new subtlety appears: the eigenphase sum (153) may be meaningless—if, for example, a given eigenphase δ_i^q is associated with one or more open channels.

To determine whether this is the case, we look at the eigenvectors resulting from the diagonalization *of the entire K matrix*. For the ith eigenphase, the squared modulus of each element of the eigenvector, $|\langle p;q \mid i\rangle|^2$, gives the (probabilistic) contribution of channel p to eigenchannel i in symmetry Λ (or total angular momentum state J). If, as often happens when more than one channel are open, a given eigenphase contains contributions from more than one energetically distinct channel, then it cannot be interpreted according to conventional eigenphase analysis.

Easy Calculation of (Approximate) Vibrational Cross Sections. To conclude this look at special and diagnostic methods, we return to low-energy vibrational excitation calculations. So many potential land mines imperil such calculations that even a crude guess at the answer may be of diagnostic value. We can obtain such a guess from the eigenphases by combining the ANV approximation for the dynamics and the SHO approximation for the vibrational target functions.

We first write the BF-FNO T-matrix elements in terms of their real and imaginary parts, as

$$T^{\Lambda}_{v\ell,v_0\ell_0} = R^{\Lambda}_{v\ell,v_0\ell_0} + i\, I^{\Lambda}_{v\ell,v_0\ell_0}. \tag{154}$$

With this definition, the integrated vibrational cross section (112) becomes

$$\sigma_{v_0\to v} = \frac{\pi}{k_0^2} \sum_{\Lambda=-\ell}^{+\ell} \sum_{\ell=0}^{\ell_{\max}} \sum_{\ell_0=0}^{\ell_{\max}} \left[\left(R^{\Lambda}_{v\ell,v_0\ell_0}\right)^2 + \left(I^{\Lambda}_{v\ell,v_0\ell_0}\right)^2 \right]. \tag{155}$$

In the ANV approximation, we can express the real part of the inelastic T-matrix element in terms of fixed-R BF-FN matrix elements as

$$R^{\Lambda}_{v\ell,v_0\ell_0} = \int_0^{\infty} \varphi_v^{(\mathrm{v})}(R)\, R^{\Lambda}_{\ell,\ell_0}(R) \varphi_{v_0}^{(\mathrm{v})}(R)\, dR, \tag{156}$$

where $R^{\Lambda}_{\ell,\ell_0}(R)$ denotes the dependence of *the real part* of the T matrix on the internuclear separation R. (The equations for the imaginary parts of the T matrix look just like those for the real parts except that R is replaced by I; we will write only the equations for the real parts.)

Now we must evaluate these integrals. Provided the excursions of the nuclei from equilibrium during the excitation are not too great, we can expand the fixed-R T-matrix elements in a Taylor series about equilibrium R_{eq} and retain only the first-derivative term to obtain

$$R^{\Lambda}_{\ell,\ell_0}(R) \approx R^{\Lambda}_{\ell,\ell_0}(R_{\text{eq}}) + \left\{ \left.\frac{d}{dR}\right|_{R_{\text{eq}}} R^{\Lambda}_{\ell\ell_0}(R) \right\} (R - R_{\text{eq}}). \tag{157}$$

This approximation reduces the ANV T-matrix elements to forms like

$$\begin{aligned} R^{\Lambda}_{v\ell,v_0\ell_0} = R^{\Lambda}_{\ell,\ell_0}(R_{\text{eq}}) \int_0^{\infty} \varphi_v^{(\text{v})}(R)\varphi_{v_0}^{(\text{v})}(R)\, dR + \\ \left\{ \left.\frac{d}{dR}\right|_{R_{\text{eq}}} R^{\Lambda}_{\ell\ell_0}(R) \right\} \int_0^{\infty} \varphi_v^{(\text{v})}(R)\,(R - R_{\text{eq}})\,\varphi_{v_0}^{(\text{v})}(R)\, dR. \end{aligned} \tag{158}$$

For vibrational excitation ($v \neq v_0$) the first of these integrals is zero by orthogonality of the vibrational wave functions. The second we can easily evaluate in the SHO approximation, since[32]

$$M(v, v_0) \equiv \int_{-\infty}^{+\infty} \varphi_v^{(\text{v})}(\xi)\, \xi\, \varphi_{v_0}^{(\text{v})}(\xi)\, d\xi = \frac{1}{\beta_v} \begin{cases} \sqrt{\dfrac{v+1}{2}} & \text{for } v = v_0 - 1 \\ \sqrt{\dfrac{v}{2}} & \text{for } v = v_0 + 1 \\ 0 & \text{otherwise} \end{cases} . \tag{159}$$

Here β_v is the usual SHO constant evaluated at the natural frequency of the vth vibrational state, $\beta_v \equiv \sqrt{\mu\omega_v/\hbar}$. So in the SHO approximation the ANV matrix elements in Eq. (158) simplify as

$$R^{\Lambda}_{v\ell,v_0\ell_0} \approx R^{\Lambda}_{\ell,\ell_0}(R_{\text{eq}})\, \delta_{v_0,v} + \left\{ \left.\frac{d}{dR}\right|_{R_{\text{eq}}} R^{\Lambda}_{\ell\ell_0}(R) \right\} M(v, v_0) \delta_{v_0,v\pm 1}. \tag{160}$$

Only the first term in these matrix elements contributes to elastic scattering and only the second to the $v_0 \to v_0 \pm 1$ excitation, the only excitation allowed in this extreme approximation.[33]

An even more extreme (and illuminating) approximation obtains if we invoke a "decoupling approximation" and set all off-diagonal elements of the BF-FN T-matrix to zero. We can write the diagonal matrix elements (including both real and imaginary parts) in terms of eigenphases as

$$T^{\Lambda}_{\ell\ell}(R) \approx e^{i\delta^{\Lambda}_{\ell}(R)} \sin \delta^{\Lambda}_{\ell}(R). \tag{161}$$

Doing this enables us to express the first derivatives resulting from the Taylor series expansion (157) in terms of the first derivative of the eigenphases. Thus for the $v_0 \to v = v_0 + 1$ excitation we have from Eq. (159) the simple approximation

$$\sigma_{v_0 \to v} \approx \frac{\pi}{k_0^2} \frac{2(v_0+1)}{\beta_v^2} \sum_{\ell,\Lambda} \left[\left.\frac{d}{dR}\right|_{R_{\text{eq}}} \delta^{\Lambda}_{\ell}(R_e) \right]^2 . \tag{162}$$

Note that *in this approximation the inelastic cross section depends only on the slope of the eigenphases (at equilibrium).* This feature reflects the importance in the actual

[32] To effect this evaluation we must change variables to $\xi \equiv R - R_{\text{eq}}$. This step results in integrals over ξ that range from $-R_{\text{eq}}$ to ∞. But we can exploit the finite extent of the vibrational wave functions to extend the domain of integration from $-\infty$ to ∞.

[33] We can evaluate the first derivatives of the real and imaginary parts of the T matrix using a finite difference approximation starting with the BF-FN T matrix at two nearby internuclear separations. Alternatively we could use a fit to, say, a quadratic equation if values at three nearly separations are available. [See Chap. 9 of Ref. 113.]

vibrational cross section of the *variation* of the interaction potential with R, one reason for the extreme sensitivity of vibrational cross sections to every aspect of this potential.

Although highly approximate, Eq. (162) is not that bad. For excitation of the symmetric stretch mode of CO_2 (with $\beta_v = 9.6042\, a_0^{-1}$) at an incident energy of 0.2 eV, inserting BF-FN Σ_g eigenphases for $\ell = 0$ near and at equilibrium into this equation gives an inelastic cross section that differs by only 2% from the result of fully converged close-coupling calculations.[114] One cannot, of course, count on this level of agreement. Still, this and similar crude approximations are invaluable guides (at least at the order-of-magnitude level), and we urge their use along with the other diagnostic tools of this section to corroborate results of more rigorous scattering calculations.

APPENDIX: CONTINUUM NORMALIZATION AND OTHER CHOICES

In setting up the machinery of this chapter we made several choices that influence many of its equations. Primary among these are (1) how to normalize the plane-wave free states $|\,\mathbf{k}'\,\rangle$ (and hence the stationary scattering states $|\,\mathbf{k}'+\,\rangle$); (2) how to normalize the angular-momentum free states $|\,E\ell m_\ell\,\rangle$ (whether to use energy normalization or wave number normalization); and (3) how to relate the scattering matrix S to the transition matrix T. There are no "standard choices" for these options; an often bewildering variety of alternatives appear in the literature of scattering theory, often generating confusion since different researchers choose different conventions. The choices implemented in this chapter are:

1. Dirac-delta function normalization of $|\,\mathbf{k}'\,\rangle$ as in Eq. (18*a*), i.e.,

$$\langle \mathbf{r}' \mid \mathbf{k}' \rangle = (2\pi)^{-3/2} e^{i\mathbf{k}'\cdot\mathbf{r}'} \implies \langle \mathbf{k} \mid \mathbf{k}' \rangle = \delta(\mathbf{k} - \mathbf{k}').$$

2. Energy normalization of the angular momentum states $|\,E\ell m_\ell\,\rangle$ so that

$$\langle E\ell m_\ell \mid E'\ell' m_\ell' \rangle = \delta(E - E')\delta_{\ell,\ell'}\delta_{m_\ell,m_\ell'},$$

where $E = \hbar^2 k^2/(2m_e)$.

3. A proportionality between the scattering amplitude and T matrix typified by Eq. (17), which leads to Eq. (31) relating the S and T matrices.

Published papers and reviews (as well as computer codes available through the CPC program library) incorporate various alternatives to these choices. To faciliate switching between various normalizations and proportionalities, we here introduce willfull arbitrariness into a few key equations of this chapter via the complex constants c and d defined below.

1. The constant c allows for arbitrary normalization of plane-wave free states, i.e.,

$$\langle \mathbf{r}' \mid \mathbf{k}' \rangle = c\,(2\pi)^{-3/2} e^{i\mathbf{k}'\cdot\mathbf{r}'} \implies \langle \mathbf{k} \mid \mathbf{k}' \rangle = |c|^2\, \delta(\mathbf{k} - \mathbf{k}').$$

To convert the equations of this chapter to arbitrary continuum normalization, merely replace $|\,\mathbf{k}'\,\rangle \to c\,|\,\mathbf{k}'\,\rangle$ and $\langle\,\mathbf{k}'\,| \to c^*\,\langle\,\mathbf{k}'\,|$. For example, the transformation matrix Eq. (19*a*) becomes

$$c_{\ell m_\ell}(\mathbf{k}') = \langle E\ell m_\ell \mid \mathbf{k}' \rangle = c\,\frac{i^\ell}{\sqrt{m_e k}}\, Y_\ell^{m_\ell *}(\hat{k}').$$

The most common alternative to Dirac-delta function normalization is no normalization, i.e., $c = (2\pi)^{3/2}$.

2. The constant d allows for arbitrary proportionality relationships between the scattering amplitude and (momentum-space) T matrix, as

$$f_{\nu,\nu_0}(\hat{r}') = \frac{d}{|c^2|}\,\frac{4\pi m_e}{\hbar^2}\,\langle \mathbf{k}'_\nu, \nu \mid \hat{T} \mid \mathbf{k}'_0, \nu_0 \rangle .$$

Introduction of d requires that we modify the relationship between the S and T matrices Eq. (31) to

$$\mathsf{S} = 1 + 2i\,d\,\mathsf{T},$$

that between the T and K matrices (33) to

$$\mathsf{T} = \frac{1}{d}\,\mathsf{K}\,(1 - i\mathsf{K})^{-1},$$

and the T-matrix boundary conditions on the radial function (28) to

$$u^q_{p,p_0}(r) \xrightarrow[r\to\infty]{} \delta_{p,p_0}\,\hat{j}_{\ell_0}(k_0 r) + d\,\sqrt{\frac{k_0}{k_p}}\,T^q_{p,p_0}\,\hat{h}^+_\ell(k_p r), \qquad \text{(open channels)}$$

Two alternative choices of d are also widely used in scattering literature:

$$\begin{aligned} d = \frac{i}{2} \quad &\Longrightarrow \quad \mathsf{S} = 1 - \mathsf{T} \\ &\Longrightarrow \quad \mathsf{T} = -2i\mathsf{K}\,(1 - i\mathsf{K})^{-1} \\ d = -\pi \quad &\Longrightarrow \quad \mathsf{S} = 1 - 2\pi i \mathsf{T} \\ &\Longrightarrow \quad \mathsf{T} = -\frac{1}{\pi}\mathsf{K}\,(1 - i\mathsf{K})^{-1} \end{aligned}$$

For example, Geltman (Ref. 61) and Lane (Ref. 29) adopt the first choice; Taylor (Ref. 51), Rodberg and Thaler (Ref. 62), and Morrison (Ref. 30) the second.

The primary remaining choice is whether to energy-normalize the angular-momentum free states, as we have done. The most common alternative found in the literature is to choose k-normalization, so that

$$\langle k\ell m_\ell \mid k'\ell' m_{\ell}' \rangle = \delta(k - k')\,\delta_{\ell,\ell'}\,\delta_{m_\ell,m_{\ell}'} .$$

Switching the equations of this chapter to k-normalized free states requires only multiplying by the prefactor in the relationship

$$|\,k\ell m_\ell\,\rangle = \left(\frac{\hbar^2 k}{m_e}\right)^{1/2} |\,E\ell m_\ell\,\rangle .$$

The crucial but tedious burden of keeping all this straight is greatly eased if one uses Dirac notation. For example, one can derive Eq. (130), which relates the scattering amplitude in the BODY frame to the FNO T matrix, from the basis proportionality relation above (using atomic units) as

$$\begin{aligned} f_{v,v_0}(\hat{k}_v, \hat{k}_0) &= \frac{d}{|c^2|}\,\frac{4\pi m_e}{\hbar^2}\,\langle \mathbf{k}_v, v \mid \hat{T} \mid \mathbf{k}_0, v_0 \rangle \\ &= \frac{d}{|c^2|}\,\frac{4\pi m_e}{\hbar^2} \sum_{\ell=0}^{\ell_{\max}} \sum_{\ell_0=0}^{\ell_{\max}} \sum_{\Lambda=-\ell}^{+\ell} \langle k_v \mid E_v \ell \Lambda \rangle \langle E_v \ell \Lambda \mid \hat{T} \mid E_0 \ell_0 \Lambda \rangle \langle E_0 \ell_0 \Lambda \mid k_0 \rangle \end{aligned}$$

with the identification $\langle E_v \ell \Lambda \mid \hat{T} \mid E_0 \ell_0 \Lambda \rangle = T^{\Lambda}_{v\ell, v_0 \ell_0}$. One can invoke alternative normalizations by simply scaling the kets appropriately before introducing explicit forms for free state functions and matrix elements. The theory underlying the present chapter is formulated in this way in Sec. II of Ref. 30.

ACKNOWLEDGEMENTS

MAM is grateful to the many students and colleagues whose collaborations over the past 17 years on a range of electron-molecule scattering problems have helped shape the perspectives offered here. We are both grateful to Dr. Grahame Danby and Dr. Wayne K. Trail for early work on our e–N_2 scattering project and for contributions to our study of exchange effects in vibrational excitation, and to Dr. Brian K. Elza for his suggestions concerning the content of this chapter and his assistance in the e–H_2 convergence studies reported herein. Finally, we owe a special debt to Dr. Trail and to Mr. William Isaacs, both of whom, in an effort that can only be described as heroic, not only read meticulously this chapter in manuscript form but checked the derivations of many of the key equations. This work was supported in part by NSF Grant No. PHY-9108890.

References

1. *Electron Molecule and Photon-Molecule Collisions*, edited by T. N. Rescigno, V. McKoy and B. I. Schneider, (New York, Plenum 1979).
2. *Electron-Molecule Collisions and Photoionization Processes*, edited by V. McKoy, H. Suzuki, K. Takayanagi, and S. Trajmar (Deerfield Beach, Florida: Verlag Chemie International, 1983).
3. *Wavefunctions and Mechanisms for Electron Scattering Processes*, edited by F. A. Gianturco and G. Stefani (Springer-Verlag, 1984).
4. *Electron-Molecule Interactions and their Applications, Volume 1*, edited by L. G. Christophorou (Academic Press, New York, 1984).
5. *Electron-Molecule Collisions*, edited by I. Shimamura and K. Takayanagi (New York, Plenum, 1984).
6. *Swarm Studies and Inelastic Electron-Molecule Collisions*, edited by L. C. Pitchford, V. McKoy, A. Chutjian, and S. Trajmar (New York: Springer-Verlag, 1986).
7. G. J. Schulz in *Principles of Laser Plasmas*, edited by G. Bekite (Wiley, New York, 1976), Chap. 2.
8. A. V. Phelps in *Electron-Molecule Scattering*, edited by S. C. Brown (Wiley-Interscience, New York, 1979), Chap. 2.
9. J. N. Bardsley, in *Electron-Molecule Collisions and Photoionization Processes*, edited by V. McKoy, H. Suzuki, K. Takayanagi, and S. Trajmar (Deerfield Beach. Fl: Verlag Chemie International, 1983), p.235.
10. G. S. Willet, *Introduction to Gas Lasers—Population Inversion Mechanisms* (New York: Pergammon, 1984).
11. S. Trajmar, D. F. Register, and A. J. Chutjian, Phys. Rept. **97**, 220 (1983).
12. J. W. McConkey, S. Trajmar, and G. C. M. King, Comments At. Mol. Phys. **22**, 17 (1988).
13. M. Kimura and M. Inokuti, Comments At. Mol. Phys. **24**, 269 (1990).

14. W. L. Morgan, Plasma Chemistry and Plasma Processing **12**, 449 (1992); JILA Data Center Report No. 34 (1991).
15. G. J. Schulz, Rev. Mod. Phys. **45**, 423 (1962).
16. D. G. Thompson, Adv. At. Mol. Phys. **19**, 309 (1984).
17. E. Enhardt and L. Frost, Comments At. Mol. Phys. **29**, 123 (1993).
18. M. A. Morrison, Adv. At. Mol. Phys. **24**, 51 (1988).
19. P. G. Burke, in *Quantum Dynamics of Molecules*, ed. by R. G. Wooley (New York: Plenum, 1980), pp. 483.
20. M. A. Morrison, in *Electron- and Photon-Molecule Collisions* edited by T. N. Rescigno, B. V. McKoy and B. I. Schneider (Plenum Press, New York, 1979).
21. B. D. Buckley, P. G. Burke, and C. J. Noble in *Electron-Molecule Collisions* edited by I. Shimamura and K. Takayanagi (Plenum, New York, 1984) page 495.
22. L. A. Collins and B. I. Schneider, in *Electron-Molecule Scattering Processes and Photoionization*, edited by P. G. Burke and J. B. West (New York: Plenum, 1988).
23. Y. Itikawa, Phys. Rept. **46**, 117 (1978).
24. L. A. Collins and D. W. Norcross, Adv. At. Mol. Phys. **18**, 341 (1983).
25. D. G. Thompson and F. A. Gianturco, Comments At. Mol. Phys. **16**, 307 (1985).
26. F. A. Gianturco and A. Jain, Phys. Rept. **143**, 347 (1986).
27. D. E. Golden, N. F. Lane, A. Temkin, and E. Gerjuoy, Rev. Mod. Phys. **43**, 642 (1971).
28. P. G. Burke, Adv. At. Mol. Phys. **15**, 471 (1979).
29. N. F. Lane, Rev. Mod. Phys. **52**, 29 (1980).
30. M. A. Morrison, Aust. J. Phys. **36**, 239 (1983).
31. M. Born and J. R. Oppenheimer, Ann. Phys. (Leipzig) **84**, 457 (1927).
32. M. A. Morrison, T. L. Estle, and N. F. Lane, *Quantum States of Atoms, Molecules, and Solids* (Prentice-Hall, Englewood Cliffs, New Jersey, 1977).
33. A primary resource for electron-molecule codes is the program library of *Computer Physics Communcations.* Information is available from Department of Applied Mathematics and Theoretical Physics, The Queen's University of Belfast, Belfast BT7 1NN, Northern Ireland.
34. M. A. Morrison and G. A. Parker, Aust. J. Phys. **40**, 465 (1987).
35. M. E. Rose, *Elementary Theory of Angular Momentum* (New York, Wiley, 1957).
36. D. A. Varshalovich, A. N. Moskalev and V. K. Khersonskii, *Quantum Theory of Angular Momentum* (World Scientific, Singapore, 1988).
37. R. N. Zare, *Angular Momentum: Understanding Spatial Aspects in Chemistry and Physics* (Wiley, New York, 1988).
38. D. M. Brink and G. R. Satchler, *Angular Momentum* (Third Edition) (New York: Oxford, 1993).
39. B. I. Schneider and L. A. Collins, J. Phys. B: At. Mol. Phys. **15**, L335 (1982); Phys. Rev. A **27**, 2847 (1983).
40. H. D. Meyer, J. Phys. B: At. Mol. Phy. **25**, 2657 (1992).
41. T. L. Gibson and M. A. Morrison, Phys. Rev. A **29**, 2497 (1984).

42. J. K. O'Connell and N. F. Lane, Phys. Rev. A **27**, 1893 (1983).
43. A. Jain and D. W. Norcross, Phys. Rev. A **34**, 739 (1986).
44. M. A. Morrison and W. K. Trail, Phys. Rev. A **48**, 2874 (1993).
45. M. J. Seaton, Comments At. Mol. Phys. **1**, 184 (1970).
46. M. A. Morrison, and L. A. Collins, Phys. Rev. A **23**, 127 (1981).
47. L. A. Collins, W. D. Robb, and M. A. Morrison, Phys. Rev. A **21**, 488 (1980).
48. B. I. Schneider and L. A. Collins, Comput. Phys. Rpt. **10**, 51 (1989).
49. E. R. Cohen and B. N. Taylor, Rev. Mod. Phys. **59**, 1121 (1987).
50. N. F. Mott and H. S. W. Massey, *The Theory of Atomic Collisions* (Third Edition) (Oxford: Clarendon Press, 1965).
51. J. R. Taylor, *Scattering Theory*, (New York: Wiley, 1972).
52. S. Hara, J. Phys. Soc. Jpn. **27**, 1592 (1969).
53. A. Temkin and K. V. Vasavada, Phys. Rev. A **160**, 190 (1967).
54. N. Chandra and A. Temkin, Phys. Rev. A **13**, 188 (1976).
55. M. A. Morrison, A. N. Feldt, and D. A. Austin, Phys. Rev. A **29**, 2518 (1984).
56. A. N. Feldt and M. A. Morrison, Phys. Rev. A **29**, 401 (1984).
57. M. A. Morrison, *Understanding Quantum Physics: A User's Manual* (Prentice-Hall Inc., Englewood Cliffs, NJ, 1990).
58. A. M. Arthurs and A. Dalgarno, Proc. R. Soc. London Ser. A **256**, 540 (1960).
59. W. K. Trail, M. A. Morrison, W. A. Isaacs, and B. C. Saha, Phys. Rev. A **41**, 4868 (1990).
60. R. G. Newton, *Scattering Theory of Waves and Particles* (Second Edition), (New York: Springer-Verlag, 1982).
61. S. Geltman, *Topics in Atomic Collision Theory* (New York: Academic Press, 1969).
62. L. S. Rodberg and R. M. Thaler, *Introduction to the Quantum Theory of Scattering* (New York: Academic Press, 1967).
63. M. A. Morrison, N. F. Lane, and L. A. Collins, Phys. Rev. A **15**, 2186 (1977).
64. D. T. Birtwistle and A. Herzenberg, J. Phys. B **4**, 53 (1971).
65. B. K. Elza, Ph. D. thesis, University of Oklahoma, 1992.
66. R. K. Nesbet, *Variational Methods in Electron-Atom Scattering Theory* (New York: Plenum, 1980).
67. U. Fano, Comments At. Mol. Phys. **1**, 140 (1970).
68. E. S. Chang and U. Fano, Phys. Rev. A **6**, 173 (1972).
69. K. A. Jerjian and R. J. W. Henry, Phys. Rev. A **31**, 585 (1985).
70. M. A. Morrison, B. C. Saha, and A. N. Feldt, Phys. Rev. A **30**, 2811 (1984).
71. A. Temkin and F. H. M. Faisal, Phys. Rev. A **3**, 520 (1971).
72. M. Shugard, and A. Hazi, Phys. Rev. A **12**, 1895 (1975).
73. M. A. Morrison, M. Abdolsalami, and B. K. Elza, Phys. Rev. A **43**, 3440 (1991).
74. A. C. Allison, Adv. At. Mol. Phys. **25**, 323 (1988).
75. W. N. Sams and D. J. Kouri, J. Chem. Phys. **51**, 4809 (1969).
76. N. F. Lane and S. Geltman, Phys. Rev. **160**, 53 (1967).
77. T. N. Rescigno and A. E. Orel, Phys. Rev. A **25**, 2402 (1982).

78. M. A. Morrison, Comput. Phys. Commun. **21**, 63 (1980).
79. L. A. Collins, D. W. Norcross, and G. B. Schmid, Comput. Phys. Commun. **79**, 63 (1980).
80. M. A. Morrison and L. A. Collins, J. Phys. B **10**, L119 (1977).
81. W. M. Huo, T. L. Gibson, M. A. P. Lima, and V. McKoy, Phys. Rev. A **36**, 1632 (1987).
82. H. -D. Meyer, Phys. Rev. A **40**, 5605 (1989).
83. L. Castillejo, I. C. Percival, and M. J. Seaton, Proc. R. Soc. London, Ser. A **254**, 259 (1960).
84. M. A. Morrison and P. J. Hay, Phys. Rev. A **20**, 740 (1979).
85. C. A. Weatherford, K. Onda, and A. Temkin, J. Phys. B **31**, 3620 (1985).
86. B. K. Elza, T. L. Gibson, M. A. Morrison, and B. C. Saha, J. Phys. B **22**, 113 (1989).
87. I. C. Percival and M. J. Seaton, Prof. Cambridge Phil. Soc. **53**, 654 (1957).
88. H. S. W. Massey, and I. C Percival, Proc. Roy. Soc. A **274**, 427 (1963).
89. G. Herzberg, *Molecular Spectra and Molecular Structure I: Spectra of Diatomic Molecules* (Second Edition) (Van Nostrand, New York, 1950).
90. K. P. Huber, and G. Herzberg, *Molecular Spectra and Molecular Structure IV. Constants of Diatomic Molecules* (Van Nostrand, New York, 1979).
91. A. A. Radzig and B. Smirnov, *Reference Data on Atoms, Molecules, and Ions* (Springer-Verlag, New York, 1986).
92. P. M. Morse, Phys. Rev. **34**, 57 (1929).
93. M. Abramowitz and I. A. Stegun, *Pocketbook of Mathematical Functions* (Frankfurt: Deutsch, 1984).
94. J. Spanier and K. B. Oldham, *An Atlas of Functions* (New York: Hemisphere, 1987).
95. G. A. Korn and T. M. Korn, *Mathematical Handbook for Scientists and Engineers* (Second Edition), (New York: McGraw Hill, 1968).
96. S. Wolfram, *Mathematica: A System for Doing Mathematics by Computer* (Second Edition) (New York: Adison-Wesley, 1991).
97. Y. P. Varshni, Rev. Mod. Phys. **29**, 664 (1957).
98. D. Steele, E. R. Lippincott, and J. T. Vanderslice, Rev. Mod. Phys. **34**, 239 (1962).
99. G. Simons, R. G. Parr, and J. M. Finlan, J. Chem. Phys. **59**, 3229 (1973).
100. J. M. Finlan and G. Simons, J. Mol. Spectrosc. **57**, 1 (1975).
101. D. W. Norcross and N. T. Padial, Phys. Rev. A **25**, 226 (1982).
102. L. A. Collins and D. W. Norcross Phys. Rev. A **18**, 467 (1978).
103. N. Chandra, Phys. Rev. A **16**, 80 (1977).
104. E. Gerjuoy, and S. Stein Phys. Rev. **97**, 1671 (1955).
105. A. Dalgarno and R. J. Moffett, Proc. Natl. Acad. Sci. India **33**, 511 (1963).
106. L. G. H. Huxley and R. W. Crompton, *The Diffusion and Drift of Electrons in Gases* (Wiley: New York, 1974).
107. U. Fano and D. Dill, Phys. Rev. A **6**, 185 (1972).
108. W. A. Isaacs and M. A. Morrison, J. Phys. B **25**, 703 (1992).
109. I. I. Fabrikant, J. Phys. B **17**, 4223 (1984).

110. M. A. Morrison, R. W. Crompton, B. C. Saha, and Z. Lj. Petrović, Aust. J. Phys. **40**, 239 (1987).

111. A. U. Hazi, Phys. Rev. A **19**, 920 (1979).

112. C. W. Clark, Phys. Rev. A **30**, 750 (1984).

113. G. Forsythe and C. B. Moler, *Computational Solutions of Linear Algebraic Systems* (Englewood Cliffs, NJ: Prentice-Hall, 1967).

114. M. A. Morrison, and N. F. Lane, Chem. Phys. Lett. **66**, 527 (1979).

THE (NON-ITERATIVE) PARTIAL DIFFERENTIAL EQUATION METHOD: APPLICATION TO ELECTRON-MOLECULE SCATTERING

A. Temkin[1] and C.A. Weatherford[2]

[1] Laboratory for Astronomy and Solar Physics, Code 680
Goddard Space Flight Center, NASA
Greenbelt, MD 20771

[2] Department of Physics and
Center for Nonlinear & Nonequilibrium Aeroscience
Florida A&M University
Tallahassee, FL 32307

INTRODUCTION

In this article, we will present a brief but, hopefully, logically consistent precis of the non-iterative partial differential equation (PDE) approach to electron-molecule scattering. Finer details of the method may be found in articles to which we shall refer.

We begin with some historical remarks to set the stage for what will follow. The non-iterative PDE approach arose in the context of electron-(atomic) hydrogen scattering. Our approach, called the nonadiabatic-theory, was introduced[1] for S-wave scattering for which the time-independent Schrödinger equation can rigorously be reduced to a 3-dimensional (3d) PDE for the S-wave function $\Psi(r_1, r_2, \theta_{12}$ [cf. Ref. 2]. From the asymptotic form of Ψ one can determine the phase shift and hence the scattering cross section. The first aspect of the nonadiabatic theory[1] was motivated by expanding Ψ in Legendre polynomials of $cos\theta_{12}$:

$$\Psi(r_1, r_2, \theta_{12}) = \frac{1}{r_1 r_2} \sum_{l=0}^{\infty} \sqrt{2l+1}\ \Phi_l(r_1, r_2)\ P_l(cos\theta_{12}) \tag{1.1}$$

Here r_1 and r_2 are the distances of the scattered and orbital electrons from the nucleus (assumed fixed) and θ_{12} is the angle between the vectors r_1 and r_2. Substitution of (1.1) into the S-wave Schrödinger equation results in a set of coupled (2-d) PDE's for the function $\Phi_l(r_1, r_2)$.

The chief idea of the nonadiabatic theory was to define a model gotten by eliminating the coupling terms associated with the very first (i.e., $l = 0$) equation of this set. The resultant PDE is

Computational Methods for Electron-Molecule Collisions
Edited by W.M. Huo and F.A. Gianturco, Plenum Press, New York, 1995

$$(-\frac{\partial^2}{\partial {r_1}^2}-\frac{\partial^2}{\partial {r_2}^2}-\frac{2}{r_<}-E)\Phi_0^{(0)}(r_1,r_2)=0 \quad (1.2)$$

Note that $r_<$ is the lesser of r_1 and r_2. The superscript of this function emphasizes that $\Phi_0^{(0)}$ differs from Φ_0 of (1.1). Nevertheless, it turns out that it describes much of the true physics, and the difference between Φ_0 and $\Phi_0^{(0)}$ can be described by a rapidly convergent series involving, however, the (as yet unsolved for) functions $\Phi_l(l > 0)$. These functions also contain much physics and additionally, they can be reasonably approximated.

It was in the context of these equations as well as (1.2), that E.C. Sullivan, a colleague of one of us, came forward with the idea of the non-iterative PDE method.[3] The essence of that idea will be presented in the next section.

To conclude this historical introduction, a similar point of serendipity bestrode our later work on electron-molecule scattering. Serendipity struck when we realized the numerical method referred to above for $e^- - H$ scattering was applicable to a molecular target [second paper of Ref. 4]. There, in an attempt to understand and calculate the resonant structures of the 2.3 eV Π_g resonance in electron-N_2 vibrational excitation, one of us introduced a method called the "hybrid theory" whereby the total wavefunction of the $e - N_2$ system was expanded in a vibrational close-coupling expansion.[4] In a somewhat symbolic notation, that expansion of the wavefunction reads

$$\Psi_m = \Phi_{N_2} \sum_v F_v^{(m)}(r,\theta)\, e^{im\phi}\, \chi_v^{(N_2)}(R) \quad (1.3)$$

In (1.3), the spherical coordinates r, θ, ϕ of the scattered electron are defined with respect to the internuclear axis between the two (nitrogen) atoms (constituting the N_2 molecule at a distance R apart), and $\chi_v^{(N_2)}(R)$ is the vth vibrational function of the N_2 molecule. Φ_{N_2} is the electronic wavefunction of N_2 (which is itself a parametric function of R). Because of the cylindrical symmetry of N_2 about its internuclear axis, the ϕ dependence of the function describing the scattered electron can be factored as indicated in Eq. (1.3). The result is a set of (2d) PDE's for the functions $F_v^{(m)}(r,\theta)$.

In essence, the bulk of our work on the PDE method is concerned with the solution of these resultant PDE's. This work will be outlined and summarized in the subsequent sections of this article, but to conclude this introduction, we[4] originally expanded the function $F_v^{(m)}$ for $m = 1$, l even ($\Rightarrow \Pi_g$ partial wave) in associated Legendre polynomials ($P_l^m(\theta)$), deriving a doubly coupled set of ordinary differential equations from which, notwithstanding the necessity of severe truncations, almost all of the qualitative information and much of the quantitative information about this complicated resonance emerged. However, it is with the PDE form of these equations that we shall here be concerned.

ESSENTIALS OF THE PDE METHOD

The method which we shall describe below was given its first definitive presentation in Ref. 5. To explain the essentials of the method, it is preferable to avoid the R-dependence implicit in the vibrational close-coupling expansion. In other words, for the present purposes, we assume R is a fixed parameter. In that case, the PDE's are no longer coupled and a typical PDE (now 2d) is indicated in Fig. 1. The PDE is broken into a set of difference equations as shown in Fig. 1, and it is the solution of that set which is sought. For pedagogical purposes, assume that there are only nine interior points in the $r-\theta$ plane as shown in Fig. 2. Note that they are numbered in a sequential array and this allows the desired solution to be represented by a vector $f_v \equiv (f_1, f_2, \ldots, f_n)$ (i.e., a one-dimensional array) where n is the number of interior points ($n = 9$ in this example). Realize also that representing the solution as a vector can be done no matter how many dimensions the original PDE contains.

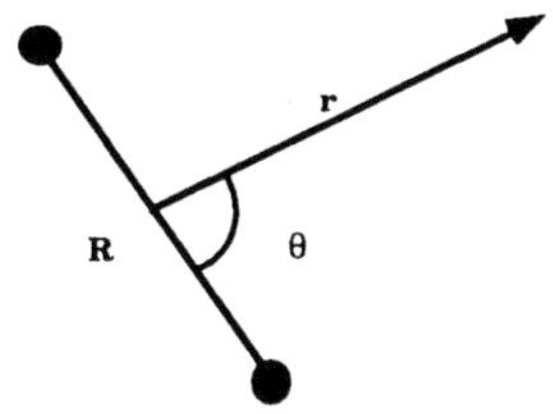

PDE

$$\left\{\frac{\partial^2}{\partial r^2} + \frac{1}{r^2}\left[\frac{1}{sin\theta}\frac{\partial}{\partial\theta}sin\theta\frac{\partial}{\partial\theta} - \frac{m^2}{sin^2\theta}\right] - V(r,\theta) + k^2\right\}f(r,\theta) = 0$$

Difference formulae

$$\frac{\partial^2 f}{\partial r^2} \approx \frac{f(r+\Delta r,\theta) - 2f(r,\theta) + f(r-\Delta r,\theta)}{(\Delta r)^2}$$

$$\frac{\partial^2 f}{\partial \theta^2} \approx \frac{f(r,\theta+\Delta\theta) - 2f(r,\theta) + f(r,\theta-\Delta\theta)}{(\Delta\theta)^2}$$

$$\frac{\partial f}{\partial \theta} \approx \frac{f(r,\theta+\Delta\theta) - f(r,\theta-\Delta\theta)}{2(\Delta\theta)}$$

Difference equations

$$\alpha_{ij}^{+}f_{i+1,j} + \alpha_{ij}^{-}f_{i-1,j} + \beta_{ij}^{+}f_{i,j+1} + \beta_{ij}^{-}f_{i,j-1} + \gamma_{ij}f_{ij} = 0$$

Figure 1. The 2d-PDE (top) is transformed into a set of difference equations using, say, the difference formulae indicated for the partial derivatives. Explicit forms for $\alpha_{ij}^{\pm}, \beta_{ij}^{\pm}, and \gamma_{ij}^{\pm}$ are given in Ref. 5.

Thus the difference form of the PDE may always be represented as an $(N \times N)$ matrix operating on an N-dimensional vector. The right hand side represents the values of the solution (or perhaps its derivative) on the boundary domain within which the PDE holds. The assumption that all the boundary values are known is the numerical equivalent of saying that we are dealing with (and the method applies only to) elliptic PDE's. The set of equations can therefore be written as a matrix equation

$$\underline{\mathbf{A}}\,\vec{f} = \vec{p} \tag{2.1}$$

Note that the matrix $\underline{\mathbf{A}}$ is block tridiagonal and the solution vector can be written as a group of three subvectors (Fig. 3). Treating the blocks and subvectors as units, the key to the non-iterative method is to break the matrix into the product of an upper times a lower tridiagonal matrix. How this is done is illustrated in Fig. 4. Note that the blocks, of which $\underline{\mathbf{L}}$ and $\underline{\mathbf{U}}$ are constructed, are obtained sequentially from the blocks of which $\underline{\mathbf{A}}$ is composed. Note also that the size of the inverses required are the same as the size of the block units. This is the key reason why the method works pragmatically. Because knowing the number of operations, and hence the machine time, to invert an $N \times N$ matrix goes as N^3, then the total time to invert a matrix which can be broken up into n independent matrices of size m ($\Rightarrow N = n + m$), is

$$time \propto n \times m^3 \ll N^3$$

The second major aspect of the non-iterative PDE method is the recovery of the desired solution vector $\vec{f}$ from $\underline{\mathbf{L}}$ and $\underline{\mathbf{U}}$. This is indicated in Fig. 5. Here it is to be noted that inversion requires only the inverses of the diagonal elements of $\underline{\mathbf{L}}$ to get the intermediate vector $\vec{\zeta}$, but once $\vec{\zeta}$ is known, no further inverses are required to evaluate the desired solution vector $\vec{f}$ (only multiplication of matrices times vectors, which is an N^2 operation). This is shown in Fig. 6.

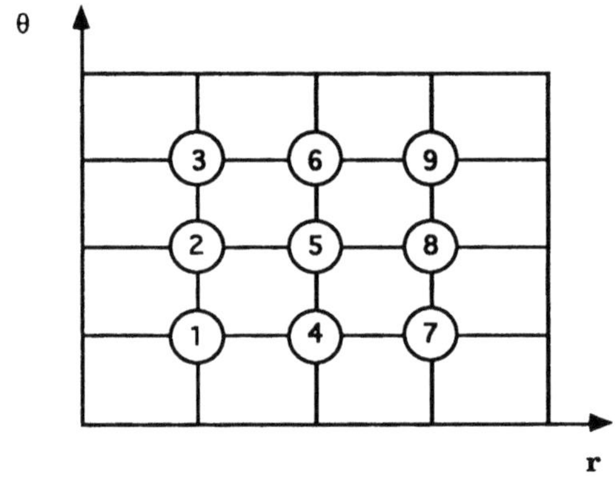

$$f(r_1, \theta_1) = f_1$$
$$f(r_2, \theta_1) = f_2$$
$$\vdots$$
$$f(r_3, \theta_3) = f_9$$

Figure 2. Enumeration of the gridpoints.

$$\left[\begin{array}{ccc|ccc|ccc}
x & x & 0 & x & & \mathbf{0} & & & \\
x & x & x & & x & & & \mathbf{0} & \\
0 & x & x & \mathbf{0} & & x & & & \\
\hline
x & & \mathbf{0} & x & x & 0 & x & & \mathbf{0} \\
 & x & & x & x & x & & x & \\
\mathbf{0} & & x & 0 & x & x & \mathbf{0} & & x \\
\hline
 & & & x & & \mathbf{0} & x & x & 0 \\
 & \mathbf{0} & & & x & & x & x & x \\
 & & & \mathbf{0} & & x & 0 & x & x
\end{array}\right]
\left[\begin{array}{c} f_1 \\ f_2 \\ f_3 \\ \hline f_4 \\ f_5 \\ f_6 \\ \hline f_7 \\ f_8 \\ f_9 \end{array}\right]
=
\left[\begin{array}{c} p_1 \\ p_2 \\ p_3 \\ \hline p_4 \\ p_5 \\ p_6 \\ \hline p_7 \\ p_8 \\ p_9 \end{array}\right]$$

Figure 3. The matrix form of the set of difference equations representing the PDE. The x's are known numbers from the PDE and the vector $\vec{p}$ on the right-hand-side is known from the boundary conditions.

Write formally

$$\begin{bmatrix} A_{11} & A_{12} & 0 & \cdots \\ A_{21} & A_{22} & A_{23} & \cdots \\ 0 & A_{32} & A_{33} & \cdots \\ \vdots & \vdots & \vdots & \ddots \end{bmatrix}$$

$$= \begin{bmatrix} L_{11} & 0 & 0 & \cdots \\ L_{21} & L_{22} & 0 & \cdots \\ L_{31} & L_{32} & L_{33} & \cdots \\ \vdots & \vdots & \vdots & \ddots \end{bmatrix} \begin{bmatrix} 1 & U_{12} & U_{13} & \cdots \\ 0 & 1 & U_{23} & \cdots \\ 0 & 0 & 1 & \cdots \\ \vdots & \vdots & \vdots & \ddots \end{bmatrix}$$

Formally multiply $\underline{\mathbf{L}}\,\underline{\mathbf{U}}$ and match to right-hand-side.

For example

$$\left.\begin{array}{l} A_{11} = L_{11} \\ A_{12} = L_{11} U_{12} \\ A_{21} = L_{21} \\ A_{22} = L_{21} U_{12} + L_{22} \end{array}\right] \Rightarrow \left[\begin{array}{l} A_{11} = L_{11} \\ U_{12} = L_{11}^{-1} A_{12} \\ L_{21} = A_{21} \\ L_{22} = A_{22} - L_{21}^{-1} + U_{12} \end{array}\right.$$

Figure 4. Decomposition of difference matrix into upper times lower triangular matrices.

Difference form of PDE: $\underline{\mathbf{A}}\,\vec{f} = \vec{p}$

Having evaluated $\underline{\mathbf{L}}$ and $\underline{\mathbf{U}}$ from

$$\underline{\mathbf{A}} = \underline{\mathbf{L}}\,\underline{\mathbf{U}} \Rightarrow \underline{\mathbf{L}}\,\underline{\mathbf{U}}\,\vec{f} = \vec{p}$$

Let $\vec{\zeta} \equiv \underline{\mathbf{U}}\,\vec{f}$ **and solve** $\underline{\mathbf{L}}\,\vec{\zeta} = \vec{p}$

for (intermediate) vector $\vec{\zeta}$

$$\begin{bmatrix} L_{11} & 0 & 0 \\ L_{21} & L_{22} & 0 \\ 0 & L_{32} & L_{33} \end{bmatrix} \begin{bmatrix} \zeta_1 \\ \zeta_2 \\ \zeta_3 \end{bmatrix} = \begin{bmatrix} p_1 \\ p_2 \\ p_3 \end{bmatrix}$$

$$\left.\begin{array}{r} L_{11}\zeta_1 = p_1 \\ L_{12}\zeta_1 + L_{22}\zeta_2 = p_2 \\ L_{32}\zeta_2 + L_{32}\zeta_3 = p_3 \end{array}\right] \Rightarrow \left[\begin{array}{l} \zeta_1 = L_{11}^{-1} p_1 \\ \zeta_2 = L_{22}^{-1}(p_2 - L_{12}\zeta_1) \\ \zeta_3 = L_{33}^{-1}(p_3 - L_{32}\zeta_2) \end{array}\right.$$

Figure 5. First part of difference equation corresponding to PDE: breaking $\underline{\mathbf{A}}$ into $\underline{\mathbf{LU}}$ and solving $\vec{\zeta}$ sequentially and noniteratively.

In summary, the basics of the non-iterative PDE method ($\underline{\mathbf{A}}\,\vec{f} = \vec{p}$) requires four major steps:

(a) evaluation of the matrix $\underline{\mathbf{A}}$ from the PDE;
(b) evaluation of the elements of $\underline{\mathbf{L}}$ and $\underline{\mathbf{U}}$ out of the elements of which $\underline{\mathbf{A}}$ is composed ($\underline{\mathbf{A}} = \underline{\mathbf{L}} \times \underline{\mathbf{U}}$);
(c) evaluation of the intermediate vector $\vec{\zeta}$ (from $\underline{\mathbf{L}}\,\vec{\zeta} = \vec{p}$); and
(d) evaluation of the solution vector $\vec{f}$ ($\underline{\mathbf{U}}\,\vec{f} = \vec{\zeta}$)

Once $\underline{\mathbf{A}}$ is known, each the steps which follow it are accomplished sequentially and non-iteratively. The method can be extended to include exchange (including a multi-configuration target wavefunction), and the method can be generalized to higher than 2d PDE's. These developments will be reviewed (again from a pedagogical point of view) in the next section.

In the penultimate section, we shall review various applications (all in the context of $e-N_2$ scattering) we have made of the method up to and including a new calculation [see Ref. 21] incorporating most of the present theoretical developments. In the final section, we discuss an as yet not applied extension of the method to e-polyatomic molecule scattering.

OTHER ELEMENTS OF THE PDE METHOD

A. Incorporation of Exchange (SCF Target)

At the outset, it would appear that the non-iterative technique would not apply when exchange is included because exchange is described by a full kernel, which is to say exchange (i.e., antisymmetry of the total wave including the incident electron with respect to the exchange of the coordinates of any two electrons) gives rise to an integro-differential equation for a scattering orbital, in which the kernel of the integral is non-zero for every value of $\vec{x}$ and $\vec{r}$. This is illustrated in Fig. 7, even in the case that the target is assumed to be a closed-shell (N_2) molecule which can be approximated by a single-determinant, each of whose electron orbitals are naturally described by a vector difference from each nucleus ($\vec{R}_1$ and $\vec{R}_2$). Even here the integral terms $W_\alpha^{(m)}(\vec{r})$ are full matrices if the kernel were to be discretized at the outset.

However, a simple trick, based on the fact that the non-separable part of the kernel is the Coulomb repulsion between the incident electron ($\vec{r}$) and an orbital electron ($\vec{x}$), utilizes the property that such a potential is also the Green's function of the Laplacian. This means, as indicated by the second equation in Fig. 8, that when W_α is acted upon by the ∇^2 operator, it gives a delta function; thus the integral representing W_α can be trivially carried out and reduces to a simple product.

In other words, W_α satisfies a PDE which is of the same type as $f(r,\theta)$ satisfies with a local potential. Thus one now has a set of non-integral PDE's to solve, to which the non-iterative method as described above can completely and straightforwardly be extended. Finally, to complete this subsection, the (3d) PDE's can be reduced to (2d) equations which come from exploiting the cylindrical symmetry of both the bound orbitals and the scattered partial waves. This, together with the method of dealing with exchange, is detailed in our paper, Ref. 6, and the resultant (2d) PDE's are presented at the bottom of Fig. 8.

B. Incorporation of Exchange (MCSCF Target)

Calculations utilizing the full formalism that we have thus far developed, have been carried out and will be reviewed in the next section. We would like to conclude

Recall

$$\underline{\mathbf{A}}\vec{\mathbf{f}} = \underline{\mathbf{L}}\,\underline{\mathbf{U}}\vec{\mathbf{f}} = \underline{\mathbf{L}}\vec{\zeta} = \vec{\mathbf{p}}$$

Having solved for $\vec{\zeta}$ **, solve** $\underline{\mathbf{U}}\vec{\mathbf{f}} = \vec{\zeta}$ **for** $\vec{\mathbf{f}}$

$$\begin{bmatrix} I & U_{12} & 0 & \cdots \\ & I & U_{23} & \cdots \\ \mathbf{0} & & I & \cdots \\ \vdots & \vdots & \vdots & \ddots \end{bmatrix} \begin{bmatrix} f_1 \\ f_2 \\ f_3 \\ \vdots \end{bmatrix} = \begin{bmatrix} \zeta_1 \\ \zeta_2 \\ \zeta_3 \\ \vdots \end{bmatrix}$$

Solve in reverse order

$$\left.\begin{matrix} f_3 = \zeta_3 \\ f_2 + U_{23} f_3 = \zeta_2 \\ f_1 + U_{12} f_2 = \zeta_1 \end{matrix}\right] \Rightarrow \left[\begin{matrix} f_3 = \zeta_3 \\ f_2 = \zeta_2 - U_{23} f_3 \\ f_1 = \zeta_1 - U_{12} f_2 \end{matrix}\right.$$

Figure 6. Second part of solution of difference equation corresponding to PDE: with $\vec{\zeta}$ known, solving for solution $\vec{f}$ sequentially. Note that no inverses of blocks are required in this step.

this subsection with the formal extension of the method of including exchange (not previously published) to the situation where the target is described by a multiple configuration self-consistent field (MCSCF) wavefunction as opposed to a single configuration self-consistent field (SCF) wavefunction. Then the target is represented by a linear superposition of determinants,

$$\Phi_{N_2} = \sum_n C_n \, det(\phi_n) \tag{3.1}$$

wherein each determinant is treated as in Fig. 7, but carries a label n to signify which configuration it refers to. In particular, the exchange terms in the scattering equation now carry an additional index n and an addtional sum over n,

$$\sum_\alpha W_\alpha^{(m)} \, \phi_\alpha \rightarrow \sum_{\alpha,n} W_{\alpha,n}^{(m)} \, \phi_{\alpha,n}, \tag{3.2}$$

where

$$W_{\alpha,n}^{(m)}(\vec{r}) = C_n \int d\vec{x} \; \phi_{\alpha,n}(\vec{x}) \, \frac{1}{|\vec{x} - \vec{r}|} \, F^{(m)}(\vec{x}). \tag{3.3}$$

Operating with $\nabla_{\vec{r}}^2$ gives an equation for each n:

$$\nabla_{\vec{r}}^2 \, W_{\alpha,n}^{(m)}(\vec{r}) = -4\pi \, C_n \, \phi_{\alpha,n}(\vec{r}) \, F^{(m)}(\vec{r}) \tag{3.4}$$

In other words, the set of equations that must be simultaneously solved now includes an additional set of terms n, for as many terms as there are in the sum (3.2). The non-iterative method is in principle, equally applicable to this set.

C. Extension of PDE Method to Three (and Higher) Dimensions

The extension to 3d (and higher dimensions) is based on the remark made above that a solution can be represented by a vector no matter how many dimensions it involves. Each dimension multiplies the length of the vector by the number of points one uses to approximate the additional dimension. The extension was given in Ref. 7 with the specific application to $e - N_2$ scattering, including vibrational excitation, in mind (a detailed exposition of the method as well as model results is contained in Ref. 8, to which the reader is particularly referred).

Static exchange ansatz:

$$\Psi^{(m)} = \sum_{i=1}^{15} (-1)^{p_i} F^{(m)}(x_i)\Phi_{N_2}(x^{(i)}; R)$$

where x_i are the space and spin coordinates of the ith electron and $x^{(i)}$ is the collection of coordinates of the remaining (fourteen) electrons. Assume a single configuration self-consistent field (SCF) approximation, Φ_{N_2}, for the ground state of the N_2 molecule. Φ_{N_2} is made out of orbitals ϕ_α, $(\alpha = 1\sigma_g, \ldots, 1\pi_{uy})$

$$\Phi_{N_2} = det\ \left[\phi^{\uparrow}_{\alpha_1}(1)\phi^{\downarrow}_{\alpha_1}(2)\phi^{\uparrow}_{\alpha_2}(3)\phi^{\downarrow}_{\alpha_2}(4)\cdots\phi^{\uparrow}_{\alpha_7}(13)\phi^{\downarrow}_{\alpha_7}(14)\right]$$

where, for example, $\uparrow$ signifies a spin-up orbital:

$$\phi^{\uparrow}_{\alpha}(j) = \phi_\alpha\left[(\vec{r}_j - \vec{R}_1), (\vec{r}_j - \vec{R}_2)\right]\chi_{\frac{1}{2},\frac{1}{2}}(j)$$

Derive the integro-differential equation for $F^{(m)}$

$$-(\nabla^2 + k^2)F^{(m)}(\vec{r}\,) = V(\vec{r}\,)F^{(m)}(\vec{r}\,) - 2\sum_{\alpha=1}^{7} W^{(m)}_\alpha(\vec{r}\,)\phi_\alpha(\vec{r}\,)$$

where

$$W^{(m)}_\alpha(\vec{r}\,) = 2\int d\vec{x}\ \phi^*_\alpha(\vec{x}\,)\frac{1}{|\,\vec{x} - \vec{r}\,|}F^{(m)}(\vec{x}\,)$$

Figure 7. First part of inclusion of exchange in PDE formalism; the matrix corresponding to $W^{(m)}_\alpha$ is full.

The basic form of the PDE is given at the top of Fig. 9. The new dimension, R (for comparison cf. Fig. 1) occurs as a second partial derivative as well as in the potential $V = V(r, \theta, R)$. For the purposes of this discussion, let this second derivative be represented by a 3-point formula analogous to that used for r in Fig. 1. The whole set of difference formulae now is represented by a block pentagonal matrix. This is illustrated in more detail in Fig. 10, but referring first to Fig. 9, we see how equating $\underline{\mathbf{A}}$ to $\underline{\mathbf{L}} \times \underline{\mathbf{U}}$, one can go through–row-by-row–to find the elements (blocks) of $\underline{\mathbf{L}}$ and $\underline{\mathbf{U}}$ sequentially from the elements of $\underline{\mathbf{A}}$. In Fig. 10, the block structure is given in greater detail showing its pentagonal form, and in particular, the manner in which the triple indices $i, j, k(i', j', k')$ are related to the single index $n(n')$. (Fig. 10 is an extended version of a figure presented in Ref. 7).

Utilize

$$\nabla_{\vec{r}}^2 \left(\frac{1}{| \vec{r} - \vec{x} |}\right) = -4\pi \; \delta(\vec{r} - \vec{x})$$

so that

$$\nabla_{\vec{r}}^2 \; W_\alpha^{(m)}(\vec{r}\,) = -4\pi \; \phi_\alpha(\vec{r}\,) F^{(m)}(\vec{r}\,) \qquad , \alpha = 1, \ldots, 7$$

To reduce integro-differential equations to simultaneous (coupled), *but nonintegral*, PDE's ($\underline{z} = (r, \theta)$), using

$$\phi_\alpha(\vec{r}\,) = \frac{1}{r} \; \phi_\alpha(\underline{z}\,) \frac{(-1)^{m_\alpha}}{\sqrt{2\pi}} \; e^{im_\alpha \phi}$$

$$F^m(\vec{r}\,) = \frac{1}{r} \; f^{(m)}(\underline{z}\,) \frac{(-1)^m}{\sqrt{2\pi}} \; e^{im\phi}$$

to obtain

$$\begin{cases} \left[\Delta(m) + k^2\right] f^{(m)}(\underline{z}) = V(\underline{z}) f^{(m)}(\underline{z}) - \dfrac{2}{r} \displaystyle\sum_{\alpha=1}^{7} \phi_\alpha(\underline{z}) w_\alpha(\underline{z}) \\ \Delta(m - m_\alpha) w_\alpha(\underline{z}) = -\dfrac{2}{r} \phi_\alpha(\underline{z}) f^{(m)}(\underline{z}) \end{cases}$$

where

$$\Delta(\mu) \equiv \frac{\partial^2}{\partial r^2} + \frac{1}{r^2} \left[\frac{\partial^2}{\partial \theta^2} + \frac{1}{sin\theta} \frac{\partial}{\partial \theta} sin\theta \frac{\partial}{\partial \theta} - \frac{\mu^2}{sin^2\theta}\right]$$

Figure 8. Second part of inclusion of exchange: the kernel itself satisfies a similar PDE, and the two PDE's are solved simultaneously.

In addition to higher dimensional PDE's, the use of higher order difference formulae will also lead to banded (block) matrices of greater band width than the basic tridiagonal form. These issues are also dealt with, and model results presented, in Ref. 8.

D. Boundary Conditions

Boundary conditions arise from (a) initial conditions, (b) asymptotic conditions, and/or (c) symmetry conditions. For all these, one must have sufficient knowledge in advance so that the PDE equations can be integrated. Essentially this amounts to knowing the vector on the right-hand-side of the corresponding difference equation, which we labeled $\vec{p}$ in the various foregoing figures.

To show how this goes, consider the (3d) case. The relevant wavefunction is called $\Psi^{(m)}(r, \theta, R)$. From the general theory of quantum mechanics, one knows $\Psi^{(m)}$ must be finite (but not necessarily zero) at $r = 0, R = 0$. But its value at $r = 0$ and for $R = 0$ is unknown. However, by defining a new function $s^{(m)}(r, \theta, R)$

$$\left\{\frac{\partial^2}{\partial r^2}+\frac{1}{r^2}\left[\frac{\partial^2}{\partial\theta^2}+\frac{1}{sin\theta}\frac{\partial}{\partial\theta}sin\theta\frac{\partial}{\partial\theta}-\frac{m^2}{sin^2\theta}\right]+\frac{m_e}{M_{nu}}\frac{\partial^2}{\partial R^2}+V(r,\theta,R)+k^2\right\}f^{(m)}(r,\theta,R)=0$$

Difference equation corresponding to PDE above is $\underline{\mathbf{A}}\vec{f}=\vec{p}$,
where the $\underline{\mathbf{A}}$ matrix is block pentagonal; it is decomposed as shown below.

$$\begin{bmatrix} A_{11} & A_{12} & 0 & A_{14} & 0 & 0 & 0 \\ A_{21} & A_{22} & A_{23} & 0 & A_{24} & 0 & 0 \\ 0 & A_{32} & A_{33} & 0 & 0 & A_{36} & 0 \\ A_{41} & 0 & 0 & A_{44} & A_{45} & 0 & A_{47} \\ 0 & A_{52} & 0 & A_{54} & A_{55} & A_{56} & 0 \\ 0 & 0 & A_{63} & 0 & A_{65} & A_{66} & 0 \\ 0 & 0 & 0 & A_{74} & 0 & 0 & A_{77} \end{bmatrix} = \begin{bmatrix} L_{11} & 0 & 0 & 0 & 0 & 0 & 0 \\ L_{21} & L_{22} & 0 & 0 & 0 & 0 & 0 \\ L_{31} & L_{32} & L_{33} & 0 & 0 & 0 & 0 \\ L_{41} & L_{42} & L_{43} & L_{44} & 0 & 0 & 0 \\ L_{51} & L_{52} & L_{53} & L_{54} & L_{55} & 0 & 0 \\ L_{61} & L_{62} & L_{63} & L_{64} & L_{65} & L_{66} & 0 \\ L_{71} & L_{72} & L_{73} & L_{74} & L_{75} & L_{76} & L_{77} \end{bmatrix} \times \begin{bmatrix} 1 & U_{12} & U_{13} & U_{14} & U_{15} & U_{16} & U_{17} \\ 0 & 1 & U_{23} & U_{24} & U_{25} & U_{26} & U_{27} \\ 0 & 0 & 1 & U_{34} & U_{35} & U_{36} & U_{37} \\ 0 & 0 & 0 & 1 & U_{45} & U_{46} & U_{47} \\ 0 & 0 & 0 & 0 & 1 & U_{56} & U_{57} \\ 0 & 0 & 0 & 0 & 0 & 1 & U_{67} \\ 0 & 0 & 0 & 0 & 0 & 0 & 1 \end{bmatrix}$$

$$\begin{bmatrix} A_{11}=L_{11} \\ A_{12}=L_{11}U_{12} \\ A_{13}=L_{11}U_{13} \\ A_{14}=L_{11}U_{14} \end{bmatrix} \Rightarrow \begin{bmatrix} L_{11}=A_{11} & A_{21}=L_{21} \\ U_{12}=L_{11}^{-1}A_{12} & A_{22}=L_{21}U_{12}+L_{22} \\ U_{13}=L_{11}^{-1}A_{13} & A_{23}=L_{21}U_{13}+L_{22}U_{23} \\ U_{14}=L_{11}^{-1}A_{14} & \vdots \end{bmatrix} \Rightarrow \begin{bmatrix} L_{21}=A_{21} \\ L_{22}=A_{22}-L_{21}U_{12}=A_{22}-A_{21}(L_{11}^{-1}A_{21}) \\ L_{23}=L_{22}^{-1}(A_{23}-L_{21}U_{13})=L_{22}^{-1}(A_{23}-A_{21}L_{11}^{-1}A_{13}) \\ \vdots \end{bmatrix}$$

Figure 9. Description of the manner in which the matrices $\underline{\mathbf{U}}$ and $\underline{\mathbf{L}}$ are generated from the block pentagonal matrix $\underline{\mathbf{A}}$. General formulae for a banded matrix of any width are given in Ref. 8.

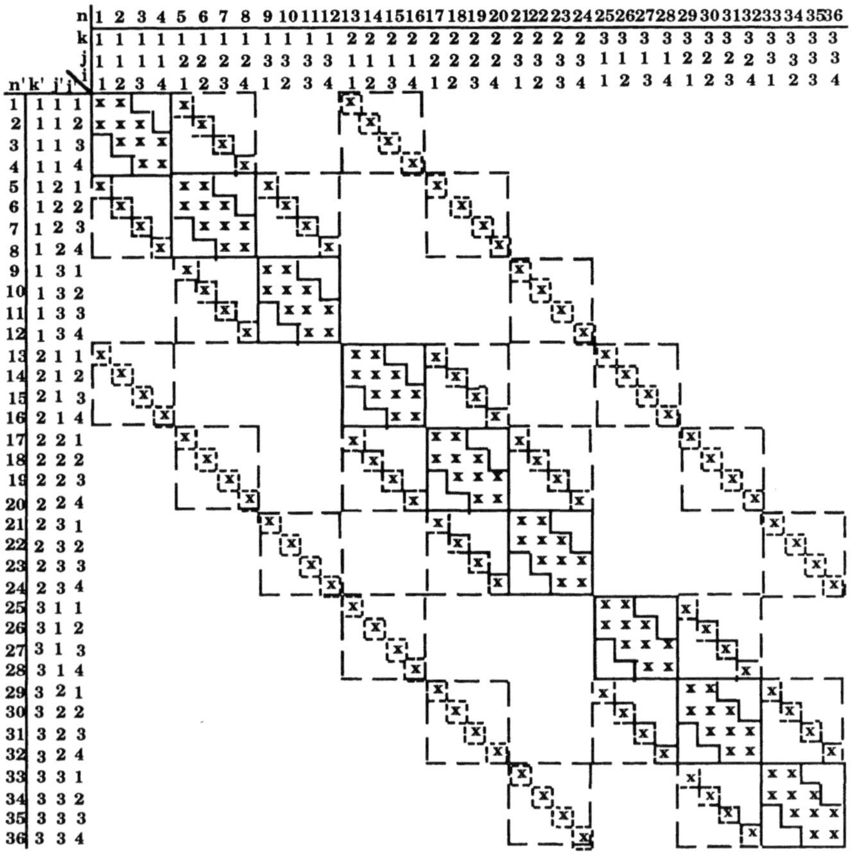

Figure 10. A matrix for example in text: $\underline{\mathbf{A}}_{n',n} = \underline{\mathbf{A}}_{i'j'k',ijk}$ are specified by $\mathbf{X}$ =nonzero elements, unfilled entries are zero. Note that $\mathbf{X}$ occurs only when one of the entries (i', i), (j', j),(k', k) differ by zero or one unit, the others remaining diagonal. This allows at most 5 $\mathbf{X}$'s per row or column for this example.

$$\Psi^{(m)} = e^{im\phi}\ \frac{s^{(m)}(r,\theta,R)}{rR}, \tag{3.5}$$

the PDE for $s^{(m)}(r,\theta,R)$

$$\begin{aligned}\{\frac{\partial^2}{\partial r^2} + \frac{2m_e}{M_n}\frac{\partial^2}{\partial R^2} + \frac{1}{r^2}(\frac{1}{sin\theta}\frac{\partial}{\partial\theta}sin\theta\frac{\partial}{\partial\theta} - \frac{m^2}{sin^2\theta})\\ + \epsilon_{N_2}(R) + V(R,r,\theta) - E\}\ s^{(m)}(r,\theta,R) = 0\end{aligned} \tag{3.6}$$

must necessarily have the boundary conditions

$$0 = s^{(m)}(0,\theta,R) = s^{(m)}(r,\theta,0) \tag{3.7}$$

Note that in (3.6), $M_n = 14m_p$ for N_2, where m_p is the mass of a proton. If the energy E is below the dissociation threshold, then dissociation cannot take place: that means

$$lim_{R\to\infty}s^{(m)}(r,\theta,R) = 0$$

This is another boundary condition. But when the electron coordinate (r) goes to infinity, the wavefunction does not vanish, because that asymptotic form must describe elastic scattering [as well as rotational and vibrational scattering within the ground electronic state, assuming (as we eventually do) that the energy is also below the threshold of the first electronically excited state]. Without going into further details here (cf. Ref. 7) the physical solution can be expressed as a linear combination of solutions $s_{vl}^{(m)}$, where each $s_{vl}^{(m)}$ is identified with a specific vibrational state and a specific spherical harmonic at large $r \equiv \rho$. This implies [$S_v(R)$ is a vibrational wavefunction of the target (N_2)].

$$lim_{r\to\rho} s_{vl}^{(m)}(\rho,\theta,R) = const. \times S_v(R)P_{lm}(\theta) \tag{3.8}$$

Note that the right-hand-side of (3.8) actually solves the PDE in the limit, because as $r \to \infty$, $V(r,\theta,R) \to 0$. Therefore, from a linear superposition of $s_{vl}^{(m)}(r,\theta,R)$, for values of $r \leq \rho$, one can actually construct the physical solution and in particular, the desired scattering matrix.

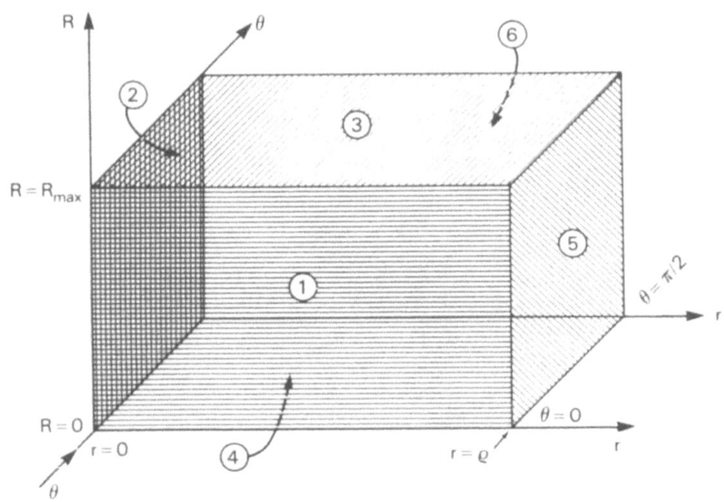

Figure 11. Integration volume of 3d-PDE. Boundary conditions are: surfaces 2,3,4, $s_{vf}^{(m)}(0,\theta,R) = s_{vf}^{(m)}(r,\theta,R_{max}) = s_{vf}^{(m)}(r,\theta,0) = 0)$; surface 5, $s_{vf}^{(m)}(\rho,\theta,R) = P_{lm}(\theta) \times S_v(R)$; surfaces 1 and 6, $s_{vf}^{(m)}(r,\theta,R)|_{\theta=0,\pi/2}$ and/or $\left[\frac{\partial}{\partial\theta}s_{vf}^{(m)}(r,\theta,R)\right]_{\theta=0,\pi/2} = 0$, (same as b.c. for $P_{lm}(\theta)$.

Putting all this together, we show in Fig. 11, the region and the boundary conditions for the 3d problem. The 3d problem is presently being investigated, however, for the 2d problem (for which the boundary conditions are correspondingly simpler, both in the vibrational close-coupling and the fixed-nuclei approximations), calculations have been carried out, and thus some of those results will be shown in the next section.

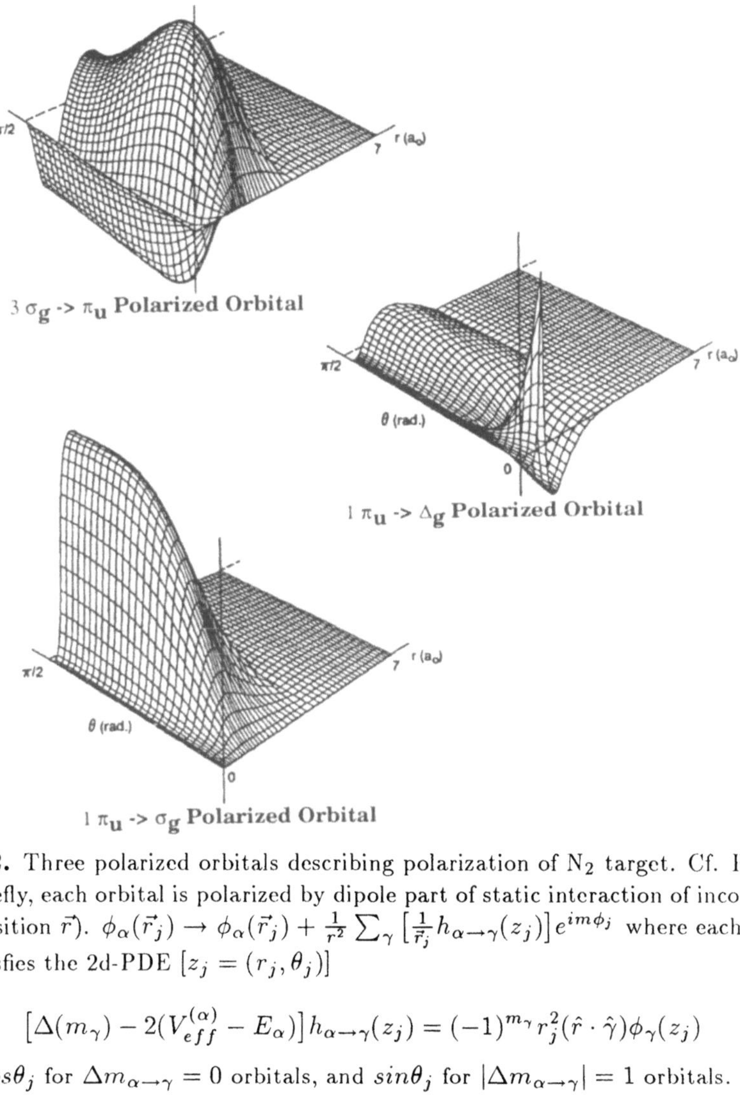

Figure 12. Three polarized orbitals describing polarization of N_2 target. Cf. Ref. 10 for details: briefly, each orbital is polarized by dipole part of static interaction of incoming electron (at position $\vec{r}$). $\phi_\alpha(\vec{r}_j) \rightarrow \phi_\alpha(\vec{r}_j) + \frac{1}{r^2}\sum_\gamma \left[\frac{1}{\vec{r}_j} h_{\alpha\rightarrow\gamma}(z_j)\right] e^{im\phi_j}$ where each polarized orbital satisfies the 2d-PDE $[z_j = (r_j, \theta_j)]$

$$\left[\Delta(m_\gamma) - 2(V_{eff}^{(\alpha)} - E_\alpha)\right] h_{\alpha\rightarrow\gamma}(z_j) = (-1)^{m_\gamma} r_j^2 (\hat{r}\cdot\hat{\gamma})\phi_\gamma(z_j)$$

$\hat{r}\cdot\hat{\gamma}$ is 2 $cos\theta_j$ for $\Delta m_{\alpha\rightarrow\gamma} = 0$ orbitals, and $sin\theta_j$ for $|\Delta m_{\alpha\rightarrow\gamma}| = 1$ orbitals.

APPLICATIONS

Although the PDE method, thus far, has been discussed in the context of scattering, its first application was in fact to a bound-state problem: evaluating the polarized orbitals from which, in particular, the polarization potential can be derived.

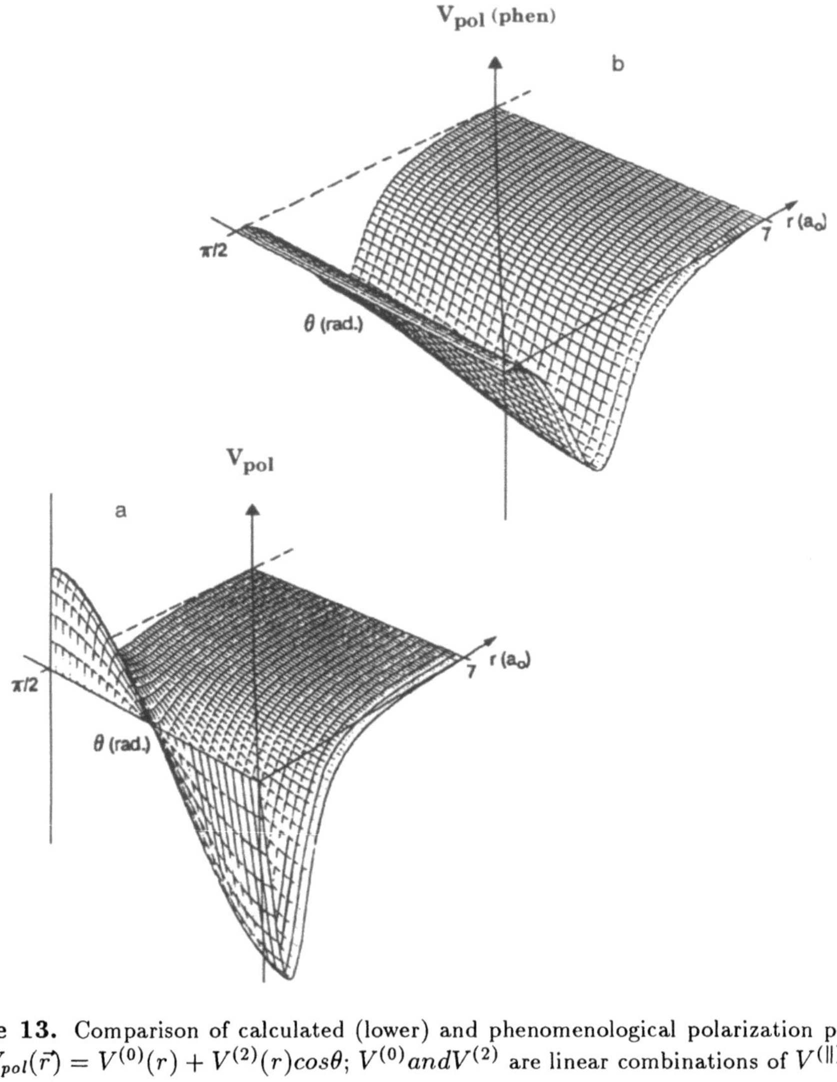

Figure 13. Comparison of calculated (lower) and phenomenological polarization potentials: $V_{pol}(\vec{r}) = V^{(0)}(r) + V^{(2)}(r)cos\theta$; $V^{(0)} and V^{(2)}$ are linear combinations of $V^{(\|)}$ and $V^{(\perp)}$ given by

$$V^{\{\|,\perp\}}(r) = \frac{4}{r^2}\int_0^r dx\, x^2 \int_0^{\pi/2} d\overline{\theta} sin\overline{\theta} \sum_{\alpha,\gamma_n} \phi_\alpha(x,\overline{\theta})\, h_{\alpha\rightarrow\gamma_n}(x,\overline{\theta})\, \{cos\overline{\theta}, sin\overline{\theta}\}$$

Polarization describes the effect of the distortion of the target system by the incoming electron. The method of polarized orbitals,[9] applied originally to atomic targets, was extended to molecular problems in Ref. 10. Briefly, each orbital in the target gets perturbed by the incident particle with the perturbed functions satisfying PDE's corresponding to a stationary electron at a distance $\vec{r}$ from the center of the molecule. The polarization potential is calculated as integrals over these functions together with the unperturbed functions. Some equations and partial results are given in Fig. 12, but the main result, the polarization potential (at $R = R_0$) is given in Fig. 13 together with a comparison with the phenomenological polarization potential, which was used in the first "*ab initio*" calculations of $e - N_2$ scattering in the fixed-nuclei approximation.[11] Whereas, in the direction of the internuclear axis ($\theta \cong 0$), they are quite similar to each other throughout the range of r, they are entirely different from each other in the perpendicular direction ($\theta \cong \pi/2$) for the interior values of r. Indeed, our calculated V_{pol} becomes somewhat repulsive there.

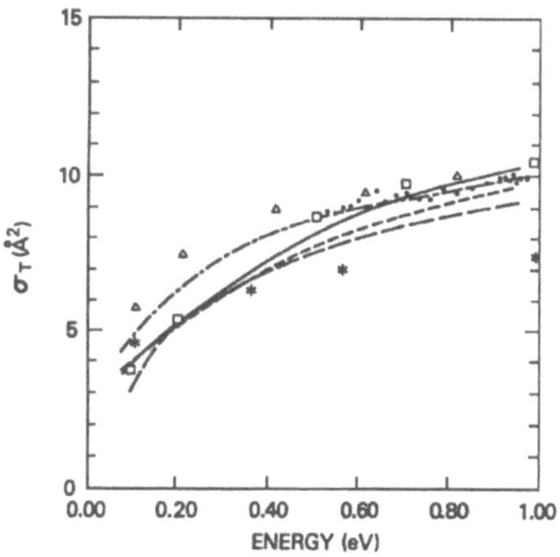

Figure 14. Total cross section at very low energy; vib close-coupling with six states in $^2\Sigma_g^+$ symmetry using MCSCF target (Ref. 12) (□□□); Hybrid Theory with fixed nuclei $^2\Sigma_g^+$ symmetry and MCSCF target (Ref. 13) (———); Hybrid Theory with fixed nuclei $^2\Sigma_g^+$ symmetry and SCF target (Ref. 14) (— — —); Padial and Norcross theory (Ref. 15) (- - -); Morrison et. al. theory (Ref. 16) (—-—); Kennerly experiment (Ref. 17) (• • •); Sohn et. al. experiment (Ref. 18) (∗ ∗ ∗); Jost et. al. experiment (Ref. 19) (△ △ △).

The second application we wish to discuss is a scattering application (in the fixed-nuclei approximation[20]) of electrons from N_2.[13] Although the most prominent low-energy feature of this process is the Π_g resonance with all of its substructure (at $k^2 \cong 2.4eV$), it turns out that even lower energy ($k^2 < 1eV$) discrepancies—experimental and calculational—had arisen, both in the normalization and shape of

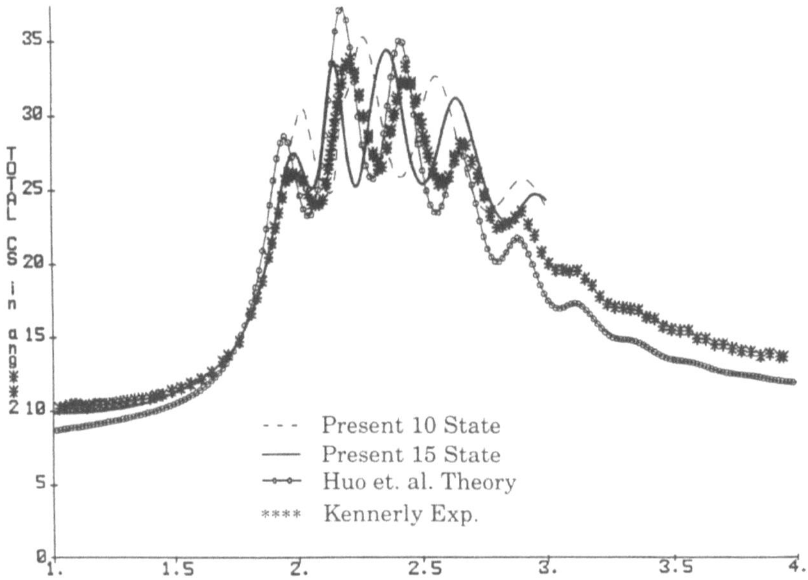

Figure 15. Total cross section in the Π_g resonance region in two vibrational close-coupling calculations (Refs. 12 and 21) versus the calculation of Huo et. al. (Ref. 23) and the experiment of Kennerly (Ref. 17).

the total scattering cross section ($\sigma_T \Rightarrow$ summed over all vibrational states). The energy in question is sufficiently below the Π_g resonance that only the lowest symmetry (i.e., Σ_g^+ partial wave) contributes to the scattering. The equations are just those given in Fig. 9, but in this calculation, we first introduced the idea of using a multiconfigurational (MCSCF) target wavefunction to calculate the static part of the electron-N_2 interaction while at the same time, using the single configuration (SCF) target to calculate the effect of exchange. In doing so however, we added to the static potential a polarization potential generated in Ref. 7, but containing an additional parameter (r_0) which is tuned to give the Π_g resonance at the correct energy. Referring to Fig. 9, this means for V in Fig. 9, we use

$$V = V_{static}(MCSCF) + \left[\, 1 - e^{-(\frac{r}{r_0})^2} \right] V_{pol}^{(OT)}$$

where $V_{pol}^{(OT)}$ in the polarization potential generated in Ref. 10 (cf. Ref. 21 for further discussion of this, particularly in view of a criticism of Meyer et. al.[22]). With regard to the fixed-nuclei calculation, the essential results at very low energy are given in Fig. 14. One sees at the outset that there are quantitative differences among (some of) the experiments. Until that is settled, one can not say anything definitive about the various calculations, or more importantly, the methodologies involved. One thing that seems certain is that one can use the fixed-nuclei approximation,[20] particularly because only the Σ_g partial wave is important in this energy range, and it is non-resonant. From a slightly less quantitative point of view, however, it is significant that there is general agreement of both the magnitude of the cross section and its upward trend with energy. With respect to our own results, note those marked (□) refer to a vibrational close-coupling calculation containing six vibrational states and an MCSCF target. The fact that the results agree with our fixed-nuclei results (—) at scattering energies 0.1 and 0.2 eV, and are slightly higher than the latter above 0.5 eV suggests that the 6-state vibrational close-coupling expansion is not quite converged at the "higher energy." Thus we tend to trust our fixed-nuclei results there, and the circumstance again points to the power of the fixed-nuclei approximation,[20] when resonant structures are not invloved.

The final application we shall discuss is the resonant cross section at slightly higher energy,[14] in the heart of the resonance region. The salient results on the total cross sections are presented in Fig. 15. Experiment is represented by Kennerly's results,[17] while not the first to show the substructure in σ_T, have been confirmed many times and are therefore considered to be quite accurate both in shape and magnitude. Our own results are given for a 10 and 15 state vibrational close-coupling expansion so as to display the convergence of this expansion in vibrational states. We consider the 15 state result as reasonably converged up to the second peak ($E \leq 2.1eV$). It is gratifying that both the position and height of the first two peaks are more accurate than any other calculation. Particularly in earlier calculations (not shown here), a much larger first peak than is seen here, is uniformly predicted. It is also significant that our calculation has required an MCSCF target to give this result. On the other hand, a (necessarily limited) vibrational close-coupling expansion begins to fail to give the higher structure correctly and fully. Here the calculation of Huo et. al.,[23] which appears to be the best of the *ab initio* calculations, starts from a point of view first embodied in the "boomerang model," [24] does very well on the substructure as a whole (cf. Schneider et. al.[25] for the inception of the *ab initio* approach as a whole).

The main thrust of our calculation,[21] however, is the shape of the angular distribution in the immediate vicinity of the major (second) substructure peak at $E \cong 2.1eV$. The results are illustrated in Fig. 16, wherein the angular distribution is

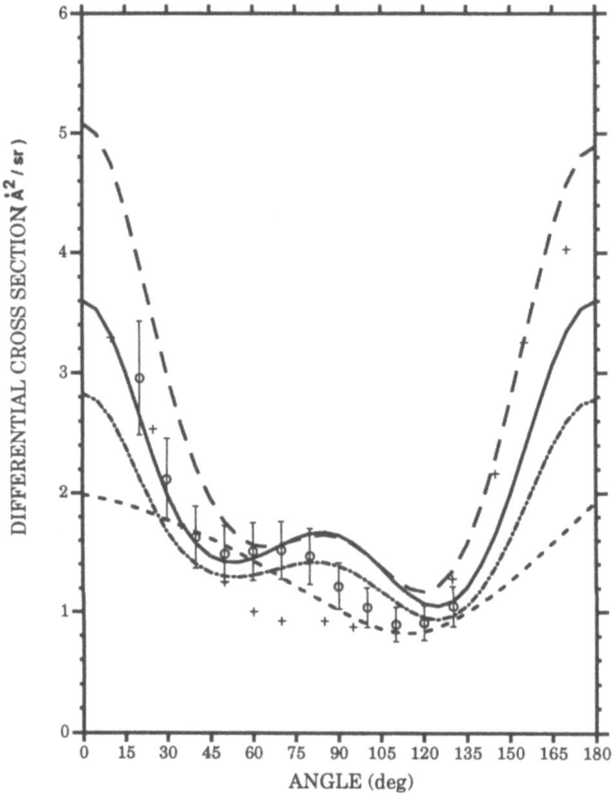

Figure 16. Angular distribution in vibrational elastic scattering. Present (15 state close-coupling results, Ref. 21) at 2.05 eV (— - —); at 2.10 eV (———); at 2.15 eV (—— —). Experimental results are (Ref. 26) (ΦΦΦ); (Ref. 27) (+++); (Ref. 4) (- - -).

given (for vibrationally elastic scattering—similar variation is found in vibrational excitation as well[21]). One sees that the shape of the angular distribution varies markedly over an energy range as narrow as 0.1 eV. But that is the experimental width of even the best of current differential measurements;[26] it is not surprising that there is disagreement among the various measurements[26,27] (other calculations have not explored this narrow energy range, but they would very likely give a similar variation, in our opinion). The message is that the experimental exploration of this variation in angular distribution is a worthwhile goal, demanding a substantial increase in energy resolution.

CONCLUSION

The substantive results are intended to illustrate the power of the non-iterative PDE method, applied to 2d equations. As we have indicated, the method has been generalized to 3d equations and a model problem has been solved with notable success.[8] However, the on-going application to the adiabatic $e - N_2$ problem is proving to be difficult.

We would like to conclude this article with a restatement of what we believe to be the most promising application of the 3d-PDE method: the application to electron scattering from polyatomic molecules[7] in the fixed-nuclei approximation. It is implicit in this article that the fixed-nuclei approximation[20] is the basic idealization for electron-molecule scattering. That is because it relies on the two fundamental aspects of the problem: the light mass of the electron relative to the nuclei of the atoms which constitute the molecule, and nevertheless, the high speed of the electron relative to the motion (vibration and rotation) of the target molecule. The latter implies that one must avoid threshold behavior, a point particularly emphasized in recent times by Morrison and collaborators.[28]

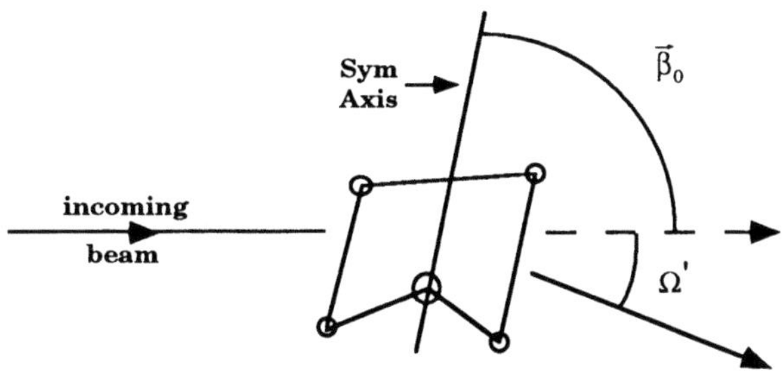

Incoming Wave Transformed To Body Frame

$$Y_{l_1}^0(\Omega') = \sum_m Y_{l_1}^m(\Omega) D_{0m}^{(l_1)^*}(\vec{\beta}_0)$$

Scattering Of $Y_l^m(\Omega)$ By Polyatomic In Body Frame

$$\hat{S}Y_l^m(\Omega) \rightarrow \sum_{l_2}\sum_{m_2} a_{mm_2}^{(l_1 l_2)} Y_{l_2}^{m_2}(\Omega)$$

Transform $Y_{l_2}^{m_2}(\Omega)$ Back To Lab Frame

$$Y_{l_2}^{m_2}(\Omega) = \sum_{m'} D_{m_2 m'}^{(l_2)}(\vec{\beta}_0) Y_{l_2}^{m'}(\Omega')$$

In Total

$$\hat{S}Y_{l_1}^0(\Omega') = \sum a_{mm_2}^{(l_1 l_2)} D_{0m}^{(l_1)^*}(\vec{\beta}_0) D_{m_2 m'}^{(l_2)}(\vec{\beta}_0) Y_{l_2}^{m'}(\Omega')$$

General Form Of Fixed Nuclei Scattering Amplitude:

$$f(\vec{\beta}_0; \Omega') = \sum D_{0m}^{(l_1)^*}(\vec{\beta}_0) a_{mm_2}^{(l_1 l_2)} D_{m_2 m'}^{(l_2)}(\vec{\beta}_0) Y_{l_2}^{m'}(\Omega')$$

Figure 17. Short derivation of the form of the scattering amplitude for e-polyatomic scattering in the fixed-nuclei approximation.

One must also use caution in applying this approximation to resonances. In particular, as we have found in the $e - N_2$ Π_g resonant application presented here, the appearance of the substructure depends critically on the vibrational motion of the target, which therefore goes beyond the fixed-nuclei (and adiabatic-nuclei[29,30]) approximation. Nevertheless, the envelope of this substructure can be described by a single resonance and this shows up in the fixed-nuclei calculations starting with the early realistic (fixed-nuclei) calculations of the $e - N_2$ system by Burke and Chandra.[31] Thus even here and many (if not most) electron-molecule scattering situations some kind of resonances arise, yet the fixed-nuclei still provides a useful approximation in cases where one needs an accurate estimate of the gross structure of the total cross section. Applications of such cross sections are numerous: a good example of the utility of fixed-nuclei calculations is scattering from sometime exotic molecules that are formed in interstellar space, such as occur in the Orion nebula, that cannot easily be produced in the laboratory, thus theoretical calculations provide the only reasonable alternative.

Expand

$$\Psi_{Total} = \frac{1}{r} \sum_{lm} \psi_{lm}(\vec{r}\,)\ \Phi_{Target}$$

Partial differential equation for ψ_{lm}

$$\left[-\frac{\partial^2}{\partial r^2} - \frac{1}{r^2}\left(\frac{1}{sin\theta}\frac{\partial}{\partial\theta} sin\theta \frac{\partial}{\partial\theta} + \frac{1}{sin^2\theta}\frac{\partial^2}{\partial\phi^2} \right) + V(\vec{r}\,) - k^2 \right] \psi_{lm}(\vec{r}\,) = 0$$

where $k^2 = E - E_{Target}$ and

$$V(\vec{r}\,) = \left\langle \Phi^*_{mol} \left| V_{e-mol} \right| \Phi_{mol} \right\rangle$$

The $a^{(ll')}_{mm'}$ are deduced from the asymptotic form of the solutions $\psi_{lm}(\vec{r}\,)$

$$\lim_{r\to\infty} \psi_{lm}(\vec{r}\,) = i^l r j_l(kr)\delta_{m0} + e^{ikr} \sum_{l'm'} a^{(ll')}_{mm'} y^{m'}_{l'}(\hat{r})$$

Figure 18. Basic form of the 3d-PDE for e-polyatomic scattering in the fixed-nuclei approximation.

The basic formulae of the application to polyatomic molecules[7] are given in Figs. 17 and 18. Fig. 17 contains a short but complete derivation of the form of the fixed-nuclei amplitude which is a minimal generalization of the diatomic target result[20] [$a^{l_1 l_2}_m \to a^{l_1 l_2}_{mm'}$]. Fig. 18 gives the basic form of the 3d PDE for e-polyatomic scattering in the fixed-nuclei approximation. The following features are noted:

(a) (Fig. 17)the scattered amplitude $f(\vec{\beta}_0, \Omega')$, when expressed in terms of the angles in the lab frame (that is, the frame which is defined by the direction of the incident electron) as well as $\vec{\beta}_0$, which are a set of Euler angles needed to rotate a coordinate defined by the (fixed) molecular target into the laboratory system of f;

(b) (Fig. 18) the basic PDE is a 3d equation in which the third coordinate, ϕ, is the azimuthal angle of the scattered electron relative to a coordinate system fixed by the target molecule (cf. Fig. 18);

(c) (not shown) the form of $f(\vec{\beta}_0, \Omega')$ may be further decomposed into irreducible representations of the additional symmetries describing the (polyatomic) target.[32] Effectively this means that linear combinations of the $a^{l_1 l_2}_{mm'}$ are diagonal in those symmetry labels. The main point, however, is that at low energy, the set of $a^{l_1 l_2}_{mm'}$

converge (rapidly) in l, l', and one doesn't have to make the further irreducible decomposition;

(d) (Fig. 18)the scattering parameters are deduced from the asymptotic form of the solutions $\psi_{lm}(r, \theta, \phi)$ of the PDE, where l, m refers to the $Y_{lm}(\theta, \phi)$ defining the boundary condition of a particular solution (at large r).

In summary, the ability to solve such PDE's can be expected to have useful application over the whole range of (low energy) electron-molecule problems. We close by expressing the hope that the community at large will avail itself of this methodology.

ACKNOWLEGEMENTS

Work of A.T. was done under NASA RTOP 432-36-58-01. C.A.W. was partially supported by NASA grant NAG-5307 and NASA contract NAGW-2930; he would also like to acknowledge a grant of computer time by Florida State University on a CRAY-YMP.

REFERENCES

1. A. Temkin, "Nonadiabatic Theory of the Scattering of Electrons from Atomic Hydrogen," Phys. Rev. 126:130(1962).
2. A.K. Bhatia and A. Temkin, "Symmetric Euler Angle Decomposition of the 2-Electron Fixed Nucleus Problem," Rev. Mod. Phys. 36:1050(1964).
3. A. Temkin and E.C. Sullivan, "Nonadiabatic Theory of Electron-Hydrogen Scattering, Part II," NASA Technical Note D-1702, (1963).
4. N. Chandra and A. Temkin, "Hybrid theory and calculation of $e - N_2$ scattering," Phys. Rev. A 13:188(1976); A. Temkin, PDE approach to hybrid theory of e-molecule scattering, in: " Symposium on Electron-Molecule Collisions," I. Shimamura and M. Matsuzawa, eds., U. of Tokyo, Tokyo (1979).
5. E.C. Sullivan and A. Temkin, "A Non-iterative Method for Solving PDE's Arising in Electron Scattering," Comp. Phys. Comm. 25:97(1982).
6. C.A. Weatherford, K. Onda, and A. Temkin, "Inclusion of exact exchange in the non-iterative PDE method," Phys. Rev. A 31:3620(1985).
7. A. Temkin, C.A. Weatherford, and E.C. Sullivan, "Extension of the PDE Method to 3 Dimensions: Application to e-Molecule Scattering," A.I.P. Conf. Proc. 204 (Amer. Inst. Phys., New York, 1990) pp. 133-139.
8. E.C. Sullivan and A. Temkin, "Further Developments in the Non-iterative Method of Solving PDE's in Electron Scattering," Comp. Phys. Comm. 71:319(1992).
9. A. Temkin, "Polarization and Exchange Effects in the Scattering of Electrons from Atoms," Phys. Rev. 107:1004(1957).
10. K. Onda and A. Temkin, "Calculation of the polarization potential for $e - N_2$ collisions," Phys. Rev. 28:621(1983).
11. P.G. Burke and N. Chandra, "Electron-Molecule Interactions III: $e - N_2$ Scattering," J. Phys. B 5:1696(1973).
12. C.A. Weatherford and A. Temkin, Very low energy e-N_2 scattering and interim hybrid calculations of the Π_g resonance, in: "Electron-Molecule Scattering and Photoionization," P.G. Burke and J.B. West, eds., Plenum, N.Y. and London (1988).
13. C.A. Weatherford, F.B. Brown, and A. Temkin, "Inclusion of electron correlation for the target wave function in low- to intermediate-energy $e - N_2$ scattering," Phys. Rev. A 35:4561(1987).

14. C.A. Weatherford, K. Onda, and A. Temkin, "Inclusion of exact exchange in PDE method: application to $e - N_2$," Phys. Rev. A 31:3620(1985).
15. N.T. Padial and D.W. Norcross, "Parameter-free model of the correlation-polarization potential for electron-molecule collisions," Phys. Rev. A 29:1742(1984).
16. M.A. Morrison, B.C. Saha, and T.L. Gibson, "Electron-N_2 scattering with a parameter-free model polarization potential," Phys. Rev. A 36:3682(1987).
17. R.E. Kennerly, "Absolute total electron scattering cross sections for N_2 between 0.5 and 50 eV," Phys. Rev. A 21:1876(1980).
18. W. Sohn, K.-H. Kochem, K.-M. Scheuerlein, K. Jung, and H. Ehrhardt, "Near-threshold vibrational excitation and elastic electron scattering from N_2," J. Phys. B 19:4017(1986).
19. K. Jost, P.G.F. Bisling, F. Eschen, M. Felsmann, and L. Walther, Total cross sections for electron scattering from N_2, Xe, Kr and Ar, in: "Abstracts of Contributed Papers, Thirteenth International Conference on the Physics of Electronic and Atomic Collisions," J. Eichler, I.V. Hertel, and N. Stolterfoht, eds., North-Holland, Amsterdam (1983).
20. A. Temkin and K.V. Vasavada, "Scattering of Electrons from H_2^+," Phys. Rev. 160:137(1978); 186:57(1969).
21. C.A. Weatherford and A. Temkin, "Completion of a hybrid theory calculation of the Π_g resonance in electron-N_2 scattering," Phys. Rev. A 49:2580(1994).
22. H. -D. Meyer, S. Pal, and U. Riss, "Inclusion of electron correlation for the target wavefunction in $e - N_2$ scattering," Phys. Rev. A 46:186(1992).
23. W.M. Huo, M.A.P. Lima, T.L. Gibson, and V. McKoy, "Correlation effects in elastic $e - N_2$ scattering," Phys. Rev. A 36:1642(1987).
24. L. Dubé and A. Herzenberg, "Absolute cross sections from the 'boomerang model' for resonant electron-molecule scattering," Phys. Rev. A 20:194(1979).
25. B.I. Schneider, M. Le Dourneuf, and Vo Ky Lan, "Resonant Vibrational Excitation of N_2 by Low-Energy Electrons: An *Ab Initio* R-Matrix Calculation," Phys. Rev. Lett. 43:1926(1979).
26. M.J. Brennan, D.T. Alle, P. Euripides, S.J. Buckman, and M.J. Brunger, "Elastic electron scattering and rovibrational excitation of N_2 at low incident energies," J. Phys. B 25:2669(1992).
27. T.W. Shyn and G.R. Carignan, "Angular distribution of electrons scattered from gases: 1.5-400 eV on N_2. II," Phys. Rev. A 22:923(1980).
28. Cf. for example M. Morrison, A.N. Feldt, and D.A. Austin, "Adiabatic Approximation for Excitation of Molecules by Low-Energy Electron Impact: Rotational Excitation of H_2," Phys. Rev. A 29:2518(1984).
29. D.M. Chase, "Adiabatic Approximation for Scattering Processes," Phys. Rev. 104:838(1956).
30. E.S. Chang and A. Temkin, "Rotational Excitation of Diatomic Molecules by Electron Impact," Phys. Rev. Lett. 23:399(1969).
31. P.G. Burke and N. Chandra, "Electron-molecule interactions III: A pseudopotential for e-N_2 scattering," J. Phys. B 5:1696(1972).
32. F.A. Gianturco and D.G. Thompson, Computational models for e-polyatomic low-energy scattering, in: "Electron-Molecule Collisions," J. Hinze, ed., Plenum, N.Y. (1983).

AN R-MATRIX APPROACH TO ELECTRON-MOLECULE COLLISIONS

Barry I. Schneider

Physics Division, National Science Foundation, Arlington, Va 22230

1. HISTORICAL INTRODUCTION

The R-matrix formalism has a long and venerable history. The method was introduced into nuclear physics by Wigner[1] and Wigner and Eisenbud[2] in the late 1940's to enable a unified treatment of nuclear reactions dominated by compound state formation. However, there are earlier sources, [3-4] which developed quite similar approaches to resonant nuclear reactions. All of these theories utilize the short-range character of the nuclear force to define a reaction zone of finite radius but differ in the mathematical details of the treatment of the wavefunction within that reaction zone. By enclosing the scattering partners within this sphere of radius $r = a$ (the R-matrix surface), where a is chosen to be the range of the nuclear force, it should be possible to characterize the system using energies and wavefunctions computed within the sphere. By matching to the known asymptotic solutions, which in the nuclear problem are simply free waves, one can easily extract the relevant scattering parameters. The connection between the internal and external solutions is provided by the R-matrix, which is a sum over quantities related to the overlap integrals (level widths) of the internal and external wavefunctions evaluated on the surface of the sphere, and the energies of the internal states.[5-8] Since the low-energy nuclear scattering problem is dominated by the formation of resonances which can be identified fairly easily with the internal states, the method is a natural one for the parametrization of nuclear cross sections. Thus the R-matrix method becomes a systematic framework for understanding and characterizing large amounts of data in terms of energy levels and widths obtained from experimental measurement. In addition, once these energies and level widths are obtained, the R-matrix provides a vehicle for predicting new results which may be difficult or impossible to obtain from experiment.

Another important point is that the internal R-matrix energies and wavefunctions may be computed using the techniques developed for nuclear structure studies, i.e., using bound-state variational methods. In the case of nuclear structure little was known about the basic nuclear force for many years and the R-matrix method remained an empirical tool for describing nuclear reactions. This began to change in the early 1960's

Computational Methods for Electron-Molecule Collisions
Edited by W.M. Huo and F.A. Gianturco, Plenum Press, New York, 1995

and there were attempts to use the approach as a more *ab-initio* tool.[9] However, by that time formal scattering theory based on the Lippmann-Schwinger [10] formalism had been developed, and the R-matrix approach lost favor because of the need to define a nuclear radius to characterize the strong-interaction zone. In addition, there was the problem of dealing with the case of long-range forces such as those between charged particles. Formally, this may be solved but again the division of configuration space into a strongly and weakly interacting region seemed somehow inelegant.

The R-matrix method was introduced into atomic physics by Burke and his associates [11-15] in the early 1970's. In contrast to the nuclear scattering problem, the forces are known in atomic and molecular physics and the problem is one of computing the electronic states and scattering cross sections of the constituent particles. Here what is needed is a robust *ab initio* method capable of providing insight and information about the atomic many-body problem. Burke recognized that the R-matrix method had many advantages over other approaches in that, in principle, one could employ bound-state methodologies to compute the R-matrix energies and wavefunctions *once and for all* and then extract the scattering information quite cheaply for many incident energies. Such an approach has considerable value when trying to characterize a certain cross section over broad energy ranges or when one is studying resonances where the cross section is changing rapidly over small energy ranges. From the computational viewpoint, the calculation of atomic bound state wavefunctions and energies was well developed and it seemed foolish not to utilize all that technology in attacking the atomic continuum.

In the atomic and molecular problem there are long-range forces and it is necessary to squarely face that question at the outset. The proposed solution was to divide space into two regions. In region one, the internal region, the particles are close to one another and it is necessary to accurately account for all short-range interactions, including exchange. In the second or external region the particles are still interacting but the forces are direct and multi-polar in character. Thus it is possible to reduce the scattering problem in the external region to the solution of a set of *coupled, ordinary differential equations* which are much easier to handle than when the particles are close together and the interactions non-local. The matching of internal and external wavefunctions may be easily carried out using a generalization of the approach for a single channel; all of the regular and irregular external solutions to the set of coupled differential equations are computed and appropriate linear combinations taken to match logarithmic derivatives at the surface. Eventually a more efficient approach to the matching was found but the essential features remain intact. Burke and his co-workers applied the R-matrix method to many problems in the electron scattering and photoionization of simple and complex atoms.[16] The method was generalized and applied first by Schneider,[17-19] Hay and Schneider[20] and Schneider and Morrison[21] and then by Burke, Mackey and Shimamura,[22] Buckley, Burke and Lan[23] and Noble, Burke and Salvini[24] in a slightly different form, to treat electron-molecule collisions. The generalization and its subsequent applications form the subject of this and the following five chapters of this book.

2. R-MATRIX THEORY FOR A SIMPLE, ONE-DIMENSIONAL, S-WAVE, RADIAL POTENTIAL

In order to set the framework for the general discussion to follow, I present here a review of the derivation of the R-matrix method for a simple, one-dimensional, s-wave, radial potential. Readers familiar with the basic content of the R-matrix theory may simply skip to the next section.

Let us consider the Schrödinger equation, in atomic units, for a one-dimensional, s-wave radial potential, $v(r)$.

$$-\frac{1}{2}\frac{d^2\Psi}{dr^2} + v(r)\Psi - E\Psi = 0 \tag{1}$$

where $v(r)$ is assumed to vanish outside of $r = a$. In order to completely specify the solutions to eq (1) it is necessary to make some mathematical statement concerning the boundary conditions satisfied by Ψ at $r = a$ (it is assumed that Ψ is regular at $r = 0$). For solutions to eq (1) for E$\geq$ 0 and $r \geq a$, it is known that $\Psi(r) = sin(kr + \delta)$, where δ, the phase shift, completely specifies the scattering information. Clearly, the problem is to determine δ. If we were dealing with a bound state problem, on an infinite interval, we could expand the unknown function Ψ in some convenient set of known functions which exponentially decayed at infinity, and determine the coefficients using the Rayleigh-Ritz variational principle.

One would like to have a similar procedure for the scattering problem but we are seemingly prevented from doing so by the boundary condition at $r = a$. To proceed, we resort to an artifice; we add and subtract an operator,

$$L_b = \frac{1}{2}\delta(r-a)\left(\frac{d}{dr} - b\right)$$

to eq (1) which, for any real b, has the effect of making $\left(-\frac{1}{2}\frac{d^2}{dr^2} + v(r) + L_b\right)$ a hermitian operator. Thus we obtain,

$$-\frac{1}{2}\frac{d^2\Psi}{dr^2} + v(r)\Psi + L_b\Psi - E\Psi = L_b\Psi \tag{2}$$

and it is now possible to proceed in a very similar fashion to the bound state problem. We introduce the expansion,

$$\Psi = \sum_n c_n\phi_n$$

where the expansion functions ϕ_n are arbitrary and are assumed to be capable of representing the scattering wavefunction for $r \leq a$, over the energy range of interest in the problem. The expansion functions need not be orthonormal, nor satisfy any fixed boundary condition at $r = a$. If the expansion functions are not orthonormal, they may be orthonormalized by the Gram-Schmidt or symmetric orthogonalization method. For the purposes of the subsequent discussion, let us assume the ϕ_n are orthonormal.

We may calculate the expansion coefficients, c_n by projecting eq (2) into the space spanned by the ϕ_n. This yields,

$$\sum_m (\phi_n \mid (-\frac{1}{2}\frac{d^2}{dr^2} + v(r) + L_b - E) \mid \phi_m)c_m = \frac{1}{2}\left[\left(\frac{d\Psi}{dr}\right)_{r=a} - b\Psi(a)\right]\phi_n(a) \tag{3}$$

If we transform to the representation which diagonalizes $(\phi_n \mid (-\frac{1}{2}\frac{d^2}{dr^2} + v(r) + L_b) \mid \phi_m)$, we get,

$$(E_i - E)\,d_i = \frac{1}{2}\left[\left(\frac{d\Psi}{dr}\right)_{r=a} - b\Psi(a)\right]\varphi_i(a)$$

$$(-\frac{1}{2}\frac{d^2}{dr^2} + v(r) + L_b - E_i)\varphi_i = 0 \tag{4}$$

where d_i is the expansion coefficient in term of the new basis. Substituting the value of d_i into

$$\Psi = \sum_i d_i \varphi_i$$

yields,

$$\Psi(r) = \frac{1}{2}\left[\left(\frac{d\Psi}{dr}\right)_{r=a} - b\Psi(a)\right]\sum_i \frac{\varphi_i(r)\varphi_i(a)}{(E_i - E)} \tag{5}$$

Setting $r = a$ on both sides of eq (5) and substituting $\Psi = sin(ka + \delta)$ and $\frac{d\Psi}{dr} = kcos(ka + \delta)$, gives us the necessary relationship to solve for δ.

3. R-MATRIX THEORY FOR ELECTRON-MOLECULE SCATTERING

3.1. General Formal Theory

The dynamical system under consideration consists of N_e electrons and N_p nuclei interacting via Coulomb potentials. We assume for the present discussion that only one of the N_e electrons, the incident or scattering electron, can be found at infinite distances from the other electrons. This excludes from consideration explicit treatments of impact ionization or other three-body breakup processes but does not exclude problems involving dissociation of the molecule *as long as the incident electron remains attached to one of the dissociating fragments.* Processes such as dissociative attachment or dissociative recombination fall into this latter category. Even some problems involving the three-body process of electron impact dissociation can be treated provided the reaction can be assumed to proceed in two independent steps.[25] In later chapters the restriction to only a single electronic continuum will be partially removed. In order to simplify the notation for the subsequent discussion, I will assume the target to be a diatomic molecule with only a single degree of internal freedom.

Let us begin by considering the Schrödinger equation defined within a restricted R-matrix hypersphere of electronic radius $r = a_e$ and nuclear radius $R = a_p$. Outside of a_e the dynamical system may be regarded as consisting of a well separated molecule and scattering electron. Similarly, outside of a_p, the system consists of two molecular fragments with the incident electron attached to one of the fragments. The problem is to construct a solution of the equation,

$$(\, H_e + H_N - E\,) \mid \Psi) = 0 \tag{6}$$

where

$$H_e = \sum_{i=1}^{n_e} h_i \; + \sum_{i<j=1}^{n_e} \frac{1}{\mid r_i - r_j \mid} \;\; ; \quad H_N = -\sum_{i=1}^{n_p} \frac{1}{2M_i}\nabla_i^2 + \sum_{i<j=1}^{n_p} \frac{Z_i Z_j}{\mid R_i - R_j \mid}$$

and

$$h_i = -\frac{1}{2}\nabla_i^2 - \sum_{j=1}^{n_p} \frac{Z_j}{\mid r_i - R_j \mid}$$

In Eq (6) and what follows round brackets are used to denote integrations over the internal region. Eq (6) is an elliptic partial differential equation whose solution is undefined without the specification of a boundary condition on the spherical hypersurfaces $\{a_e, a_p\}$. Since we have assumed that the dynamical system of electrons and nuclei interact by purely long-range, electrostatic forces outside of $\{a_e, a_p\}$, it is only necessary to be able to ennumerate the possible fragmentation modes or channels on the hypersurface to begin to specify the necessary conditions. Since we have excluded three-body breakup from consideration, only elastic and inelastic electronic excitation, vibrational excitation, and dissociative attachment (neutral target) or dissociative recombination (positive ion target) are allowed.

To complete the specification of the solution it is necessary to determine the logarithmic derivative of the wavefunction describing the relative motion of these fragments on the R-matrix hypersurface. In fact, it is just these logarithmic derivatives which are needed to completely determine the scattering cross sections. From a knowledge of the solutions of the Schrödinger equation in the external region and the complete set of internal R-matrix eigenstates, which satisfy a different, but *known* boundary condition on the hypersurface, the required logarithmic derivatives may be found by matching internal and external wavefunctions on the hypersurface between the two regions. Thus the major computational task of solving the scattering problem is shifted to determining the complete, discrete set of internal R-matrix eigenvalues and eigenfunctions which can then be used to describe the scattering wavefunction *at any energy* within the hypersphere. Any complete set will suffice. In the original R-matrix derivation, as well as in much of the subsequent work, the set was chosen to satisfy fixed and real logarithmic derivative conditions on the hypersphere, i.e.,

$$\left[\frac{\partial \Psi}{\partial r}\right]_{r=a_e} = b_e \Psi\Big|_{r=a_e} \quad ; \quad \left[\frac{\partial \Psi}{\partial R}\right]_{R=a_p} = b_p \Psi\Big|_{R=a_p} \tag{7}$$

where $\{b_e,b_p\}$ are real and identical for all {electronic,nuclear} channels. Other choices of these parameters are possible and for many of these there are particular reaction theories[3-4] associated with that choice. It would take us too far away from the main subject to examine this question now but the interested reader may consult the previous references and Schneider[26] for further information. The advantage of the Wigner and Eisenbud choice of logarithmic derivative is that the internal spectrum is not only discrete but also real.

Once we have chosen values of $\{a_e,a_p\}$ and $\{b_e,b_p\}$, the eigenvalue problem,

$$(\, H_e + H_N - E_i \,) \mid \Psi_i\rangle = 0 \tag{8}$$

becomes well defined and subject to numerical treatments using techniques developed for ordinary bound-state problems with one caveat: *the matrix elements are defined over only the internal region.* Since many of the techniques developed for the evaluation of the one-and-two-electron matrix elements required for the practical solution of eq (8) depend on having an infinite volume of integration, care must be taken before leaping ahead blindly. We shall return to this point later. Another problem, which is especially difficult for electron-molecule collisions, is the fixed boundary condition. In general this would require all of the functions describing the relative motion of the scattering fragments to satisfy that boundary condition. Fortunately, there is another, equivalent derivation, due to Bloch,[6, 27] which eliminates this requirement and has the advantage of better convergence properties as well. Let us define an operator,

$$\begin{aligned} L_b = L_b^e + L_b^p \;&=\; \frac{1}{2}\sum_{t=1}^{n_c^e} \mid \Phi_t\rangle \delta(r-a_e)\left(\frac{\partial}{\partial r} - b_e\right)\langle \Phi_t \mid \\ &+\; \sum_{d=1}^{n_c^p} \frac{1}{2M_d} \mid \Phi_d\rangle \delta(R-a_p)\left(\frac{\partial}{\partial R} - b_p\right)\langle \Phi_d \mid \end{aligned} \tag{9}$$

where

$$\begin{aligned} | \Phi_t\rangle &= | \Theta_q(1,2,\cdots,n_T;R)\, Y_{lm}(\Omega_e)\, \chi_{q\alpha}(R)\rangle \\ | \Phi_d\rangle &= | \Theta_a(1,\cdots,n_a)\Theta_b(n_{a+1},\cdots,n_{T+1})\, Y_{LM}(\Omega_p)\rangle \end{aligned}$$

and

$$\begin{aligned} | \Theta_q(1,2,\cdots,n_T;R)\rangle &= \text{electronic target wavefunction} \\ Y_{lm}(\Omega_e) &= \text{spherical harmonic of scattering electron} \\ | \chi_{q\alpha}(R)\rangle &= \text{vibrational wavefunction of target} \\ | \Theta_a(1,\cdots,n_a)\Theta_b(n_{a+1},\cdots,n_{T+1})\rangle &= \text{bound wavefunction of dissociated fragments} \\ Y_{LM}(\Omega_p) &= \text{spherical harmonic for dissociated fragments} \end{aligned}$$

where the sum over $\{t,d\}$ runs over the {electronic,nuclear} channels. The L operator projects the full wavefunction onto all of the possible products on the R-matrix hypersurface. This projection is itself a wavefunction in the relative co-ordinate of the scattered fragments. The δ function instructs us to evaluate the projection (and its derivative) at distances such that all exchange interactions between the fragments are, for practical purposes, negligible. From a mathematical standpoint, as shown by Bloch,[6] the addition of the L operator to the Hamiltonian, makes the sum of $H + L_b$ a Hermitian operator. This would not be true of the unmodified Hamiltonian because of the surface terms due to the kinetic energy operators acting on the unbound particle co-ordinates.[28] The L operator has the effect of cancelling these surface terms. We now proceed by adding and subtracting the L operator to eq (6) and then rearranging to get,

$$(\, H_e + H_N + L_b - E\,)\,|\,\Psi) = L_b\,|\,\Psi) \tag{10}$$

A formal solution of eq (10) may be obtained as,

$$|\,\Psi) = (\, H_e + H_N + L_b - E\,)^{-1} L_b\,|\,\Psi) \tag{11}$$

Utilizing the solutions of the R-matrix eigenvalue equation,

$$(\, H_e + H_N + L_b - E_i\,)\,|\,\Psi_i) = 0 \tag{12}$$

we obtain,

$$|\,\Psi) = \sum_{i=1}^{\infty} \frac{|\,\Psi_i)(\Psi_i\,|\,L_b\,|\,\Psi)}{(E_i - E)} \tag{13}$$

with

$$\begin{aligned} (\Psi_i\,|\,L_b\,|\,\Psi) &= \frac{1}{2}\sum_{t=1}^{n_c^e} F_{i,t}(a_e)\left(\frac{\partial F_{E,t}(r)}{\partial r} - b_e F_{E,t}(r)\right)_{r=a_e} \\ &+ \sum_{d=1}^{n_c^p} \frac{1}{2M_d} F_{i,d}(a_p)\left(\frac{\partial F_{E,d}(R)}{\partial R} - b_p F_{E,d}(R)\right)_{R=a_p} \end{aligned}$$

where $\{F_i, F_E\}$ are the partial wavefunctions in the relative scattering co-ordinate evaluated on the hypersphere for the {R-matrix,scattering} wavefunctions. To completely determine the scattering wavefunction it is necessary to specify $\{F_{E,t}(a_e), F_{E,d}(a_p)\}$ and the corresponding derivatives on the hypersphere. To accomplish this we project eq (13) onto the set $\{| \Phi_t\rangle, | \Phi_d\rangle\}$ and evaluate the resulting equations at $\{a_e, a_p\}$. This yields,

$$
\begin{aligned}
F_{E,t}(a_e) &= \frac{1}{2}\sum_{t'=1}^{n_c^e} R_{t,t'} \left(\frac{\partial F_{E,t'}(r)}{\partial r} - b_e F_{E,t'}(r)\right)_{r=a_e} \\
&+ \sum_{d=1}^{n_c^p} R_{t,d}\frac{1}{2M_d}\left(\frac{\partial F_{E,d}(R)}{\partial R} - b_p F_{E,d}(R)\right)_{R=a_p} \\
F_{E,d}(a_p) &= \frac{1}{2}\sum_{t=1}^{n_c^e} R_{d,t}\left(\frac{\partial F_{E,t}(r)}{\partial r} - b_e F_{E,t}(r)\right)_{r=a_e} \\
&+ \sum_{d'=1}^{n_c^p} R_{d,d'}\frac{1}{2M_{d'}}\left(\frac{\partial F_{E,d'}(R)}{\partial R} - b_p F_{E,d'}(R)\right)_{R=a_p} \qquad (14)
\end{aligned}
$$

where

$$
R_{t,t'} = \sum_{i=1}^{\infty} \frac{F_{i,t}(a_e)F_{i,t'}(a_e)}{(E_i - E)} \; ; \; R_{t,d} = \sum_{i=1}^{\infty} \frac{F_{i,t}(a_e)F_{i,d}(a_p)}{(E_i - E)} \; ; \; R_{d,d'} = \sum_{i=1}^{\infty} \frac{F_{i,d}(a_p)F_{i,d'}(a_p)}{(E_i - E)}
$$

and provides the desired relationship between scattering wavefunction and derivative on the surface of the hypersphere. In order to solve this set of matrix equations it is necessary to know the functional form of the scattering functions $\{F_{E,t}(a_e), F_{E,d}(a_p)\}$. If there was no interaction in the external region each $\{F_{E,t}(a_e), F_{E,d}(a_p)\}$ could be expressed as an unknown linear combination of regular and irregular Bessel functions. This would then be substituted into eq (14) and the resultant set of linear equations, whose dimension is $(n_c^e + n_c^p)$, be solved for the unknown coefficients. Subsequent manipulation of that matrix of coefficients provides the K, T or S matrices of standard scattering theory.

If the interaction does not vanish outside the hypersurface but is local, it is possible to reduce the scattering problem to the solution of the following set of coupled *differential* equations in the external region;

$$
\begin{aligned}
\left(-\frac{1}{2}\nabla_e^2 - \epsilon_t\right) F_{E,t} + \sum_{t'=1}^{n_c^e} V_{t,t'}F_{E,t'} + \sum_{d=1}^{n_c^p} V_{t,d}F_{E,d} &= 0 \qquad (15) \\
\left(-\frac{1}{2M_d}\nabla_R^2 - \epsilon_d\right) F_{E,d} + \sum_{t=1}^{n_c^e} V_{d,t}F_{E,t} + \sum_{d'=1}^{n_c^p} V_{d,d'}F_{E,d'} &= 0
\end{aligned}
$$

These differential equations may be solved by any convenient numerical method (i.e., Numerov method) for all of the regular and irregular solutions. The procedure that follows is identical to the free particle case except for the replacement of the regular and irregular Bessel functions with the numerically calculated regular and irregular functions of eq (15). Another more efficient approach is to propagate the R-matrix from the hypersphere to distances sufficiently large that the interaction may be assumed to be zero. This may be accomplished using the R-matrix propagation technique developed by Light and Walker[29, 31] and/or the continued fraction approach of Noble and Nesbet.[30] A matching to free waves at these distances completes the specification of the scattering information.

Before concluding this section I would like to mention a related approach to treating the scattering problem within a finite volume, the so-called eigenchannel method.[32–33] The eigenchannel method replaces the R-matrix eigenvalue problem by an eigenvalue equation at fixed E for the parameter b in the Bloch L operator. This requires a rediagonalization of the Hamiltonian matrix for each incident energy but solves the formal difficulties of the non-uniform convergence of the derivative of the scattering wavefunction at the the R-matrix boundary. Just as in the Bloch formalism, arbitrary basis sets may be employed to expand the scattering wavefunction in the internal region. In fact, it has recently been shown that the two methods are equivalent for fixed, finite basis sets.[34] The eigenchannel approach has been successfully employed in atomic scattering problems by Greene and his co-workers.

3.2. Fixed Nuclei Calculations

The above theory is quite general but of limited utility in practical applications to electron-molecule collisions. In the derivation presented, there was no cognizance taken of the 2000 fold difference in mass of the electrons and nuclei. This difference, which is the basis of the Born-Oppenheimer (BO) approximation, enables the molecular physicist to adiabatically separate these two motions. The nuclei are first regarded as fixed in space and a dynamical problem for the electronic motion is formulated regarding the R variables as a parameter. The nuclei are then moved to a new position and the solution of the electronic problem repeated. Once the entire electronic surface has been mapped out, it may be considered a potential for the dynamical motion of the nuclei.

For most spectroscopic problems in molecular physics the BO separation works extremely well. One talks about potential curves, i.e., the fixed nuclei, electronic eigenvalue problem and the ro-vibrational levels, the eigenvalues for the rotating/vibrating molecular framework which are derived from these curves. The BO separation basically decouples these motions and provides an important simplification of the molecular dynamics. However, it does not always work. When there are degeneracies or near degeneracies of the electronic potential surfaces one may expect even small perturbations to cause this picture to break down. When this does occur, it is necessary to couple the two degrees of freedom to remove the degeneracy. Here again, one speaks about curve crossings (or avoided crossings), and the computation of coupling matrix elements between electronic levels via ∇_R^2. The meaning of the BO approximation when one is discussing the electronic continuum is more problematic. When an electron is at energies comparable to the vibrational (or rotational) energy spacing, there is good reason to question the validity of the separation of electronic and nuclear motion. However, this does not necessarily signify a *complete* breakdown of the BO approximation. It may indeed be possible to restore the essential features of the BO separation without too much discomfort. We shall return to this in the next section and chapters 9 and 12 are devoted specifically to such questions. For now, let us assume its validity and proceed to the fixed nuclei electronic scattering problem.

$$(H_e + L_b^e - E_i(R)) \mid \Psi_i\rangle = 0 \tag{16}$$

Here it must be remembered that the nuclear co-ordinates are parameters not dynamical variables. As a consequence, it is possible to show that

$$L_b^e = \frac{1}{2} \sum_{q l_q m_q} \mid \Theta_q(1,2,\cdots,n_T;R)\, Y_{l_q m_q}(\Omega_e)\rangle \delta(r-a_e)\left(\frac{\partial}{\partial r} - b_e\right) \langle \Theta_q(1,2,\cdots,n_T;R)\, Y_{l_q m_q}(\Omega_e) \mid \tag{17}$$

where the apparent dependence of L_b^e on the vibrational basis has been removed by closure. The operator should be multiplied by $\delta(R-R')$ but this has been eliminated for notational convenience. The eigenvalue equation is solved by expanding $|\Psi_i)$ in an *arbitrary* basis, calculating the necessary one-and-two-electron matrix elements over the basis set and then constructing and diagonalizing the resulting matrix representation of $H + L_b^e$. The expansion of the wavefunction is expressed as,

$$|\Psi_i) = \sum_{q\alpha} |\Theta_q(1,2,\cdots,n_T;R)\,\phi_\alpha(n_{T+1};R))\,C^i_{q\alpha} + \sum_{\mu} |\Xi_\mu(1,2,\cdots,n_{T+1};R))\,D^i_\mu \quad (18)$$

where

$$\begin{aligned} \phi_\alpha(n_{T+1};R) = F_{\alpha l_\alpha m_\alpha}(r;R)\,Y_{l_\alpha m_\alpha}(\Omega) &= \text{open channel orbital} \\ \Xi_\mu(1,2,\cdots,n_{T+1};R) &= \text{correlation term} \end{aligned}$$

A number of features of this expansion which make it particularly suited to electron-molecule scattering should be stressed. First, as a consequence of the Bloch operator formalism, it is not necessary to enforce any fixed boundary condition on the one-electron functions used to expand the scattering orbitals.[6] In practice that means any convenient basis set may be used for $\phi_\alpha(r;R)$ *as long as it is flexible enough to represent the scattering functions inside the R-matrix hypersphere.* Second, although we have explicitly chosen to write the open channel orbital with a specific dependence on (l,m), this does not necessitate the inclusion of the very large (l,m) components characteristic of single-center expansion techniques. The great flexibility of the second term in the summation enables the use of *multi-center expansions* to represent those portions of the scattering wavefunction near the atomic nuclei and to compactly account for the polycentric character of the molecular interaction potential. Stated differently, much of the power of traditional bound-state quantum chemistry is now at our disposal for the scattering problem. By choosing the open channel orbitals to be the only members of the expansion set having a non-negligible amplitude at a_e, the matching to the external solution of the Schrödinger equation becomes much simplified.[36] At these distances only a limited number of small (l,m) components are able to penetrate the large angular momentum barrier for low-energy collision processes.

In addition, it is possible to simplify the problems associated with the calculation of the one-and-two-electron matrix elements having a non-negligible contribution outside the internal region to those involving only these orbitals. In practice, only a subset of these integrals need to be corrected for external contributions by subtracting the contribution to the integrals from a_e to infinity. Since the charge distributions have attained their long-range form outside a_e, it is possible to make considerable use of multi-polar expansions to perform the necessary correction. This is discussed in more details in chapters 10 and 13. Once the Hamiltonian matrix is constructed and the matrix eigenvalues and eigenfunctions determined by diagonalization, the matching procedure can be restricted to only the open channel terms. The matching condition may be written in a compact matrix notation as,

$$\mathbf{\Psi} = \mathbf{R}\,(\dot{\mathbf{\Psi}} - \mathbf{b_e}\mathbf{\Psi}) \quad (19)$$

where

$$\begin{aligned} R_{ql_qm_q,q'l_{q'}m_{q'}} &= \sum_{i=1}^{\infty} \frac{\phi_{ql_qm_q,i}(a_e;R)\,\phi_{q'l_{q'}m_{q'},i}(a_e;R)}{(E_i - E)} \\ \phi_{ql_qm_q,i}(a_e;R) &= \sum_{\alpha} C^i_{q\alpha} F_{l_\alpha m_\alpha}(a_e;R) \end{aligned}$$

The most important feature of this procedure is that the bulk of the work is restricted to a single, energy independent step; the formation and diagonalization of the Hamiltonian matrix. The construction of the R-matrix and the matching of internal and external solutions via eq (19) is computationally far less demanding than the construction and diagonalization of the Hamiltonian matrix. When the Hamiltonian matrix becomes too large to obtain all of its roots, it is possible to reformulate the problem as the solution of a large set of linear algebraic equations. There are well known iterative methods to solve such linear equations but the disadvantage is that this procedure needs to be repeated at each incident energy and much of the elegance of the original R-matrix approach is lost.

There are a few practical matters which need to be discussed before leaving the fixed nuclei problem. The essential computational difficulty in the application of R-matrix methods to electron-molecule scattering is the calculation of the multi-center electron repulsion integrals involving open-channel orbitals. For linear molecules, all of the two-electron integrals can be reduced to the calculation of certain special functions and one-dimensional quadratures. In addition, the one-dimensional quadrature can be performed almost as easily inside a finite volume as over all space. The key feature that enables the treatment of these linear molecules to be simplified is that it is possible to use the rotational symmetry of the molecule and prolate spheroidal co-ordinates to remove one of the angular co-ordinates analytically. For non-linear molecules no such simplification exists and one is faced with multi-dimensional quadratures for integrands without special structure. In bound-state problems, the use of Cartesian or spherical Gaussians enables all integrals to be performed analytically over the infinite volume. This is not the case for finite volume integration. However, if the entire set of bound and R-matrix open-channel orbitals were expanded in a Gaussian basis, the integrals could still be simplified by subtracting the external contributions to the integrals outside the hypersphere, using multi-polar expansions, from the analytic values over the entire space. Some early work along these lines was attempted[35] and chapter 13 considers such an approach in detail. However the bulk of the R-matrix calculations being done today are on linear molecules using numerically determined R-matrix open-channel orbitals with the integrations performed numerically as described above.

3.3. Beyond the Fixed Nuclei Approximation

The effects of the excitation of various rotational or vibrational modes of the target molecule as well as its dissociation into neutral or ionic fragments may be incorporated into the R-matrix formalism with varying levels of sophistication. The most primitive as well as the most widely used approach, the adiabatic-nucleus (AN) approximation.[37] This computes the electronic S, T or R-matrix at each geometry and then calculates a matrix element such as,

$$\mathrm{S}_{q\alpha_q,q'\alpha_{q'}} = \langle \chi_{q\alpha_q}(R) \mid \mathrm{S}(R) \mid \chi_{q'\alpha_{q'}}(R) \rangle \tag{20}$$

to describe the vibrational transition. A similar procedure can be used for rotational excitation. At the heart of this approximation is the assumption that both the energetics and the dynamics of the nuclear response to the electron is adiabatic. This will fail when 1) the small energy differences between the electronic energy and the ro-vibrational energies have a critical effect on the cross section, 2) when there is a significant time for the nuclei to respond to the additional electron or 3) when degeneracies and level crossings cause breakdown of the BO approximation. Small energy differences are always important near thresholds and it is obvious that the AN approximation will not give the correct threshold behavior for ro-vibrational transitions. There are simple methods to correct the problem when energetics is the primary cause of the AN breakdown. However, this is not always the case and it is often necessary to calculate off-shell versions of the fixed nuclei, electronic wavefunctions to get agreement with experiment.[38-40] But even here one is not completely abandoning the separation of electronic and nuclear degrees of freedom. The off-shell matrix elements only require the calculation of electronic scattering (fixed nuclei) wavefunctions *at the proper energy values of the ro-vibrational transition.* No matrix elements involving the derivative of these scattering wavefunctions with respect to nuclear co-ordinates is required, see chapter 12 for a fuller discussion..

When the nuclear dynamics are critical to a more rigorous treatment of the scattering problem, as for example near resonances, a more accurate approach is needed. Such an approach was developed by Schneider, LeDourneuf and Burke[41] in 1979 and applied successfully to resonant processes in electron + N_2 scattering shortly thereafter.[42] It has become the basis for the proper treatment of nuclear dynamics within the R-matrix formalism and remedies the first and second problems with the AN approximation listed above.[43] To my knowledge there have been no applications of any scattering formalism which have required the calculation of derivatives of electronic wavefunctions with respect to nuclear co-ordinates; this problem is considered further in the next chapter.

Let us turn our attention to the formalism developed in reference 41. The primary assumption made by those authors was that within the strong interaction zone (R-matrix internal region), an adiabatic separation of electronic and nuclear motion was still possible even when the asymptotic kinetic energy of the electrons was low. The basis of this contention is that the electrons are attracted by the short-range forces to the nuclei and sufficiently accelerated that a local BO approximation is quite realistic. This can be easily verified mathematically by examining the computed electronic R-matrix levels for curve crossings or near degeneracies. As long as the internal electronic R-matrix levels remain well separated it is likely that the BO approximation will be valid. As the electrons recede into the outer region it is necessary to return to a description appropriate to the target, i.e. a rotating/vibrating molecule in order to compute the required transition matrix elements. This frame transformation between an internal representation dominated by the physics of the molecular frame to an external representation dominated by the laboratory frame is not unique to the R-matrix method.[44] In fact, the two frames are equivalent descriptions of the dynamical system if the entire Hamiltonian is retained in both frames. The use of the molecular frame in the internal region, *coupled with the neglect of derivatives of the internal electronic wavefunctions with respect to nuclear co-ordinates* in that frame, constitutes what ref. 41 call the BO approximation for the internal R-matrix levels. The formalism then naturally leads to a description in which it is possible to extract detailed ro-vibrational cross sections by matching internal BO dominated R-matrix wavefunctions to proper asymptotic wavefunctions for the isolated molecule.

The essence of the procedure is to expand the total scattering wavefunction as a linear combination of the complete set of internal R-matrix electronic eigenstates.

$$| \Psi) = \sum_i | \Psi_i(1, 2, \cdots, n_{T+1}; R) \eta_i(R)) \tag{21}$$

By projecting eq (10) onto the $| \Psi_i)$ (integrating only over the electronic co-ordinates) and *neglecting all derivatives of* $| \Psi_i)$ *with respect to* R (the BO approximation) one obtains the following equation for $\eta_i(R)$

$$(E_i(R) + H_N + L_b^p - E) \mid \eta_i(R)) = (\Psi_i \mid L_b \mid \Psi) \tag{22}$$

where

$$\begin{aligned}(\Psi_i \mid L_b \mid \Psi) &= \frac{1}{2} \sum_{c\nu_c l_c m_c} F_{i,cl_cm_c}(a_e; R)\, \chi_{c\nu_c}(R) \left(\frac{\partial F_{E,c\nu_c l_c m_c}(r)}{\partial r} - b_e F_{E,c\nu_c l_c m_c}(r) \right)_{r=a_e} \\ &+ \sum_{abLM} \frac{1}{2M_{ab}} F_{i,abLM}(R)\, \delta(R - a_p) \left(\frac{\partial F_{E,abLM}(R)}{\partial R} - b_p F_{E,abLM}(R) \right)_{R=a_p}\end{aligned}$$

It should be noted that if the energy is below the heavy particle breakup threshold, the second term in the above equation may be dropped.

If we define

$$G_i(R \mid R') = \left\langle R \mid (E_i(R) + H_N + L_b^p - E)^{-1} \mid R' \right\rangle \tag{23}$$

then the formal solution to eq (22) becomes

$$\begin{aligned}\eta_i(R) &= \frac{1}{2} \sum_{c\nu_c l_c m_c} \int dR\, G_i(R \mid R') F_{i,cl_cm_c}(a_e; R')\, \chi_{c\nu_c}(R') \\ &\qquad \left(\frac{\partial F_{E,c\nu_c l_c m_c}(r)}{\partial r} - b_e F_{E,c\nu_c l_c m_c}(r) \right)_{r=a_e} \\ &+ \sum_{abLM} \frac{1}{2M_{ab}} G_i(R \mid a_p) F_{i,abLM}(a_p) \left(\frac{\partial F_{E,abLM}(R)}{\partial R} - b_p F_{E,abLM}(R) \right)_{R=a_p}\end{aligned} \tag{24}$$

Substituting eq (24) into eq (21) and then projecting onto the channel functions yields,

$$\begin{aligned}F_{E,c\nu_c l_c m_c}(a_e) &= \frac{1}{2} \sum_{c'\nu_{c'} l_{c'} m_{c'}} R_{c\nu_c l_c m_c, c'\nu_{c'} l_{c'} m_{c'}} \left(\frac{\partial F_{E,c'\nu_{c'} l_{c'} m_{c'}}(r)}{\partial r} - b_e F_{E,c'\nu_{c'} l_{c'} m_{c'}}(r) \right)_{r=a_e} \\ &+ \sum_{abLM} R_{c\nu_c l_c m_c, abLM} \frac{1}{2M_{ab}} \left(\frac{\partial F_{E,abLM}(R)}{\partial R} - b_p F_{E,abLM}(R) \right)_{R=a_p}\end{aligned}$$

$$\begin{aligned}F_{E,abLM}(a_p) &= \frac{1}{2} \sum_{c\nu_c l_c m_c} R_{abLM, c\nu_c l_c m_c} \left(\frac{\partial F_{E,c\nu_c l_c m_c}(r)}{\partial r} - b_e F_{E,c\nu_c l_c m_c}(r) \right)_{r=a_e} \\ &+ \sum_{a'b'L'M'} R_{abLM,a'b'L'M'} \frac{1}{2M_{a'b'}} \left(\frac{\partial F_{E,a'b'L'M'}(R)}{\partial R} - b_p F_{E,a'b'L'M'}(R) \right)_{R=a_p}\end{aligned} \tag{25}$$

where

$$\begin{aligned}R_{c\nu_c l_c m_c, c'\nu_{c'} l_{c'} m_{c'}} = \sum_{i=1}^{\infty} \int dR \int dR' \chi_{c\nu_c}(R) F_{i,cl_cm_c}(a_e; R)\, G_i(R \mid R') \\ F_{i,c'l_{c'}m_{c'}}(a_e; R') \chi_{c'\nu_{c'}}(R')\end{aligned}$$

$$\begin{aligned}R_{c\nu_c l_c m_c, abLM} &= \sum_{i=1}^{\infty} \int dR\, \chi_{c\nu_c}(R) F_{i,cl_cm_c}(a_e; R)\, G_i(R \mid a_p)\, F_{i,abLM}(a_p) \\ R_{abLM,a'b'L'M'} &= \sum_{i=1}^{\infty} F_{i,abLM}(a_p)\, G_i(a_p \mid a_p)\, F_{i,a'b'L'M'}(a_p)\end{aligned} \tag{26}$$

It is important to note that there is a rather clean separation of electronic and nuclear variables in eq (26). That part of the generalized R-matrix dealing with discrete electronic/vibrational transitions can be given a rather simple interpretation if one is willing to introduce a complete set of vibrational levels, $\theta_{iq}(R)$ of the ith electronic R-matrix state.

$$R_{c\nu_c l_c m_c, c'\nu_{c'} l_{c'} m_{c'}} = \sum_{iq} \frac{\Gamma^{iq}_{c\nu_c l_c m_c} \Gamma^{iq}_{c'\nu_{c'} l_{c'} m_{c'}}}{\left(\epsilon_{iq} - E\right)} \tag{27}$$

$$\Gamma^{iq}_{c\nu_c l_c m_c} = < \chi_{c\nu_c}(R) \mid F_{i,cl_c m_c}(a_e; R) \mid \theta_{iq}(R) > \tag{28}$$

The numerator of this expression can be interpreted as the product of two generalized Franck-Condon factors which relate the overlap of the levels of the initial and final vibrational levels of the target and that of the intermediate R-matrix vibrational states. These Franck-Condon factors are divided by an energy denominator which vanishes at the intermediate state vibrational energies, ϵ_{iq}. Thus one would expect that in situations dominated by well defined vibrational resonances, peaks would appear in the cross section at or near these intermediate energies. The strength of the cross section would be governed by the generalized Franck-Condon factors. In fact, as will be described the next chapter on applications of the R-matrix method to resonant vibrational excitation and dissociative attachment, this is precisely what is observed experimentally.

This concludes the discussion of formal R-matrix theory. The following chapters will present detailed applications of the method to electronically elastic, and inelastic scattering, vibrational excitation, dissociative attachment as well as the related problem of molecular photoionization.

References

[1] Wigner, E. P., Phys. Rev. **70**, 606 (1946), **73**, 1002 (1948)

[2] Wigner, E. P. and Eisenbud, L., Phys. Rev. **72**, 29 (1947)

[3] Kapur, P. L. and Peierls, R. E., Proc. Roy. Soc. (London), **A 166**, 277 (1938)

[4] Siegert, A. J. F., Phys. Rev. **56**, 750 (1939)

[5] Thomas, R. G., Phys. Rev. **88**, 1109 (1952)

[6] Bloch, C., Nucl. Phys. **4**, 5039 (1957); An elegant treatment which unifies all of the reaction theories in one formalism

[7] Lane, A. N. and Thomas, R. G., Rev. Mod. Phys. **30**, 257 (1958); A very important early reference, although the notation is cumbersome

[8] Lane, A. N. and Robson, D., Phys. Rev. **3**, 774 (1966); Shows how the framework of reference 6 can used to systematize the theories developed after 1957.

[9] Buttle, P. J. A., Phys. Rev. **160**, 719 (1967)

[10] Lippmann, B. A. and Schwinger, J., Phys. Rev. **79**, 469 (1950)

[11] Burke, P. G., Hibbert, A. and Robb, W. D., J. Phys. B **4**, 1153 (1971)

[12] Burke, P. G. and Seaton, M. J., Methods Comput. Phys. **10**, 1 (1971)

[13] Burke, P. G., Comput. Phys. Comm. **6**, 288 (1973)

[14] Burke, P. G. and Robb W. D., J. Phys. B **5**, 44 (1972)

[15] Burke, P. G. and Robb W. D., Adv. At. Mol. Phys. **11**, 143 (1975)

[16] Burke, P.G. and Berrington, K. A., "R-matrix Theory of Atomic and Molecular Processes" (IOP Publishing, Bristol, 1993)

[17] Schneider, B. I., Chem. Phys. Lett. **2**, 237 (1975)

[18] Schneider, B. I., Phys. Rev. **A11**, 1957 (1975)

[19] Schneider, B. I., Invited Paper. Proceedings of X ICPEAC, Paris, France (1977)

[20] Schneider, B. I. and Hay, P. J., Phys. Rev. **A13**, 2049 (1976)
[21] Schneider, B. I. and Morrison, M. A., Phys. Rev. **A16**, 1003 (1977)
[22] Burke, P. G., Mackey, I. and Shimamura, I., J. Phys. B **10**, 2497 (1977)
[23] Buckley, B. D., Burke, P. G., and Vo Ky Lan, Comput. Phys. Commun. **17**, 175 (1979)
[24] Noble, C. J., Burke, P. G., and Salvini, S., J. Phys. B **15**, 3779 (1979)
[25] An example of this situation can be found in the dissociation of a number of molecules by electron impact. The electron dynamics must be computed at each geometry to high precision but once the adiabatic nuclei wavefunctions are known a semiclassical or reflection approach to the nuclear breakup will often suffice.
[26] Schneider, B. I., Phys. Rev. **A24**, 1 (1981)
[27] Nesbet, R. K., "Variational Methods in Electron-Atom Scattering Theory", P. G. Burke and H. Kleinpoppen Eds., (Plenum Press, New York and London, 1980) pointed out that very early work on variational approaches to scattering such as Kohn, W., Phys. Rev. **74**, 1763 (1948) and Jackson, J. L., Phys. Rev. **83**, 301 (1951) had implicity recognized that it was not necessary to impose a fixed boundary condition on the trial scattering wavefunction. However, neither of these authors couched their discussion in terms of the usual R-matrix theory and more importantly did not propose the diagonalization of a modified Hamiltonian as in reference 6 to simplify the computational effort over many energies
[28] A simple application of Greens's theorem to the volume integral involving the coordinate of the scattered electron illustrates the situation quite simply
[29] Light, J. C. and Walker, R. B., J. Chem. Phys. **65**, 4272 (1976)
[30] Noble, C. J. and Nesbet, R. K., Comput. Phys. Comm. **33**, 399 (1984)
[31] Schneider, B. I. and Walker, R. B., J. Chem. Phys. **70**, 2466 (1979)
[32] Greene, C. H and Longhuan, K., Phys. Rev. **A38**, 5953 (1988)
[33] Le Rouzo, H. and Raseev, G., Phys. Rev. **A29**, 1214 (1984)
[34] Robicheaux, F. To Be Published
[35] V. R. Saunders of Daresbury Laboratory has programmed and used such a procedure with Gaussian orbitals to compute photoionization cross sections
[36] The word open channel is here meant to mean any channel with non-negligible amplitude on the R-matrix surface. In practice it may be convenient to include certain closed channels in this definition, especially if they are Rydberg in character, in order to keep the size of the internal region at a manageable level
[37] Temkin, A. and Vasavada, K. V., Phys. Rev. **160**, 109 (1967)
[38] Shugard, M. and Hazi, A. U., Phys. Rev. **A12**, 1895 (1975)
[39] Morrison, M. A., J. Phys. B **19**, L707, (1986); Morrison, M. A., Abdolsalami, M. and Elza, B. K., Phys. Rev. **A43**, 3440 (1991)
[40] See the part on the Complex Kohn Variational Method in this book
[41] Schneider, B. I., LeDourneuf, M. and Burke, P. G., J. Phys. B **12**, L365 (1979)
[42] Schneider, B. I., LeDourneuf, M. and Vo Ky Lan, Phys. Rev. Lett. **43**, 1926 (1979)
[43] There are earlier as well as later approaches to resonant scattering which recognize the importance of an internal, fixed nuclei, intermediate electronic level. A representative sample includes: Birtwistle, D. T. and Herzenberg, A., J. Phys. B **4**, 53 (1972); Dube, L. and Herzenberg, A., Phys. Rev. **A20**, 195 (1979); Domcke, W. and Cederbaum, L. S., Phys. Rev. **A16**, 1465 (1977); Berman, M., Estrada, H., Cederbaum, L. S. and Domcke, W., Phys. Rev. **A28**, 1363 (1983); Greene, C. H and Jungen, Ch., Adv. Atom. Mol. Phys. **21**, 51 (1985)
[44] Chang, E. S. and Fano, U., Phys. Rev. **A6**, 173 (1972)

NON-ADIABATIC EFFECTS IN VIBRATIONAL EXCITATION AND DISSOCIATIVE RECOMBINATION

Lesley A. Morgan

Computer Centre, Royal Holloway, University of London, Egham, Surrey TW20 0EX, England

1. INTRODUCTION

In this chapter, we focus on applications of the non-adiabatic approximation[1] to nuclear vibration and dissociative attachment. The formalism has been described in the preceding chapter and, to avoid repetition, we will refer to its equations and adopt the same notation. In this chapter we will give particular emphasis to the practical problems encountered. The molecular targets for which calculations using this method have been carried out include N_2,[2,3] HF,[4] HCl,[5] HBr,[6] CO,[7] and HeH^+[8,9] and we will use the last two systems as illustrative examples. A simple extension allow for coupling between the R-matrix poles is also described.[10]

2. METHOD

2.1. General Considerations

The success or failure of the R-matrix methods described in the preceding chapter rests on our ability to chose compact basis set expansions, which give good representations of the system of target molecule plus scattered electron, in the inner region of configuration space ($r \leq a_e$ and $R \leq a_p$). Since the nuclear motion part of the calculation is built on a sequence of fixed nuclei calculations, we must first ensure that these are of adequate quality before proceding further.

The most important choice is that of the electronic part of the target wavefunction $\Theta_q(1, 2, \cdots, n_T; R)$. An SCF wavefunction may be adequate for low energy vibrational excitation, but dissociative attachment clearly requires a more sophisticated, CI, target which dissociates to the correct fragments.

The scattered electron is described by single particle functions $\phi_\alpha(n_{T+1}; R)$. These are chosen to form an orthonormal set on the range $0 \leq r \leq a_e$, complete up to some scattering energy E_{max}. Apart from these requirements, they are quite arbitrary. A good choice, since they are independant of the nuclear coordinate R, is spherical Bessel

Computational Methods for Electron-Molecule Collisions
Edited by W.M. Huo and F.A. Gianturco, Plenum Press, New York, 1995

functions (or Coulomb functions for charged targets) with wavenumbers chosen such that they are mutually orthogonal.

The remaining part of the electronic wavefunction given in Eq 18 above, $\Xi_\mu(1,2,\cdots,n_{T+1};R))$, is constructed from same set of molecular orbitals used to construct the target wavefunction and chosen to represent correlation and short range polarization effects.

The calculations described below were all carried out using a heavily modified version of the ALCHEMY molecular structure package[11, 12] which is described in the next chapter.

The first stage is sequence of fixed nuclei calculations on a suitably chosen grid R_k. These will provide the potentials $E_i(R)$ and boundary amplitudes $F_{i,cl_cm_c}(a_e;R)$ of Eq 24. It should be noted that it is not necessary for us to explicitly extract resonance positions and widths before calculating vibrationally resolved cross-sections, since the mechanism is implicitly included in the non-adiabatic formalism.

The generalized R-matrices, Eq 26, are obtained by integration over the nuclear coordinate R. In order to carry out the necessary interpolations, it is important to ensure that the functions involved are smoothly varying. The most immediate problem is due to the fact that the boundary values F are constructed from eigenvalues of the electronic hamiltonian. These have arbitrary signs which will change randomly on the grid. If CI targets are used, these will also have random phases for the same reason. Orthogonalization of the continuum orbitals to the target MOs is another source of arbitary sign changes. It is imperative, therefore, that having carried out a benchmark calculation at one geometry, we enforce a consistent sign convention at each stage in the calculations at other grid points.

A second, and more fundamental, cause of discontinuities in the boundary amplitudes are avoided crossings between the R-matrix poles considered as functions of geometry. If these crossings are quite sharp and at geometries which are peripheral to the transitions of interest, they can be treated diabatically, simply by relabelling the poles and amplitudes to one side of the crossing. If the interaction is weak, the diabatic representation described below can be used. Other instances of strong interaction are outside of the scope of the present method.

In the outer regions of configuration space, short range potentials may be neglected and the scattering equations have the form given in Eq 15. These are identical in structure to the equations which arise in scattering from atomic targets and so we can solve them using numerical methods already developed for these problems.[13] In all calculations carried out to date, any long range couplings between the electronic and dissociating channels have been ignored.

2.2. Diabatic Transformation

The non-adiabatic formalism of the preceding chapter was based on the diagonalization of the full hamiltonian using a basis of the form,

$$|\Psi) = \sum_i |\Psi_i(1,2,\cdots,n_{T+1};R)\,\eta_i(R)) \tag{29}$$

where $\Psi_i(1,2,\cdots,n_{T+1};R)$ is the electronic R-matrix eigenstate at internuclear distance R. In order to obtain Eq 22 all derivatives of these functions with respect to R were neglected. There are cases where this is too drastic an approximation even though the R

dependance is weak. An example is the case of the dissociative attachment of electrons to HeH^+, see below, where the non-adiabatic coupling between the electronic states is the only mechanism for the process.

Since the electronic eigenstates form a complete set at each geometry, we can use the set of eigenfunctions at some specific geometry $R = R_0$ to expand the electronic part of the wavefunction at any other geoemtry. We can therefore replace the expansion above with the diabatic basis

$$| \Psi) = \sum_i | \Psi_i(1, 2, \cdots, n_{T+1}; R_0) \eta_i(R)) \tag{30}$$

If we now project the basic R-matrix equation (Eq 10) on to the $| \Psi_i)$ and integrate over the electronic co-ordinates, we obtain the equation, analogous to Eq 22, for $\eta_i(R)$

$$\sum_j (E_{ij}(R) + H_N + L_b^p - E) | \eta_j(R)) = (\Psi_i | L_b | \Psi) \tag{31}$$

where

$$E_{ij}(R) = \sum_k \langle \psi_i(R_0) | \psi_k(R) \rangle E_k(R) \langle \psi_k(R) | \psi_j(R_0) \rangle \tag{32}$$

and $< | >$ denotes integration over electronic coordinates. Note that, since the diabatic basis, $\Psi_i(1, 2, \cdots, n_{T+1}; R_0)$, is independent of R, we have not had to neglect any derivatives.

The rest of the analysis follows as before, but with the boundary amplitudes now defined by,

$$\Gamma^{iq}_{c\nu_c l_c m_c} = \sum_j \int dR \langle \Psi_i(R_0) | \Psi_j(R) \rangle \chi_{c\nu_c}(R) F_{j,cl_cm_c}(a_e; R) \eta_{jq}(R) \tag{33}$$

The diabatic transformation has the effect of replacing the problem of the evaluation of derivatives of the internal electronic wavefunctions with respect to the nuclear coordinate, with that of evaluating the overlaps $\langle \Psi_i(R_0) | \Psi_j(R) \rangle$. They therefore represent the non-adiabatic coupling between electronic states. If we can assume that the target molecular wavefunctions Θ_q and correlation term Ξ_μ are very slowly varying functions of R over the range of interest, and that the R dependance is entirely contained in the coefficients $C_{q\alpha,i}$ and $D_{\mu,i}$, then these integrals become inner products of the form $\sum_k B_{i,k}(R_0) B_{j,k}(R)$ where we have combined the $C_{q\alpha,k}$ and $D_{\mu,k}$ into a single coefficient matrix $B_{j,k}$.

Rather than use the Green's function directly, we use its spectral form obtained by solving the eigenvalue problem:

$$\sum_{iji'j'} \int dR (\eta_j(R) | E_{ii'}(R) + T_R \delta_{ii'} + L_b^p | \eta_{j'}(R)) c_{ijk} c_{i'j'k'} = E_k \delta_{kk'} \tag{34}$$

We have adopted the usual convention of including the nuclear repulsion term in the electronic hamiltonian and hence H_N is replaced by the nuclear kinetic energy operator T_R. As a consequence, the eigenvalues $E_i(R)$ (and diagonal elements $E_{ii}(R)$ of the diabatic coupling matrix), increase exponentially as R tends to zero. For this reason we restrict the range of R to $a_o \leq R \leq a_p$ as shown in Figure 1 below.

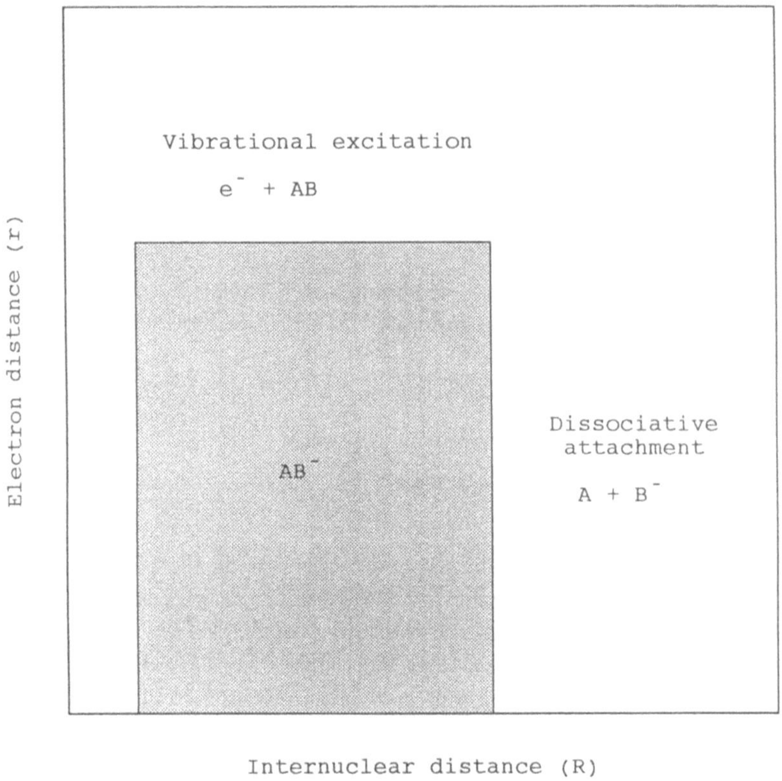

Figure 1. The partitioning of configuration space.

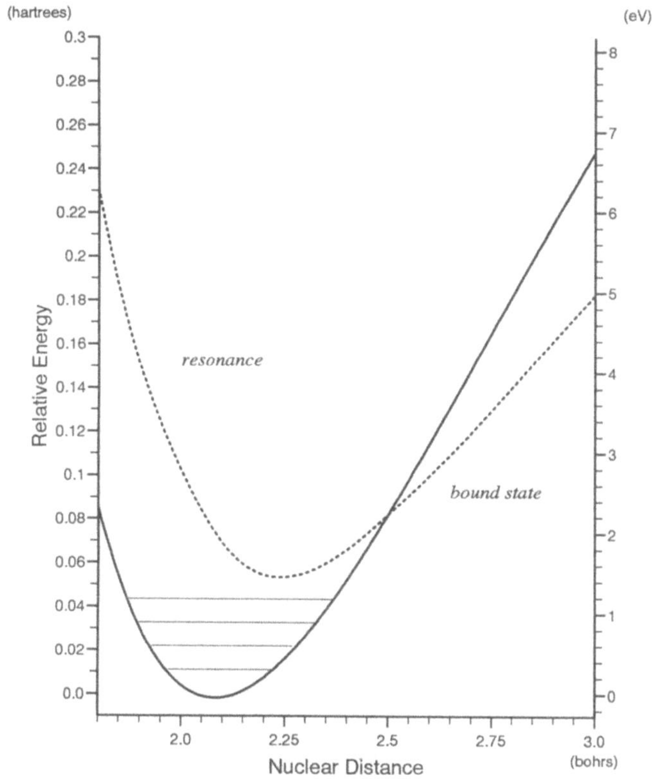

Figure 2. Potential energy curves for ground state of CO and the $^2\Pi$ resonance.

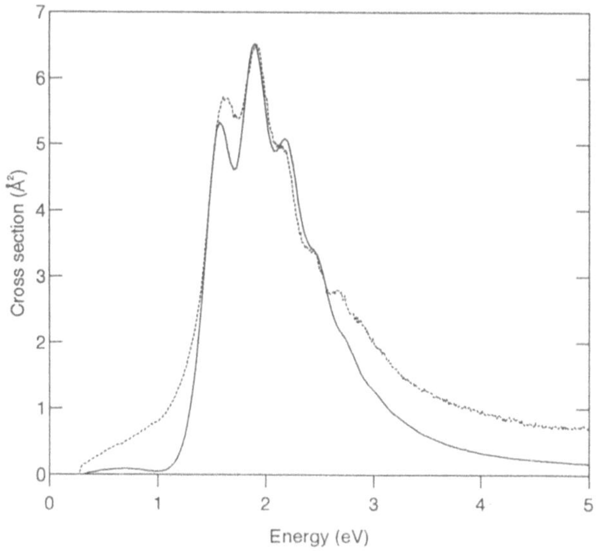

Figure 3. Cross section for the excitation of the v=1 level. Solid line $^2\Pi$ contribution from SEP approximation; dotted line, experiment (Allan 1989).

3. VIBRATIONAL EXCITATION OF CO

Experimentally measured cross sections for vibrational excitation of CO by low energy electron impact show a pronounced $^2\Pi$ resonance. An overview of the small amount of theoretical work on CO which preceded the calculations described here, can be found in reviews by Lane,[14] Norcross and Collins[15] and Morrison.[16] The R-matrix method, as described above and in other chapters of this volume, had already been used successfully to study low energy electron scattering by N_2, which has a similar shape resonance in the $^2\Pi_g$ symmetry. [2, 3] The work on CO, to a large extent, parallels this, the main difference being that the target is heteronuclear.

As noted above, the quality of the target wavefunctions is the most important single factor in the calculation. We used the STO basis given by Kirby-Docken and Liu[17] to build wavefunctions for the $X\,^1\Sigma^+$ ground state. Since this basis contains some rather diffuse orbitals. we first checked that there were no spurious effects due to the finite R-matrix radius (our usual value of 10 a_0+). Spherical Bessel functions for $l = 0, 1...6$ were found to provide adequate representation of the scattered electron.

Several fixed nuclei models were investigated before nuclear motion was introduced. The first set of models used the SCF wavefunction obtained from the Slater basis as their target wavefunction. The functions Ξ_i in Eq 18 were of two types. The first, or 'correlation', terms are obtained by allowing the extra electron to occupy any of the virtual orbitals allowed by the overall symmetry of the scattering problem. We will refer to any calculation which includes only a single target state and the corresponding correlation terms as 'static exchange' (SE). The second set of terms are 2-particle, 1-hole (2-p,1-h) configurations and represent short range polarization effects. Single state calculations which include this type of term will be refered to as 'static exchange plus polarization' (SEP). Long range polarization effects can be included by the addition of polarized pseudo-states (PS) chosen to give a good representation of the static polarizabilies of the target.

The second set of models used wavefunctions obtained from a complete active space CI (CASCI) calculation in which six electrons were distributed amongst the 5σ, 6σ, 1π and 2π orbitals in all possible ways. This generated a ground state consisting of 30 CSFs. Much better CI wavefunctions are of course available. However the size of the scattering calculation constrains us to use very compact representations of the target. It is difficult to postulate criteria for the choice of target wavefunctions for use in scattering calculations. It seems reasonable to assume that highly accurate absolute target energies are not required since only relative energies are involved. In vibrationally resolved calculations the shape of the target potential energy curve must however be accurate. Other properties which clearly affect scattering are the dipole, quadrupole and static polarizabilities of the target. It is well known that, for CO, the SCF approximation gives the wrong polarity for the equilibrium dipole moment. In fact the dipole moment changes sign near to the Equilibrium geometry and the SCF approximation over estimates it (assuming positive values correspond to polarity C^+O^-) at all geometries. Our CI target did much better, giving $D=-0.049$ au at equilibrium compared with the experimental value of -0.048 au. In models based on CI targets, the L^2 part of the fixed nuclei trial wavefunction was restricted to 'correlation' type terms, since the inclusion of 2-particle, 1-hole terms can have the effect of over-correlating the N+1 electron state relative to the N electron target, see the next chapter for a discussion of this.

Further fixed nuclei calculations were carried out on a grid of 10 geometries in the range $1.8 \leq R \leq 3.0$ a_0. No avoided crossings between the potential curves were observed and since the scattering is dominated by the shape resonance, the non-adiabatic method described in the preceding chapter was used unmodified. In order to keep the model entirely *ab initio*, we used target vibrational wavefunctions obtained by fitting the target potential energy curves to Morse potentials. This should be adequate for the low energies and restricted range of geometries considered. Orthogonal polynomials were used for the basis $\eta_j(R)$, introduced in Eq 21.

Here we will only consider the $^2\Pi$ symmetry, since this is the only one to display any structure at low energies. It was clear from a comparison of SE and SEP approximations that the short range polarization effects provided by the 2-p,1-h terms in the SEP approximation were crucial in getting an accurate representation of the resonance. In the published work,[7] the results obtained using the CI target were found to be inferior to those obtained from the models based on the SCF target. This was subsequently traced to be a side effect of the Lagrange orthogonalization[18] used to avoid linear dependence between the target and continuum orbitals. New results obtained using the CI target and Schmidt orthogonalization only are very similar to the published SEP cross sections.

The correct resonance width is crucial to getting the correct structure in vibrationally inelastic cross sections. If the width is too large then the resonance lifetime is too short for the 'boomerang' effect to occur and no substructure will be observed in the resonance peak. Under-correlation, as in a SE calculation, is the most common cause of this problem. However, as noted above, care must be taken in chosing the L^2 basis when using a truncated CI target. The single excitation, 'correlation', terms should produce no spurious effects, though linear dependence between diffuse virtual and continuum orbitals can manifest itself here. Two-particle one-hole terms can lead to over correlation and were omitted.

At internuclear separations greater than about 2.5 a.u. all calculations predicted that CO^- would be bound. Resonance positions and bound state energies obtained using the SEP model are shown in Figure 2.

In Figure 3 we show the R-matrix results for the excitation of the v=1 level, compared with the measurements of Allan.[19] The theoretical results show only the $^2\Pi$ contribution since the contribution from other symmetries is negligible on the scale of the graph. The experimental data is relative, therefore we have normalized it to give the same maximum peak height for the v=0→1 cross section as the corresponding theoretical data. In the vicinity of the resonance there is excellent agreement between theory and experiment. The theoretical v=0→1 cross-section appears to lie below experiment below the resonance, but this could be attributed to an underestimate of the $^2\Sigma$ contribution. This symmetry is very sensitive to polarization effects. Although pseudo-states were introduced to model these effects, they were constructed somewhat arbitrarily from the molecular orbitals used to construct the ground state wavefunction and no attempt was made to optimize them as has been suggested by Malegat et al.[20] Cross sections for the excitation of higher levels do not show any low energy shoulders and theory and experiment are in good agreement. Above the resonance the discrepancies between theory and experiment can be attributed primarily to the truncation of the number of vibrational channels which we retained in the outer region of configuration space.

4. DISSOCIATIVE RECOMBINATION OF HeH

Dissociative recombination (DR) conventionally proceeds by either or both of the following mechanisms. In the direct mechanism, an incident electron is captured into a repulsive electronic state of the neutral molecule and the nuclei move apart in this potential. The molecule is stabilised against the competing autoionisation process when the internuclear separation is such that the potential energy curve of the neutral lies below the ionic ground state potential energy curve. In the indirect mechanism,[21] the electron is captured into a vibrationally excited Rydberg state, AB^*, of the neutral molecule, which then predissociates with the repulsive state. Of the two, the first is normally the dominant process and the interference between the two processes generally results in complicated resonance structure in the cross section.

Both mechanisms depend critically on the presence of a curve crossing between the ionic and neutral potential energy curves. For systems without suitable curve crossings, such as HeH, DR cross sections have normally been assumed to be negligible. Yousif and Mitchell [22] measured the DR cross section for HeH^+ in a merged beam experiment and found it to be of a magnitude comparable with systems where there is a curve crossing. These results have been confirmed by other workers using similar techniques.[23, 24]

Since previous theoretical models of dissociative recombination are all dependant on the presence of curve-crossings, they are inapplicable to HeH. They neglect the direct coupling of the electronic and nuclear continua which arises from the nuclear kinetic energy operator and, in this sense, can be regarded as depending on 'indirect' coupling effects.

The generalization of the non-adiabatic R-matrix theory, as described above, has been applied to HeH by Sarpal et al.[9] Curve crossings are not necessary in this model, but it is important that there is coupling between the R-matrix electronic curves at some internuclear separations, not necessarily close to the equilibrium separation of the ion. This coupling allows for flux to go from higher states, which describe the electronic continuum, into the lower states which correspond to states involved in the DR process.

The method was first applied to the vibrational excitation of HeH^+.[8] A graphic illustration of the importance of non-adiabatic effects is provided by their results. No resonances were found in any of the fixed geometry calculations, nor in an adiabatic approximation of nuclear motion. Large numbers of narrow resonances appeared once nuclear motion was treated non-adiabatically. The eigenphase sum, obtained by diagonalizing the non-adiabatic K-matrix, is compared with that obtained in the fixed nuclei approximation at the equilibrium geometry, for a small energy range near the elastic threshold, in Figure 4.

In order to obtain DR cross sections it was necessary to include the lowest thirteen R-matrix poles in the diabatic formalism described above since the curves were observed to interact strongly at small bond lengths. An additional seven poles were treated in the non-coupling formalism.

Figure 5 compares the summed DR cross section to all three $^2\Sigma$ states with the experimental total cross section of Yousif and Mitchell.[22] The cross section to the ground state, X $^2\Sigma$, was found to be the largest in magnitude with the cross section to the A $^2\Sigma$ being an order of magnitude smaller. The cross section for the C $^2\Sigma$ state was smaller still. The structure in all three cross sections is very similar.

The agreement between theory and experiment is good; in particular, the window resonances around 0.028 eV and 0.05 eV are reproduced quite accurately and the cross section displays a sudden decrease around 0.1 eV in both sets of results. This drop

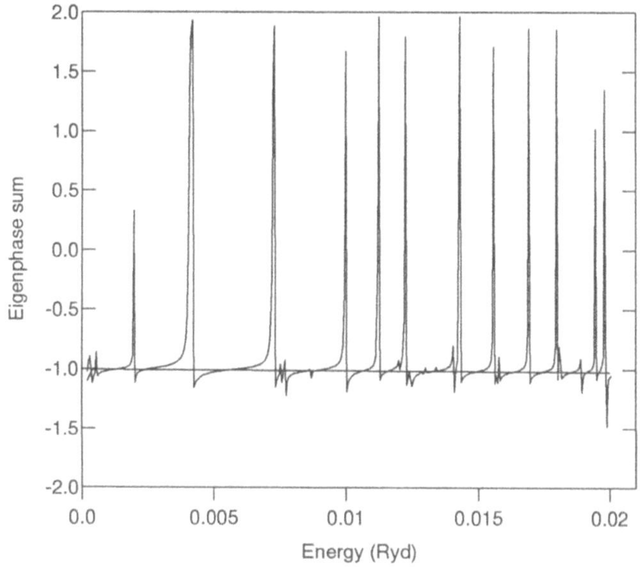

Figure 4. Eigenphase sums for HeH; solid line, non-adiabatic approximation; dashed line, fixed geometry.

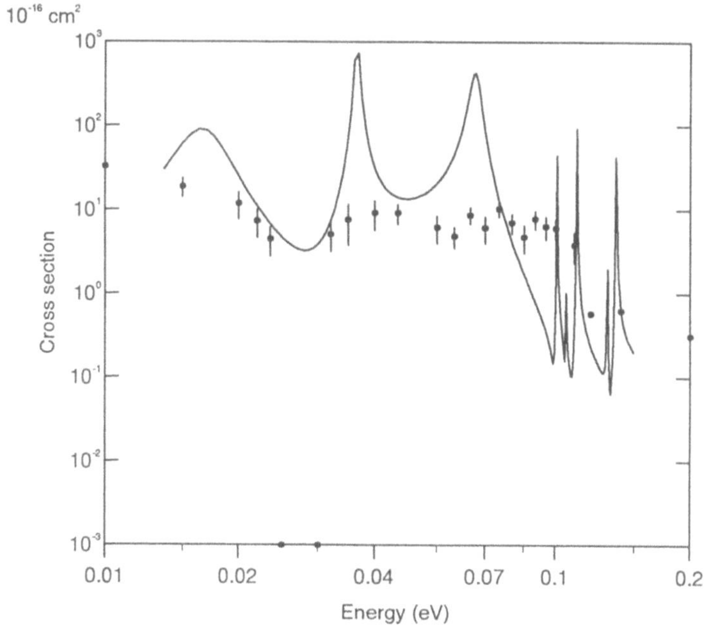

Figure 5. Dissociative recombination cross sections for HeH; solid line, theory; dashed line, Yousif and Mitchell.

in cross section for energies greater than 0.1 eV is attributed to the large number of Rydberg states converging to the $v = 1$ threshold. The theoretical results do not take account of instrumental effects on the cross section and ignore the effects of rotational motion. Both of these are expected to smear out some of the sharper resonant features in the theoretical cross section. The contribution from the B $^2\Pi$ state was also not included, but it is expected that this will be much smaller than the $^2\Sigma$ contributions since only weak interaction had been observed for the $^2\Pi$ symmetry.[26] The fact that DR is due to non-adiabatic coupling was verified by neglecting the couplings between states, i.e. reverting to formalism of Schneider et al ,[1] when the cross section was found to be effectively zero.

These results suggest that DR cross section for species which do not have curve crossings between the dissociating neutral state and the ionic ground state may sometimes be larger than conventionally assumed. In these cases DR may proceed by the non-adiabatic coupling between the electronic and nuclear continua.

5. SUMMARY

The calculations described above are representative of work currently being carried out with the UK R-matrix codes. The most serious short-comings are in the description of the targets in the fixed geometry calculations. Recent developments will enable us to use somewhat better target wavefunctions but problems still remain with the inclusion of the large number of, often diffuse, states which lie close above the lowest valence states of most molecules. The diabatic representation introduced in this chapter has, to date, only been tested on HeH where there are no avoided crossings between electronic R-matrix eigenstates, except at small bond lengths. Calculations for other processes, such as photoionization, with nuclear motion included in this non-adiabatic approximation, are in progress.

References

[1] B.I. Schneider, M. Le Dourneuf and P.G. Burke, *J. Phys. B: At. Mol. Phys.* **12**, L365 (1979).

[2] B.I. Schneider, M. Le Dourneuf and Vo Ky Lan, Phys. Rev. Lett. **43**, 1926 (1979)

[3] L.A. Morgan *J. Phys. B: At. Mol. Phys.* **19** L439 (1986).

[4] L.A. Morgan and P.G. Burke *J. Phys. B: At. Mol. Opt. Phys.*, **21**, 2017 (1988).

[5] L.A. Morgan, P.G. Burke and C.J. Gillan *J. Phys. B: At. Mol. Opt. Phys.*, **23**, 99 (1990).

[6] R. Fandreyer, P.G. Burke, L.A. Morgan and C.J. Gillan *J. Phys. B: At. Mol. Opt. Phys.*, **26**, 3625 (1993).

[7] L.A. Morgan *J. Phys. B: At. Mol. Opt. Phys.* **24** 4649 (1991).

[8] B.K. Sarpal, J. Tennyson and L.A. Morgan, *J. Phys. B: At. Mol. Opt. Phys***24**, 1851 (1991).

[9] B.K. Sarpal, J. Tennyson and L.A. Morgan, submitted to *J. Phys. B: At. Mol. Opt. Phys.*, 1994.

[10] C.J. Gillan, O. Nagy, P.G. Burke, L.A. Morgan and C.J. Noble, *J. Phys. B: At. Mol. Phys.* **20**, 4585 (1987).

[11] A.D. McLean " Conference on Potential Energy Surfaces in Chemistry ", W.A. Lester Jr., ed., IBM Research Laboratory, San Jose, p87 (1971).

[12] C.J. Noble, Daresbury Laboratory Technical Memorandum DL/SCI/TMT33T (1982).

[13] P.G. Burke and K.A. Berrington, "R-matrix Theory of Atomic and Molecular Processes" (IOP Publishing, Bristol, 1993).
[14] N.F. Lane *Rev. Mod. Phys.*,**52**, 29 (1980)
[15] D.W. Norcross and L.A. Collins *Adv. At. Mol. Phys.*,**18**, 341 (1982)
[16] M. Morrison *Adv. At. Mol. Phys.*,**24**,51 (1988)
[17] K. Kirby-Docken and B. Liu, *J. Chem. Phys.* **66** 4309 (1977).
[18] J. Tennyson, P.G. Burke and K.A. Berrington,*Comput. Phys. Commun.*, **47**, 207 (1987).
[19] M. Allan, *J. Electron. Spectrosc. Rel. Phenomena* **48** 219 (1989).
[20] L. Malegat, M. Vincke, C.G. Gillan and M. Le Dourneuf *J. Phys. B: At. Mol. Opt. Phys.* **25**, 727 (1992).
[21] J.N. Bardsley, *J. Phys. B: At. Mol. Phys.* **1**, 365 (1968).
[22] F.B. Yousif and J.B.A. Mitchell, *Phys. Rev.* A **40** 4318 (1989).
[23] M. Larsson, Private communication (1993)
[24] T. Tanabe et al,*Phys. Rev. Lett.* **70**, 422 (1993).
[25] B.K. Sarpal, J. Tennyson and L.A. Morgan, Dissociative Recombination: Theory, Experiment and Applications, ed B R Rowe, J B A Mitchell and A Canosa (Plenum, New York, 1993).
[26] B.K. Sarpal, S.E. Branchett, J. Tennyson and L.A. Morgan, *J. Phys. B: At. Mol. Opt. Phys.* **24**, 3685 (1991).

THE UK MOLECULAR R-MATRIX SCATTERING PACKAGE: A COMPUTATIONAL PERSPECTIVE

Charles J. Gillan[1], Jonathan Tennyson[2], and Philip G. Burke[1]

[1] Department of Applied Mathematics and Theoretical Physics, Queen's University of Belfast, N. Ireland BT7 1NN, United Kingdom
[2] Department of Physics and Astronomy, University College London, London WC1E 6BT, United Kingdom

1. INTRODUCTION

The R-matrix theory has been presented in the previous chapters of this book from a theoretical standpoint. Such is its complexity however that a separate chapter, that is this one, must be devoted to the implementation of the theory as a set of computer programs. Obviously if the R-matrix theory or any other method of solving the electron molecule scattering problem is to be viable it must be economic in the sense that it can be used routinely to produce numeric data for comparison with experimental measurements. Today, this means that the theory can be coded as a set of one or more computer codes which can be run in realistic time scales to produce data. This chapter attempts to show how the R-matrix method has been implemented, by a collaboration in the UK, in order to meet this objective. The reader should remember that the program suite reported here has been, and continues to be, developed and maintained by many people including overseas visitors and not just these authors.

Readers of R-matrix theory papers, of which there are now many in the literature, are used to seeing the R-matrix basis expansion of eq 18 as the starting point for any discussion of the theory. For the sake of completeness, however, the salient details of R-matrix theory and its approach to wavefunction expansion are given here. The reader should note that when using the R-matrix computer programs the wavefunction is in fact the last item to be built; a feature shared by the, closely related, configuration interaction method in atomic and molecular bound state studies.

We seek the quantum mechanical description of a scattering event where we know, or assume, that the system is described by the non-relativistic Schrödinger Hamiltonian which we denote by H_{N+1}; in this system there are N electrons which constitute the target and one other particle, the projectile, which will also be an electron in this article. The R-matrix codes can, it should be stated, handle the case of a positron incident on a molecule too. Basic quantum mechanics states that when the system has energy E

Computational Methods for Electron-Molecule Collisions
Edited by W.M. Huo and F.A. Gianturco, Plenum Press, New York, 1995

then it has a wavefunction, Ψ_E, given by

$$(H_{N+1} - E)\,\Psi_E = 0. \tag{35}$$

In fixed nuclei R-matrix theory we divide configuration space into two regions, the inner and outer, as shown in figure 1 and discussed by Gillan and co-workers,[1, 2] and apply

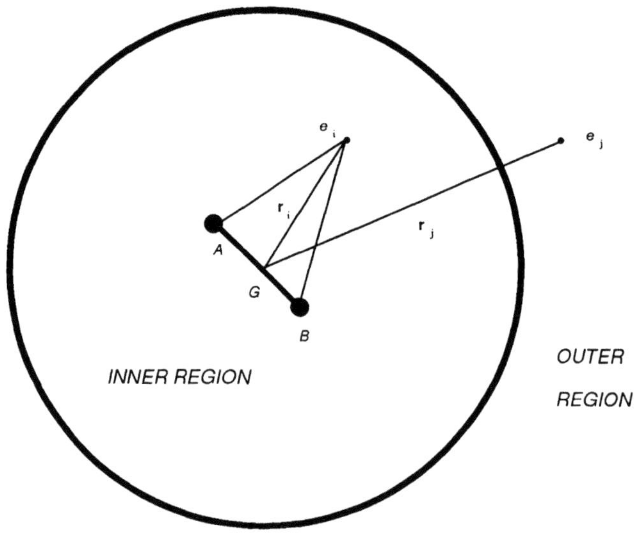

Figure 1. Division of configuration space in fixed nuclei R-matrix theory

the following expansion of the wavefunction in the inner region,

$$\Psi_E = \sum_k A_{Ek}\psi_k \tag{36}$$

where the ψ_k are energy independent basis functions and the A_{Ek} are coefficients whose value changes as we change the total energy E. The power of R-matrix theory lies in the fact that these ψ_k are independent of energy and need be constructed only once for all scattering energies.

The next section discusses the *morphology* of the ψ_k from the point of view of a fixed nuclei calculation leaving consideration of nuclear motion to chapters 2 and 5. A precis of the outer region part of a fixed nuclei calculation is then the subject matter of the subsequent section. Following these discussions there is a section outlining the steps that have to be performed when carrying out a full R-matrix study of any target and complimenting this is a section describing the algorithms used at various stages in the computer program implementation. The chapter closes with a prospective look at new developments taking place in the R-matrix code suite.

2. CHOOSING BASIS FUNCTIONS AND CONFIGURATIONS

The inner region R-matrix basis functions, ψ_k, are expansions over configuration state functions, CSFs, which in turn are composed of linear combinations of Slater determinants; these determinants involve spin orbitals with the spatial part of these, the orbitals ϕ_i, being linear combinations of functions defined on the nuclei and on the centre of gravity of the molecule. The strong connection with quantum chemistry, specifically configuration interaction studies of molecular bound states, should now be obvious. Thus in studying the fixed nuclei inner region problem one must consider two features:

1. The choice of the nuclear centred basis functions, the form of the functions defined on the scattering centre and their combination into a molecular orbital set.
2. The building of a vector space of CSFs by distribution of electrons, and their quantum number couplings, in the molecular orbital set.

Ultimately any fixed nuclei R-matrix calculation is judged by the specific form it employs for each of these.

2.1. Basis Functions and Orbitals

Clearly the expression of molecular orbitals, ϕ_i, as linear combinations of basis functions, U_j, may be best represented as a matrix, c_{ij}, where

$$\phi_i = \sum_j c_{ji}\, U_j. \tag{37}$$

The structure of the matrix, **C**, is illustrated in figure 2 where it is shown that the orbital set is composed of two distinct parts. In the present scheme Slater Type Functions

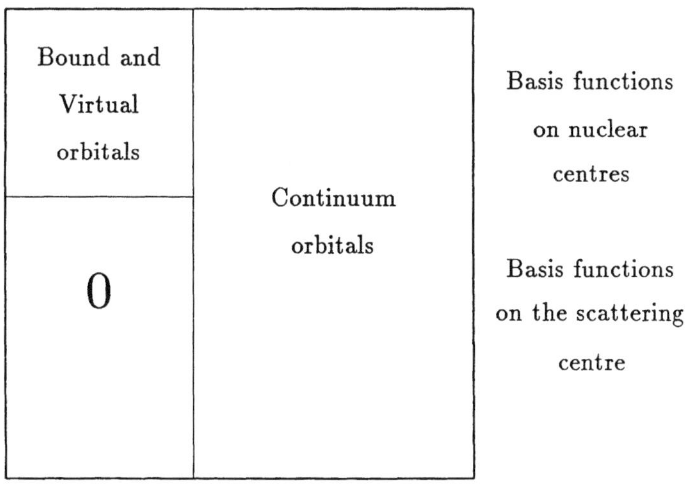

Figure 2. Structure of orbital set used in an R-matrix calculation showing that the one particle basis is composed of two distinct types of functions.

(STFs, also known as Slater Type Orbitals STOs) χ_i, centered on each of the atomic nuclei are used to span the bound space while numerical functions defined on the centre of gravity of the molecule comprise the continuum part. The STFs have the form,

$$\chi_i\,(n,l,m,k,\zeta) = \sqrt{\frac{(2\zeta_i)^{2n+1}}{(2n)!}}\; r_k^{n-1}\, e^{-\zeta_i r}\; Y_{lm}(\hat{\mathbf{r}}_k) \tag{38}$$

where k defines the nuclear centre and numerical functions have the form,

$$\frac{1}{r_G}\, u_{ij}\,(r_G)\, Y_{l_i m_i}(\hat{\mathbf{r}}_G) \tag{39}$$

the Y_{lm} being complex spherical harmonics.[3] The radial parts, u_{ij}, are obtained by solving the model, single channel scattering problem

$$\left[\frac{\mathrm{d}^2}{\mathrm{d}r^2} - \frac{l_i(l_i+1)}{r^2} + 2V_0(r) + k_j^2\right] u_{ij}(r) = 0 \tag{40}$$

subject to the fixed boundary conditions

$$u_{ij}(0) = 0, \qquad \left[\frac{a}{u_{ij}(r)}\frac{\mathrm{d}u_{ij}(r)}{\mathrm{d}r}\right]_{r=a} = \mathrm{b} \tag{41}$$

Here a is the R-matrix boundary radius, b is an arbitrary constant frequently set at zero and V_0 is an arbitrary spherical potential which is often chosen to be the spherically symmetric part of the target ground state static potential. The numerical functions are therefore solutions of a Stürm-Liouville problem with homogeneous boundary conditions and the radial part of the scattering wave function is ultimately a generalized Fourier series expansion on these functions. Any finite expansion over these functions will exhibit the Gibbs phenomenon near the R-matrix boundary and so a correction term, proposed by Buttle[4] must be added to the R-matrix on the boundary to compensate for the finite expansion length. Note that only these numerical functions are non-zero on the R-matrix boundary and therefore they alone provide the link between the inner and outer regions.

It is important to note that the orbital set is chosen to be orthogonal; this is because such a choice simplifies greatly the evaluation of the Hamiltonian matrix elements (see later). The target occupied and virtual orbital part of the matrix **C** is computed first using standard quantum chemical techniques; numerical functions representing the continuum are then added to the basis set. Initially zero matrices are inserted on the target – continuum off-diagonal blocks and a unit matrix is placed in the lower right hand corner. At this stage an orthogonalization procedure is applied, typically the Schmidt process. This means that the continuum orbital space picks up a target orbital component a feature that can cause practical linear dependence problems at the four index transformation stage if the coefficients produced are too large.

The final structure of the orbital matrix **C** is illustrated in figure 2 where its morphology is clearly visible. It is, of course, imperative that the compound basis set, Slater functions and numerical functions, is linearly independence but it is necessary to distinguish between the mathematical and computational nature of this attribute. Computer arithmetic is performed using fixed size, floating point numbers a feature which imposes limits on the accuracy of the real numbers that can be represented. The numerical integration technique employed will add a further source of error. This means that it is possible to have basis sets which exhibit numerical linear independence on a thirty two or sixty four bit computer architecture but which would be linearly independent if more accuracy were available.

Having chosen the expansion basis for the molecular orbitals, the fundamental quantities on which the rest of the calculation depends are atomic integrals so called because they involve atomic centered functions (in scattering also the centre of gravity). In R-matrix theory integrals are required over a finite range of the radial co-ordinate not the infinite range needed in quantum chemistry and it is this feature which introduces a major distinction between bound state calculations and scattering evaluations. For a more detailed discussion of the classification of atomic integrals including a description of the breakdown of two electron integrals into Coulomb, hybrid and exchange types the reader should consult Shavitt.[5]

Bound state Slater integral codes perform their task using combinations of numerical quadrature over the different co-ordinates and hence they are easily modified to include functions which are defined numerically as in our scattering problem; it is for this reason that our integrals program has been developed from a Slater bound state package.[6] It should be noted however that Slater functions are only useful for linear molecules as noted in chapter 1.

2.2. Configurations for e-N_2 with an SCF Target Representation

Having defined a molecular orbital set we can now make the R-matrix basis functions ψ_k concrete thereby illustrating their relationship to the physics and chemistry of the problem. It will be shown, in particular, that the form chosen for the ψ_k is closely related to the close coupling expansion;[7] the R-matrix calculation reported by Gillan and co-workers[1] is an appropriate vehicle for this purpose.

The simplest possible representation of the ground state wavefunction is the Hartree Fock or single configuration model which is, following the aufbau principle,[8]

$$1\sigma_g^2\ 1\sigma_u^2\ 2\sigma_g^2\ 2\sigma_u^2\ 3\sigma_g^2\ 1\pi_u^4. \tag{42}$$

When an SCF evaluation of the $X^1\Sigma_g^+$ state of N_2 is performed, a basis of Slater type will be used which is much larger than just having one function for each atomic orbital. This leads to several extra orbitals being evaluated and these are known as virtual orbitals; chemically, they are approximately representing an extra electron outside an N_2 core. In the basis of Nesbet[9] there are seven σ type and three π type Slater functions on each atom yielding a molecular orbital space which may be classified as follows

- Occupied Space

$$1\sigma_g, 1\sigma_u, 2\sigma_g, 2\sigma_u, 3\sigma_g, 1\pi_u \tag{43}$$

- Virtual Space

$$(4,\ldots,7)\sigma_g,\ (3,\ldots,7)\sigma_u,\ (2,3)\pi_u,\ (1,\ldots,3)\pi_g \tag{44}$$

Although the virtual orbitals are more diffuse than the occupied at a radius of 7 bohr from the center of gravity they have decayed to negligible amplitude whereas the continuum orbitals have not. Consider scattering in the $^2\Sigma_g^+$ symmetry, as reported by Gillan and collaborators,[1] where there are thirty continuum molecular orbitals of σ_g type; these can be numbered as orbitals $(8,\ldots,37)\sigma_g$ in our set. We may now make a convenient description of the R-matrix basis functions, ψ_k, in terms of orbital configurations. Thus,

1. The first part of each ψ_k corresponds to the configurations

$$1\sigma_g^2\ 1\sigma_u^2\ 2\sigma_g^2\ 2\sigma_u^2\ 3\sigma_g^2\ 1\pi_u^4\ (^1\Sigma_g^+)\ \{8\sigma_g,\ldots,37\sigma_g\}^1\ (^2\Sigma_g^+) \tag{45}$$

These are the configurations which account for one electron in a continuum orbital with the rest of the electrons in the target state.

2. The second part of each ψ_k consists of two types of configurations.
   - One particle terms with the form:

$$1\sigma_g^2\, 1\sigma_u^2\, \left\{2\sigma_g^2\, 2\sigma_u^2\, 3\sigma_g^2\, 1\pi_u^4\, (^1\Sigma_g^+)\, \{4,5,6,7\sigma_g\}^1\right\}\, (^2\Sigma_g^+) \tag{46}$$

   where the scattering electron drops down into the virtual orbitals of the same symmetry but the target electrons remain as before
   - Two particle - One hole terms with the form:

$$1\sigma_g^2\, 1\sigma_u^2 \otimes \left\{2\sigma_g^2\, 2\sigma_u^2\, 3\sigma_g^2\, 1\pi_u^4\right\}^{-1} \otimes \{4,\ldots,7\sigma_g, 3,\ldots,7\sigma_u, 2,3\pi_u 1,2,3\pi_g\}^2\, (^2\Sigma_g^+) \tag{47}$$

   in which there is one hole in the target and two particles in the virtual space.

   Both of these types are L^2 configurations because all the electrons are in short range orbitals but we distinguish the two types. We call the former type orthogonality configurations because they compensate for the fact we have imposed orthogonality of the continuum molecular orbitals to the virtual space, a restriction that simplifies the computer programs immensely but is in fact not required by the quantum theory. These configurations also allow the continuum electrons to occupy high angular momentum states in the region of the nuclear singularities. The latter configurations are correlation configurations, sometimes referred to as (short range) polarization configurations, which introduce correlation effects between target and projectile. Note that these are correlation effects within the radial range of the STFs and not long range effects such as dipole polarization, a feature which can be modelled by using dipole pseudo-states.[10]

In passing, we note that a calculation in which the correlation configurations are omitted is referred to as being at the static exchange level (SE) while the addition of correlation configurations is a static exchange plus polarization approximation (SEP).

2.3. Configurations for e-N_2 with a CI Target Representation

In multistate scattering studies it is almost always necessary to use a configuration interaction representation of the target states, usually in the valence orbital space,[11] in order to obtain good relative energy separations. Here e-N_2 elastic scattering is reconsidered, using the same basis as above, as a simple example to illustrate the general situation for a CI target. The N_2 ground state is now represented by a CSF expansion too. Thus its wavefunction Φ is given by

$$\Phi = \sum d_i\, \Delta_i \tag{48}$$

where the Δ_i are a set CSFs, the choice of which defines different types of target state approximation, and the d_i are coefficients which are obtained by diagonalizing the target state Hamiltonian in the basis Δ_i. Adopting the valence configuration interaction representation the CSF space is defined by the electron distribution,

$$1\sigma_g^2\, 1\sigma_u^2\, \{2\sigma_g 2\sigma_u 3\sigma_g 1\pi_u 1\pi_g 3\sigma_u\}^{10}. \tag{49}$$

Thus we freeze the lowest two orbitals and distribute the remaining ten electrons in all possible ways over the rest of the orbital set; the dimension of this CSF space, for the ground state of N_2.

In the work of Gillan *et al* [1] there were twenty seven continuum molecular orbitals of symmetry π_g which we number $4, \ldots, 30$. Proceeding as before, but now looking at the resonant $^2\Pi_g$ scattering symmetry instead, the ψ_k can be analysed into parts as follows:

1. The first part of each R-matrix basis function

$$1\sigma_g^2\, 1\sigma_u^2\, \{2\sigma_g 2\sigma_u 3\sigma_g 1\pi_u 1\pi_g 3\sigma_u\}^{10}\, (^1\Sigma_g^+)\, \{4\pi_g \ldots 30\pi_g\}^1\, (^2\Pi_g) \tag{50}$$

consists of the direct product of the Δ_i defining the target space with one electron in the continuum space. Note that the target eigenvector coefficients, d_i, associated with each CSF have not yet been accounted for. They are in fact introduced prior to diagonalization of the Hamiltonian matrix.

2. The L^2 part of each ψ_k may be further divided into three pieces:
   - Orthogonality CSFs

$$1\sigma_g^2\, 1\sigma_u^2\, \{2\sigma_g 2\sigma_u 3\sigma_g 1\pi_u 1\pi_g 3\sigma_u\}^{10}\, (^1\Sigma_g^+)\, \{2, 3\pi_g\}^1\, (^2\Pi_g) \tag{51}$$

   accounting for the fact that the continuum molecular orbitals are also orthogonal to the unoccupied virtual orbitals.
   - Correlation CSFs

$$1\sigma_g^2\, 1\sigma_u^2\, \{2\sigma_g 2\sigma_u 3\sigma_g 1\pi_u 1\pi_g 3\sigma_u\}^{11} (^2\Pi_g). \tag{52}$$

   where the scattering electron enters the charge cloud of the target state.
   - Two particle - One hole Additional Correlation CSFs

$$\begin{aligned} 1\sigma_g^2\, 1\sigma_u^2 \otimes \quad & \{2\sigma_g 2\sigma_u 3\sigma_g 1\pi_u 1\pi_g 3\sigma_u\}^9\, (Z) \otimes \\ & \{4 \ldots 7\sigma_g, 4 \ldots 7\sigma_u, 2, 3\pi_u, 2, 3\pi_g\}^2\, (^2\Pi_g). \end{aligned} \tag{53}$$

   These CSFs are analagous to the two particle one hole terms for the SCF target but are vastly increased in number with a CI target. The quantity Z indicates that the nine remaining target electrons must have their coupling restricted to represent a target *hole*. For the N_2 ground state Z takes the values $^2\Sigma_g^+$, $^2\Sigma_u^+$, $^2\Pi_u$, $^2\Pi_g$ etc.

In summary then, with a CI target it is not possible to delineate the role of specific configurations as clearly as in the case of an SCF target.

3. THE OUTER REGION

In the external region of configuration space, a single centre, no exchange, close coupling expansion of the scattering wavefunction is adopted, reflecting the fact that physical interactions, inherently different to those of the inner region, now dominate the scattering process.[12] A body fixed frame of reference is chosen and the expansion adopted is

$$\Psi_E = \sum_{i=1}^{n} \bar{\Phi}_i(x_1, \ldots x_N, \sigma_{N+1})\, \frac{1}{r_{N+1}}\, F_i(r_{N+1})\, Y_{l_i m_{l_i}}(\hat{\mathbf{r}}_{N+1}) \tag{54}$$

where the $x_i = (\hat{\mathbf{r}}_i, \sigma_i)$ represent the angular and spin co-ordinates of electron i and the functions $\bar{\Phi}_i$ are formed by coupling the scattering electron spin with the target state spins to yield eigenfunctions of S^2 and S_z for the N+1 electron system. The channel functions for the problem, involving all but the radial co-ordinate of the scattering electron, are $\left[\bar{\Phi}_i Y_{l_i m_{l_i}}\right]_\Lambda$ and are eigenfunctions of the operator L_z with eigenvalue Λ as well as of the operators S^2 and S_z for the N+1 electron system.

On substituting this form of Ψ_E into equation (35) and projecting onto the channel functions one obtains

$$\left\{\frac{d^2}{dr^2} - \frac{l_i(l_i+1)}{r^2} + k_i^2\right\} F_i(r) = 2\sum_j V_{ij}(r)\, F_j(r) \tag{55}$$

which is a set of n, coupled, homogeneous, differential equations for the reduced radial wavefunctions F_i. The quantity k_i^2 is defined as

$$k_i^2 = 2(E - \epsilon_i) \tag{56}$$

ϵ_i being the target state eigenenergies. The matrix V_{ij}, which is a function of the radial variable r_{N+1} is the potential matrix which describes the coupling between the channels of interest. It has the form

$$V_{ij} = <\bar{\Phi}_i Y_{l_i m_{l_i}} \mid \sum_{k=1}^{N} \frac{1}{\mid \mathbf{r}_{N+1} - \mathbf{r}_k \mid} - \frac{Z_A}{\mid \mathbf{r}_{N+1} - \mathbf{R}_A \mid} - \frac{Z_B}{\mid \mathbf{r}_{N+1} - \mathbf{R}_B \mid} \mid \bar{\Phi}_j Y_{l_j m_{l_j}} > \tag{57}$$

where $Z_A, \mathbf{R}_A$ and $Z_B, \mathbf{R}_B$ are the charges and position vectors of the two nuclei respectively and N is the number of electrons in the target. Since the radial co-ordinate of the scattered electron is greater than the radial co-ordinates of all of the target electrons, for this particular region of configuration space, it is possible to expand the matrix elements of the direct potential in inverse powers of r_{N+1}. Applying the spherical harmonic addition theorem yields the expression[2]

$$V_{ij} = \sum_{s=1}^{\infty} \frac{a^{(s)}_{il_i,jl_j}}{r_{N+1}^{s+1}} - \frac{Z_A + Z_B - N}{r_{N+1}} \delta_{ij}\, \delta_{l_i l_j}. \tag{58}$$

Evaluation of the asymptotic potential coefficients, $a^{(s)}_{il_i,jl_j}$ is straightforward and requires data on the properties and transition moments of the target states included in the R-matrix expansion.

From a mathematical perspective the set of equations (55) have $2n^2$ solutions at every scattering energy each of the n channels yielding n regular and n irregular solutions. Physically, we may distinguish between open and closed channels, using equation (56), because the channels may be ordered contiguously in the same fashion as their associated target eigenstates. The ground state energy defines the zero level for scattering so that at all positive scattering energies those channels associated with the ground state are open. At scattering energies above the highest target state threshold all channels will be open. Between these extremes there will be, in general, n_a open channels ($n_a < n$) and n_b closed channels ($n_b = n - n_a$).

In the limit that the electron-molecule distance tends to infinity we may define, within each open channel i, n_a different, linearly independent, standing wave solutions j, of the form

$$F_{ij} \sim \frac{1}{\sqrt{k_i}} \left(\sin(k_i r - \frac{1}{2} l_i \pi)\delta_{ij} + \cos(k_i r - \frac{1}{2} l_i \pi) K_{ij}\right). \tag{59}$$

The $n_a \times n_a$ coefficients K_{ij} define the K-matrix which embodies the information on the scattering process. From it, or more precisely the related T-matrix, we may calculate physical observables for the scattering process. In the n_b closed channels, the radial function decays exponentially,

$$F_{ij} \sim e^{-|k_i|r} \tag{60}$$

indicating that there is no flux lost at infinity.

4. SCATTERING COMPUTATIONS: A TOP DOWN VIEW

The best way to present an overview of the computational steps that one carries out in applying the R-matrix method is to list those steps in a top down fashion. Thus we have:

1. Generation of target orbitals and target states.
   In fact this encompasses the entire field of molecular electronic structure theory and is not a trivial problem! In early work it was sufficient to use a simple single configuration representation of the lowest electronic state of the molecule. Therefore an SCF calculation in a chosen basis of STFs was sufficient to obtain the orbital expansion coefficients. In recent years coupled electronic state calculations have become more important thereby requiring more complicated evaluations. In R-matrix calculations all electronic states are expanded in one common set of target orbitals and since we are forced to use only moderate CI expansions for these states, it is often quite difficult to obtain an optimal common orbital set.
2. Evaluation of the numerical basis functions.
3. Evaluation of integrals between atomic and/or centre of mass functions.
4. Continuum orbital generation. Lagrange and/or Schmidt orthogonalization of the continuum orbitals to the bound state orbitals.
5. Four index transformation from atomic integrals to molecular orbital integrals using the results of the previous step.
6. Creation of the Hamiltonian matrix and its diagonalization to yield eigenvalues and eigenvectors at fixed geometry. This step, which has to be repeated for each scattering symmetry under study, actually may be divided into sequential substeps as follows:
   (a) Configuration state function generation
   (b) Hamiltonian matrix element evaluation in terms of coefficients and integral labels; otherwise known as the formula tape construction.
   (c) Formula tape reindexing and reordering for efficient Hamiltonian matrix construction
   (d) Unification of formula tape and actual integrals to construct the Hamiltonian matrix which is then diagonalized to yield eigenvalues and eigenvectors.
7. Data Management
   The previous steps are carried out for each scattering symmetry and with the nuclei held fixed in space. It is necessary then to collate all of the data from a range of internuclear geometries into an ordered set before proceeding to the nuclear motion stage.
8. Nuclear Motion
   When nuclear motion is being considered explicitly in the R-matrix calculations, it is necessary to set up and diagonalize a Hamiltonian incorporating the nuclear kinetic energy operator.

9. External Region
   This changes depending upon the problem in question. The possibilities in our present package are,
   (a) Electron/Positron Scattering
   (b) Dissociative Attachment/Recombination
   (c) Photoionization
   (d) Bound States
   Each of these possibilities must be considered separately. Of these, (a)-(c) produce a K-matrix, at each scattering energy, which is then used in the next step.
10. Once a K-matrix is obtained, several physical observables can be calculated from it, such as total and differential cross sections. It is these physical observables which are then compared with experiment.

This list, in itself, presents a coarse grain modularization of the computational task. In the next section we discuss the actual program modules which implement these steps and in so doing consider the fine grain modularization of the task. For each module we indicate where further information on the coded algorithms can be found.

5. DESCRIPTION OF INDIVIDUAL MODULES

Each step in the previous section is implemented by one or more programs in our package thus modularizing the task further. Much of the inner region code has been developed from the Alchemy I program suite written by scientists at the IBM Almaden Research Laboratory, San José California. The interested reader should consult an article by McLean[13] who has described the earliest versions of the Alchemy suite. Alchemy I and its modern counterpart Alchemy II[14] have been obtained by our group under license from IBM.

5.1. Atomic Integrals

This is the integral generator, for STFs, contained in the Alchemy suite adapted to generate all of the integrals for a scattering calculation. The algorithms used to evaluate the bound state integrals[3, 13] have been modified to restrict the range of the radial co-ordinate and to include functions defined numerically.[6, 15]

5.2. Hartree Fock Self Consistent Field

This is the self consistent field program from the Alchemy suite which solves the matrix Hartree Fock procedure.[16, 17]

5.3. Numerical Basis Function Generator

This program, described by Salvini,[18] solves the Stürm Liouville problem, equation (40), for the numerical continuum basis functions u_{ij} and evaluates the Buttle corrections[4] mentioned earlier.

5.4. Orthogonal Molecular Orbital Generator

This module reads a set of molecular orbital expansion coefficients and associated basis function overlap integrals and proceeds to generate an orthogonal orbital set. For bound state studies only the Schmidt option is used while for scattering situations an option exists to Lagrange orthogonalize the continuum molecular orbitals instead.[19]

Once the orthogonal continuum orbitals are available this code computes the magnitude of the continuum orbitals on the R-matrix boundary and then writes this data to disk; it is used by the external region codes to build the R-matrix elements at each scattering energy.

5.5. Four Index Transformation from Atomic to Molecular Integrals

This is essentially the unmodified Alchemy I four index transformation program which has been discussed by McLean;[13] it takes atomic integrals and multiplies them by combinations of molecular orbital coefficients to produce molecular integrals for Hamiltonian construction.

5.6. Configuration State Function Generator

This module generates CSFs with the appropriate spin and symmetry couplings for the system under study. In scattering work it is necessary to use a genealogical scheme to generate the configurations because of the form of the vector space of CSFs required; Nesbet has discussed a similar program for atomic scattering studies.[20] Specifically, it is imperative that one can specify exactly the coupling of the N electron target as well as the N+1 particle system. Orel *et al* [21] have given a detailed description of the method of pseudo-(N+1) electron CI calculation in order to circumvent the phase consistency problems inherent in CSF generation; this technique is also used in the R-matrix codes.

5.7. Formula Tape Evaluation and Symbolic Expansion

This code produces expressions for the Hamiltonian matrix elements in terms of coefficients and integrals labels. Assuming that molecular orbitals are orthonormal the Hamiltonian elements, H_{ij} may be written as linear combinations of one and two electron integrals[22] between these orbitals (transformed integrals),

$$
\begin{aligned}
< \Delta_i \mid H \mid \Delta_j > \quad &= \sum_{kl} A_{kl}^{ij} < \phi_k(\mathbf{r}_1) \mid h_1 \mid \phi_l(\mathbf{r}_2) > \\
&+ \sum_{klmn} B_{klmn}^{ij} < \phi_k(\mathbf{r}_1)\, \phi_l(\mathbf{r}_1) \mid \frac{1}{r_{12}} \mid \phi_m(\mathbf{r}_2)\, \phi_n(\mathbf{r}_2) > \\
&+ \delta_{ij} \frac{Z_A Z_B}{\mid \mathbf{R}_A - \mathbf{R}_B \mid}.
\end{aligned} \tag{61}
$$

The structure of the CSF space for scattering readily lends itself to the use of a symbolic approach to formula tape evaluation. Morgan and Tennyson[23] have implemented a new section of code to embrace that symbolic technique first reported by Liu and Yoshimine[24] in a fashion appropriate for scattering calculations.

5.8. Formula Tape Reordering

Given the formula tape and the dataset containing the molecular integrals, the Hamiltonian matrix could be constructed straightforwardly by reading the formulae sequentially and then carrying out the summation in equation (61). This, however, would become a bottleneck in the calculation as it would require, for each matrix element, successive reading and re-reading of the sequentially organized, integrals dataset. Instead, Yoshimine,[25] has devised a three stage algorithm which leads to faster set up of the Hamiltonian. At the heart of the algorithm is the idea that the lengthy integrals tape should only be read once and a fundamental step in achieving this goal is the re-ordering of the energy (Hamiltonian) expressions.

5.9. Hamiltonian Matrix Construction and Diagonalization

The operations carried out by the CI program are:

1. Read the sorted energy expression tape
2. Read the transformed integrals tape
3. Build the lower half triangle of the real symmetric Hamiltonian matrix in core
4. Carry out a contraction with target eigenstate vectors, if needed.
5. Diagonalize the matrix to find *all* eigenvalues and eigenvectors using the Householder procedure and store these for use by the external region program.

The first three steps above are accomplished by use of stages two and three of the Yoshimine procedure.[25] The contraction process has been discussed by Orel *et al* [21] with regard to application of the Kohn method to electron molecule scattering; the same operation is carried out within the R-matrix codes.

5.10. Density Matrix and Property Analysis Code

Wavefunction properties provide, in general, a more stringent test of the charge density than the computed energy and so most molecular structure suites include a module to evaluate these quantities. Fortunately, most of the wavefunction properties that one needs to calculate are for one electron operators. The transition property, or transition moment, T_{IJ} between the pair of N-electron wavefunctions Ψ_I and Ψ_J is defined as

$$\mathrm{T}_{IJ} = < \Psi_I \mid \sum_{i=1}^{N} f(\mathbf{r}_i) \mid \Psi_J > \tag{62}$$

where $f(\mathbf{r})$ is a one electron property operator for example $z = r\cos\theta$. When $\Psi_I = \Psi_J$ then T_{II} is a wavefunction property or target state moment.

It is possible to write down expressions for T_{II} and T_{IJ} in terms of the density and transition density matrices respectively.[26, 27] Once a density matrix is available *any* one electron property may be evaluated by contracting that density matrix with a matrix of appropriate property integrals. The two most common types of expectation value required in R-matrix scattering studies are the target state properties needed for building the potential matrix in the asymptotic region[2] and transition dipoles needed to evaluate the contribution of dipole polarized pseudostates;[10] transition dipoles are also required to compute photoionization cross sections.

5.11. Nuclear Motion Inner Region

In essence this code sets up and diagonalizes the nuclear motion Hamiltonian matrix in the inner region[1] implementing the theory which is described in the previous chapter. The program is controlled by a driving routine which,

1. Reads all of the fixed nuclei eigenvectors, eigenvalues and primitive boundary amplitudes.
2. Generates or reads target vibrational wavefunctions.
3. Sets up the nuclear motion Hamiltonian which involves the nuclear kinetic energy operator.
4. For each scattering energy evaluates the generalized R-matrix.
5. Interfaces with external region programs.

Clearly, there is a very large amount of input scratch file data required, the exact amount depending on the number of internuclear separations used. Target vibrational wavefunctions, of the Morse variety, can be generated by the code if the required parameterization of the Morse potential energy curve is input. On the other hand, numerical vibrational wavefunctions can be read in. These are certainly required for when studying excitation to high vibrational levels where the Morse prescription is less accurate.

The bulk of the computation occurs in the third step where several integrals are evaluated. However, it should be stressed that, for diatomics, the elements of the nuclear motion Hamiltonian are much more straight forward to evaluate than those needed for the electronic part of the calculation. The input data is first fitted to a bicubic spline over the range of internuclear separations used. Then, the integrals are carried out using Simpson's rule although there is also an option to use Gauss Legendre quadrature. The Householder method is again used for the diagonalization. In summary the set up and diagonalization of the nuclear motion Hamiltonian is a microcausm of the electronic situation.

5.12. External Region Codes and Utility Programs

The end product of any scattering calculation is a set of calculated physical observables which allow comparison between different computational methods and with experiments. The solution of the external region equations produces initially a K-matrix at each energy as discussed previously. The eigenphases and eigenphase sum are easily obtained by diagonalizing the K-matrix and yield important information on resonances and on partial wave mixing; additionally multichannel quantum defects may be evaluated. The S and T-matrices are obtained from the definitions,

$$S = \frac{1 + \mathrm{i}K}{1 - \mathrm{i}K} \quad \text{and} \quad T = S - 1. \tag{63}$$

Given the T-matrix and the definition of the scattering channels in the problem, simple summation formulae yield the experimentally measurable values of the total and differential scattering cross sections.

Since the K-matrix serves as an intermediate to the evaluation of other data it is seldom necessary in practice to examine the actual matrix itself. Obviously, in any given situation one must always solve the coupled equations and obtain the K-matrices at each scattering energy of interest, but the user is free to choose the physical observables calculated by calling further subroutines from the library.

Library member INTERF reads a user defined input deck and combines the files of Buttle corrections, boundary amplitudes and CI vectors produced by the inner region codes. The information is passed to library member RSOLVE which solves the external region problem using a combination of R-matrix propagation[28] and Gailitis[29] asymptotic expansion techniques.[30] When wavefunctions are also required, they are propagated inwards using direct numerical integration.[31] The output of RSOLVE is a file of K-matrices from which eigenphase sums and other properties are calculated. The complete list of library members and their functions is as follows:

GETCOR System dependent dynamic core allocation.

INTERF Reads and formats the inner region and input data for use by the rest of the library members.

VIBRMT Solves the nuclear motion R-matrix problem.

RSOLVE Solves the external region coupled differential equations and obtains the K-matrix at each scattering energy.

EIGENP Diagonalizes the K-matrices and calculates the eigenphases and eigenphase sum.

RESON Locates resonances and fits the associated eigenphase sum to a Breit-Wigner form.[32]

MQDT Calculates multichannel quantum defects.[33]

TMATRX Computes the T-matrix from the K-matrix at each energy.

IXSECS Evaluates integrated cross sections.

BOUND Locates bound states using an adaptation[31] of the method of Seaton.[34]

TDIP Transition dipoles among bound states obtained using BOUND.[31]

RATES Integration of collision strengths over a Maxwellian temperature distribution.

DCS The differential cross section evaluation from T-matrices using a previously published program.[35]

The use of the outer region code has been extensively adapted to study nuclear motion effects which go beyond the Born-Oppenheimer approximation. This is important for vibrational excitation near threshold or a resonance, dissociative recombination and dissociative attachment; these processes are discussed in chapter 2.

6. FUTURE DIRECTIONS

This article has described a large and complex software package which may be used to carry out state of the art, *ab initio* calculations on electron, positron and photon scattering by linear molecules. Inherent in all computer packages of this size is the need to develop and maintain it. Obviously this task may be divided into two sub-tasks as follows:

- more efficient algorithms must be found, where possible, and implemented as modules within the existing structure
- code must be kept up to date, transformed as needed, in order to exploit new and better FORTRAN compilers which in turn exploit new and better computer architectures.[36, 37]

The R-matrix method may be applied to areas of molecular physics beyond those discussed in this article and so the computer package is perpetually being expanded to encompass a larger problem domain. The refinements necessary to introduce Breit-Pauli type relativistic corrections into the theory have recently been discussed.[38] Until recently the R-matrix suite was restricted to linear systems, in practice only to diatomic molecules, because of the integral generator used. Working with the author of the Gaussian integrals code MOLECULE[39] new code is being added to the R-matrix suite to evaluate bound-continuum integrals for scattering from polyatomic targets. The R-matrix program suite has been radically enhanced from that which was used to study electron molecule scattering a decade ago and, clearly, the suite that has been discussed in this chapter is merely an outline of that which will hopefully exist at the end of the decade.

ACKNOWLEDGEMENTS

The authors wish to thank the International Business Machines Corporation who have provided copies of both the Alchemy I suite and their MOTECC-90 compilation of chemistry software under Joint Study agreements. Additional thanks are due to Prof. Jan Almlöf for providing a version of the program MOLECULE which has been optimized for vector processing on Cray architectures.

References

[1] C.J. Gillan, O. Nagy, P.G. Burke, L.A. Morgan and C.J. Noble, *J. Phys. B:At. Mol. Phys.* **20**, 4585 (1987)
[2] P.G. Burke, I. Mackey and I. Shimamura, *J. Phys. B:At. Mol. Phys.* **10**, 2497 (1977)
[3] M. Yoshimine and A.D. McLean, *IBM Journal of Research and Development* **12**, 206 (1968)
[4] P.J.A. Buttle, *Phys. Rev.* **A160**, 719 (1967)
[5] I. Shavitt, *Methods of Computational Physics* **1** (1963)
[6] C.J. Noble, "The Alchemy linear molecule integral generator", *Daresbury Laboratory Report DL/SCI/TM33T*, SERC Daresbury Laboratory, (1982)
[7] H.S.W. Massey and C.B.O. Mohr, *Proc. Roy. Soc.(London)* **A136**, 289 (1932)
[8] J.C. Morrison, A.W. Weiss, K. Kirby and D.Cooper, "Structure of atoms and molecules", *in*: "Encyclopedia of Applied Physics Volume 6", VCH Publishers Inc, (1993)
[9] R.K. Nesbet, *J. Chem. Phys*, **40**, 3619 (1964)
[10] C.J. Gillan, C.J. Noble and P.G. Burke, *J. Phys. B:At. Mol. Phys.* **21**, L53 (1988)
[11] W.C. Ermler, A.D. McLean and R.S. Mulliken, *IBM Research Report RJ3243* (1981)
[12] I. Shimamura, "Rotational excitation of molecules by slow electrons", *in*: "Electron Molecule Collisions", I. Shimamura and K. Takayanagi, ed., Plenum Press, New York (1985)
[13] A.D. McLean, "Potential energy surfaces from ab-initio computation: current and projected capabilities of the Alchemy computer program", *in*: "Proc. Conf. on Potential Energy Surfaces in Chemistry", W A Lester Jr., ed., IBM San Jose (1971)
[14] A.D. McLean, M. Yoshimine, B.H. Lengsfield III, P.S. Bagus and B. Liu, "ALCHEMY II: A research tool for molecular electronic structure interactions", *in*: "MOTECC Modern Techniques in Computational Chemistry", Clementi E ed., ESCOM Science BV, Leiden (1991)
[15] J. Kendrick and B.D. Buckley, "An adaptation of the Alchemy atomic and molecular integrals packages for R-matrix electron-molecular collisions", *Daresbury Laboratory Report DL/SCI/TM22T*, SERC Daresbury Laboratory, (1980)
[16] C.C.J. Roothaan, *Rev. Mod. Phys.* **23**, 69 (1951)
[17] P.S. Bagus, *IBM Research Report RJ1077* (1972)
[18] S. Salvini, *Ph.D Thesis* Queen's University of Belfast (1984)
[19] J. Tennyson, P.G. Burke and K.A. Berrington, *Comp. Phys. Commun.* **47**, 207 (1987)
[20] R.K. Nesbet, "Variational Methods in Electron Atom Scattering Theory,"Plenum, New York (1980)

[21] A.E. Orel, T.N. Rescigno and B.H. Lengsfield III, *Phys. Rev.*, **A44**, 4328 (1991)
[22] U. Fano, *Phys. Rev.*, **A67**, 140 (1965)
[23] L.A. Morgan and J. Tennyson, *J. Phys. B:At. Mol. Opt. Phys.*, **26**, 2429 (1993)
[24] B. Liu and M. Yoshimine, *IBM Research Report RJ2849* (1980)
[25] M. Yoshimine, *J. Comp. Phys.*, **11**, 449 (1973)
[26] R. McWeeny and B.T. Sutcliffe, "Methods of Molecular Quantum Mechanics", Academic Press, London (1969)
[27] I. Shavitt, "The method of configuration interaction in electronic structure theory", *in:* "Methods of Electronic Structure Theory", H.F. Schaefer III, ed., Plenum Press, New York (1977)
[28] K.L. Baluja, P.G. Burke and L.A. Morgan, *Comput. Phys. Commun.* **27**, 299 (1982)
[29] M. Gailitis, *J. Phys. B:At. Mol. Phys.* **9**, 843 (1976)
[30] C.J. Noble and R.K. Nesbet, *Comput. Phys. Commun.* **33**, 399 (1984)
[31] S.E. Branchett and J. Tennyson, *J. Phys. B:At. Mol. Opt. Phys.* **25**, 2017 (1992)
[32] J. Tennyson and C.J. Noble, *Comput. Phys. Commun.* **33**, 421 (1984)
[33] B.K. Sarpal and J. Tennyson, *J. Phys. B:At. Mol. Opt. Phys* **25**, L49 (1992)
[34] M.K. Seaton, *J. Phys. B:At. Mol. Phys.* **18**, 2111 (1985)
[35] L. Malégat, *Comput. Phys. Commun.* **60**, 391 (1990)
[36] W. AbuSufah, D. Kuck and D. Lawrie, "Automatic program transformations for virtual memory computers", *in:* "National Computer Conference", AFIPS Conference Proceedings (1979)
[37] B. Liu and N. Strother, "Peak performance from VS Fortran", *in:* "Proceedings of SHARE 67," (1986)
[38] R. Fandreyer, P.G. Burke, L.A. Morgan and C.J. Gillan, *J. Phys. B:At. Mol. Opt. Phys.* **26**, 3625 (1993)
[39] J. Almlöf, "The Program System MOLECULE: Integrals", *USIP Report 74-29*, Stockholm, (1974)

ELECTRON COLLISIONS WITH THE He_2^+ CATION

Brendan M. McLaughlin[1] and Charles J. Gillan[2]

[1] Institute for Theoretical Atomic and Molecular Physics,
Harvard-Smithsonian Center for Astrophysics
60 Garden Street, Cambridge, MA 02138, USA

[2] Theoretical and Computational Physics Research Division,
Department of Applied Mathematics and Theoretical Physics.
The Queen's University of Belfast, Belfast BT7 1NN, UK

1. INTRODUCTION

This chapter illustrates one application of the R-matrix approach to electron collisions with diatomic molecules. The He_2^+ cation is the chosen target and the scattering equations are solved within a two state approximation in which those target states are represented by a complete configuration interaction (CI) expansion. Our work is one of the latest in a long line of applications of R-matrix theory to the field of electron-diatomic molecule collisions; the field began with the paper by Schneider [1] on $e^- - H_2$, work which was soon extended to the more complex $e^- - F_2$ interaction.[2] Only the details of applying R-matrix theory to the $e^- - He_2^+$ system are presented here because the theoretical aspects of the method have already been outlined in chapter 1 and the specific computational implementation that was used in the previous chapter.

The study of electron collisions with He_2^+ molecular ions has a wide variety of applications and interests as it provides information necessary for a doorway to related theoretical problems. For collisional ionization between pairs of metastable $He(2\,^1S, 2\,^3S)$ atoms, detailed dynamical calculations for the heavy particle motion require the knowledge of the electronic properties such as the resonance energies and widths, and the potential energy curves of both the neutral and ionic species. Using the molecular R-matrix approach the initial state of such processes appears naturally as a resonance in this form of calculation and the resonance energy and width can be obtained from an analysis of the eigenphase sum. For example the $^{1,3}\Sigma_u^+$ and $^{1,3}\Sigma_g^+$ autoionizing resonances widths as a function of the helium dimer bond separation variable R are required to fully understand the energy spectrum of electrons produced by ionizing collisions between pairs of metastable $He(2\,^1S, 2\,^3S)$ atoms.[3] This enables the study of Penning and associative ionization processes[4] in field-free and photon-assisted He^* - He collisions.

Computational Methods for Electron-Molecule Collisions
Edited by W.M. Huo and F.A. Gianturco, Plenum Press, New York, 1995

The dissociative recombination process is the main mechanism for removal of ions in any laboratory, atmosphere, or astrophysical plasma cool enough to contain a molecular complex.[5] This process is of interest in the helium afterglow as it proceeds only with the participation of vibrationally excited ions.[6] Estimates of the rate coefficient α_{DR} for the dissociative recombination process in the He_2^+ complex,

$$e^- + He_2^+(1\sigma_g^2 1\sigma_u \, X^2\Sigma_u^+) \rightarrow He^* + He \tag{64}$$

are in the range (10^{-8} to 10^{-10} cm^3s^{-1}).[7] No detailed theoretical calculations exist for this process,[8] and it may be studied within the context of the non-adiabatic R-matrix approach to electron molecular ion collisions, see chapter 2. *Ab initio* studies of electronically excited vibrational transitions in the ground state of the He_2^+ cation,

$$e^- + He_2^+(1\sigma_g^2 1\sigma_u \, X^2\Sigma_u^+ v) \rightarrow e^- + He_2^+(1\sigma_g^2 1\sigma_u \, X^2\Sigma_u^+, v') \tag{65}$$

requires one to incorporate nuclei motion into the theory, thereby providing a wealth of new information on this complex. Photoionization of the excited electronic states of the helium dimer into vibrationally resolved states of the He_2^+ complex may also be investigated using the R-matrix approach and have direct comparison with experiment.[9]

Detailed structure calculations for the transition dipole moment[10, 11] coupling the X and A states of the He_2^+ molecular ion, have allowed the radiative association and the photodissociation cross section to be determined.[12, 13] These processes are of interest in the supernovae ejecta (SN 1987A), planetary nebulae (NGC 7027), the early universe and the interstellar medium. The radiative association process

$$He^+ + He(A^2\Sigma_g^+) \rightarrow He_2^+(1\sigma_g^2 1\sigma_u \, X^2\Sigma_u^+) + h\nu \tag{66}$$

may have been an important process for the removal of He^+ ions in the CO region of the ejecta of SN 1987A. Recent calculations by Dalgarno and co-workers indicate removal of He_2^+ ions by radiative association is unimportant in the SN 1987A compared to removal by recombination and charge transfer however this process leads to the production of He_2^+ molecular ions at an earlier time in the early universe than H_2^+. Its formation does not have any consequence though as its weak chemical bond (≈ 2.5 eV) will be broken by the photodissociation process,

$$He_2^+(1\sigma_g^2 1\sigma_u \, X^2\Sigma_u^+) + h\nu \rightarrow He^+ + He \tag{67}$$

the dissociative recombination, or by charge transfer processes. The photodissociation process in He_2^+ may also be important in cool helium rich white dwarfs.[13]

The work reported here is confined to *ab initio* studies of the elastic scattering of low energy electrons from ground state He_2^+ molecular ions and the computations of bound states of the helium dimer from our collision work. In the work presented here we emphasis the application of the R-matrix method to electron collisions with the He_2^+ cation using complete CI target wave functions, obtained from natural orbital analysis of multiconfiguration self-consistent-field complete-active-space (MCSCF-CASSCF)[14, 15] calculations on the X and A states of He_2^+ complex. The collision work on the He_2^+ cation[16] and the H_2[17] molecule were the first to use fully correlated target wave functions within the R-matrix context, and this approach has subsequently been used in detailed *ab initio* electron collision work on HeH^+. For electron collisions with larger systems such as N_2, CO and O_2, target wave functions at only the valence-configuration-interaction (VCI) level are routinely used.

2. TARGET WAVE FUNCTIONS

All of the *ab initio* calculations were performed with a two-state approximation in the R-matrix basis for the target He_2^+ molecular wave functions using the ground X $^2\Sigma_u^+$ and first excited A $^2\Sigma_g^+$ state, which were obtained from a set of orthogonal molecular orbitals. In all cases we used the (4s,2p,2d) double ζ plus polarization (DZ+P), Slater type orbital basis set of Reagen et al.[18] The sensitivity of the scattering results to the target wave functions representations at SCF and CI level has previously been demonstrated[15, 19] and will not be reproduced here. Linear dependence forced us to delete one of the σ_g and one of the σ_u basis functions in the SCF procedure. This yields the orbital set,

$$1\sigma_g \rightarrow 6\sigma_g, \quad 1\sigma_u \rightarrow 6\sigma_u, \quad 1\pi_u \rightarrow 3\pi_u, \quad 1\pi_g \rightarrow 3\pi_g, \quad 1\delta_g, 1\delta_u. \tag{68}$$

We then Schmidt-orthogonalized the open shell virtual orbitals to the bound orbitals and used the resulting orbital set for the expansion of both target states. Initially we used single configurations for the X $^2\Sigma_u^+$ and the A $^2\Sigma_g^+$ states, and then proceeded to carry out full CI and MCSCF-CASSCF calculations on these target states in this common orbital set. The dimension of the configuration state function (CSF) space was 562 and to make the target wave functions manageable in our codes it was necessary to reduce the number of CSF's. We performed this by calculating the first-order spin-reduced density matrix for the X $^2\Sigma_u^+$ state using the full CI approximation, diagonalized this and transformed the initial molecular orbital set to natural orbitals.[20] The natural orbitals serve as a means for analyzing the contribution of the successive members of the space of molecular orbitals to the total wave function. The new orbital set was truncated by omitting all natural orbitals with occupation numbers less than 10^{-3}. The truncated set consisted of the following orbitals,

$$1\sigma_g \rightarrow 3\sigma_g, \quad 1\sigma_u \rightarrow 2\sigma_u, \quad 1\pi_u, \quad 1\pi_g\,. \tag{69}$$

Performing a full CI calculation with the truncated natural orbital set (6) yielded 30 CSF's per state with a negligible degradation in absolute energies.[15]

Note that the use of natural orbitals obtained from the ground state introduces an extra approximation into the expansion of the excited A $^2\Sigma_g^+$ state of the target He_2^+ cation. This approximation may be improved using state averaged natural orbitals, which yield a better representation of the excited target states of the complex.

3. SCATTERING CALCULATIONS

We use complete configuration interaction[21] target wave functions for the X and A states of the He_2^+ cation generated from the truncated natural orbital space. To solve the inner region scattering problem we used the modified, machine portable, version of the IBM Alchemy quantum chemistry program suite[22, 23] as outlined in the previous chapter. The continuum molecular orbitals, η_j, representing the scattered electron, are formed by taking suitable linear combinations of the target molecular orbitals and additional continuum basis functions, the radial parts of which are obtained by numerically solving the model scattering problem subject to the fixed boundary conditions with the spherical part of the static potential of the SCF target ground state of He_2^+. The use of the fixed boundary condition requires the introduction of the Buttle correction[24] when computing the R-matrix. The R-matrix radius we chose to be 10 a_0, a value large enough that the target state charge distributions are enveloped by the hypersphere as in our previous work.[15, 19] Four partial waves were used for each continuum symmetry and convergence was achieved with three partial waves per continuum symmetry. We chose the continuum basis to be complete up to 11 Rydbergs, a condition that required the use of about nine continuum functions per partial wave.

The outer region collision problem was solved using a single center, no exchange, close coupling expansion of the wave function with R-matrix propagator[25] and accelerated Gailitis expansion methods employed. For the scattering process the external region was solved for positive energies and the R-matrix matched onto outgoing Coulomb wave solutions. As in our previous studies the direct potential in the outer region calculations contains, in addition to the dominant Coulomb term, contributions from the ground and excited state quadrupole moments as well as the dipole moment coupling the X and A states. The T-matrices obtained by this procedure were then employed in standard formulae to produce differential and integral cross sections.

For scattering calculations with a full CI representation of the target states the truncated natural orbital set of equation(6) was used. To ensure consistency with the use of full CI the $\mathcal{L}^2$ terms consisted of all possible arrangements of four electrons in the natural orbital space. Our scattering calculation was therefore completely correlated within this truncated orbital space.

4. R-MATRIX TECHNIQUE FOR BOUND STATES

To evaluate bound states of He_2^* we employed the method of Seaton[26] as adapted and applied first to molecules by Sarpal et al[27] for HeH* and Branchett and Tennyson for H_2.[28, 29] Bound states may be obtained by solving the external region scattering problem with all channels closed that is at energies below the ground state of the He_2^+ target. It must be stressed that this requires no re-evaluation of the inner region. This technique is ideal for the computation of diffuse Rydberg states of the He_2 molecule, that is states in which there is one diffuse electron outside a He_2^+ core. Its advantage over the traditional linear combination of atomic orbitals approach in quantum chemistry (LCAO) is simply that the atomic basis set does not have to contain very diffuse functions; their role has been replaced by the numerical solutions to the coupled, homogeneous, ordinary differential equations in the outer region. The absence of numerical linear dependence problems associated with diffuse basis functions not with standing, it must be remembered that considerable development of integrals codes has been necessary to solve the inner region problem.

All the electronic states converging to a particular ionization threshold can in principle be determined once a suitable **R** - matrix has been constructed. We note that

Table 1. Comparison of the energies in Hartrees for the A $^1\Sigma_u^+$, bound state of He_2 at selected bond separations with the work of quantum chemists.

R(a_0)	1.5	2.0	2.5	3.0	3.5	4.0
SOCI[36]	-5.0873816	-5.1363949	-5.1151796	-5.0863092	-5.0653736	-5.0535918
MCSCF[37]	–	-5.1286836	-5.1080124	-5.0794566	-5.0587398	-5.0470075
This work	-5.066586	-5.121300	-5.102514	-5.073907	-5.051675	-5.017428

the He_2 $(1\sigma_g^2\, 1\sigma_u\, ns\sigma, nd\sigma, ng\sigma\ ^{1,3}\Sigma_u^+)$ and He_2 $(1\sigma_g^2\, 1\sigma_u\, np\sigma, nf\sigma, nh\sigma\ ^{1,3}\Sigma_g^+)$ Rydberg series in the helium dimer have been detected and analyzed using this method.[27, 28]

There is a countably infinite number of Rydberg states for each quantum number s, p, d, ... all of which converge onto the ground state energy. In practice the numerical procedures used, coupled with the finite size of a computer word, enforce a limit on the highest n value obtained in any series.

In our original work to evaluate the bound states of the helium dimer from $e^- - He_2^+$ collisions[19] we reported on the $He_2(a^3\Sigma_u^+)$ state using the bound state method of Ohja and Burke.[30] This was then extended to higher lying states of this same symmetry[15] and to the $^{1,3}\Sigma_g^+$ symmetries,[31] using the modified method of Seaton[27, 28] to evaluate bound states of the helium dimer. The power of the R-matrix technique is in calculating a large number of diffuse Rydberg states of the molecular system. We have previously illustrated this for the $^3\Sigma_u^+$ and $^{1,3}\Sigma_g^+$ symmetries in the helium dimer, and presently now for the $^1\Sigma_u^+$ symmetry. We note that Tennyson and co-workers,[27, 32, 33] Norcross and Gorczyca,[34] and McLaughlin et al[15] have illustrated the power of the quantum defect[35] and close-coupling method in evaluating Rydberg bound states of molecules such as H_2, HeH, He_2, H_2^-, HF^- and HCl^-.

5. RESULTS

All the calculations reported here were performed with the nuclei held fixed in space. The resonances in the cross sections and eigenphase sums $\eta(\epsilon)$ were detected and fitted to the Breit-Wigner form for several overlapping resonances,

$$\eta(\epsilon) = \eta_0(\epsilon) + \sum_{i=1}^{M} \tan^{-1} \frac{\Gamma_i}{2(\epsilon_n^i - \epsilon)} \, . \tag{70}$$

Here $\eta_0(\epsilon)$ is the background phase shift at scattering energy ϵ, ϵ_n^i the position of the $i^{\rm th}$ resonance, Γ_i the width, and M the number of overlapping resonances. The resonances were fitted automatically using a modified version of the computer code of Tennyson and Noble,[38] with a quadratic form for the background phase shift $\eta_0(\epsilon)$.

In our studies on electron collisions with the He_2^+ cation we have indicated that the following resonance processes occur in the elastic scattering process for the $^{1,3}\Sigma_g^+$ and $^{1,3}\Sigma_u^+$ symmetries:

$$\begin{array}{l} e^- + He_2^+(1\sigma_g^2 1\sigma_u \; X\,^2\Sigma_u^+) \\ \qquad \downarrow \\ He_2(1\sigma_g 1\sigma_u^2 ns\sigma, nd\sigma, \ldots \; ^{1,3}\Sigma_g^+, \text{ or } 1\sigma_g 1\sigma_u^2 np\sigma, nf\sigma, \ldots \; ^{1,3}\Sigma_u^+) \\ \qquad \downarrow \\ e^- + He_2^+(1\sigma_g^2 1\sigma_u \; X\,^2\Sigma_u^+) \end{array} \tag{71}$$

These process correspond to a rearrangement of the He_2^+ cation core with the continuum electron attaching itself to the repulsive A $^2\Sigma_g^+$ unstable excited state. The quasi-bound states consist therefore of an unstable excited He_2^+ core surrounded by, and interacting with, a diffuse Rydberg orbital.

Table 1 compares our $^1\Sigma_u^+$ bound state results, obtained from the adapted method of Seaton for molecules[27, 28] with results from quantum chemistry calculations as an indication of their accuracy. From Table 1 it is seen that our absolute energies are in satisfactory agreement with the second-order-configuration-interaction (SOCI) calculations of Yarkony[36] and the multiconfiguration self-consistent-field (MCSCF) calculations of Sunil et al[37] for the range of bond lengths considered.

Table 2. The lowest sixteen bound states for the He_2 $^1\Sigma_u^+$ symmetry at the fixed bond separation of 2.0625 a_0. All energies are in Hartrees. The effective quantum number ν and quantum defect μ of these states are included.

State	Energy	ν	μ
A $^1\Sigma_u^+$	-5.098742	2.078411	-0.078411
$3d\sigma$ $^1\Sigma_u^+$	-5.037923	3.017101	-0.017101
$3s\sigma$ $^1\Sigma_u^+$	-5.028448	3.316715	-0.316715
$4d\sigma$ $^1\Sigma_u^+$	-5.013689	4.036102	-0.036102
$4s\sigma$ $^1\Sigma_u^+$	-5.010100	4.295043	-0.295043
$5g\sigma$ $^1\Sigma_u^+$	-5.002873	5.015358	-0.015358
$5d\sigma$ $^1\Sigma_u^+$	-5.002662	5.042271	-0.042271
$5s\sigma$ $^1\Sigma_u^+$	-5.001006	5.268924	-0.268924
$6g\sigma$ $^1\Sigma_u^+$	-4.996789	6.020779	-0.020779
$6d\sigma$ $^1\Sigma_u^+$	-4.996679	6.044777	-0.044777
$6s\sigma$ $^1\Sigma_u^+$	-4.995793	6.250604	-0.250604
$7g\sigma$ $^1\Sigma_u^+$	-4.993129	7.024364	-0.024364
$7d\sigma$ $^1\Sigma_u^+$	-4.993067	7.045986	-0.045986
$7s\sigma$ $^1\Sigma_u^+$	-4.992538	7.238404	-0.238404
$8g\sigma$ $^1\Sigma_u^+$	-4.990756	8.026796	-0.026796
$8d\sigma$ $^1\Sigma_u^+$	-4.990718	8.046647	-0.046647

Tables 2 gives the numerical values for several of the Rydberg electronic states $He_2^*(1\sigma_g^2 1\sigma_u ns\sigma$, $nd\sigma$ and $ng\sigma\ ^1\Sigma_u^+)$ at the fixed internuclear separation of 2.0625 a_0 to illustrate the power of this quantum defect technique. Figures 1 and 2 respectively illustrate $ns\sigma\ ^1\Sigma_u^+$ and $nd\sigma\ ^1\Sigma_u^+$ Rydberg states of the helium dimer as a function of bond separation. Finally in Table 3 we present our results for the $np\sigma\ ^1\Sigma_u^+$ and the $nf\sigma\ ^1\Sigma_u^+$ resonance states, obtained from our two-state e^- – He_2^+ scattering calculations. The results are for the autoionization widths of the $He_2(1\sigma_g 1\sigma_u^2 np\sigma\ ^1\Sigma_u^+)$ and $He_2(1\sigma_g 1\sigma_u^2 nf\sigma\ ^1\Sigma_u^+)$ quasi-bound resonance states at 2.0625 a_0. The resonance positions and widths were obtained by fitting the eigenphase sum from our two-state scattering calculations to a Breit-Wigner form for multiple overlapping resonances, similar to our recent work on the $^3\Sigma_u^+$ symmetry[15] in e^- – He_2^+ collisions.

Table 3. Autoionizaton width Γ, position ϵ_r relative to the $He_2^+(1\sigma_g^2 1\sigma_u\ X\ ^2\Sigma_u^+)$ state. The real and imaginary parts, α and β respectively, of the complex quantum defect $\alpha+i\beta$, for several $1\sigma_g 1\sigma_u^2 np\sigma\ ^1\Sigma_u^+$ and $1\sigma_g 1\sigma_u^2 nf\sigma\ ^1\Sigma_u^+$ resonance He_2^* states are included, n being the principal quantum number. The results are for the fixed bond separation of 2.0625 a_0.

Resonance	Γ (Ry)	ϵ_r (Ry)	$n-\alpha$	β
$2p\sigma$ $^1\Sigma_g^+$	0.21422×10^{-2}	0.574345	2.627218	0.97116×10^{-2}
$3p\sigma$ $^1\Sigma_g^+$	0.10435×10^{-2}	0.636866	3.484551	1.10376×10^{-2}
$4f\sigma$ $^1\Sigma_g^+$	0.14894×10^{-3}	0.659398	4.088408	0.25446×10^{-2}
$4p\sigma$ $^1\Sigma_g^+$	0.42628×10^{-3}	0.669065	4.465006	0.94864×10^{-2}
$5f\sigma$ $^1\Sigma_g^+$	0.10049×10^{-3}	0.808951	5.107805	0.33480×10^{-2}
$5p\sigma$ $^1\Sigma_g^+$	0.21603×10^{-3}	0.685714	5.462719	0.88039×10^{-2}
$6f\sigma$ $^1\Sigma_g^+$	0.10093×10^{-3}	0.692500	6.117064	0.57758×10^{-2}
$6p\sigma$ $^1\Sigma_g^+$	0.12559×10^{-3}	0.695283	6.462904	0.84759×10^{-2}

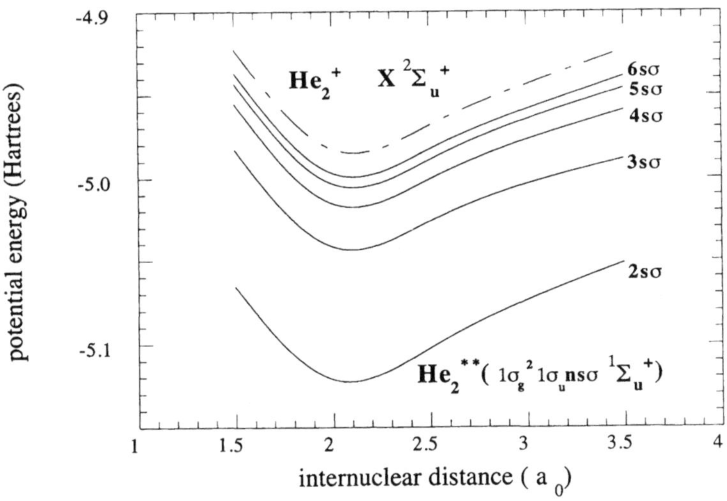

Figure 1. The Rydberg $ns\sigma$ bound states of the helium dimer as a function of bond separation for the symmetry $^1\Sigma_u^+$.

6. CONCLUSIONS

We have carried out low energy elastic scattering calculations of electrons from the He_2^+ cation for the $^1\Sigma_u^+$ symmetry. Results have been presented at several bond separations to illustrate the power of the R-matrix approach in calculating many bound states of the helium dimer. This work on bound states is an extension of the SCF and Frozen-Orbital (FO) calculations of Guberman and Goddard[39] to higher lying $(n\ell\sigma, n'\ell')$ Rydberg states of He_2^*. Our scattering results, using complete CI target state representations obtained from the natural orbitals generated by MCSCF – CASSCF calculations, provide new information on bound and continuum states of the $e^- - He_2^+$ collision complex. The low lying bound state calculations from our work show suitable agreement with quantum chemistry results. Calculations on several other symmetries within the fixed nuclei approximation are in progress and will be extended to incorporate nuclei motion using a non-adiabatic R-matrix approach,[40] discussed in chapter 2.

Our *ab initio* work on the He_2^+ cation has permitted the calculation of the position and widths of several super-excited resonance states of the helium dimer. Results for the resonances widths calculated with correlated full CI targets indicate they are an order of magnitude larger than those calculated at the static exchange plus polarization level.[15, 19] Resonances widths calculated as a function of the helium dimer bond separation variable R are then suitable for use in dynamical studies of collisional ionization in He^*–He and He^*–He^* collisions.

Finally one may apply this state of the art method of generating target state wave functions to other small molecules such as LiH, Li_2, Be_2, ..., where use of complete CI targets can be made. In future work we intend to use state averaged natural orbitals, which provide greatly improved representations of excited target states of the complex.

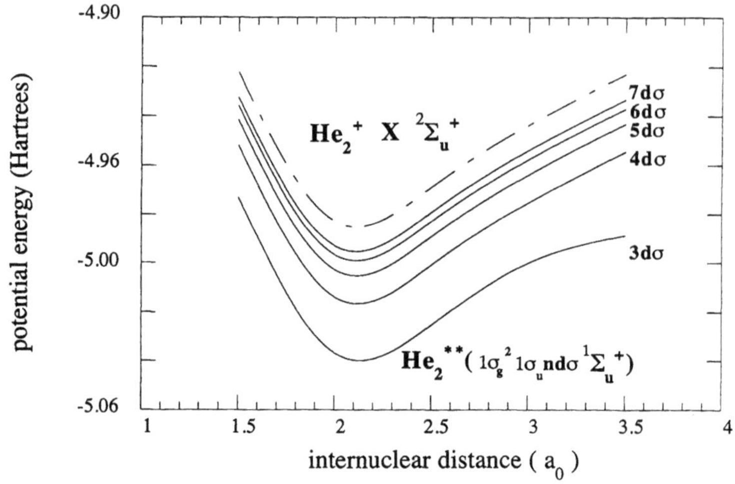

Figure 2. The Rydberg $nd\sigma$ bound states of the helium dimer as a function of bond separation for the symmetry $^1\Sigma_u^+$.

ACKNOWLEDGMENTS

We thank the IBM Corporation for provision of the Alchemy I and II suites under Joint Study agreements with The Queen's University of Belfast. The Institute for Theoretical Atomic and Molecular Physics at The Harvard-Smithsonian Center for Astrophysics is supported by a grant from the National Science Foundation.

References

[1] B. I. Schneider, Phys. Rev. A, **11**, 1957 (1975), Chem. Phys. Lett., **31** 237 (1975).

[2] B. I. Schneider and P. J. Hay, Phys. Rev. A, **13**, 2049 (1976).

[3] M. W. Muller, W. Bassert, M. W. Ruf, H. Hotop, W. Meyer and M. Movre, Z. Phys. D., **21**, 89 (1991), R. J. Bieniek, M. W. Muller and M. Movres, J. Phys. B., **23**, 4521 (1990).

[4] P. E. Siska, Rev. Mod. Phys., **65**, 337 (1993), A. Niehaus, Phys. Rep. **186**, 149 (1990), A. N. Klucharev and V. Vujnovic, Phys. Rep. **185**, 55 (1990) and J. Weiner, F. Masonou-Seews and A. Giusti-Suzor, Adv. At. Mol. Opt. Phys., **26**, 209 (1990), P. Monichicourt, P. Pradel, D. Dubreuil and J. J. Laucagne, Phys. Rev. A, **40**, 1147 (1989), Phys. Rev. A, **40**, 6706 (1989).

[5] A. Dalgarno, Planet. Space Sci., **40**, 1583 (1993).

[6] V. A. Ivanov and Yu. E. Skoblo, Opt. Spectrosc. (USSR), **65**, 445 (1988).

[7] J. B. A. Mitchell, Phys. Rep., **186**, 240 (1990).

[8] S. L. Guberman, *Physics of Ion-Ion and Electron-Ion Collisions*, ed. by F. Brouillard and J. W. McGowan, (Plenum Press, New York and London 1983) 167.

[9] D. C. Lorents, N. Bjerre and H. Helm, Bull. Amer. Phys. Soc., **35**, 1168 (1990).

[10] A. Metropoulos, Y. Li, G. Hirsch and R. J. Buenker, Chem. Phys. Lett., **198**, 266 (1992).

[11] B. M. McLaughlin, C. J. Gillan, P. G. Burke and J. S. Dahler, *Atomic and Molecular Physics*, ed. C. Cisneros, T. J. Morgan and I. Alvarez , (World Scientific, New Jersey and London 1991) 161.
[12] P. Stancil, J. Babb and A. Dalgarno, Ap. J., **414**, 672 (1993).
[13] P. Stancil, Ap. J., *in press* (1994), and references therein.
[14] A. C. Wahl and G. Das, *Modern Theoretical Chemistry : Vol. 3, Methods of Electronic Structure Theory*, Edited by H. F. Schaefer III, (Plenum, New York and London, 1977) 51.
[15] B. M. McLaughlin, C. J. Gillan, P. G. Burke and J. S. Dahler, Phys. Rev. A, **47**, 1967 (1993).
[16] C. J. Gillan, B. M. McLaughlin, P. G. Burke and J. S. Dahler, Bull. Amer. Phys. Soc., **35**, 1140 (1990), B. M. McLaughlin, C. J. Gillan, P. G. Burke and J. S. Dahler, *Minnesota Supercomputer Institute Report*, **90/182** (1990).
[17] S. E. Branchett and J. Tennyson, Phys. Rev. Lett. **64**, 2889 (1990).
[18] P. N. Reagen, J. C. Browne and F. A. Matsen, Phys. Rev., **132**, 304 (1963).
[19] B. M. McLaughlin, C. J. Gillan, P. G. Burke and J. S. Dahler, Nucl. Instrum. & Methods B, **53**, 518 (1991).
[20] R. McWeeny and B. T. Sutcliffe, *Methods of Molecular Quantum Mechanics*, (Academic Press: New York 1978) p94.
[21] I. Shavitt, *Modern Theoretical Chemistry : Vol. 3, Methods of Electronic Structure Theory*, Edited by H. F. Schaefer III, (Plenum, New York and London, 1977) p189.
[22] A. D. McLean, *Proceedings of the Conference on Potential Energy Surfaces in Chemistry*, ed. by W. A. Lester Jr., (IBM, San Jose, 1971) 87.
[23] C. J. Noble, Daresbury Laboratory Report, DL/SCI/TM33T (1982).
[24] P. J. A. Buttle, Phys. Rev., **162**, 719 (1967).
[25] L. A. Morgan, Comput. Phys. Commun., **31**, 419 (1984).
[26] M. J. Seaton, J. Phys. B., **18**, 2111 (1985).
[27] B. K. Sarpal, S. E. Branchett, J. Tennyson and L. Morgan, J. Phys. B., **24**, 3685 (1991).
[28] S. E. Branchett, *PhD Thesis, University College London* (1991).
[29] S. E. Branchett and J. Tennyson, J. Phys. B., **25**, 2017 (1992).
[30] P. Ohja and P. G. Burke, J. Phys. B., **16**, 3513 (1983).
[31] C. J. Gillan, B. M. McLaughlin and P. G. Burke, *Dissociative Recombination: Theory, Experiment and Applications*, (Plenum: New York 1993) 155.
[32] S. E. Branchett and J. Tennyson, J. Phys. B., **25**, 2017 (1992).
[33] B. K. Sarpal and J. Tennyson, J. Phys. B., **25**, L49 (1992).
[34] T. Gorczyca and D. W. Norcross, Phys. Rev. A, **45**, 140 (1992), Phys. Rev. A, **42**, 5132 (1990).
[35] M. J. Seaton, Rep. Prog. Phys., **46**, 167 (1983).
[36] D. Yarkony, J. Chem. Phys. **90**, 7164 (1989).
[37] K. K. Sunil, J. Lin, H Siddiqui, P. E. Siska, K. D. Jordan and R. Shephard, J. Chem. Phys., **78**, 6190 (1983), K. D. Jordan and P. E. Siska, J. Chem. Phys., **80**, 5027 (1984).
[38] J. Tennyson and C. J. Noble, Comput. Phys. Commun., **33**, 421 (1984).
[39] S. L. Guberman, *PhD Thesis, Caltech* (1973), University Microfilms, Ann Arbor, Michigan, S. L. Guberman and W. A. Goddard III, Phys. Rev. A, **12**, 1203 (1975).
[40] B. K. Sarpal, J. Tennyson and L. Morgan, J. Phys. B., **24**, 1851 (1991).

ROVIBRATIONAL EXCITATION BY ELECTRON IMPACT

Helmar T. Thümmel[1,2] *, Thomas Grimm-Bosbach[1], Robert K. Nesbet[3], and Sigrid D. Peyerimhoff[1]

[1] Institut für physikalische und theoretische Chemie, Universität Bonn, Wegelerstraße 12, 53115 Bonn, Germany
[2] NASA Ames Research Center, Moffett Field, California 94035-1000, USA
[3] IBM Almaden Research Center, San José, California 95120, USA

1. INTRODUCTION

This chapter covers theoretical methods for rotational and vibrational coupling in electron-molecule scattering in the region where non-adiabatic effects are important. So far, they have been applied only to diatomic molecules, but should be applicable for polyatomic molecules as well. Here we assume that the fixed-nuclei electronic R-matrix is already determined over a grid of energy and nuclear coordinates using the methods described in other chapters. The problem considered here is the computation of ro-vibronic scattering cross sections from the fixed-nuclei electronic R-matrix. Note that ro-vibronic coupling inherently involves a breakdown of the Born-Oppenheimer separation whenever electronic kinetic energy is converted into vibrational or rotational energy. The treatment of ro-vibronic motion in the R-matrix formulation can be separated into two steps: (1) how to deduce the ro-vibronic R-matrix from the fixed-nuclei electronic R-matrix determined at some electronic radius r_1 where the molecular charge density is negligible and (2) how to propagate the resulting ro-vibronic R-matrix to determine the cross sections.

In recent years, analytic methods have become available for solving the multichannel close-coupling equations for electrostatic multipole and polarization potentials.[1, 2, 3] Large angular momentum quantum numbers, which imply close-coupling equations of high dimension, can be taken into account by use of closure formulas based on the Born approximation.[4] With these methods, the ro-vibronic R-matrix, defined in a basis of vibrational and rotational eigenfunctions, can be matched to asymptotic functional forms to determine cross sections. In section 2, we describe a recently developed method for deducing the vibronic R-matrix from the fixed-nuclei electronic R-matrix. At scattering energies outside the rotational threshold region, it is generally a good approximation

* Present address: NASA Ames Research Center, Moffett Field, California 94035-1000, USA

Computational Methods for Electron-Molecule Collisions
Edited by W.M. Huo and F.A. Gianturco, Plenum Press, New York, 1995

to average over rotational levels. Also most experimental cross sections are rotationally averaged except for very precise, low-energy measurements and those are confined to small molecules. However rotational averaging cannot always be done at the level of the vibronic close-coupling equations. In particular, the external potential of a molecular dipole moment is dynamically screened by molecular rotation. Furthermore, even without a dipole potential, rotational averaging in the threshold region may not give an adequate description of rapidly varying cross sections, such as threshold excitation peaks, that are observed as a superposition of rotational substructures. Sec. 3 discusses a full treatment of molecular rotation. The ro-vibronic R-matrix is obtained from the vibronic R-matrix by a rotational frame transformation, with additional propagation of the vibronic R-matrix to smaller radii than the electronic r_1, if necessary, to stabilize the frame transformation. Standard analytic methods are then used to solve external close-coupling equations in the laboratory reference frame. In this chapter, our emphasize is on non-adiabatic rotational effects. The breakdown of adiabatic-nuclei rotational methods near ro-vibrational thresholds is discussed, and the threshold laws which are expected for ro-vibrational state-to-state transitions are compared with those implied by ANR methods.

Section 4 discusses our recent results on the highly polar systems, using $e - HF$ as an example. By solving the ro-vibrational close-coupling equations, non-adiabatic rotational coupling has been fully accounted for in these calculations. Cross sections for rotational excitation and ro-vibrational state-to-state transitions are calculated. For the hydrogen halides and a few other molecules experimental results indicate narrow and intense threshold peaks in vibrational excitation cross sections, first discovered by Rohr and Linder[5] some time ago. The present rotationally 'resolved' calculations provide the first analysis of rotational fine-structures near these vibrational thresholds.

Section 5 presents new results for the case of resonant scattering. The non-adiabatic phase-matrix method (NADP), described in Section 2, is applied to $e - N_2$, in the energy range of the well known $^2\Pi_g$ shape resonance.

2. VIBRONIC R-MATRIX THEORY

A fixed-nuclei R-matrix is assumed to be available for a range of electronic energies ϵ and internuclear distances q. Vibronic coupling occurs both through the explicit dependence of the R-matrix on the coordinate q and because the energy available to the continuum electron depends on the kinetic energy T_q of nuclear motion. The energy-modified adiabatic (EMA) approximation[6] was introduced in order to correct adiabatic nuclei (AN) theory in the vibrational threshold region, where neglect of T_q gives qualitatively incorrect results, and to provide a basis for describing the effect of fixed-nuclei resonances on vibrational resonant structure observed in molecules such as N_2. Fixed-nuclei electronic close-coupling equations are converted into vibrational close-coupling equations by introducing the operator T_q and by projecting the equations into a basis of vibrational eigenfunctions. Formally, this replaces the numerical electronic energy parameter ϵ by the operator

$$\epsilon_{op} = E - H_q, \tag{72}$$

which acts on functions of the vibrational coordinates. Here E is the specified total energy and H_q is the Hamiltonian of the nuclear motion. The EMA approximation is based on the postulate that a correspondence principle is valid for matrix elements of functions of H_q. It is assumed that fixed-nuclei scattering matrices can be converted to their vibronic (or ro-vibronic) counterparts by replacing functional forms that depend on the parameter ϵ by matrix expressions valid in certain limiting cases. In the present work, this postulate is applied to the R-matrix rather than to the usual asymptotic scattering matrices.

Standard AN theory is expected to be valid for scattering energies well above the vibrational thresholds, such that the vibrational kinetic energy can be neglected. Then ϵ_{op} is only a function of q. The vibronic R-matrix elements are

$$< v \mid R \mid v' >= \int dq \chi_v^{\dagger}(q) R^{FN}(\epsilon_{op}, q) \chi_{v'}(q), \tag{73}$$

reducing in this limit to the AN formula,

$$< v \mid R \mid v' >= \int dq \chi_v^{\dagger}(q) R^{FN}(\epsilon(q), q) \chi_{v'}(q). \tag{74}$$

In the present treatment, the effective vibrational channel energy is chosen in a different manner. Because the R-matrix does not distinguish between open and closed channels, there is no characteristic threshold structure, and the parameter ϵ can have positive as well as negative values. A consistent choice of $\epsilon_{vv'}$ is the arithmetic mean of the diagonal values of ϵ

$$\epsilon_{vv'} \hat{=} E - (E_v + E_{v'})/2. \tag{75}$$

Although the R-matrix does not have branch points as a function of the electronic energy parameter ϵ, it necessarily has a sequence of poles. Integration of such a function over the vibrational coordinate q, or interpolation in ϵ, is not expected to give meaningful results. To avoid this difficulty, the real symmetric matrix R^{FN} is replaced by an equivalent phase matrix Φ, defined by

$$\tan \mathbf{\Phi} = \mathbf{R^{FN}}. \tag{76}$$

The matrix R^{FN} is diagonalized and Φ is constructed from the eigenvectors and inverse tangents of the eigenvalues. These eigenphases are determined only modulo π. A phase map is constructed showing each eigenphase as a function of ϵ and q, and multiples of π are added to give smooth variation for adiabatically connected eigenchannels on the correct branch of the arctan. The resulting EMA vibronic phase matrix is

$$\Phi_{vl,v'l'}(E) \hat{=} < v \mid \Phi_{ll'}^{FN}(\epsilon_{vv'}, q) \mid v' > . \tag{77}$$

Defining $\epsilon_{vv'}$ as discussed above, the vibronic R-matrix is constructed from the matrix definition tanΦ. This procedure characterizes the energy-modified adiabatic phase (EMAP) method.

This construction is not expected to be valid when matrix Φ varies rapidly with energy. It can be shown from the variational expression for the fixed-nuclei R-matrix that the trace of the corresponding adjusted phase matrix (the eigenphase sum) must increase monotonically with scattering energy ϵ. If the analytic continuation of the R-matrix to complex energy values has a single pole near the real axis, standard Breit-Wigner analysis can be applied. This situation defines a "precursor resonance", associated with a true scattering resonance or bound state as the R-matrix boundary is extended to large electron radii. The practical criterion for a precursor resonance is a relatively sharp local maximum of the energy derivative of the eigenphase sum. Clearly the vibrational kinetic energy can no longer be neglected. The basic strategy for determining the vibronic R-matrix in this case is to separate the Φ matrix into a rapidly-varying part Φ_1, describing precursor states for which a non-adiabatic analysis must be used, and a smoothly-varying part Φ_0 which can be converted to a vibronic matrix by the EMAP method.

Separation of a rapidly varying foreground phase matrix from a slowly varying background matrix is done following Smith.[7] Since a resonance corresponds to a peak in the energy derivative of the eigenphase sum, the most meaningful projection of a resonance onto eigenchannels should correspond to that eigenvector of the matrix $d\Phi/d\epsilon$ that has maximum eigenvalue. In the present work, a precursor resonance energy ϵ_{res} is defined by a local maximum of $\frac{d}{d\epsilon}\mathrm{Tr}\Phi$, for fixed q. The foreground phase matrix Φ_1 takes the form

$$\Phi_1(\epsilon) = y\phi_{res}(\epsilon)y^\dagger, \tag{78}$$

where y is a unit vector and ϕ_{res} is defined by a Breit-Wigner function such that

$$\frac{d}{d\epsilon}\phi_{res}(\epsilon_{res}) = \frac{2}{\gamma}, \tag{79}$$

where γ is the width of the resonance. With the requirement that

$$\frac{d}{d\epsilon}\left(y^\dagger\Phi y\right)|_{res} = \frac{2}{\gamma} \tag{80}$$

the projection of the background derivative onto the one-dimensional subspace defined by y will vanish at ϵ_{res}. This determines both y and γ through the eigenvalue equation

$$\left(\frac{d\Phi}{d\epsilon}\right)|_{res}\, y = y\left(\frac{2}{\gamma}\right). \tag{81}$$

The maximum eigenvalue gives γ and y is the corresponding eigenvector. The residual background phase matrix Φ_0 is defined as

$$\Phi_0 = \Phi - \Phi_1. \tag{82}$$

The resonant vibronic phase matrix $\Phi_1^{vv'}$ is generated by the non-adiabatic phase-matrix (NADP) method, outlined here. If a precursor resonance energy is given as a function of q, it defines an effective vibrational Schrödinger equation of the form

$$\left(H_q + \epsilon_{res}(q)\right)\chi_s(q) = \chi_s(q)E_s. \tag{83}$$

Any eigensolution of this equation has the property that the residual electronic energy operator ϵ_{op} takes on the definite numerical value $E - E_s$. This number is substituted into the Breit-Wigner formula for the precursor resonance, and the resonant vibronic phase matrix is constructed as a sum over these eigenstates. Given the eigenfunctions χ_s, vibronic phase matrix elements are determined by

$$\begin{aligned} < v \mid \Phi_1 \mid v' > \;&=\; \sum_s < v \mid y(q) \mid s >< s \mid \tan^{-1}\left[\frac{\gamma(q)}{2(E_s - E)}\right] \mid s > \\ &\quad < s \mid y^\dagger(q) \mid v' > . \end{aligned} \tag{84}$$

The evaluation of this sum requires a complete set of eigenfunctions $\chi_s(q)$, in general including the vibrational continuum. A discrete but complete set of functions $\chi_s(q)$ is provided by a basis of spline-delta functions.[8] The effective completeness of these basis functions depends on the accuracy of representation of vibrational functions by a cubic spline fit over a finite range of q, and has been verified in test cases. The resulting vibronic R-matrix can be reconstructed as

$$R = tan\left(\Phi_1^{vv'} + \Phi_0^{vv'}\right) \tag{85}$$

where $\Phi_0^{vv'}$ is the vibronic background phase matrix determined by the EMAP method.

3. ROTATIONAL COUPLING

The method most widely used for treating rotation is the adiabatic-nuclei rotation[9] (ANR) approximation. It has been extensively applied in electron scattering from diatomic molecules. (See for example the reviews of Morrison[10] and Lane.[11]) The essential criterion for the validity of the ANR approximation is that the rotational Hamiltonian H_{rot} can be neglected in the vibrational close-coupling equations. This requires the classical collision time of an electron ($t = L/v$, where L is the effective range of the scattering potential and v the electron velocity) to be short compared with a typical rotational period ($\approx 10^{-14}$ sec). It is assumed in the ANR approximation that electronic motion is uncoupled from nuclear rotation, that H_{rot} commutes with the long-range interaction potentials, and that the energies in the entrance and exit channels are independent of the rotational quantum number j, implying target state degeneracy. The ANR method is not expected to be accurate near thresholds or for highly excited rotational states. Rotational energies cannot be neglected in comparison with a static dipole potential at large distances and are comparable to quadrupole and polarization potentials at intermediate distances, so the ANR method fails when the kinetic energy of an incident or scattered electron is small. For polar molecules, it has been shown[12, 13] that the ANR differential cross sections for the transfer of one unit angular momentum diverge in the forward direction. The corresponding integrated cross sections are infinite at all scattering energies. Nevertheless, the ANR approach may be adequate to obtain accurate cross sections, except for forward scattering, at scattering energies large compared with rotational level spacings. There is little quantitative information about the magnitude of non-adiabatic corrections to the ANR approximation. Prior to work described here, no direct calculation of rotational state-selected cross sections near threshold, using a fairly realistic interaction potential from ab initio calculations, has been published. One step beyond the ANR is the "multipole-extracted adiabatic-nuclei" (MEAN) method of Norcross and Padial[4]. It permits the inclusion of the effects of the rotational Hamiltonian for those large impact parameters for which the ANR breaks down, but for which the laboratory-frame first Born approximation (FBA) may be valid. The MEAN approach has been applied to obtain differential cross sections for ro-vibrational state-to-state transitions for e-CO,e-HCl,[4, 14, 15] e-HCN [16] and e-HF.[17]

As described in the last section, the vibronic R-matrix is constructed in the molecular or body frame of reference (BF) from variational R-matrices computed for a range of specified nuclear coordinates. Thus the method includes the electron-vibrational interaction as well as an exact treatment of electronic exchange in the strong interaction region of small electronic radii. Given the vibronic R-matrix the calculation can be completed either in the body-frame, treating nuclear rotation by the adiabatic-nuclei rotation (ANR) approximation or by the multipole-extracted adiabatic-nuclei (MEAN) approximation, or, after transformation to the laboratory frame of reference (LF), by direct analytic solution of the rotational close-coupling equations. All three of these alternatives are considered here.

The BF vibronic K-Matrix is computed by matching the vibronic R-matrix to solutions of the asymptotic vibrational close coupling equations. The BF T-matrix can be obtained from the K-matrix using the matrix equation

$$\mathbf{T} = 2i\mathbf{K}/(1 - i\mathbf{K}). \tag{86}$$

In order to obtain approximate cross sections for ro-vibrational state-to-state transitions the T-matrix in the lab-frame representation indexed by ro-vibrational channels can be obtained by applying a radial frame transformation [18] to the body BF T-matrix as

$$
\begin{aligned}
{}^{\mathrm{ANR}}\mathrm{T}^{J\eta}_{vjl,v_0j_0l_0} &= [(2j_0+1)(2j+1)]^{1/2} \\
& \sum_{\Lambda=-l_{min}}^{l_{min}} \begin{pmatrix} j_0 & l_0 & J \\ 0 & \Lambda & -\Lambda \end{pmatrix} {}^{\mathrm{BF}}\mathrm{T}^{\Lambda}_{vl,v_0l_0} \begin{pmatrix} j & l & J \\ 0 & \Lambda & -\Lambda \end{pmatrix}
\end{aligned} \tag{87}
$$

where l_{min} is the smallest of (l_0, l, J). Λ is the projection of total electronic angular momentum on BF z-axis, and the conserved quantum numbers J and η describe total angular momentum and parity. (For given J the lab-frame scattering equations split into two different sets of η depending upon whether $(-1)^{l+j} = (-)^{l_0+j_0}$ is either positive or negative.) ${}^{ANR}T$ is the central quantity of the adiabatic-nuclei rotation method and can be substituted into standard expressions for the calculation of differential and integral cross sections for ro-vibrational state-to-state transitions [19]:

$$
\frac{d\sigma_{v_0j_0\to vj}}{d\Omega} = \frac{k^{-2}_{v_0j_0}}{4(2j_0+1)} \sum_L A_L^{v_0j_0vj}\, P_L(cos\theta) \tag{88}
$$

$$
\begin{aligned}
\sigma_{v_0j_0\to vj} &= \sum_{J,\eta} \sigma^{\eta J}_{v_0j_0\to jl} \\
&= \frac{k^{-2}_{v_0j_0}}{4(2j_0+1)} \quad A_0^{v_0j_0vj}
\end{aligned} \tag{89}
$$

The coefficients $A_L^{v_0j_0,vj}$ depend explicitly only on the transition matrix $T^{J\eta}_{(vjl,v_0j_0l_0)}$ and on algebraic factors. (The precise definition is not of interest here, but it involves, in principle, infinite sums over angular momenta l_0, l as well as a summation over J and η.) The $\sigma^{J\eta}_{v_0,j_0\to vj}$ are partial cross sections in J for ro-vibrational state-to-state transitions.

$$
\sigma^{\eta J}_{v_0j_0\to jl} = \frac{k^{-2}_{v_0j_0}}{4(2j_0+1)} \sum_{l_0l} (2J+1)\, |T^{\eta J}_{vjlv_0j_0l_0}|^2 \tag{90}
$$

In the ANR, $k^2_{v_0,j_0} = k^2_{v_0}$. The energies in the entrance and exit channels are independent of the rotational quantum number j. It has been shown[9] that the integral cross section

$$
\sigma^{BN}_{v_0\to v} = \sum_j \sigma^{ANR}_{v_0j_0\to vj} \tag{91}
$$

is independent of the initial rotational state j_0.

3.1. Threshold laws for ro-vibrational state-to-state transitions

Since it is generally not experimentally possible to resolve rotational state-to-state cross sections, it is important to understand the form of rotational threshold laws predicted by exact theory. The multichannel theory of threshold effects as presented by Bransden[20] was applied to both long- and short-range potentials by Bardsley and Nesbet.[21] This theory can be expressed in a form that includes the effects of resonance, bound or virtual state poles of the scattering matrix close to an excitation threshold.[22] It is convenient to parametrize threshold behavior in terms of the real and symmetric K-matrix for open channels above a given threshold. Consider the general case of a threshold with M open channels below threshold, labeled as α, β .., and n new channels labeled p,q,... opening at the threshold. If there is no unscreened coulomb interaction in the asymptotic region, but long-range components of the form $\sim r^{-s}$ with s greater than 2 are present, the asymptotic single-particle wave function $\psi_{\alpha,\beta}(r)$ for each channel α can be expressed in terms of spherical Bessel functions:

$$\psi_{\alpha,\beta}(r) = u_{\alpha,\beta} w_{0\alpha}(r) + v_{\alpha,\beta}(r) w_{1\alpha}(r). \tag{92}$$

Here β denotes the incident channel and

$$\begin{aligned} w_{0\alpha}(r) &= k_\alpha^{1/2} r j_{l_\alpha}(k_\alpha r) \overset{r\to\infty}{\sim} \sin(k_\alpha r - 1/2 l_\alpha \pi) \\ w_{1\alpha}(r) &= k_\alpha^{1/2} r y_{l_\alpha}(k_\alpha r) \overset{r\to\infty}{\sim} \cos(k_\alpha r - 1/2 l_\alpha \pi) \end{aligned} \tag{93}$$

for open channels. Solving the coupled equations, subject to the boundary conditions, eq.(93) defines the K-matrix

$$K_{\alpha,\beta} = v_{\alpha,\beta} u_{\alpha,\beta}^{-1} \tag{94}$$

For $K_{a,p}$, corresponding to inelastic scattering, we obtain from the K-matrix Born approximation

$$K_{\alpha,p} \sim k_\alpha^{s-l_p-5/2}\, k_p^{l_p+1/2}, \tag{95}$$

and for scattering in the new channels opening at threshold

$$K_{qp} \sim \begin{cases} k_p^{s-2} & \text{for } l_p + l_q > s-3 \\ k_p^{l_p+l_q+1} & \text{for } l_p + l_q \le s-3 \end{cases}, \tag{96}$$

valid for small k_p.

As will be discussed below, if two or more degenerate channels are coupled by an interaction $\sim r^{-2}$, the asymptotic form of the channel wave functions changes, and these threshold laws are not valid. Such a situation occurs for example for inelastic scattering of electrons from atomic hydrogen, if the relativistic splitting of the 2s and 2p orbitals is neglected, and in the case of electron-molecule scattering if ro-vibronic levels in a dipolar molecule are accidentally degenerate. Due to conservation of total angular momentum J and parity, long-range dipole and quadrupole couplings generally do not occur between degenerate levels labeled by the same rotational quantum number j. For scattering of electrons from a diatomic molecule, channels with $\Delta j = 0$ are coupled by the spherical polarizability $\sim r^{-4}$, those with $\Delta j = 2$ are coupled by the quadrupole interaction $\sim r^{-3}$ and, for polar molecules, channels with $\Delta j = 1$ are coupled by the long-range dipole potential $\sim r^{-2}$.

¿From eq.(96) it follows for inelastic scattering that

$$\begin{aligned} K^{J\eta}_{v_0 j_0 l_0, vjl} &\sim k_{vj}^{l+1/2} \\ T^{J\eta}_{vjl, v_0 j_0 l_0} &\sim k_{vj}^{l+1/2} \\ \sigma^{J\eta}_{v_0 j_0 \to vj} &\sim k_{vj}^{2l_{min}+1}, \end{aligned} \tag{97}$$

where k_{vj} is the wave vector in a threshold channel and l_{min} is the lowest partial wave l that contributes in eq. (90). For elastic scattering, K and T-matrix elements behave as $\sim k_{vj}$ for s-waves, implying finite values of the corresponding threshold cross-sections. These matrix elements are $\sim k_{vj}^2$ for higher partial waves. The largest rotational transitions are those with an s-wave threshold channel. For a diatomic molecule there generally is one s-wave channel (l=0,j=J,v) for given J, opening at the threshold $E_{v,j=J}$. We will call this channel the dominant channel. For ro-vibrational transitions into the dominant channel we have

$$\sigma^J_{v0,j0\rightarrow vj=J} \sim k_{vJ} \tag{98}$$

and for elastic scattering in the dominant channel

$$\sigma^J_{v,j=J\rightarrow v,j=J} \sim const. \tag{99}$$

As was discussed previously[22, 23], eq.(98) may be further modified if there is a pole in the complex k−plane at $k = -i\beta$ for a virtual state or at $k = +i\beta$ for a resonance or bound state on the real energy axis below threshold, where β is a positive real number. In either case the excitation cross section varies as

$$\sigma^J_{v0,j0\rightarrow vj=J} \sim k_{vJ}/(k^2_{vJ} + \beta^2) \tag{100}$$

for positive real k_{vJ} immediately above a threshold. This gives the general form of a threshold excitation peak, rising as k_{vJ}, then falling as k^{-1}_{vJ}. Given the K-matrix as a function of energy, the parameter β can be determined by fitting to appropriate analytic forms. The energy range in which the lowest-order terms dominate may be rather small, and there may be large second-order terms. For example, ladder coupling of a long-range dipole potential may strongly affect rotational elastic scattering or transitions with $\Delta j = 2$.[13, 24] In principle, a ro-vibrational close-coupling calculation is required for quantitative treatment of the small-energy behavior of state-to state cross sections. The practical difficulties of such calculations are formidable, and provide motivation for extracting as much information as possible from parametrized formal theory and from systematic approximations.

3.2. Breakdown of the ANR approximation

In this section we derive the ro-vibrational threshold laws implied by the ANR approximation, and compare them with the full ro-vibrational theory outlined above. In the ANR method it is assumed that all rotational sublevels of a given vibronic state are degenerate, i.e. $k_{vj} = k_v$. Contrary to the formula (97), this assumption implies for nonpolar molecules, using eq.(96), that all T-matrix elements for rotational excitation have the threshold scaling

$$T^{J\eta}_{vjl,vj_0l_0} \sim \begin{cases} k_v^{s-2} & \text{for } l + l_0 > s - 3 \\ k_v^{l_p+l_q+1} & \text{for } l + l_0 \le s - 3 \end{cases} \tag{101}$$

The accuracy of the ANR approximation has been investigated by Morrison[25] for various ro-vibrational state-to-state transitions in e-H_2 scattering. Identical model potentials were used for ANR and for exact ro-vibrational close coupling calculations. For $\sigma_{vj=00\rightarrow 02}$ the error is smaller than 10% for impact energies larger than 0.1eV.

For polar molecules, degenerate channels are connected by a long-range static dipole potential $\sim r^{-2}$. The implied threshold laws and scattering quantities such as eigenphase sums and cross sections have analytic behavior different from the formulas valid for nonpolar molecules. This qualitative difference is in fact nonphysical, since a static dipole field occurs in nature only for a nonrotating molecule. It is important to understand the analytic theory of this behavior in order to motivate the need to introduce a dynamical theory of the rotational screening of long-range potentials in molecules. Clark[26] investigated the static dipole theory within the framework of the fixed-nuclei (FN) approximation for electron scattering from polar molecules. His derived threshold laws for the FN eigenphase sum have been verified by recent ab initio calculations for the $^2\Sigma^+$, $^2\Pi$ and $^2\Delta$ states of $e-HF$.[24] The basic idea is[27, 26] to expand the wave function in the asymptotic region as

$$\Psi_\beta(\boldsymbol{r}) = r^{-1}\sum_\alpha \Omega_{N_\alpha}(\theta,\phi)\psi^N_{\alpha,\beta}(r) \tag{102}$$

where Ω_N are the eigenfunctions of an effective angular momentum operator $\hat{O}$

$$(\boldsymbol{l}^2 - D\cos\theta)\Omega_N = N(N+1)\Omega_N. \tag{103}$$

Here D is the dipole moment connecting the degenerate channels and θ is the angle between the internuclear axis and $\boldsymbol{D}$. For the function Ω an expansion in spherical harmonics can be used:

$$\Omega_N(\theta,\phi) = \sum_{l\geq|\Lambda_\alpha|} A^{(\Lambda)}_{Nl} Y_{l,\Lambda}(\theta,\phi). \tag{104}$$

The eigenvalues $N(N+1)$ of $\hat{O}$ define generalized (generally nonintegral) quantum numbers, which replace l in the usual partial wave expansion. For $N(N+1) > -1/4$, N is real; for $N(N+1) < -1/4$, N= $-1/2 \pm i\mu$ is complex. The associated radial Schrödinger equation

$$\left(\frac{d^2}{dr^2} - \frac{N_\alpha(N_\alpha+1)}{r^2} + k^2\right)\bar{w}_\alpha(r) = 0 \tag{105}$$

has a set of fundamental asymptotic solutions $\bar{w}_{0.\alpha}, \bar{w}_{1,\alpha}$, given by Bessel functions of general (complex) order which replace the spherical Bessel functions in eq.(93). A distorted-wave $\bar{K}$-matrix in the dipole-adopted basis is obtained by solving the coupled equations subject to the boundary conditions of these "dipole functions".

Threshold laws are obtained using the $\bar{K}$-matrix Born approximation for real N. ¿From the small argument expansion of the dipole functions it follows, for inelastic scattering into the new (degenerate) channels p,q opening at threshold, that

$$\bar{K}_{\alpha,p} \sim k^{N_p+1/2} \tag{106}$$

and for elastic scattering

$$\bar{K}_{p,q} \sim k^{N_p+N_q+1}, \tag{107}$$

valid if no long-range interaction is present other than the dipole interaction connecting the degenerate states. Transforming $\bar{K}$ to l-representation i.e. comparing the asymptotic forms of the dipole functions $\bar{w}_{0.\alpha}, \bar{w}_{1,\alpha}$ with the free asymptotic forms at large r (eq.93), it follows[26] that all matrix elements $K_{p,q}$ for transitions between the degenerate channels have finite values c_1 at threshold that only depend on the dipole moment which couples the degenerate channels. Thus the threshold scaling law for elastic scattering in degenerate channels coupled by the dipole potential takes the form

$$K_{p,q} = c_1(l_p, l_q, D) + c_2 k^{2N_{min}+1} + ... \tag{108}$$

For inelastic scattering into the degenerate channels,

$$K_{\alpha,p} \sim k^{N_{min}+1/2} + ... \tag{109}$$

Table 1. Threshold behavior of K_{l_p,l_q} and $\bar{K}_{N_p,N_q}$ matrix elements for the $^2\Sigma^+$-scattering-state of e-HF from R-matrix calculation using dipole adopted basis functions as a function of energy, E. An overcritical dipole moment D=0.764 a.u. couples the degenerate channels. The element $\bar{K}_{N_0,N_0}$ corresponding to the complex eigenvalue N_0 dominate at small E. The values of $\bar{K}^a_{N_0,N_0}$ calculated from the analytic formula are included in the second column. No element of the corresponding K-matrix goes to zero at threshold.

E/Ry	$\bar{K}^a_{N_0,N_1}$	$\bar{K}_{N_0,N_0}$	$\bar{K}_{N_0,N_1}$	$\bar{K}_{N_1,N_1}$	$K_{S,S}$	$K_{S,P}$	$K_{P,P}$
0.000001	0.5505	0.5469	0.0000	0.0000	3.7590	0.2531	-0.1117
0.000005	0.3320	0.3298	0.0000	0.0000	2.2323	0.3140	-0.1141
0.00001	0.2624	0.2605	0.0001	0.0000	1.9257	0.3263	-0.1146
0.00005	0.1281	0.1265	0.0002	0.0000	1.4667	0.3448	-0.1153
0.0001	0.0770	0.0754	0.0003	0.0000	1.3255	0.3507	-0.1156
0.0005	-0.0367	-0.0383	0.0011	0.0000	1.0601	0.3625	-0.1162
0.001	-0.0863	-0.0881	0.0020	-0.0002	0.9614	0.3677	-0.1166
0.005	-0.2119	-0.2158	0.0085	-0.0021	0.7466	0.3843	-0.1200
0.008	-0.2536	-0.2595	0.0134	-0.0038	0.6841	0.3925	-0.1225
0.01	-0.2746	-0.2820	0.0167	-0.0047	0.6539	0.3975	-0.1239
0.05	-0.4614	-0.5238	0.0678	-0.0148	0.3847	0.4664	-0.1005
0.1	-0.5734	-0.7182	0.0669	-0.0350	0.2073	0.4637	-0.0734

Instead of eqs.(95,96) these are power-law scalings with noninteger exponents. For complex N, $k^{N_{min}+1/2}$ has a complex phase factor, and the situation is more complicated. One can show[27, 28] that all matrix elements in the vicinity of a threshold oscillate as a function of ln k. For example, the low energy expansion for the the eigenphase sum associated with the K-matrix connecting the degenerate states for the case of one complex $N_0 = 1/2 + i\mu$ can be written as $\delta(E) \sim \delta_D(E) + \text{const}(D)$ where the first energy dependent term is given by [26]:

$$\bar{K}_{N_0,N_0} = \tan\delta_D(E) \sim \tanh\left(\frac{\pi\mu}{2}\right)\tan\left[\mu\ln\left(\frac{k_{N_0}r_0}{2}\right) - \phi + \xi\right]. \qquad (110)$$

Here $\xi = \arctan\,(b/\mu)$ is a phase shift introduced by the short-range interactions, which can be calculated from the R-matrix,[24] and $\phi = \arg[\Gamma(1+i\mu)]$. This threshold law has been verified by recent *ab initio* calculations for the scattering state $^2\Sigma^+$ of e-HF and has been found to describe the low energy behavior of K-matrix elements within a few percent for scattering energies less than 1eV (table 1).

¿From the observation that all matrix elements $^{BF}K^{\Lambda}_{vl_0,vl}$ with the same v have finite values at threshold it follows that all $^{ANR}T^{J\eta}_{vjl,v j_0 l_0}$ have finite values at threshold. The corresponding cross sections $\sigma^{J\eta}$ diverge as k_v^{-2}. Cross sections for simultaneous rotational and vibrational excitation have power law scalings with noninteger exponents $\sim k_v^{2N_{min}+1}$. Terms that oscillate as a function of $\ln k_v$ are present for both vibrationally elastic and inelastic scattering if the dipole moment D connecting the degenerate states is overcritical i.e. lead to eigenvalues $N(N+1) < -1/4$. So far we have shown that the low-energy behavior of ANR scattering quantities such as $T^{\eta,J}$ and $\sigma^{\eta,J}$ for ro-vibrational excitation is different from that implied by an exact treatment of rotation. In order to calculate the total integrated (eq. 89) and differential cross sections (eq. 88) for ro-vibrational state-to-state transitions, summations must be carried out over J and η. For polar molecules, it has been shown[12, 13] that within the ANR approach these sums do not converge for transitions with angular momentum transfer of one unit $\Delta j = 1$. In fact differential cross sections diverge in the forward direction ($\theta = 0$) and the corresponding integrated cross sections are infinite at all impact energies.

3.3. The Multipole extracted adiabatic-nuclei approximation

Breakdown of the ANR approximation for forward scattering is due to partial waves with high electronic angular momenta. For low angular quantum numbers, the difference between the T-matrix obtained from a lab-frame calculation including H_{rot} and the ${}^{ANR}T$-matrix is not significant for polar targets except near rotational thresholds.[29] In contrast, for large impact parameters (high l) where the ANR breaks down, the lab-frame first order Born approximation (LF FBA), indexed by ro-vibrational channels including the effects of rotational motion and long-range interactions, may be valid. This is the basis of the "multipole-extracted adiabatic-nuclei approximation".[4] Angular momentum transfer analysis[30] is used to modify the LF FBA differential cross section by removing its lowest order angular momentum terms, replaced by the corresponding partial waves terms from the ANR-T matrix. This Born-closure procedure is also appropriate for molecules with significant quadrupole moments or polarizabilities. Differential cross sections are calculated as

$$\frac{d\sigma_{v_0 j_0 \to vj}}{d\Omega} = \frac{d\sigma^{FBA}_{v_0 j_0 \to vj}}{d\Omega} + \Delta\frac{d\sigma_{v_0 j_0 \to vj}}{d\Omega} \tag{111}$$

where

$$\Delta\frac{d\sigma_{v_0 j_0 \to vj}}{d\Omega} = \frac{k^2_{vj}}{k^2_{v_0 j_0}} \sum_{l_t} C(j_0 j l_t) \Delta\frac{d\sigma^{l_t}_{v_0 \to v}}{d\Omega} \tag{112}$$

and

$$\Delta\frac{d\sigma^{l_t}_{v_0 \to v}}{d\Omega} = \frac{1}{k_v k_{v_0}} \sum_{\lambda} [B_\lambda(v_0 \to v, l_t) - B^{FBA}_\lambda(v_0 \to v, l_t)] P_\lambda(\cos\theta) \tag{113}$$

The C(...) are Clebsch-Gordon coefficients and l_t is the angular momentum transferred during the collision. The coefficients B_λ and B^{FBA}_λ similar to the A_L of eq. (89) depend on matrix elements of ${}^{BF}T$, the FBA of ${}^{BF}T$ for the dominant long-range interaction and on algebraic factors. The first term of eq. (111) is the FBA for the same long-range potentials in the laboratory-frame, and can be evaluated in closed form[31]. The sum in eq.(112) converges for $\theta = 0$, since high l contributions that cause difficulties in the body frame cancel in eq.(113). Their net contributions are summed in closed form in the laboratory-frame through the first term of eq.(111).

3.4. Rovibrational Close Coupling with Born closure

The nonphysical behavior of ANR cross sections near threshold can only be avoided by including molecular rotation in the equations of motion. Given a vibronic R-matrix in the body-frame (eq.87), ro-vibrational coupling can be treated exactly in the asymptotic region outside the R-matrix radius r_1 by applying a radial frame transformation. This converts the vibronic R-matrix into the laboratory frame (LF) of rotation. The LF K-matrix is obtained by matching the LF R-matrix to solutions of the asymptotic ro-vibrational close-coupling equations.[32] For polar molecules, the expansion (88) in Legendre polynomials may converge very slowly. Laboratory-frame Born closure, using the analogue of eq.(111), provides an efficient alternative method.[31, 32] Because the FBA for the LF T-matrix is exact in the limit of large l, Born closure in this case eliminates the computational cost of solving close-coupling equations with a very large number of coupled channels for a large number of values of total J.

4. APPLICATION to $e - HF$

4.1. Preliminary Remarks

Electron collisions with polar molecules play a fundamental role in various fields from interstellar space and upper atmosphere of planets to gaseous discharges and radiation interactions with matter. The cross sections for polar molecules exceed those of non-polar molecules by several orders of magnitude and cross sections are large for rotational excitation.[13] For example in a gas, which may consist of molecules as CN,CH,OH, CO, H_2O, NH_3 and H_2CO which are found in interstellar space, the electron temperature, T_e, often exceeds the temperature of the molecular gas T_m. The relaxation T_e-T_m, i.e the thermal balance, occurs mainly through rotational excitations of polar molecules where T_e is below several thousand Kelvin degrees (see Itikawa 1978[12] and references therein).

The understanding of the vibrational excitation peaks observed in the hydrogen halides and a few other molecules as H_2O, H_2S, CO_2 CH_4 and SF_6 continues to be one of the most interesting problems in low-energy electron-molecule scattering. Recent reviews covering this subject have been written by Domcke,[33] Morrison[10] and Cvejanović.[34] We do not intend to give a complete overview of the history but simply to summarize some important aspects.

For HF and HCl, threshold structures where first observed by Rohr and Linder,[5] who found an isotropic angular dependence of the scattered electrons in the whole region of the peaks. Refined and more extended measurements by the Kaiserlautern group[35, 36, 37, 38] have confirmed the earlier results and have provided more detailed information on rotational effects and angular distributions. Vibrational excitation of HF and HCl was found to be associated with strong rotational transitions $\Delta j = \pm 1, \pm 2, \pm 3$ and anisotropy at all energies, especially in the regions of the threshold peaks. Several theoretical models have been proposed to explain these threshold peaks. In the last decade there has been much controversy about the relation of threshold peaks to virtual states, bound states or resonances as well as the role of the long-range dipole potential (see review of Domcke[33] and references therein). If there is a resonance above an excitation threshold, there are two poles of the S-matrix S(k) at $k = \pm\alpha - i\beta$ in the complex k-plane, with $\beta > 0$. If β is defined as minus the imaginary part of k at a pole associated with an excitation threshold, positive β characterizes a virtual state, while negative β characterises a resonance or bound state below the threshold. Dubé and Herzenberg[39] explained the main features observed in HCl by the concept of a virtual state in the s-partial wave, which was shown to yield large and narrow threshold peaks. Another concept which seems most relevant to an explanation of the peaks is that of nuclear excited Feshbach resonances.[40] A Feshbach resonance has a parent state and therefore an interference structure should be visible below the parent state in the open channel, i.e. a narrow structure in the excitation functions $v = 0 \rightarrow v'$ should occur just below the vibrational threshold $E_{v'+1}$. For a nuclear-excited Feshbach resonance the parent states are just quasibound vibrational levels of the molecular negative ion. For HF the existence of quasibound HF^- levels $v = 2, 3$ and 4 has been experimentally confirmed.[38] For $v = 1$, however, there is no indication of an interference structure in the cross section $v = 0 \rightarrow 0$ below the threshold $E_{v=1}$, within an experimental energy resolution of 20meV.

There are two recent theoretical publications on $e + HF$. Morgan and Burke[41] applied an R-matrix method in which non-adiabatic effects of the vibrational motion are approximated using the method proposed by Schneider et al,[42] see chapter 2. Their results on integrated vibrational excitation cross sections, which have been obtained without resolving the rotational subchannels, are in qualitative agreement with the measurements. They found that the peaks correspond to poles in the S-matrix which can be interpreted as nuclear-excited Feshbach resonances. In this first ab initio calculation, inclusion of effects of the permanent dipole moment as well as the polarizability at intermediate and short range were found to be necessary in order to reproduce the observed threshold structures.

Snitchler et al[43] have performed vibrational close-coupling calculations. Exchange effects were included using a separable parameter-free correlation polarization potential. The effects of rotational motion were approximated in the body-frame using the MEAN method described in a section above. They found interference structures below the vibrational thresholds $v = 2,3$ and 4. Differential cross sections are similar to the measurements but differ in the location of the peaks and in detailed shape.

Contrary to the measurements of the Kaiserslautern group all theoretical calculations so far for HF and HCl predict essentially vertical onsets of the peaks. The theoretical peaks occur within a few meV above threshold. It has been argued that proper inclusion of rotation could possibly close the gap between theory and experiment.[33, 41] At this point, it must be mentioned that for HCl recently new aspects have been added to the enigma of threshold excitation peaks: In comparison with earlier data, recent measurements of Cvejanović et al[34, 44] (see also Schafer and Allan,[45]) using a threshold sensitive electron spectrometer, show sharper onsets, peak locations closer to threshold and more intense threshold peaks for the excitation functions $v = 0 - 1$ and $v = 0 - 2$. Well-characterized oscillations around the v=2 threshold have been observed, which were not found in the earlier measurements.[5, 35] Experiments close to threshold are difficult and one has to await further experimental clarification of these points.

We have given the technical details for generating vibronic R-matrices using the EMAP method in previous publications,[24, 46] in the following quoted as I and II. We here will emphasize the effects of rotationally selected transitions for a molecule with a large dipole moment, using HF as an example. The ro-vibronic close-coupling equations where solved for total angular momentum up to $J \leq 9$ including three vibrational channels. The standard asymptotic code has been modified to include molecular rotation. The phase matrix Φ was found to change smoothly with energy if the eigenchannels are connected diabatically as a function of internuclear distance H-F (subsection 2). The rotational basis set consists of functions

$$(l,j)\begin{cases} \leq 7 & \text{for } J \leq 5 \\ \leq J+2 & \text{for } J > 5 \end{cases} \tag{114}$$

Contributions of higher angular momenta have been included using Born closure. The rotational basis set for J=2,$\eta = +1$ is listed in table 2 as an example.

Table 2. Rotational basis set for J=2, $\eta = +1$ for electron scattering from HF in its $^1\Sigma^+$ electronic ground state. Basis functions (jl=20) relate to s-wave ro-vibrational channels, which we call "dominant" channels; functions (jl=11,31) to p-wave channels.

j	0	1	1	2	2	2	3	3	3	4	4	4	...
l	2	1	3	0	2	4	1	3	5	2	4	6	...

4.2. Near Threshold Rotational Excitation

In this sections results are presented for pure rotational excitation of HF i.e. for electron energies well below the thresholds for vibrational excitation. In figure 1 the largest cross sections $\sigma^J_{vj=0,j\to 0,j'}$ for the example J=2 and $\eta = +1$ are displayed as a function of E_0 in the vicinity of the threshold $E_{v=0,j=2}$ i.e. the dominant s-wave channel. E_0 is the energy of the electron incident on the ground ro-vibronic state of molecule with energy ϵ_0. Thus $E_0 = E_T - \epsilon_0$ differs from the total energy E_T of the colliding particles by the constant ϵ_0. The corresponding incident energy of the electron for the target in an arbitrary initial state $v = 0, j$ is given by $E_{inc} = E_0 - Bj(j+1) = k^2_{0j}/2m$, where B=2.59eV [47] is the molecular rotational constant of HF and m the electron mass.

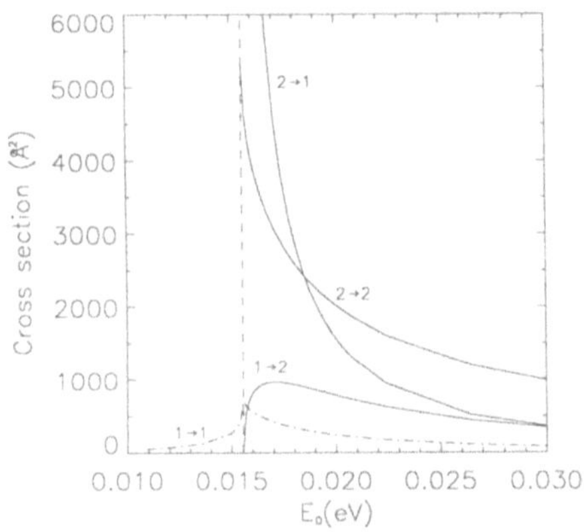

Figure 1. Energy dependence of the electron–HF cross sections $\sigma^J_{v=0j_0\to v=0,j}$ for total angular Momentum J=2 and $\eta = +1$. The threshold $E_{v=0,j=2}$ is indicated by a vertical broken line.

The curves reveal the characteristics of a multichannel threshold structure in the presence of a strong long-range dipole potential $\sim r^{-2}$. The cross section $j = 1 \to 1$ corresponding to background scattering in the old channel $vj = 01$ (predominantly p-wave) shows a Wigner cusp[48] structure at threshold. In contrast, for $j = 2 \to 2$, the cross section for elastic scattering in the new s-wave (dominant) channel opening at threshold has a large finite value at the threshold $E_{v=0,j=2}$.

The cross section for $j = 1 \to 2$ corresponding to rotationally inelastic scattering $(p \to s)$ shows a typical threshold excitation peak rising $\sim k_{02}$ consistent with the general scaling law for scattering with an out-state s-wave channel. The maximum is only 2 meV above threshold. The cross section for a de-excitation process or superelastic transition in which the electron gains energy is generally related to that of the inverse process (excitation), through the law of detailed balances.[49] For ro-vibronic transitions, remembering the (2j+1)-fold degeneracy for a target-state with total angular momentum j we have:

$$(2j_0+1)k_{v_0j_0}^2\sigma_{v_0,j_0\to v,j} = (2j+1)k_{vj}^2\sigma_{j,v\to v_0,j_0} \tag{115}$$

It follows that the de-excitation cross sections diverge as k_{02}^{-1} for small wave numbers k_{02}. As can be seen from figure 1 for the highly polar HF-molecule the de-excitation cross section with $\Delta j = -1$ is dominant for electron energies $E_{inc} < 5meV$. Thus extremely slow electrons will effectively gain energy thereby cooling the molecule. When $k_{v_0j_0}/k_{v,j} \sim 1$ the exact fraction is

$$\sigma_{v_0,j_0\to v,j}/\sigma_{v,j\to v_0,j_0} = (2j+1)/(2j_0+1) \tag{116}$$

indicating that excitation processes generally dominate de-excitation processes at larger electron energies.

Fitting the corresponding K-matrix elements to an analytic form suggested by the threshold scaling laws it follows from the multichannel theory of threshold effects that the multichannel threshold structure at $E_{v=0,j=2}$ is due to a virtual state with $\beta = +0.03a_0^{-1}$. Similar structures are present (see I for details) for each total angular momentum J at each threshold $E_{v=0,j=J}$.

In many experiments, except for very light molecules such as H_2 with large rotational level spacings,[12, 11] it is difficult to achieve sufficient energy resolution to measure rotational state-to-state transitions directly. Furthermore no measurement preparing the target molecule in a defined initial ro-vibronic state has been reported up to now. When rotational states cannot be resolved, cross sections summed over all possible final rotational states are useful:

$$\sigma_{j_0} = \sum_j \sigma_{v_0,j_0\to v,j} \tag{117}$$

In figure 2 we give the curves for $j_0 = 0,1,2,3$ as a function of E_0. All of these curves share a common high energy limit. The overall shape of each curve near threshold can be explained by the dominant partial cross sections $\sigma_{j\to j-1}^{J=j}$, $\sigma_{j\to j}^{J=j}$, $\sigma_{j\to j+1}^{J=j}$, $\sigma_{j\to j}^{J=j+1}$ (cusp) and $\sigma_{j\to j+1}^{J=j+1}$ (excitation peak), where the latter two partial cross sections are responsible for rotational structure at the threshold E_{j+1}. If the initial rotational states are not known, cross sections averaged over the initial states may be calculated for a direct comparison with measurements:

$$\sigma(T) = \sum_{j_0} N_{j_0}(T)\sigma_{j_0} \tag{118}$$

In eq. (118), $N_{j_0}(T)$ is the fraction of the molecules in its rotational state j_0 and is usually specified by the temperature T of the molecular gas. Supposing that the gas is in rotational equilibrium the rotational distribution is given by [50]

$$N_{j_0}(T) \propto (2j_0+1)\exp[-Bj_0(j_0+1)/\kappa T] \tag{119}$$

where κ is the Boltzmann constant. In figure 3, $\sigma(T)$ is displayed as a function of the electron energy E_{inc} for Temperatures of 0K, 100K and 400K. T=400K corresponds to the typical temperatures of the target gas for beam experiments[37] implying a maximum of population for j=2 (23.4%) and a population below 1% for rotational states of HF $j >$ 7. The rotational fine structures tend to be smoothed out by increasing temperature. For incident electron energies larger 0.05 eV the three curves coincide indicating the energy range, above which the rotationally averaged cross sections are independent of the initial rotational state distribution. This gives a lower limit where the ANR-approximation might be valid. Unfortunately no measurements close enough to threshold are available up to now to verify the fine structures predicted here.

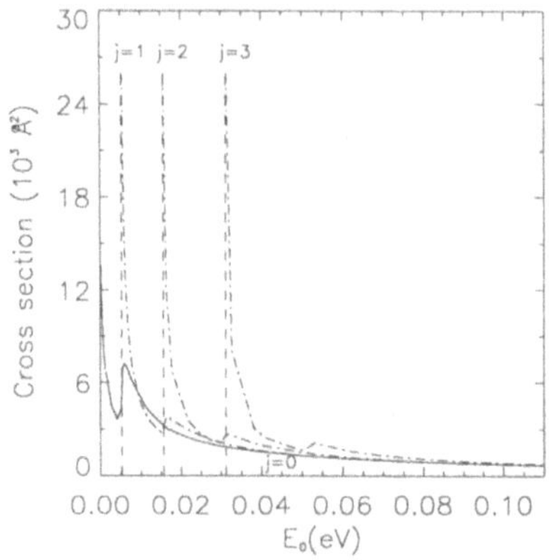

Figure 2. Summed integrated cross sections over final rotational states summed as a function of energy E_0 for vibrational elastic scattering of HF. Curves are displayed for the lowest rotational levels j=0,1,2 and 3. The corresponding thresholds are indicated by vertical dashed lines.

4.3. Rovibrational Excitation

In this subsection we give the results for ro-vibronic excitation cross sections in the energy region of the famous vibrational excitation peaks observed for the excitation of the lowest vibrational levels of HF. To begin with consider partial cross sections J=2 and $\eta = +1$ as an example. As can be seen from figures 4 and 5, there are basically two types of transitions. Transitions from an in-state to an out-state in the dominant channel (vj=22) show threshold excitation peaks above threshold. Transitions to the out-state p-wave channel (vj=21) show narrow cusps. The most striking feature is the strong, sharp cusp for the rotational de-excitation $\Delta j = -1$ ($s \rightarrow p$ transition), which is similar in magnitude to the threshold excitation peak for rotational elastic (s-wave) scattering. In order to study the importance of the asymptotic potential of the HF target molecule outside the R-matrix sphere of ten bohr, we have repeated the calculations neglecting the long-range polarization (b), neglecting polarization and quadrupole potentials (c) and ignoring the potential outside the R-matrix sphere (d). For $vj = 02 \rightarrow 22$ and $vj = 02 \rightarrow 21$ these results are also included in figures 4 and 5. Especially for the latter transition there is only a small effect due to the quadrupole and static polarizability. When the long-range dipole potential is omitted, the cusp structures vanish and the threshold excitation peak for the rotational elastic function gets broader and more rounded. Thus the cusp structures are clearly a phenomena related to the permanent dipole moment of the target molecule.

The largest state-to-state integrated cross sections (eq.89) from the thermally most populated in-state $vj = 02$ for excitations of $v = 2$ ro-vibrational out-states are summarized in figure 6. Since the integral cross section is simply the sum over J (eq. 89) shoulders, peaks and the detailed shape in the state-to-state cross sections can be explained by decomposition into partial cross sections σ^J, which have the cusps and peaks

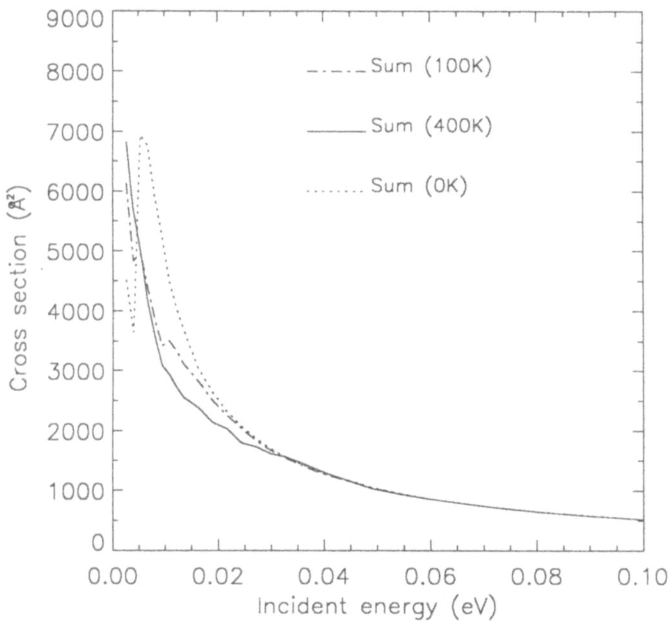

Figure 3. Temperature dependence of rotationally averaged, summed integrated cross sections for vibrationally elastic scattering of HF.

discussed above for the example of J=2. In table 3 we have summarized the most important of these structures together with the partial waves involved, and the energetic location of the relevant threshold. Similar structures are present for transitions from other in-states. Note that the $\Delta j = -1$ cusp structure decreases rapidly to 0.05Å^2, which is of the same magnitude as the cross section for the $\Delta j = +1$ process.

The summed cross sections (eq. 117) for $j_0 = 0, 1$ and 2 are shown in figure 7a for the excitation of v=2 rotational levels and in figure 7b for excitations of v=1. All curves for different j_0 share a common high energy limit, indicating the energy region where independence of the initial rotational state is reached and ANR-methods might be valid. The basic features remain the same on varying the vibrational quantum number. However, as was shown in II, there are interference structures in the v=0→1 cross sections with a sharp minimum below the threshold v=2 due to a phase shift of π in the open channels but no structures were found in the $v = 0 \to 0$ function below the threshold v=1. Indeed threshold analysis of the present rotationally resolved data indicates, see II, that the multichannel threshold structure at $v = 1$ can be characterized as a consequence of an underlying virtual state (for example $\beta = +0.025a_0^{-1}$ for J=2). In contrast, for $v = 2$, β is negative, suggesting a pole position of the S-matrix consistent with a nuclear-excited Feshbach resonance below threshold ($\beta = -0.021a_0^{-1}$ for J=2). Thus, despite the fact that the main features of the threshold peaks discovered for $v = 0 \to 1$ and $v = 0 \to 2$ are similar, the underlying pole structure of the S-matrix $S^{J,\eta}_{jv,j_0,v_0}(k)$ clearly moves in the complex k-plane with vibrational quantum number.

The rotationally averaged, total integrated cross section $\sigma(T)$ (eq. 117) for $v = 0 \to 1$ at a temperature of 400K is shown in figure 8 in comparison with the theoretical results of Snitchler et al[43] and the earlier R-matrix calculation of Morgan and Burke.[41] The overall shape agrees well with the vibrational close-coupling (VCC) calculation, where rotation has been approximated by the MEAN method (Note however, that in

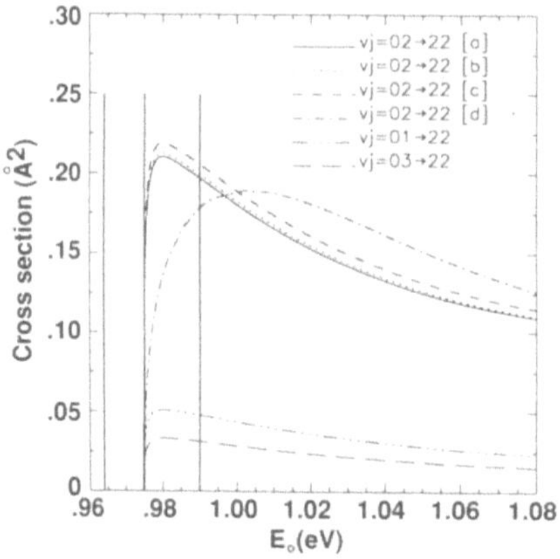

Figure 4. The largest partial cross sections $\sigma^{J=2}_{v=0,j_0\to v=1,j=2}$ (η = +1) for e+HF in the near threshold energy region. The ro-vibrational thresholds vj=21,22 and 23 are indicated by vertical broken lines. (a) Inclusion of dipole, quadrupole moment and polarizability; (b) neglect of polarizability; (c) pure dipole; (d) complete neglect of potential outside the R-matrix sphere.

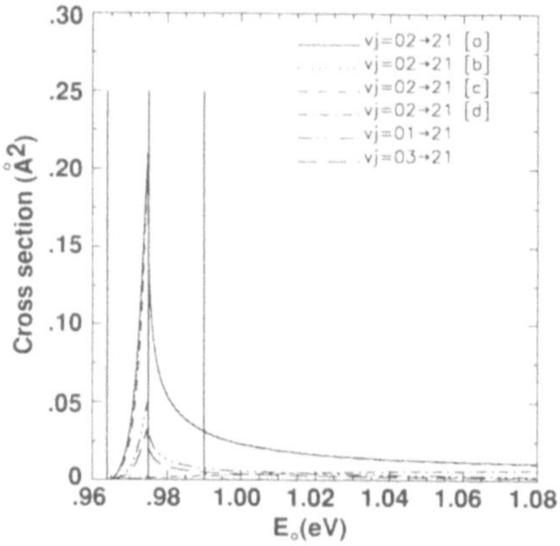

Figure 5. Same as figure 4 for the out-state vj=21.

Table 3. Largest cross sections $\sigma^J_{v_0j_0\to vj}$ for the in-state $v_0j_0 = 02$ in the energy region of the vibrational excitation peak $v = 0 \to 2$ of e+HF. Processes with an out-state in a dominant (s-wave) channel lead to threshold excitation peaks (TEP) with maxima a few meV above the threshold $E_{v=2,j=J}$. Other lead to narrow cusps at the threshold indicated.

$v_0j_0 \to vj$	J	partial in	wave out	structure	threshold
02 → 22	2	s	s	TEP	$E_{v=2,j=2}$
" "	3	p	p	cusp	$E_{v=2,j=3}$
02 → 21	2	s	p	cusp	$E_{v=2,j=2}$
" "	1	p	s	TEP	$E_{v=2,j=1}$
02 → 23	3	p	s	TEP	$E_{v=2,j=3}$
02 → 20	2	s	d	cusp	$E_{v=2,j=2}$
" "	1	p	p	cusp	$E_{v=2,j=1}$

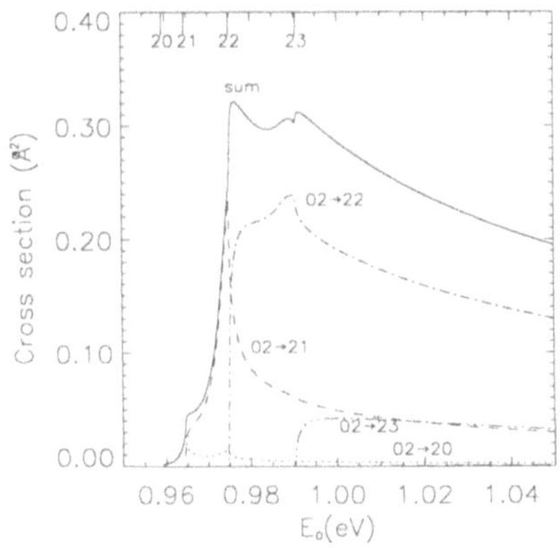

Figure 6. Rovibrational state-to-state cross sections as a function of E_0 from the in-state vj=02 for excitation of v=2 rotational levels. The contribution of higher rotational out-states in the sum is smaller than 0.01Å. The thresholds vj=20,21,22 and 23 are indicated at the top of the figure.

the VCC calculation there is a critical dependence on the choice of the polarization potential at intermediate distances. Absolute values get smaller by a factor of two with a polarization cutoff beyond 3.5 a_0). Contrary to the earlier theoretical results, the onset of the present curve is less steep. Rotational de-excitations are responsible for non-zero cross section at electron energies E_{inc} below the threshold $E_{v=1,j=0}$. Moreover, the rotational fine structures of the σ_{j_0} curves superimpose to form pronounced shoulders on the right-hand side of the $v = 0 \rightarrow 1$ vibrational excitation peak. The MEAN approximation, using the laboratory-frame Born approximation for high-order partial waves, corrects the divergence of forward differential cross sections that occurs in the ANR approximation for vibrational elastic scattering from a polar molecule. MEAN does not directly correct the breakdown of the ANR near threshold but, as indicated by the results of Snitchler et al, the MEAN approximation is useful much closer to thresholds. The computed rotationally averaged cross section at 90° is compared with the measurement of Knoth et al[35] in figure 9 for T=400K, corresponding to the experimental conditions. Theoretical and experimental results in this case differ in shape as well as in the location of the peak. The present results indicate that this large difference between theory and experiment cannot be explained by effects of the rotational motion. Rotational non-adiabatic effects are only important in a relatively small energy range, say 0.05eV from threshold. Absolute values agree better for larger energies. The angular dependence of ro-vibrational state-to-state cross sections at 0.9 and 1.2 eV shows excellent agreement with the rotationally resolved experimental data,[35] especially for the rotationally summed curves (figures 10a,b). Vibrational excitation is accompanied by a high degree of rotational excitation. The angular dependence of the scattered electrons is anisotropic. Knoth et al, using a deconvolution procedure based on the ANR-method, found the elastic contribution to be of the same order or even smaller than the contri-

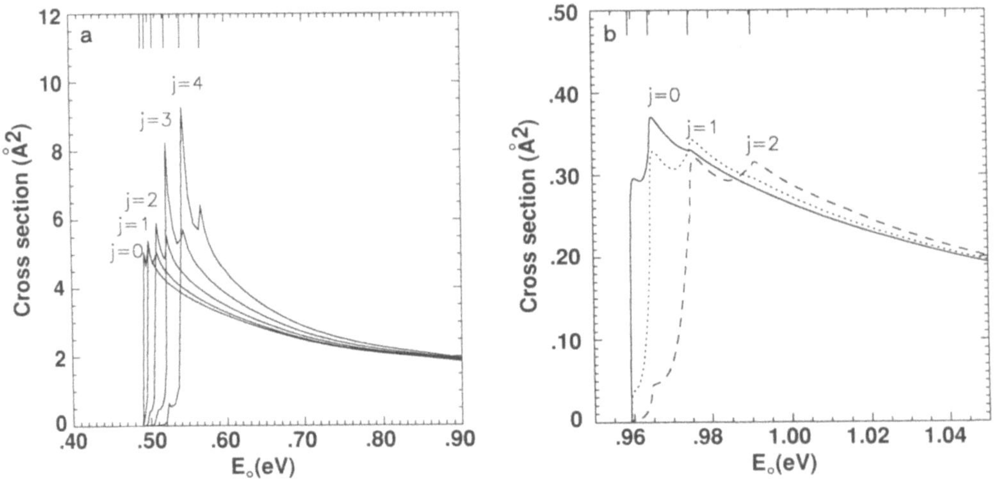

Figure 7. Cross sections rotationally summed for in-states $v_0 j_0 = 0j$. (a) Excitation of v=1 ro-vibrational levels. (b) Excitation of v=2 ro-vibrational levels. The thresholds $E_{v=1,2,j=0,1,2,3....}$ are indicated by short vertical lines at the top of the figures.

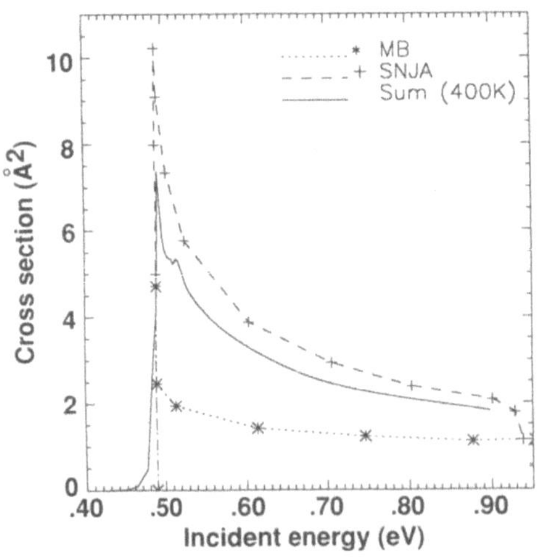

Figure 8. Rotationally averaged summed integrated cross section $v = 0 \rightarrow 1$ for e+HF as a function of electron energy E_{inc}. Present results are temperature weighted for T=400K. MB, R-matrix results of Morgan and Burke;[41] SNJA, vibrational close-coupling calculation of Snitchler et al.[43]

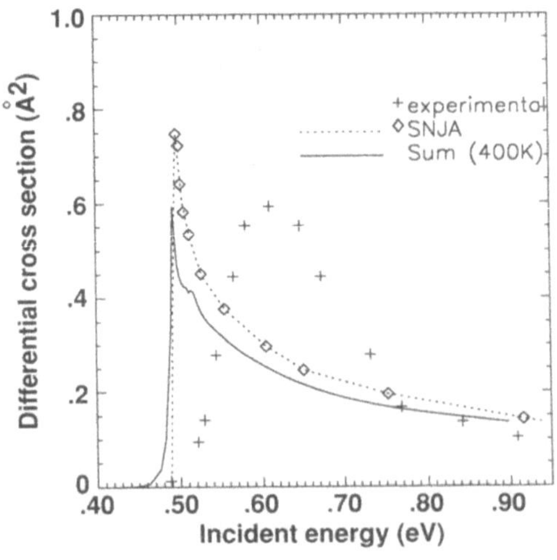

Figure 9. Rotationally averaged summed cross section at 90° for $v = 0 \rightarrow 1$ as a function of electron energy E_{inc}. Present results are temperature weighted for T=400K. SNJA, vibrational close-coupling calculation of Snitchler et al;[43] EXP, experimental results of Knoth et al.[35]

butions corresponding to angular momentum transfer of one unit ($\Delta j = 1$). In contrast, the computed rotational elastic process is larger.

5. APPLICATION to $e - N_2$

The low-energy cross section for electron scattering by N_2 is dominated by the well-known $^2\Pi_g$ shape resonance. The vibrational excitation functions exhibit a rich oscillatory structure within a broad resonance profile. This feature was first observed experimentally by Schulz,[51] who considered the fundamental mechanism to be the formation of a short-lived state of N_2^-. When an incident electron is captured into the lowest vacant electronic orbital $^1\pi_g(2p)$ of N_2, the resulting metastable state decays by emitting a π_g electron with a minimum angular momentum l=2 (d-wave). This resonance has the property that the lifetime of the scattering complex N_2^- is comparable to a molecular vibrational period. It has been a standard test case for theoretical concepts and computational methods suitable for strong vibronic coupling. The relevant literature has been reviewed by Lane[11] and by Domcke.[33]

5.1. Technical details

Our calculations were carried out including only the electronic ground state of N_2. The $^1\Sigma_g^+$ target wave function is an SCF wave function constructed by Slater type orbitals (STO) basis, whose exponents are the same as those previously used by Nesbet.[52] This gives the orbital basis

$$1\sigma_g 2\sigma_g 3\sigma_g 1\sigma_u 2\sigma_u 1\pi_u 4\bar{\sigma}_g 5\bar{\sigma}_g 6\bar{\sigma}_g 7\bar{\sigma}_g 3\bar{\sigma}_u 4\bar{\sigma}_u 5\bar{\sigma}_u 6\bar{\sigma}_u 7\bar{\sigma}_u 2\bar{\pi}_u 3\bar{\pi}_u 1\bar{\pi}_g 2\bar{\pi}_g 3\bar{\pi}_g \qquad (120)$$

where the first six orbitals are occupied and the remaining are virtual orbitals. The continuum basis for each partial wave includes eight regular spherical Bessel functions with a fixed boundary condition at the R-matrix radius, 10 a_0 here. For $l = 2,4,6$ three numerical asymptotic functions (NAFs) are added to ensure convergence inside the R-matrix radius. The continuum functions, centered on the center of mass of the target molecule, are Schmidt-orthonormalized to the target SCF-MO's. We carried out variational calculations for the symmetries $^2\Sigma_g^+, ^2\Sigma_u^+, ^2\Pi_g, ^2\Pi_u$ in the static-exchange plus polarization (SEP) approximation. In the SEP model charge and polarization effects are taken into account by including single-particle virtual dipole excitations of the target valence electrons, described by correlation functions of two-particle/one-hole character, see chapter 3.

Accurate variational calculations of electron-molecule scattering require great computational effort. However, it is clear from the physics of low-energy electron scattering that such detailed calculations are necessary only for the lowest-order partial waves. This fact is exploited in the asymptotic distorted wave (ADW) approximation.[54] In the present work, vibronic R-matrixes for the symmetries $^2\Delta_g, ^2\Delta_u, \ldots, ^2\Theta_g$ were computed using the multichannel ADW approximation, for partial waves up to $l_{max} = 6$. An effective one-electron potential for high-order partial waves is modelled in the ADW approximation by retaining the true asymptotic multipole potentials outside an inner radius r_0. Here we have included the quadrupole term and the isotropic and anisotropic polarizability. The quadrupole moments are taken from our SCF calculation, and the polarizabilities are from Morrison.[55] The target SCF-calculations are done on a grid of 16 internuclear distances from 1.668 a_0 up to 3.0 a_0. Figure 11 shows a "phase map" of the eigenvalue of the phase matrix Φ corresponding to the dominant d-wave (resonant) channel, displayed as a function of both energy and internuclear distance. For

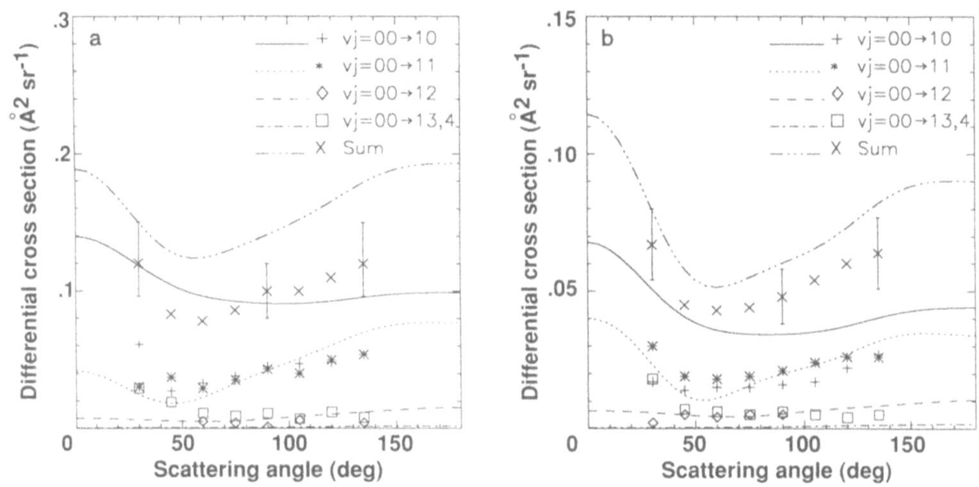

Figure 10. (a) Angular dependence of ro-vibrational state-to-state cross sections for the excitation $v = 0 \rightarrow 1$ of HF at 0.9eV. The curves refer to the present results, the symbols to the measurement of Knoth et al.[35]

(b) Same as (a) for an incident energy of 1.2eV.

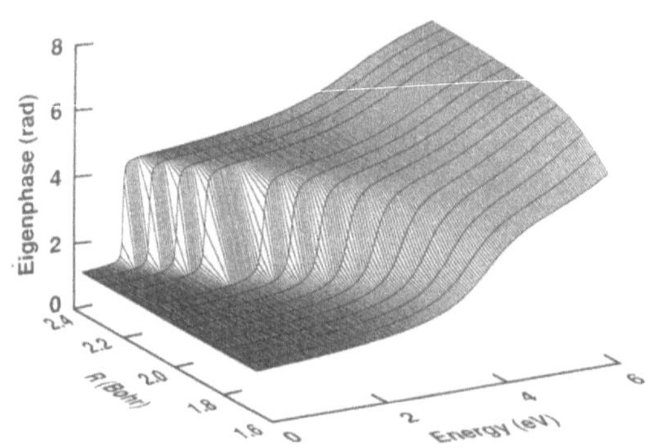

Figure 11. Phasemap for the d-wave component of the $^2\Pi_g$ state of $e + N_2$.

each internuclear separation q a precursor resonance energy $\epsilon_{res}(q)$ eq. (78) is determined as a point of inflection of TrΦ. Since the phases are only determined modulo π the smooth behaviour in q and ϵ on the correct branch of the arctan was provided by suitable addition of multiples of π. Rotational motion has been included by matching the frame-transformed vibronic R-matrix to the solutions of the ro-vibrational close coupling equations, up to total angular momentum $J \leq 8$. The rotational basis set employed consists of of all functions $\boldsymbol{l} + \boldsymbol{j} = \boldsymbol{J}$ with parity $\eta = \pm 1$. The goal of this study is to test our methodology for the case of resonant scattering. We will discuss the vibrationally inelastic cross section $v = 0 \rightarrow 1$, which is dominated by the $^2\Pi_g$-resonance, and compare the rotationally resolved data with earlier experimental and theoretical results.

5.2. Results

The total integral cross section for the $v = 0 \rightarrow 1$ excitation function is compared with experimental data in figure 12. The agreement with the four data points reported from the measurement of Jung et al[56] is excellent. The overall shape agrees well with the experimental results of Allan[57] and with the Multichannel Schwinger calculation of Huo et al,[58] in which the the Feshbach Projection operator formulation of Domcke[59, 33] was used for the treatment of nuclear dynamics. The relative size of the peaks differs somewhat from the earlier ab initio calculation of Morgan[60] based on the vibronic R-matrix method of Schneider et al.[42] In the present work a Morse potential constructed from the spectroscopic parameters of N_2[61] was employed for the target potential energy curve. Morgan[60] used an SCF curve, whose somewhat larger ω_e value gives larger vibrational level spacings. Results for the energy position and magnitude of the first peak of the $v = 0 \rightarrow 1$ function are summarized in table 4 for comparison.

As was shown in the measurements by Wong and Dubé[62] and of Jung et al 1982[56] oscillating resonance structures are present for all rotational transitions $\Delta j = 0, \pm 2$ and ± 4. The energy dependence of the differential cross section at 60° for ro-vibrational excitation with $\Delta j = \pm 4$ is displayed in figure 13 as an example. These transitions can be explained as an interference of incoming and outgoing resonant $d\pi$ partial waves, exchanging four units of angular momentum with the target. The calculated values were rotationally averaged for a temperature of $300K$, corresponding to the experimental temperature. Our theoretical curve is in good agreement with the experiment. The peaks at higher energies ($\geq 3eV$) are somewhat larger than the experimental ones. At these energies the cross sections are sensitive to the fixed-nuclei calculations for large internuclear distances q.

The angular dependence of the ro-vibrational state-to-state transition with $\Delta j = 0$ is shown in figure 14 for energies in the range of the $^2\Pi_g$ resonance. For given energy the curves are peaked in the forward and backward directions, symmetric about 90°. Thus the system $e - N_2$ corresponds primarily to d-wave scattering in the resonance region. A comparison with the rotationally resolved measurements of Jung et al[56] for an electron energy of 2.47 eV indicates (figure 15) that there is a good overall agreement in the angular dependence as well as in the relative importance of the rotationally elastic versus the inelastic processes with $\Delta j = \pm 2$ and 4. The calculated rotationally resolved curves are well within the experimental error bars of 20%. The largest difference occurs for scattering angles greater than 90°. In agreement with earlier theoretical work,[64, 65] the calculated differential cross sections for v=0-1 in the energy region of the shape resonance are essentially symmetric for all rotational transitions about 90°, even when vibrational and rotational non-adiabatic effects are included, as is done here. Contrary

Table 4. Positions and magnitudes of the maxima in the cross section $v = 0 \rightarrow 1$ in comparison with measurement of Allan et al[57], and the calculations of Morgan[60] and Huo et al.[58]

Reference	Energy/eV	Cross section/Å^2
Morgan	2.12	6.94
Huo	1.95*	5.74
Present	1.96	5.7
Allan	1.95	5.6

* Calculated second Maximum is slightly larger

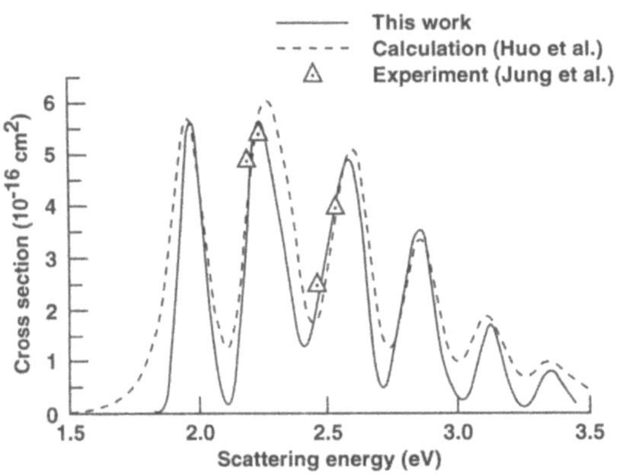

Figure 12. Integral cross section for the $v = 0 \rightarrow 1$ transition in comparison with the calculation of Huo et al 1987.[58] The symbols refer to experiments of Jung et al.[56]

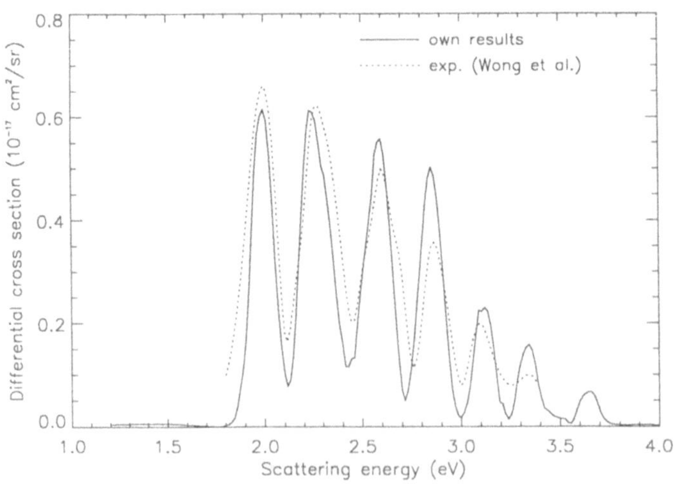

Figure 13. Energy dependence of differential cross section of the $v = 0 \rightarrow 1$ excitation of e-N_2 for $\Delta j = \pm 4$ at an angle of 60°. The dashed curve refers to the measurement of Wong and Dubé.[62]

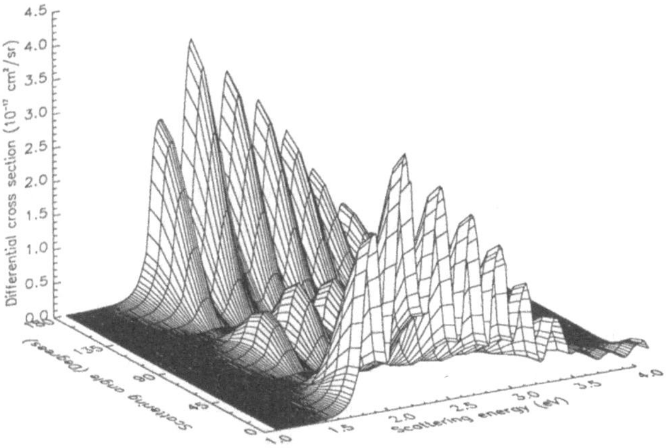

Figure 14. Angular dependence of differential cross section $v = 0 \rightarrow 1$ for the rotational elastic process $\Delta j = 0$ in the energy region of the $^2\Pi_g$ shape resonance of $e - N_2$.

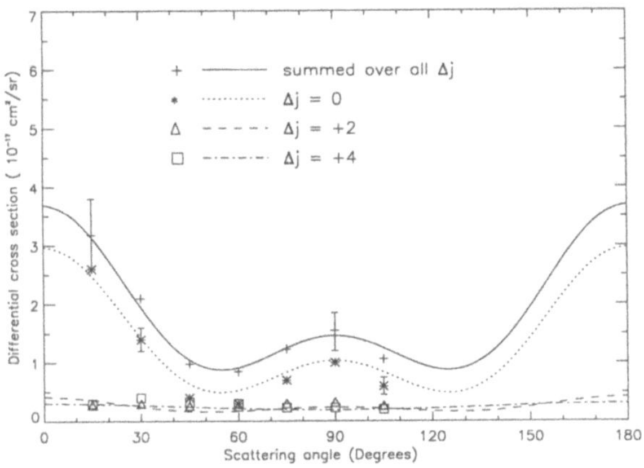

Figure 15. Differential cross section $v = 0 \rightarrow 1$ at the incident energy $2.47eV$ for rotational transitions $\Delta j = 0, 2$ and 4. The symbols refer to the measurements of Jung et al 1982.[56]

in the measurements[56, 66] the second minimum at about 120° seems to be slightly lower by a few percent in the energy region from 1.5 to 5eV.

6. CONCLUSIONS

In this section new methods for treating the effects of nuclear dynamics in electron-molecule scattering, starting from fixed-nuclei variational R-matrix calculations, have been outlined. The methodology allows treatment of vibrational and rotational coupling including non-adiabatic effects. The breakdown of the adiabatic nuclei rotational method (ANR) was discussed. In the case of a polar molecule, however weak the dipole moment may be, the ANR leads to qualitatively false results near ro-vibrational thresholds. Only proper inclusion of rotation gives the correct threshold scaling laws. A full treatment of rotation can be done by frame-transforming the vibronic R-matrix into the labo¡ratory frame and then matching to solutions of the external ro-vibronic close-coupling equations. For the highly polar system e-HF we have presented calculations and analysis of fine structures due to rotational coupling that modify vibrationally elastic and inelastic cross sections. In rotationally averaged cross sections, rotational effects lead to pronounced shoulders and to rounded onsets of the vibrational excitation peaks characteristic of dipolar target molecules. These fine structures are a consequence of underlying rotational substructures. There are threshold excitation peaks and huge narrow cusps, which are predicted for ro-vibrational state-to-state transitions with $\Delta j = -1$. The structures superimpose, depending on the rotational temperature of the target gas. The energy range of these non-adiabatic rotational effects is only 0.05eV about the vibrational thresholds, indicating a lower bound to the range where ANR methods might be valid. The multipole-extracted adiabatic nuclei (MEAN) method does not directly correct the ANR in the case of a strong long-range potential near threshold, but the inherent error in the threshold region is less severe for MEAN than for ANR.

The new NADP method has been applied to resonant vibrational scattering for the example of the $^2\Pi_g$ shape resonance of e-N_2. This resonance dominates the low-energy range from 1.0 to 4eV. The characteristic oscillating structure associated with the vibrational motion is very well reproduced. The angular dependence and relative importance of rotationally elastic versus inelastic processes with $\Delta j = \pm 2, \pm 4$ is in good agreement with rotationally resolved measurements of Jung et al[56] and Wong and Dubé[62]. In agreement with previous theoretical results for the $v = 0 - 1$ cross section, our data indicate that the angular dependence is essentially symmetric about 90° even when vibrational and rotational non-adiabatic effects are included. The cross section shows an enhancement in forward and backward direction with a minimum at about 60°. Contrary to theory, in the measurements[66, 56] the second minimum about 120° is smaller by a few percent.

References

1 K. Baluja, P.G. Burke, and L.A. Morgan, Comput. Phys. Commun., **27**, 299 (1982).
2 L.A. Morgan, Comput. Phys. Commun., **31**, 419 (1984).
3 C.J. Noble and R.K. Nesbet, Comput. Phys. Commun., **33**, 399 (1984).
4 D.W. Norcross and N. T. Padial, Phys. Rev. A, **25** 226 (1982).
5 K. Rohr and F. Linder, J. Phys. B, **9**, 2521 (1976).
6 R.K. Nesbet, Phys. Rev. A, **19**, 551 (1979).
7 F.T. Smith, Phys. Rev., **118**, 349 (1960).
8 R.K. Nesbet and T. Grimm-Bosbach, J. Phys. B, **26**, L423 (1993).
9 E.S. Chang and A. Temkin, Phys. Rev. Lett., **23**, 399 (1969).
10 M.A. Morrison, Adv. At. Mol. Phys., **24**, 51 (1988).
11 N.F. Lane, Rev. Mod. Phys., **52**, 29 (1980).
12 Y. Itikawa, Phys. Rep., **46**, 117 (1978).
13 D.W. Norcross and L.A. Collins Adv. At. Mol. Phys., **18**, 341 (1982).
14 N.T. Padial and D.W. Norcross, Phys. Rev. A, **29**, 1590 (1984).
15 N.T. Padial and D.W. Norcross, Phys. Rev. A, **29**, 1742 (1984).
16 A. Jain and D.W. Norcross, Phys. Rev. A, **32**, 134 (1985).
17 G. Snitchler, D.W. Norcross, A. Jain and S. Alston, Phys. Rev. A, **42**, 671 (1990).
18 E. S. Chang and U. Fano, Phys. Rev. A, **6**, 173 (1972).
19 A. M. Arthurs and A. Dalgarno, Proc. R. Soc. London Ser. A, **256**, 540 (1960)
20 B. H. Bransden, "Atomic Collision Theory", (New York: Benjamin) 1970.
21 J. N. Bardsley and R. K. Nesbet, Phys. Rev. A, **8** 203 (1973).
22 R. K. Nesbet, J. Phys. B, **13**, L193 (1980).
23 J. Phys. B, **14**, 4889 (1981).

[24] H. T. Thümmel, R. K. Nesbet and S. D. Peyerimhoff, J. Phys. B, **25**, 4553 (1992).
[25] M. A. Morrison, A. N. Feldt and D. Austin, Phys. Rev. A, bf 29, 2518 (1984).
[26] C. W. Clark, Phys. Rev. A, bf 30 750 (1984).
[27] M. Gailitis and R. Damburg, Proc. Phys. Soc., **82** 192 (1963).
[28] I. I. Fabrikant, J. Phys. B, **11**, 3621 (1978).
[29] L. A. Collins and D. W. Norcross, Phys. Rev. A, **18** 467 (1978).
[30] U. Fano and D. Dill, Phys. Rev. A., **6**, 185 (1972).
[31] O. H. Crawford and A. Dalgarno, J. Phys. B, **4**, 494 (1971).
[32] N. Chandra, Phys. Rev. A., **16**, 80 (1977).
[33] W. Domcke, Phys. Rep. **208,2** 97 (1991).
[34] S. Cvejanović, in "The Physics of Electronic and Atomic Collisions", AIP Conf. Proc. **295** 390 (1993).
[35] G. Knoth, M. Rädle, M. Gothe, H. Ehrhardt and K. Jung, J. Phys. B **22** 299 (1989).
[36] G. Knoth, M. Gothe, K. Jung and H. Ehrhardt, Phys. Rev. Lett. **62** 1735 (1989).
[37] M. Rädle, G. Knoth, K. Jung and H. Ehrhardt, J. Phys. B **22** 1455 (1989).
[38] H. Ehrhardt, in "Aspects of electron-Molecule Scattering and Photoionization", AIP Conf. Proc. **204**; ed A. Herzenberg.
[39] L. Dubé and A. Herzenberg, Phys. Rev. Lett. **38** 820 (1977).
[40] J. P. Gauyacq and A. Herzenberg, Phys. Rev. A **25** 2959 (1982).
[41] L. A. Morgan and P. G. Burke, J. Phys. B **21** 2091 (1988).
[42] B. I. Schneider, M. Le Dourneuf and P. G. Burke, J. Phys. B **12** L365 (1979).
[43] G. Snitchler, D. Norcross, A. Jain and S. Alston, Phys. Rev. A **42,1** 671 (1990).
[44] S. Cvejanović, J. Jureta, D. Cubrić and D. Cvejanović, to be published
[45] O. Schafer and M. Allan, J. Phys. B **24** 3069 (1991).
[46] H. T. Thümmel, R. K. Nesbet and S. D. Peyerimhoff, J. Phys. B **26** 1233 (1993).
[47] R. N. Sileo and A. T. Cool, J. Chem. Phys. **65** 117 (1976).
[48] E. P. Wigner, Phys. Rev. **73** 1002 (1948).
[49] J. R. Taylor, "Scattering Theory", John Wiley and Sons. Inc. (1972).
[50] G. Herzberg, "Spectra of Diatomic Molecules", D. Van Nostrand Company, Inc, (1950)
[51] G. J. Schulz, Phys. Rev. **135** A988 (1964).
[52] R. K. Nesbet, J. Chem. Phys. **40** 3619 (1964).
[53] C. J. Gillan, O. Nagy, P. G. Burke, L. A. Morgan, C. J. Noble, J. Phys. B, **20** 4585 (1987).
[54] R. K. Nesbet, J. Phys. B **17** L897 (1984).
[55] M. A. Morrison, P. J. Hay, J. Chem. Phys. **70** 4034 (1979).
[56] K. Jung, T. Antoni, R. Müller, K-H. Kochem, H. Ehrhardt, J. Phys. B, **15** 3535 (1982).
[57] M. Allan, J. Phys. B, **18** 4511 (1985).
[58] W. M. Huo, M. A. P. Lima, T. L. Gibson, V. McKoy, Phys. Rev. A, **36** 1642 (1987).
[59] M. Bermann and W. Domcke, Phys. Rev. A, **29** 2485 (1984).
[60] L. A. Morgan, J. Phys. B, **19** L439 (1986).
[61] K. P. Huber and G. Herzberg, "Molecular Spectra and Molecular Structure", Van Nostrand-Reinhold, New York, (1979).
[62] S. F. Wong, L. Dubé, Phys. Rev. A, **17** 570 (1978).
[63] L. Dubé, A. Herzenberg, Phys. Rev. A, **20** 194 (1979).
[64] C. J. Gillan, O. Nagy, P. G. Burke, L. A. Morgan and C. J. Noble, J. Phys. B, **20** 4585 (1987).
[65] N. Chandra, A. Temkin, Phys. Rev. A, **13** 188 (1976).
[66] M. J. Brennan, D. T. Alle, P. Euripides, S. J. Buckman, M. J. Brunger, J. Phys. B, **25** 2669 (1992).

TAILORING THE R-MATRIX APPROACH FOR APPLICATION TO POLYATOMIC MOLECULES

Kurt Pfingst, Bernd M. Nestmann and Sigrid D. Peyerimhoff

Institut für Physikalische und Theoretische Chemie, Universität Bonn, Wegelerstraße 12, D–53115 Bonn, Germany

1. PRELIMINARIES

In this chapter a concept will be presented for applying the R-matrix formalism to low-energy electron scattering off molecular systems with more than two atoms. A detailed description of this concept is given by Nestmann et al,[1] Pfingst et al[2] and Nestmann et al;[3] here we will restrict ourselves to an overview.

In polyatomic systems the number of electrons is quite large and the molecular symmetry is generally low. Thus, the numerical effort for calculating cross sections for electron scattering off such systems requires a very efficient concept to be employed. Similar problems occur in bound state calculations for large molecular systems. The aim of the present approach is to transfer, as far as possible, the high standard of efficiency which has been achieved in bound state calculations to scattering problems. For this purpose, the R-matrix formalism appears to be particularly suitable as it provides a well-defined discretization of the scattering continuum, which enables existing quantum chemistry codes to be employed with only minor changes. In the calculation presented in section 3 it will be shown that the scattering phases can be described to a large extent by only a few eigenstates of the related variational problem. These states can be handled in the same manner as excited bound states. In general, it is possible to treat the most important states at a higher level of accuracy than the less important ones. By reducing the number of states considered explicitly, one has a good chance to obtain an explanation of structures in the cross section by examining the properties of the calculated wave functions.

A second possibility to reduce the numerical effort in the standard R-matrix approach is to use exclusively Gaussian-type basis functions. It will be shown in section 2 that these functions are suitable for the representation of the continuum orbitals describing the scattered electron within the R-matrix sphere.

The following discussion of scattering processes is restricted to the fixed-nuclei approximation. The notation in the formulae refers to chapter 1, except for some minor changes. So, the $\phi_{ql_qm_q,i}(a_e;R)$ in eq. 19 will be replaced by $\omega_{qlm,i}(a_e;R)$, in order to

Computational Methods for Electron-Molecule Collisions
Edited by W.M. Huo and F.A. Gianturco, Plenum Press, New York, 1995

avoid confusion with the $\phi_\alpha(n_{T+1}; R)$ in eq. 18 and to indicate that the (l,m)-quantum numbers of the scattered electron are assumed to be independent of the target-state quantum number q.

Furthermore, the functions representing the open channels in eq. 18 will be taken in their antisymmetrized form $\mathcal{A}(\Theta_q, \phi_\alpha)$ in order to be consistent with their realization in the computer codes (see the Appendix to this chapter). One should be aware that the orthogonality of these functions does not follow automatically from the orthogonality of the Θ_q and the ϕ_α. This problem will be considered in section 2.

In the following section a technique is introduced for applying standard integral programs as used in bound-state calculations to the R-matrix approach. Section 3 describes how to obtain the R-matrix expansion, eq. 19, if the $(n_T + 1)$-particle space becomes too large for a full diagonalization. In section 4 we examine electron scattering of methane, as a first application of this concept for polyatomic molecules. Section 5 summarizes and concludes this chapter.

2. USING GAUSSIAN BASIS FUNCTIONS FOR R-MATRIX CALCULATION

From experience in bound-state calculations, Gaussian-type functions are found to be the best applicable basis functions for treating the electronic structure of polyatomic molecules. It has furthermore been shown[4, 5] that Gaussian-type basis sets are also well suited to approximate free-particle wave functions $j_l(kr)Y_{lm}(\hat{r})$ inside a spherical region Ω defined by a finite radius $r = a_e$.

Table 1. Exponents of the continuum-type Gaussian functions

No.	s	p	d
1	0.130137	0.104354	0.111252
2	0.101981	0.082488	0.089412
3	0.080524	0.065610	0.072361
4	0.063367	0.052121	0.058496
5	0.049648	0.041110	0.047005
6	0.038298	0.032122	0.037397
7	0.029151	0.024635	0.029225
8	0.021901		

An example of such an approximation is given in fig. 1 for an s-wave. In table 1 the exponents of the continuum-type Gaussian functions, as they are employed in the calculations presented within this section for representing the free electron, are given for s-, p- and d-waves. The reason for using Gaussian-type basis functions in molecular-structure calculations is the high efficiency in evaluating the integrals representing the Hamiltonian. Such evaluation, however, requires the integration to be carried out over the entire three-dimensional space $\mathcal{R}^3$. This violates the requirement of the R-matrix formalism to confine the integrals to a finite sphere Ω. A possible way out of this dilemma is suggested by the expansion of the scattering wave function Φ_i as given in eq. 18. A basic assumption of the R-matrix approach is that the electron density of the target wave functions Θ_q as well as that of the correlation terms Ξ_μ vanish outside Ω.

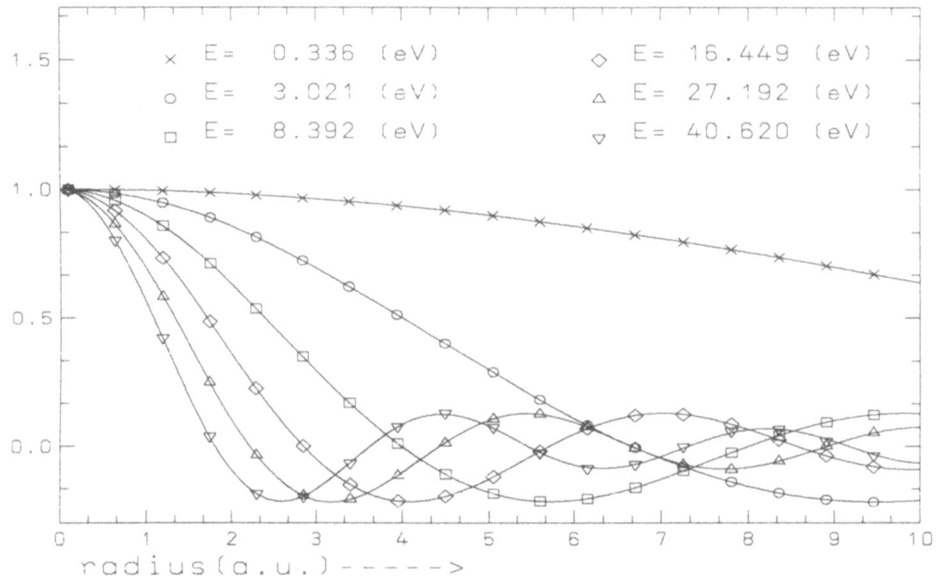

Figure 1. The lowest six Bessel functions (symbols), quantized by the boundary condition $(\frac{d}{dr}r)j_l(kr)|_{r=a_e} = 0$ and their approximation by a linear combination of Gaussians (full curves) for $l = 0$.

We now define a partitioning of the potential V occurring in the Hamiltonian for the $(n_T + 1)$-electron system:

$$V = V_\Omega + V_{\mathcal{R}^3-\Omega} , \tag{121}$$

$$V_\Omega = \begin{cases} V & \text{if } r_i \leq a, \ i = 1, \cdots, n_T + 1 \\ 0 & \text{otherwise} \end{cases} .$$

$$V_{\mathcal{R}^3-\Omega} = \begin{cases} V & \text{if } r_i > a, \text{ for some i} \\ 0 & \text{otherwise} \end{cases} .$$

The Ω-confined potential V_Ω required in R-matrix calculations can be obtained as the difference between the two components V and $V_{\mathcal{R}^3-\Omega}$. The potential V can be constructed from the standard integrals for electron-nuclei attraction and electron-electron repulsion. Under the restriction of at most one free electron, the potential for the external region can be expressed by the expansion

$$V_{\mathcal{R}^3-\Omega} = \sum_{q\alpha} \sum_{q'\alpha'} |\Theta_q > (\phi_\alpha | V_{qq'} | \phi_{\alpha'})_{\mathcal{R}^3-\Omega} < \Theta_{q'}|. \tag{122}$$

The $V_{qq'}$ denote the coupling potentials between the electronic channels affecting the scattered electron outside the sphere. They are provided by the multipole (transition) moments of the related target states. For practical purposes, the $V_{qq'}$ can be obtained from the dipole and quadrupole (transition) moments of the target wave functions Θ_q, $\Theta_{q'}$, the higher terms of the multipole expansion can be neglected. The evaluation of $V_{\mathcal{R}^3-\Omega}$ is very efficient because only one-particle integrals have to be considered.

One should note, however, that the application of formula 121 requires the projection of the scattering states onto the electronic channels.

The calculation of the integrals for the kinetic energy and the overlap make only a minor contribution to the computer time and can therefore be calculated directly confined to Ω. In the case of a charged target the contribution of the coulomb potential outside the sphere will be subtracted from the one-particle integrals.

As in other R-matrix calculations the one-particle basis functions are divided into valence-type functions, usually centered at the nuclei of the atoms and continuum-type functions located at the scattering center. By definition, only the latter have non-vanishing contributions outside the R-matrix sphere and must not be used for describing the target states Θ_q. This convention guarantees that all those functions $\mathcal{A}(\Theta_q, \phi_\alpha)$ contributing to the open channels are mutually orthogonal. In general one has

$$(\mathcal{A}(\Theta_q, \phi_\alpha), \mathcal{A}(\Theta_{q'}, \phi_{\alpha'})) = \delta_{q,q'}\delta_{\alpha,\alpha'} - d_{\Theta_q,\Theta_{q'}}(\alpha', \alpha),$$

where $d_{\Theta_q,\Theta_{q'}}$ denotes the (transition-) density matrix between the target states Θ_q and $\Theta_{q'}$.[6, 7] Due to the construction of the target, however, the $d_{\Theta_q,\Theta_{q'}}(\alpha', \alpha)$ term vanishes, if α or α' refer to a continuum-type orbital.

Dividing the one-particle basis into valence- and continuum-type functions has a further advantage. As only integrals between functions centered at the origin have to be calculated in a modified manner, the angular components can be separated from the radial one and evaluated analytically. The remaining quadrature over the radial coordinate is done numerically.

For linear molecules a standard R-matrix method based on the *ALCHEMY* package (see chapter 3) can be applied. Within this package Slater-type orbitals (STO's) are employed for the description of the molecular orbitals and Bessel and/or numerical functions for the continuum orbitals. Moreover, this code confines all integrals needed for the representation of the Hamiltonian directly to the R-matrix sphere. In the following this package is referred to as the '*STO*-Code'. For reasons, which will be given in the following section the method using the concept described above is referred to as the selected states R-matrix method (*SSRM*). In order to make a comparison between the two approaches we consider the example of $e - N_2$ scattering. This molecule has been well examined in the past[8, 9] (see also chapter 3) using the R-matrix approach for linear molecules as well as other methods.

For the moment discussion will be restricted to the well-defined static exchange (SE) level. This means, only configuration state functions (CSF's) generated by the first term of eq. 18 are included in the (n_T+1)-particle CI and the target ground state Θ_q is described by an *SCF* wavefunction. Since polarization of the target due to the scattered electron is not included at the SE level, it provides a transparent test of the approximation discussed above, namely the effective confining of the integration to a finite region of space. Technical details concerning the basis sets employed are given in the appendix of this chapter.

First the scattering states of $^2\Pi_g$ symmetry of N_2^- are considered. In table 2 the five energetically lowest-lying roots of the (n_T+1)-particle CI are shown. ΔE_i are the energy differences between the ith eigenvalue E_i of the (n_T+1)-particle CI and the respective reference energy E_{ref}, which is the energy of the target ground state. In addition the amplitudes $\omega_{qlm,i}(a_e, R_0)$ of eq. 19 are tabulated at the R-matrix radius $a_e = 10au$ for the equilibrium geometry $R_0 = 2.068\ a_0$. As can be seen from this equation, for a given collision energy $\epsilon = E - E_{ref}$ and a given radius a_e the R-matrix is determined by these two quantities: energy differences and amplitudes. For $^2\Pi_g$ symmetry only one partial wave of $d\pi_g$ symmetry is examined. The table shows good accordance between the two different approaches for both the energy differences and the amplitudes.

Table 2. The five energetically lowest-lying roots in N_2^- of $^2\Pi_g$ symmetry in the static exchange approximation. ΔE_i are the energy differences between the ith eigenvalue E_i of the (n+1)-particle CI and the respective reference energy E_{ref}, $\omega_{1lm,i}(a_e)$ are the amplitudes of eq. 18.

STO-Code				*SSRM*		
E_i (Hartree)	ΔE_i (eV)	$\omega_{1lm,i}$ l=2, m=1		E_i (Hartree)	ΔE_i (eV)	$\omega_{1lm,i}$ l=2, m=1
-108.91728	1.94	0.0861		-108.87625	1.95	0.0858
-108.83532	4.17	0.0459		-108.79848	4.07	0.0455
-108.66117	8.91	0.0743		-108.62144	8.88	0.0748
-108.33367	17.82	0.0743		-108.29272	17.83	0.0737
-107.89089	29.87	0.0695		-107.84980	29.88	0.0723

In figure 2 the $^2\Pi_g$ eigenphases of the K-matrix from both the *SSRM* and the *STO*-Code calculation are plotted. They are compared with results of Morrison et al,[8] using a model potential and Meyer,[10] based on an optical potential method. Both Morrison et al and Meyer employed Gaussian-type basis sets to represent the target ground state by an *SCF* wavefunction. The basis set used by Morrison et al is similar to the basis set in the calculation discussed here, whereas Meyer used a slightly larger basis set given by Berman and Domcke.[11] The figure shows a very good agreement between the *SSRM* eigenphases and the results of Morrison et al. There are noticeable, but still small, differences to the results of Meyer and to the *STO*-Code calculation, these can be traced to differences in the AO basis sets.[2]

We now turn to the $^2\Sigma_g^+$ symmetry as an example for non-resonant scattering. In table 3 the six energetically lowest-lying roots are shown. For this scattering symmetry two partial waves of $s\sigma_g$ and $d\sigma_g$ symmetry are considered. Again the energy differences ΔE_i and the amplitudes $\omega_{qlm,i}$ compare very well between the *SSRM* and the standard *STO*-Code calculation.

In figure 3 the corresponding $^2\Sigma_g^+$ eigenphase sums are compared with the data of Meyer,[10] Morrison et al[8] and Burke et al.[12] The curves from the present calculations are almost indistinguishable and differ only by a small amount from the referenced data over the whole energy range considered here.

3. CONSTRUCTION OF THE R-MATRIX USING A SELECTED NUMBER OF SCATTERING STATES

Including electron correlation in the theoretical treatment to an amount which is necessary for describing resonances and electronic inelastic processes, may increase the size of the (n_T+1)-particle basis to such an extent that a full diagonalization as it is required for constructing the R matrix using eq 19, is no longer possible. This situation becomes more difficult if the number of electrons in the target molecule which have to be correlated becomes larger.

However, only scattering states with non-vanishing amplitudes for $r = a_e$ contribute to the R matrix. Moreover, due to the energy denominator, states with energies

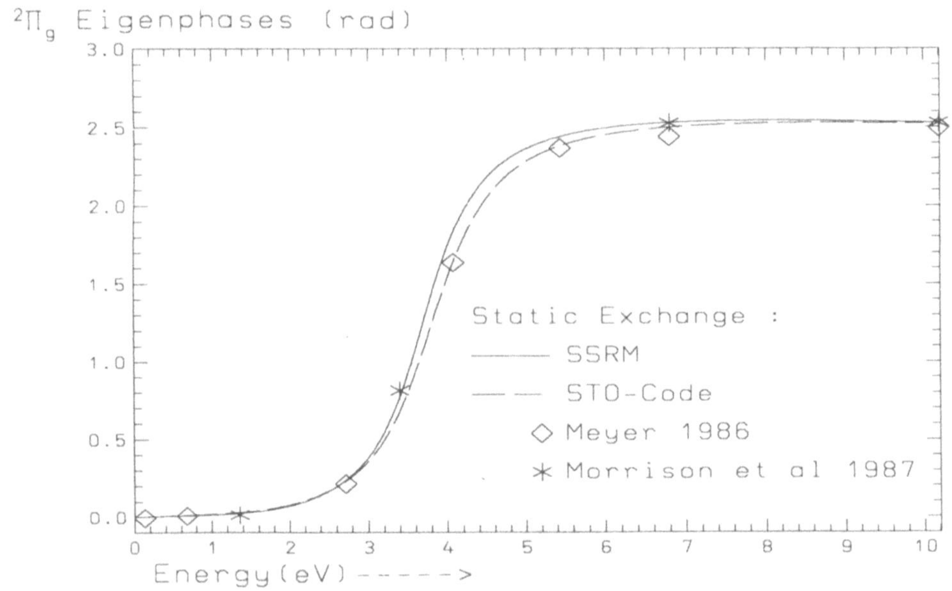

Figure 2. Eigenphases of $^2\Pi_g$ symmetry at the static exchange level.

far above the considered energy region become less important than those with energies close to this. This suggests that one can restrict the expansion of the R matrix to scattering states with noticeable amplitudes $\omega_{qlm,i}(a_e)$ and with energies E_i not too far above the ionization potential of the target which provides the upper limit of the electronic one-particle scattering problem. With respect to this approach the concept presented is called the selected states R-matrix method (SSRM).

In order to give an example, again the resonant $^2\Pi_g$ symmetry for electron scattering off N_2 shall be considered. This time, however, the $(n_T{+}1)$-scattering process is treated at the static exchange plus polarization (SEP) level. This means, the configuration space of the static exchange (SE) calculation, described in the previous section, is augmented by two-particle-one-hole configurations with respect to the configuration of the target ground state. The hole is in the $3\sigma_g$, $2\sigma_u$ or $1\pi_u$ shell and the two particles are allowed to occupy all combinations of the virtual orbitals consistent with the total symmetry. As in the SE calculation, the target ground state of N_2 is represented by the SCF wavefunction.

Comparing the *SSRM* eigenphase in figure 4, solely including the four lowest-lying SEP roots E_i with the *STO*-Code phases (dash-dotted), evaluated by taking all roots (350) of the $(n_T{+}1)$-particle CI into account, two noticeable effects can be observed:

1. The position and the width of the resonance are already determined by the four lowest roots.
2. There is a pronounced structure in the *SSRM* eigenphase curve in the energy range between 2.5 (eV) and 8.5 (eV) not occurring in the *STO*-Code curve, including all 350 roots.

If the *STO*-Code calculation is also restricted to the lowest four roots, the resulting curve coincides with the corresponding *SSRM* curve. This is a clear indication that the structure is caused by the missing higher-lying roots.

It is conspicuous that this structure does not appear in the phases at SE level. In

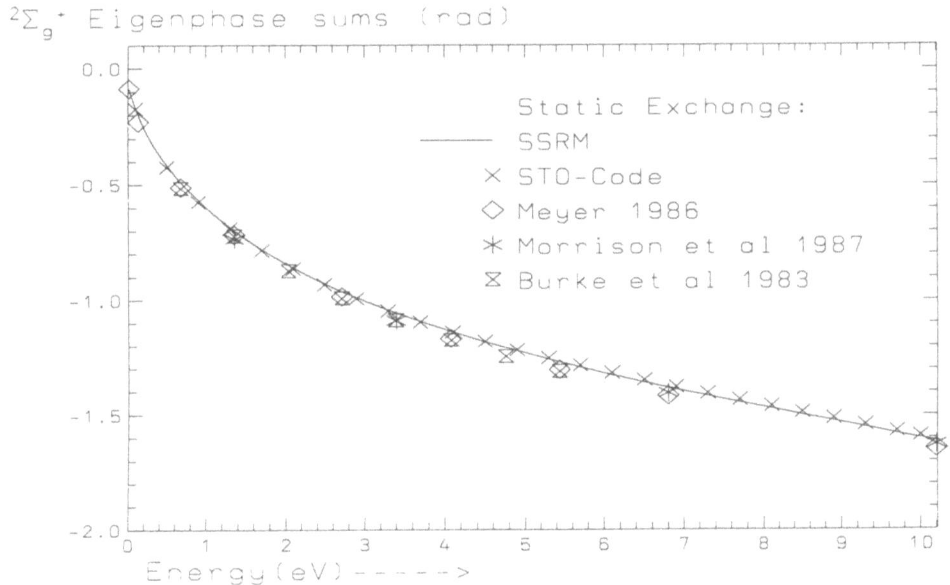

Figure 3. Eigenphase sums of $^2\Sigma_g^+$ symmetry at the static exchange level.

fact, in contrast to the SE calculation, the (n_T+1)-particle space spanned by a number of selected states Ψ_i, $i = 1, \cdots, n$, generally does not contain the functions $\mathcal{A}(\Theta_q, \phi_\alpha)$ representing the open channels completely.

In order to complete the (n_T+1)-particle space with respect to the open channels one has to find an additional set of orthonormal functions $\tilde{\Psi}_j$, $j = n+1, \cdots, m$ built up from the original space and orthogonal to the selected states, which provide the expansions

$$\mathcal{A}(\Theta_q, \phi_\alpha) = \sum_{i=1}^{n} C_{q\alpha}^i \Psi_i + \sum_{j=n+1}^{m} \tilde{C}_{q\alpha}^j \tilde{\Psi}_j \tag{123}$$

for all (q, α) considered. Formally, the remaining (unselected) eigenstates of the original problem can be taken as such a set of functions. In practice, the $\tilde{\Psi}_j$ are obtained by diagonalizing a subspace of the original one. Provided that the $\tilde{\Psi}_j$ are orthogonal to the selected Ψ_i, the matrix elements of the Hamiltonian between these states vanish and the energies and amplitudes of the $\tilde{\Psi}_j$ can be used directly in the expansion of the R-matrix eq. 19.

A set of functions $\tilde{\Psi}_j$, which approximately satisfy the condition of orthogonality to the Ψ_i and which fulfill eq. 123, can be obtained from the eigenstates of the SE approach.

As can be seen in figure 4, if the higher-lying roots are added at the static exchange level, then the *SSRM* eigenphase curve shows a perfect agreement with the best *STO*-Code curve. Thus the missing higher-lying SEP roots can be well approximated by SE roots, at least in the case of resonant scattering. The agreement with the data of Morrison et al,[8] also shown in figure 4, is not as good as it is at the static exchange level, but still satisfactory. As it is well known from the literature,[13, 8] at the SEP level

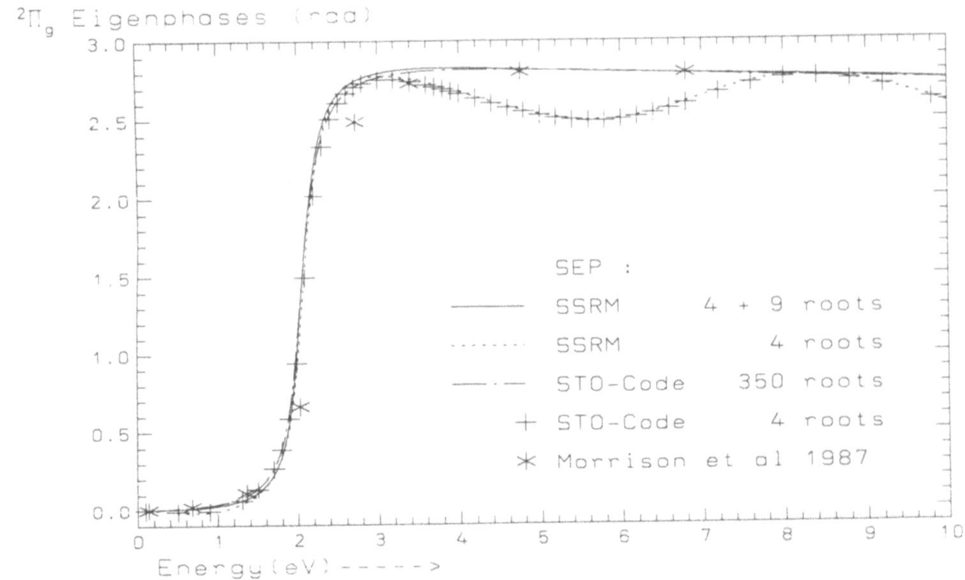

Figure 4. Eigenphases of $^2\Pi_g$ symmetry at the static exchange plus polarization level.

the width and the position of the $^2\Pi_g$ resonance depend strongly on both the AO basis set used and the way polarization is included in the calculation.

As an example of non-resonant scattering we again consider the $^2\Sigma_g^+$ symmetry of (e^-, N_2). Figure 5 shows the eigenphase sums of the *SSRM* calculations in comparison to the data of Morrison et al[8] and Burke et al.[12] Again the eigenphase sums of the two present calculations are nearly identical and located between both reference data.

4. APPLICATION TO ELECTRON-SCATTERING OF METHANE

The cross sections in electron-methane scattering show a Ramsauer-Townsend minimum at 0.4 eV and a marked increase for higher energy with a maximum at 8 eV. Both structures have been well examined by many experiments.[14–23] Very extensive studies of electron-methane scattering were published by Jain[24] using model potentials. Gianturco and Scialla[25] reproduced the Ramsauer-Townsend minimum close to the experimental values using a modified semiclassical exchange approximation including polarization by employing an effective model potential.[26–28] *Ab initio* results were obtained by the Schwinger multichannel method by Lima et al[29, 30] and by the complex Kohn method by McCurdy and Rescigno[31] and Lengsfield et al.[32] Lima et al provided differential cross sections in the static exchange (SE) approach[29] and in the static exchange plus polarization (SEP) approach.[30] For higher energies (between 7.5 and 20 eV) they achieved good agreement between the calculated differential cross sections in the SE approach and the experimental results. At the SEP level, they found a Ramsauer minimum at 0.1 eV. Similarly, McCurdy and Rescigno[31] obtained partial cross sections in A_1, T_2 and E symmetry at the SE level and the resulting integral cross sections were in good agreement with the experiments for higher energies. The calculations of Lengsfield et al[32] at the SEP level reproduced exactly the Ramsauer minimum at 0.4

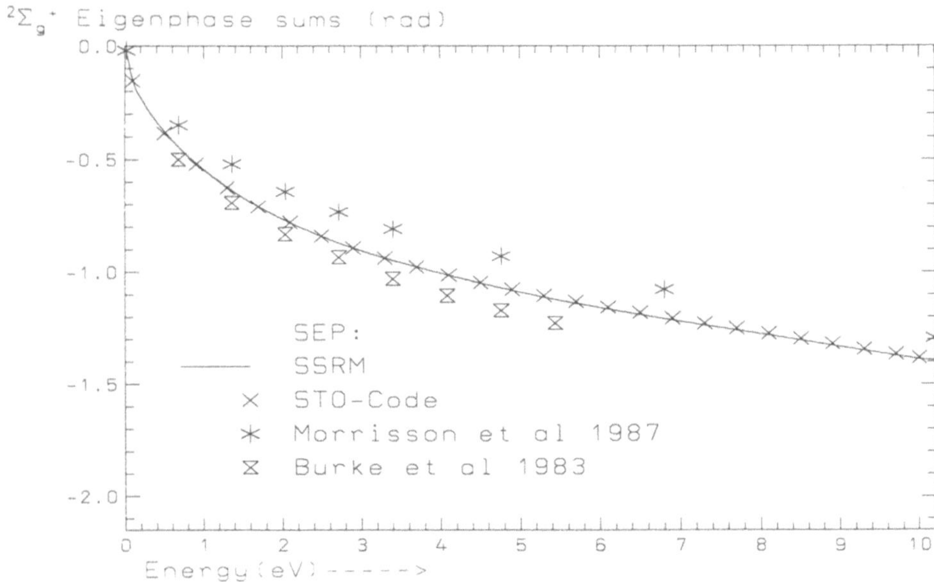

Figure 5. Eigenphase sums of $^2\Sigma_g^+$ symmetry at the static exchange plus polarization level.

eV; their calculated differential cross sections for low energies $(0.5 - 1.0eV)$ were also in good agreement with experiment. However, they also reported a high sensitivity of the low-energy results with respect to the basis set employed which could be traced to the influence of the target state polarizability.

In the presented R-matrix calculation the code employed only accepts Abelian symmetry groups. Therefore, the T_d symmetry group of methane is reduced to the D_2 subgroup. Under this reduction the t_2 orbitals split into b_1, b_2 and b_3 components and the two components of an e orbital are both in the a representation. For this reason the three-fold degeneracy of the $^2T_2$ states is lifted but the two-fold degeneracy of the ^2E states remains. In our basis representation the e components of the scattered electron are given by a superposition of a $(l,m) = (2,0)$ and a $(l,m) = (2,\pm 2)$ wave function. We remove the degeneracy of the ^2E scattering states by projecting to the component with continuum orbitals of $m = 0$.

For constructing the R-matrix along the approach described in section 3 we used four SEP and ten SE scattering states in $^2A_1$ symmetry, six SEP and eleven SE states in $^2T_2$ symmetry and three SEP and eight SE states in ^2E symmetry (after lifting the degeneracy). The selected SEP states are the energetically lowest ones with major contributions to the associated open channels.

As the methane molecule possesses neither a dipole nor a quadrupole moment we neglect $V_{qq'}$ in eq. 121. For defining the $F_{q,l,m}$ from eq. 14 a dipole polarizability of $\alpha_0 = 17.5au^{33}$ of the target is assumed for the external potential.

The energies of the lowest poles and the corresponding amplitudes are listed in table 4. The energies are given in eV relative to the energy of the target ground state. For comparison, the three lowest values of $E_i^0 = k_i^2/2$ with k_i as the solutions of

$$\frac{d\ k_i r\ j_l(k_i r)}{dr}\bigg|_{r=a} = 0$$

are also shown. These values can be interpreted as R-matrix poles of the free-particle problem.

Some information about the $e^- + CH_4$ scattering process can be obtained directly by considering table 4. For the first four scattering states in T_2 symmetry there is very little mixing of the $(l, m) = (1, 0)$ and the $(l, m) = (2, 2)$ components of the amplitudes $\omega_{qlm,i}$. This separation is still reasonable for the fifth and sixth state of this symmetry. In the energy region between 0.34 and 11 eV the s-component of the potential provided by the methane molecule is repulsive because the poles are shifted to higher energies compared to the free-particle problem. A nearly vanishing s-component can be observed at the energy of the lowest $^2A_1$ pole, which causes a Ramsauer minimum close to this energy. The $^2T_2 - p$ component of the potential is repulsive all over the energy region considered, the $^2T_2 - d$ component as well as the E-component are attractive.

In fig. 6 the results of the partial cross sections in $^2A_1$ (dashed), $^2T_2$(dash-dotted) and ^2E symmetry (dotted) and the integral cross section (full curve) in the energy region below the ionization energy of methane are shown. The partial cross sections for

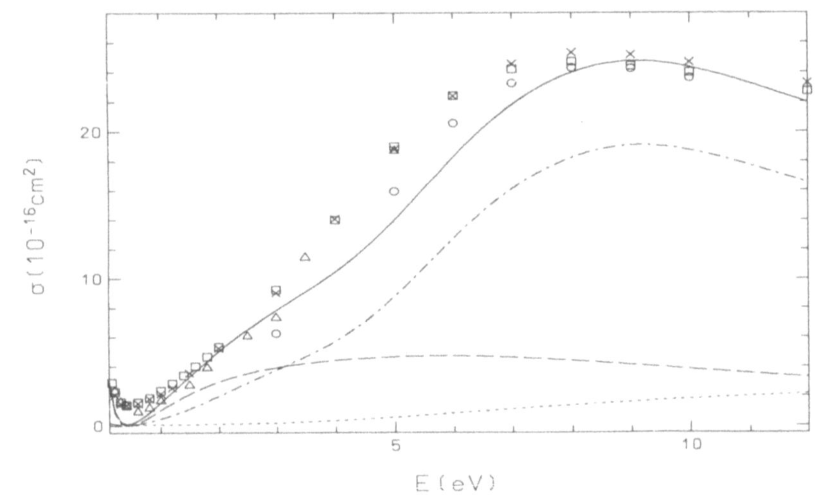

Figure 6. Calculated partial cross sections in $^2A_1$ (dashed), $^2T_2$ (dash-dotted) and ^2E (dotted) symmetry and the integral cross section (full line), integral cross sections from complex Kohn calculations,[32] (circles) experimental total cross section from Ferch et al[20] (squares), Lohmann and Buckman[22] (crosses), and integral elastic cross sections from Sohn et al[23] (triangles). The partial cross sections for $^2T_2$ and ^2E symmetry are multiplied by their degree of degeneracy.

$^2T_2$ and ^2E symmetry are multiplied by their degree of degeneracy. For comparison, experimental values for the total cross section from Ferch et al[20] and Lohmann and Buckman,[22] and integral elastic cross sections from Sohn et al[23] are added to the picture. The results of Sohn et al are obtained from differential cross sections by integration over the scattering angle, the other two experimental results are obtained directly from time-of-flight transmission experiments. In addition, theoretical values for the integral cross

Table 3. The six energetically lowest lying roots of $^2\Sigma_g^+$ symmetry in the static exchange approximation. Notation as in table 2.

STO-Code					*SSRM*			
E_i (a. u.)	ΔE_i (eV)	$\omega_{1lm,i}$ l=0	$\omega_{1lm,i}$ l=2		E_i (a. u.)	ΔE_i (eV)	$\omega_{1lm,i}$ l=0	$\omega_{1lm,i}$ l=2
-108.96788	0.56	0.1133	0.0047		-108.92752	0.56	0.1129	0.0074
-108.91320	2.05	0.0108	0.1288		-108.87165	2.07	0.0123	0.1283
-108.81471	4.73	0.1051	0.0258		-108.77532	4.70	0.1050	0.0272
-108.70829	7.63	0.0374	0.0989		-108.66475	7.70	0.0394	0.0997
-108.54304	12.12	0.0901	0.0591		-108.50507	12.05	0.0895	0.0551
-108.38476	16.43	0.0647	0.0812		-108.33978	16.55	0.0647	0.0854

Table 4. The lowest R-matrix poles ΔE_i (in eV) and corresponding amplitudes in SEP approach for the lowest scattering states of methane in the considered A_1, E and T_2 symmetry. E_i^0 denotes the poles of the free-particle problem.

A_1		E		T_2			s	p	d
ΔE_i	$\omega_{100,i}$	ΔE_i	$\omega_{120,i}$	ΔE_i	$\omega_{110,i}$	$\omega_{122,i}$		E_i^0	
0.3363	-0.3293						0.3357		
				1.0900	-0.3785	-0.0030		1.0242	
		1.9931	0.4024	1.9568	-0.0014	0.3957			2.0379
3.9077	0.3494						3.02123		
				5.8301	-0.3441	0.0062		5.0904	
		6.9464	0.3179	6.1495	-0.0024	-0.2859			7.5373
10.9766	0.3234						8.3925		
				13.7675	-0.3242	0.0613		11.8093	
		13.8163	0.3116	11.5543	0.0409	0.3040			15.6147
15.6294	0.1274						16.4492		

section from Lengsfield et al[32] are included. Fig. 7 shows the same cross sections on a larger energy scale.

Taking into account that the calculations are based on the fixed nuclei approximation we find the agreement between the calculated integral cross sections and the experimental results very satisfactory. The threshold behavior and the position of the Ramsauer minimum, as well as the broad structure around 8 eV provided by the $^2T_2$ component are well reproduced. However, some discrepancy between the calculated and measured results are obvious. The calculated cross section in the minumum is much smaller than the observed value and the maximum of the calculated cross section is shifted to higher energies. In the energy region between 4 and 7 eV the calculated values are clearly smaller than the measured ones.

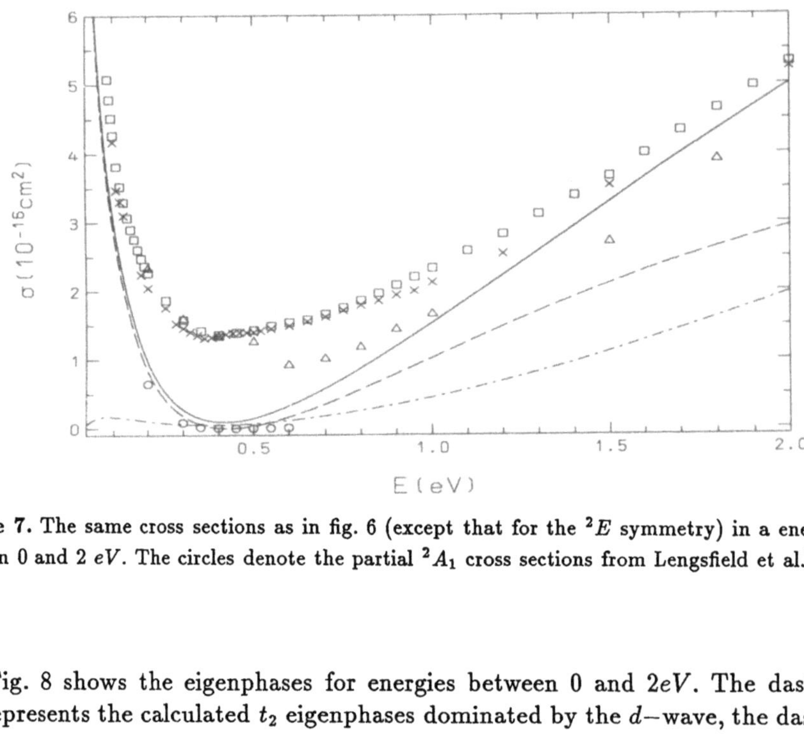

Figure 7. The same cross sections as in fig. 6 (except that for the ^2E symmetry) in a energy range between 0 and 2 eV. The circles denote the partial $^2A_1$ cross sections from Lengsfield et al.[32]

Fig. 8 shows the eigenphases for energies between 0 and $2eV$. The dash-dotted line represents the calculated t_2 eigenphases dominated by the $d-$wave, the dashed line represents the p component of t_2, the dotted line the e and the full line corresponds to the a_1 component. The crosses, squares and triangles are the experimental values for the s-, p- and d- scattering phases respectively, given by Sohn et al.[23] They were obtained by fitting to the measured values an expression for differential cross section derived for electron-atom scattering.[34–36] In this formula the scattering phases for $s-$, $p-$ and $d-$waves occur explicitly, while the phases for higher angular momenta are included by applying the Born approximation to the dipole polarizability of the target.

The agreement between theoretical and experimental values is surprisingly good for the s and the d phases in the energy region considered. The decreasing behavior of the calculated p phase for increasing energy values is also confirmed by the experimental values. The difference in behavior of the $t_2 - d$ and the e phases indicates that CH_4 does not correspond to a truly spherical system.

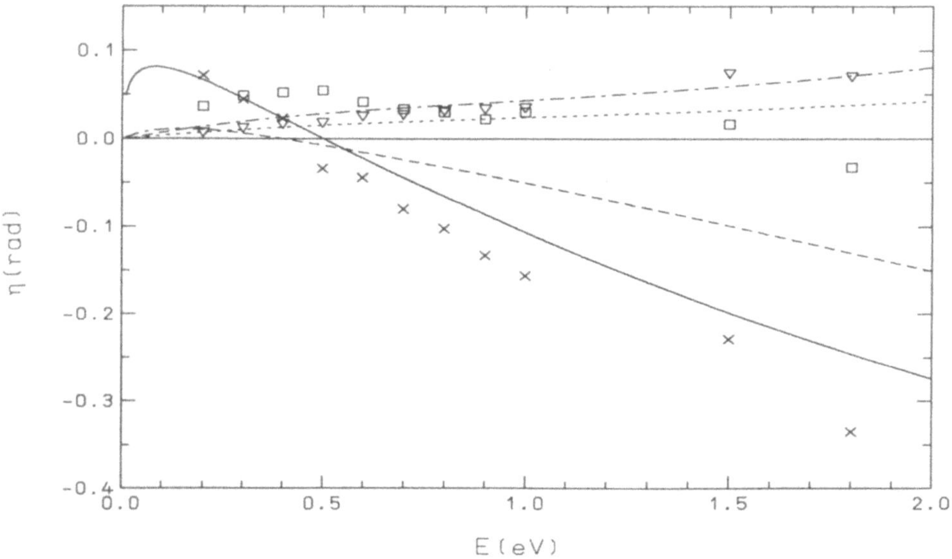

Figure 8. Calculated eigenphases for $a_1 - s$ (full line), $t_2 - p$ (dashed), $t_2 - d$ (dash-dotted), $e - d$ (dotted). Corresponding experimental values from Sohn et alSohn:1986 for s (crosses), p (squares) and d (triangles)

5. CONCLUSION

Within this section possible ways of increasing the efficiency of the R-matrix calculations are proposed. Our aim has been to apply the R-matrix approach to electron scattering processes for larger, especially polyatomic molecules. Therefore, techniques were employed which deviate from the standard approach for linear molecules. These techniques can be characterized by the following three points:

- Exclusively Gaussian-type functions are used in the one-particle basis set representing the molecular orbitals of the target system as well as the continuum orbitals describing the scattered electron.

- The integrals corresponding to the potential terms of the Hamiltonian are confined to the inside of a sphere typical in R-matrix calculations by subtracting their contributions outside the sphere from their values in the entire space using a multipole expansion for the potential. The one-electron integrals corresponding to the kinetic energy and the overlap matrix elements are calculated in a finite sphere Ω.

- In the spectral representation of the R-matrix the CI calculations are restricted to a small number of selected states. Energetically higher-lying roots are approximated very effectively by roots from a static exchange calculation.

It was shown for electron scattering of N_2 that the approximations described above do not affect the accuracy of the results. For both scattering symmetries regarded here, the resonant ($^2\Pi_g$) and non-resonant ($^2\Sigma_g^+$) symmetry, good accordance was found between the *SSRM* results, the *STO*-Code data and data from the literature.

Electron scattering of methane was considered as an application of the *SSRM* method to a polyatomic molecule. The observed broad structure in the integral cross sections with a maximum at about $8eV$, as well as the Ramsauer minimum for small collision energies could be reproduced by the present calculation. In general our results are in good agreement with experiments and recent theoretical results from other authors. Due to the importance of vibrational inelastic processes we expect a noticeable correction of the theoretical results if nonadiabatical effects are taken into consideration.

6. APPENDIX

6.1. Basis sets employed in the *SSRM* N_2 calculations

A (9s ,5p) Gaussian-type basis set for nitrogen in a [5s, 3p] contraction given by Dunning[37] augmented by one d-type function ($\alpha = 1.8846$) was used to describe the valence-type orbitals.

For the calculation in $^2\Pi_g$ symmetry the continuum orbitals are described by the seven d-type functions given in table 1. For the $^2\Sigma_g^+$ symmetry in addition the last five s-type functions of this table are employed.

The orthonormal continuum orbital basis set is obtained from the continuum basis functions by orthogonalizing them with respect to the valence basis followed by a subsequent mutual orthogonalization by diagonalizing the overlap matrix.

6.2. Basis sets employed in the *STO*-Code N_2 calculations

A Slater-type (*STO*) basis set of Cade et al[38] for nitrogen was used to represent the molecular orbitals, omitting the 4f function. This basis set contains five s-, three p- and three d-type Slater functions for the σ *MO*'s and three p- and two d-type *STO's* for the π *MO*'s.

The continuum orbitals, added to the Slater-type basis set, were represented by spherical Bessel functions with a fixed boundary condition at the R-matrix radius as well as by some functions not obeying this boundary condition, both located at the center-of-mass. The latter were numerical asymptotic functions (*NAF* 's) generated using a procedure of Nesbet et al.[39] chosen to improve convergence inside the R-matrix radius. In the present calculations for the $^2\Pi_g$ symmetry four d-type Bessel functions and one d-type *NAF* were employed to generate continuum *MO*'s of π symmetry. In the calculations for the $^2\Sigma_g^+$ symmetry we used two *NAF*'s and four Bessel functions of both s- and d-type to represent the continuum *MO*'s of σ symmetry. The continuum orbitals were Schmidt-orthogonalized to the *SCF MO*'s.

6.3. Other specifications in the N_2 calculations

For the internuclear distance R_0 a value of 2.068 a_0 is assumed. In the asymptotic region a quadrupole moment of 15.85096 au and an isotropic and anisotropic polarizability of 11.426 au and 3.359 au were assumed.

6.4. The CI Code employed

The CI calculations of the scattering states are based on the multireference single and double excitation (MRD) CI method.[40-42] In all cases the R-matrix radius a_e has been chosen to be 10 a_0.

6.5. Basis sets employed in the *SSRM* CH_4 calculations

For describing the valence type orbitals we modified the basis set given by Lengsfield et al.[32] For the calculations in $^2A_1$ symmetry two cartesian d-polarization function with exponents 0.75 and 0.09 are added to the carbon basis, in all other cases only one d function with exponent 0.75 is added. The hydrogen basis is left unchanged. In this manner we obtained a (9,6,2)/[5,4,2] and a (9,6,1)/[5,4,1] carbon basis for $^2A_1$ symmetry and the remaining symmetries, respectively, and a (5,1)/[3,1] basis for hydrogen. In order to avoid numerical linear dependences with respect to the continuum-type functions given above we have to remove the most diffuse s functions of the carbon atom.

The exponents of the Gaussian-type functions describing the continuum orbitals are taken from table 1. For the $^2A_1$ scattering states only the s-continuum functions and for the ^2E states only the d functions are included in the basis set. Calculations in $^2T_2$ symmetry require p as well as d continuum functions for representing the t_2 continuum orbitals.

The orientation of the molecule is defined by the positions of the four H atoms: (a,a,a), (-a,-a,a), (a,-a,-a), (-a,a,a), a=1.1847 a_0

References

[1] Nestmann B M, Nesbet R K and Peyerimhoff S D, J. Phys. B: At. Mol. Phys., **24**, 5133 (1991).

[2] Pfingst K, Nestmann B M and Peyerimhoff S, J. Phys. B: At. Mol. Opt. Phys., in press.

[3] Nestmann B M, Pfingst K and Peyerimhoff S D, J. Phys. B: At. Mol. Opt. Phys., in press.

[4] Kaufmann K, Baumeister W and Jungen M, J. Phys. B: At. Mol. Opt. Phys., **22**, 2223 (1989).

[5] Nestmann B M and Peyerimhoff S D, J. Phys. B: At. Mol. Phys., **23**, L773 (1990).

[6] Scheerbaum R R, Shakin C M and Thaler R M, Ann. Phys. NY, **76**, 333 (1973).

[7] Nestmann B M, Krumbach V and Peyerimhoff S D, Phys. Rev. A, **42**, 5406 (1990).

[8] Morrison M A, Saha B C and Gibson T L, Phys. Rev. A, **36**, 3682 (1987).

[9] Gillan C J, Noble C J and Burke P G, J. Phys. B: At. Mol. Opt. Phys. **21**, L53 (1988).

[10] Meyer H D, Phys. Rev. A, **34**, 1797 (1986).

[11] Berman M and Domcke W, Phys. Rev. A, **29**, 2485 (1984).

[12] Burke P G, Noble C J and Salvini S, J. Phys. B: At. Mol. Phys., **16**, L113 (1983).

[13] Schneider B I and Collins L A, Phys. Rev. A, **30**, 95 (1984).

[14] Brode R B, Phys. Rev., **25**, 636 (1925).

[15] Brüche E, Ann. Phys. Lpz, **83**, 1065 (1927).

[16] Brüche E, Ann. Phys. Lpz, **4**, 387 (1930).

[17] Ramsauer C and Collath R, Ann. Phys. Lpz, **4**, 91 (1930).

[18] Barbarito E, Basta M and Caliechio M, J. Chem. Phys., **71**, 54 (1979).

[19] Tanaka H, Okada T, Boesten L, Suzuki T, Yamanoto T and Kubo M, J. Phys. B: At. Mol. Phys, **15**, 3305 (1982).

[20] Ferch J, Granitza B and Raith W, J. Phys. B: At. Mol. Phys., **18**, L445 (1985).

[21] Jones R K, J. Chem. Phys., **82**, 5424 (1985).

[22] Lohmann B and Buckman S J, J. Phys. B: At. Mol. Phys., **19**, 2565 (1986).

[23] Sohn W, Kochem K-H, Scheuerlein K-M, Jung K and Ehrhardt H, J.Phys. B: At. Mol. Phys., **19**, 3625 (1986).

[24] Jain A, Phys. Rev. A, **34**, 954 (1986).
[25] Gianturco F A and Scialla S, J. Phys. B: At. Mol. Phys., **20**, 3171 (1987).
[26] O'Connel J K and Lane N F, Phys. Rev. A, **37**, 1893 (1983).
[27] Padial N T and Norcross D W, Phys. Rev. A, **29**, 1742 (1984).
[28] Gianturco F A, Jain A and Pantano L C, J. Phys. B: At. Mol. Phys., **20**, 571 (1986).
[29] Lima M A P, Gibson T L, Huo W M and McKoy V, Phys. Rev. A, **32**, 2696 (1985).
[30] Lima M A P, Watari K and McKoy V, Phys. Rev. A, **39**, 4312 (1989).
[31] McCurdy C W and Rescigno T N, Phys. Rev. A, **39**, 4487 (1989).
[32] Lengsfield III B H, Rescigno T N and McCurdy C W, Phys. Rev. A, **44**, 4296 (1991).
[33] Landolt Börnstein, Zahlenwerte und Funktionen (Berlin: Springer Verlag) 6. Auflage, vol 1, part 3, p 511 (1951).
[34] O'Malley T F, Spruch L and Rosenberg L, J. Math. Phys., **2**, 491 (1961).
[35] O'Malley T F, Phys. Rev., **130**, 1020 (1963).
[36] Thompson D G, Proc. R. Soc. London, **294**, 160 (1966).
[37] Dunning T H, J. Chem. Phys., **53**, 2823 (1970).
[38] Cade P E, Sales K D, Wahl A C, J. Chem. Phys., **44**, 1973 (1966).
[39] Nesbet R K, Noble C J, Morgan L A, and Weatherford C A, J. Phys. B: At. Mol. Phys., **17**, L891 (1984).
[40] Buenker R J and Peyerimhoff S D, Theor. Chim. Acta, **35**, 33 (1974).
[41] Buenker R J and Peyerimhoff S D, Theor. Chim. Acta, **39**, 217 (1975).
[42] Buenker R J and Peyerimhoff S D, New Horizons in Quantum Chemistry ed P O Löwdin and B Pullman (Dortrecht: Reidel) p 183 (1983).

R-MATRIX TECHNIQUES FOR INTERMEDIATE ENERGY SCATTERING AND PHOTOIONIZATION

C.J. Noble

DRAL Daresbury Laboratory
Warrington WA4 4AD, UK

1. INTRODUCTION

One of the most interesting challenges facing theories of electron-molecule collisions is to provide accurate descriptions of collisions at intermediate energies where large numbers of electronic states of the target are strongly coupled and where there are significant resonant processes. In this region there are an infinite number of open channels including ionizing channels. In order to obtain accurate representations of the target wavefunctions it may be necessary to use large configuration interaction (CI) expansions. In view of these difficulties it is perhaps not surprising that so far there have been few investigations of the topic. In this chapter the progress which has been made to apply R-matrix theory to intermediate energy scattering will be reviewed. Some ideas concerning possible future work are also mentioned.

When a low- or intermediate-energy electron strikes an atomic or molecular target resonant effects in which the incident electron is trapped in the vicinity of the target occur frequently and may dominate the collision process. The intermediate energy scattering range may be regarded as extending from the around the first ionization threshold to several times the ionization energy. In this region resonances are typically broader and more likely to overlap than in the low-energy region but may still play important roles in the collision process. Coupling between the channels may be important so techniques such as the Born and Distorted Wave Born approximations, which are successful at higher scattering energies where the channel coupling is relatively weak, are not applicable. As an infinite number of asymptotic channels are open it is necessary to develop approximate techniques which are capable of representing the effect of the most strongly coupled channels.

A generalized R-matrix theory designed to treat electron-atom and electron-molecule scattering at intermediate energies was introduced by Burke et al.[1] This theory is summarized in the next section. In applications of the approach to electron-atom collisions the method has been termed the Intermediate Energy R-Matrix (IERM) method.

Computational Methods for Electron-Molecule Collisions
Edited by W.M. Huo and F.A. Gianturco, Plenum Press, New York, 1995

It now seems probable that this extended R-matrix approach will be more generally applicable and should not be viewed as a limited intermediate energy technique.

Various aspects of the implementation of the theory which are crucial for efficient large-scale calculations are discussed in the following sections. A method which has been developed recently to calculate scattering Hamiltonian matrices is described. Other issues, such as the need to maintain a balanced representation of correlation in the target and in the scattering systems and the appearance of pseudoresonances, are mentioned.

Detailed *ab-initio* calculations are required to interpret the very accurate data on molecular photoionization which is now available from measurements using lasers and synchrotron radiation sources. The application of R-matrix theory to low-energy molecular photoionization is discussed in the second part of the chapter. The theoretical formalism is outlined and examples presented. One of the interesting features of molecular photoionization spectra is the appearance of autoionization structure.[2] Information about the autoionizing resonances producing the structure may be obtained by studying electron scattering by molecular ions. A complex energy R-matrix approach has been developed which may be used, for example, to obtain the position and width of the autoionizing resonances observed in photoionization spectra. More generally the approach may be used to study the analytic properties of the calculated scattering matrix. Examples of results obtained for electron scattering by the oxygen molecular ion are presented. Atomic units are assumed throughout.

2. MOLECULAR SCATTERING AT INTERMEDIATE ENERGIES

2.1. Intermediate Energy R-Matrix Theory

The scattering of an electron by an $(N+2)$-electron molecule is described by the Schrödinger equation

$$H\Psi = E\Psi \tag{124}$$

where the nonrelativistic Hamiltonian H may be written

$$H = H_T^{N+1} + K_{\text{inc}} + V_{\text{int}} \tag{125}$$

and E is the total energy. In eq. (125) H_T is the Hamiltonian of the $(N+1)$-electron target, K_{inc} is the kinetic energy of the incident electron and V_{int} the interaction between the target and the incident electron. The solution of eq. (124) is simplified by assuming the target nuclei are fixed and by working in a body fixed reference frame. This amounts to neglecting rotational and vibrational effects, but the results of this fixed-nuclei approximation may be used in subsequent calculations to solve the complete scattering calculation using the techniques described in previous chapters. The essential simplification introduced in the R-matrix approach is to divide configuration space into two or more regions which can be treated separately and then combined by matching R-matrices (inverse logarithmic derivatives) on the boundaries of the regions. The interaction is largely confined to the innermost region which is chosen to encompass the region of space where the charge cloud of the target is appreciable. In this region the collision system corresponds to an excited molecular complex. As the region is finite it may be spanned by a denumerable set of basis states which may be constructed as a single orthonormal set partitioned into bound and continuum orbitals.

Consider first the solution of the scattering problem in the internal region. The generalized R-matrix theory relaxes the usual R-matrix restriction that only a single electron may occupy the continuum orbitals by allowing both the incident and one valence electron to occupy these orbitals. The internal region is taken as a hypersphere in configuration space defined by the radial coordinates of both the incident and one valence electron assuming the values $r_{N+1} = a$, $r_{N+2} = a$ for some constant a as illustrated in Figure 1. The scattering wavefunction is written as a close-coupling sum

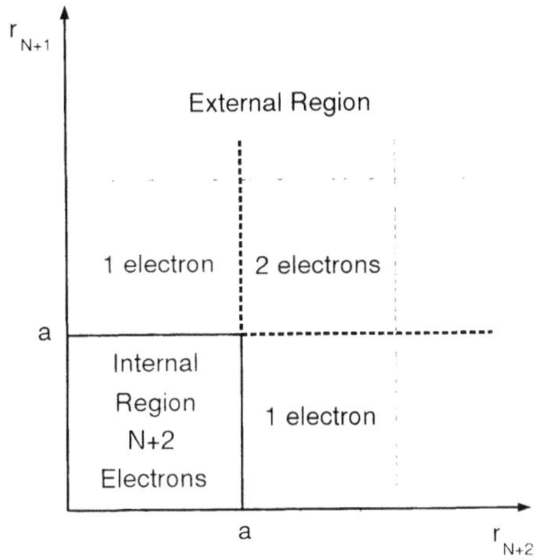

Figure 1. Division of configuration space into internal and external regions.

of the form

$$\Theta_k = \mathcal{A}\sum_{ij} \bar{\psi}_i(1 \ldots N+1; \hat{x}_{N+2}) u_j(r_{N+2}) \alpha_{ijk} + \sum_i \tilde{\chi}_i(1 \ldots N+2) \beta_{ik} \tag{126}$$

where the channel functions $\bar{\psi}_i$ are obtained by coupling the spin-angle functions of the incident electron (combined spatial and spin coordinates are denoted by $\boldsymbol{x} \equiv (\boldsymbol{r}, \sigma)$, carets indicate angular variables) to an $(N+1)$-electron R-Matrix state, ψ_k, of the form

$$\psi_k = \mathcal{A}\sum_{ij} \bar{\phi}_i(1 \ldots N; \hat{x}_{N+1}) u_j(r_{N+1}) a_{ijk} + \sum_i \chi_i(1 \ldots N+1) b_{ik}. \tag{127}$$

The channel functions $\bar{\phi}_i$ in eq. (127) represent an N-electron ion state, ϕ_i, satisfying the usual bound-state Schrödinger equation

$$H_T^N \phi_i(1 \ldots N) = E_i^T \phi(1 \ldots N) \tag{128}$$

coupled to the spin-angle functions of the $(N+1)$-th electron. The states of the ionic target, ϕ_i, are usually represented by CI wavefunctions expanded in terms of nuclear-centred single-particle basis functions. The various summations in eqs. (126) and (127) must, of course, be truncated in practice and the choice of terms to be included determines both the accuracy and the type of collisions which can be described. The functions u_j represent continuum basis functions and the operator $\mathcal{A}$ indicates that the continuum electrons are to be antisymmetrized with respect to the bound electrons as required by the Pauli Principle. We return later to a discussion of the $\mathcal{L}^2$ terms included in the second expansions in eqs. (126) and (127) which involve only nuclear centred orbitals.

The conclusion that the generalized R-matrix approach encapsulated in eqs. (126) and (127) should be generally applicable follows by recognizing that the decision to use states ψ_k in eq. (126) amounts to the introduction of a consistent set of pseudostates.[3]

Using these representations the Hamiltonian matrices for the $(N+1)$- and $(N+2)$-electron systems may be constructed. As the internal region is finite, the kinetic energy operators produce surface terms and the matrices are not Hermitian. Bloch[4] has shown that the symmetrized Hamiltonian formed by adding the operator

$$L_{N+2} = \frac{1}{2}\sum_{j=1}^{N+2}\sum_{i} |\bar{\psi}_i(r_j^{-1}) > \delta(r_j - a)\left(\frac{d}{dr_j} - \frac{b}{r_j}\right) < \bar{\psi}_i(r_j^{-1})| \tag{129}$$

to the Hamiltonian is Hermitian. In eq. (129) the channel functions are written

$$\bar{\psi}_i(r_j^{-1}) \equiv \bar{\psi}_i(1\ldots j-1, j+1\ldots N+2; \hat{x}_j). \tag{130}$$

and b corresponds to a constant but arbitrary logarithmic derivative on the boundary (usually $b = 0$). Implicit in the choice of the boundary radius a defining the internal region is the assumption that all basis functions located on nuclear centres which are used to expand the single-particle orbital set (and consequently to expand both the states ϕ_i and the $\mathcal{L}^2$ terms χ and $\tilde{\chi}$ in eqs. (126) to (128)) have become negligible on the boundary. Continuum basis functions located on the centre of gravity of the target must remain finite on the boundary in order that the scattering problems in the internal and external regions may be combined. As a result of the orthogonalization necessary to obtain a single orthonormal set of molecular orbitals, the continuum orbitals will involve both nuclear-centred and continuum basis functions. This construction facilitates the change of symmetry required in moving from the symmetry of the molecule in the internal region to the spherical symmetry of the external region at the boundary.

The R-matrix, which connects the internal and external regions, is defined by

$$f_i(a) = \sum_j R_{ij}^{N+2}\left(r\frac{df_j}{dr} - bf_j\right)_{r=a} \tag{131}$$

where the reduced radial wavefunction f_i is defined by the overlap

$$f_i(r_{N+2}) \equiv < \bar{\psi}_i(r_{N+2}^{-1})|\Theta > . \tag{132}$$

The conventional choice has been to diagonalize the symmetrized internal region Hamiltonian to obtain the constants α_{ijk}, β_{ik} appearing in eq. (126) and then to calculate the spectrally resolved form of the R-matrix.[5] The R-matrix at different scattering energies is immediately available and only the calculation in the external region needs to be repeated at each energy. This choice is preferable whenever the number of scattering energies is small compared to the dimension of the Hamiltonian matrix. For scattering at intermediate energies or other very large cases this option needs to be reassessed as it is likely that linear algebraic methods will be more efficient.[6] In this case the R-matrix is given in terms of the symmetrized Hamiltonian matrix obtained in the representation of $(N+2)$-electron basis states θ_k defined in eq. (126) by the relation

$$\mathbf{R} = \mathbf{W}(\mathbf{H}_{N+2} + \mathbf{L}_{N+2} - \mathbf{1}E)^{-1}\mathbf{W}^T \tag{133}$$

where the elements of the reduced-width amplitude matrix $\mathbf{W}$ are given by

$$W_{ik} = \left\langle \bar{\psi}_i(r_{N+2}^{-1}|\theta_k\right\rangle\Big|_{r_{N+2}=a} . \tag{134}$$

The truncation of the close-coupling sums represented by the first expansions in eqs. (126) and (127) effectively determines the scattering model. The second summations in each of these equations are primarily included to account for terms in which the incident electron occupies unoccupied or partially filled orbitals of the same symmetry as the symmetry of the orbitals u_j. These "orthogonality" terms are essential to ensure the completeness of the many-electron wavefunctions. It is customary, however, to augment the orthogonality contributions with additional terms which are supposed to compensate for the truncation of the close-coupling sums and to account for charge polarization effects. Using the Feshbach formalism[7] it can be shown these terms act as an optical potential. It is well known that these additional terms must be chosen carefully to avoid inconsistent treatments of correlation effects in the target and the scattering complex. The terms also give rise to unphysical pseudoresonances.[1] If the extended R-matrix procedure is implemented as an expansion in terms of pseudostates and only orthogonality terms are retained in addition to the close-coupling sums these ambiguities are reduced and may be eliminated. There will be pseudothresholds associated with each pseudostate. However, Bray and Stelbovics[8] have demonstrated that as the number of pseudostates is increased the effect of the pseudothresholds decreases and eventually disappears. The close-coupling expansion is known to be slowly convergent so in this ansatz it is necessary to treat long close-coupling expansions.

In following sections it will be shown that it is possible to devise computationally efficient procedures taking advantage of the particular form of the scattering expansion. These considerations suggest that in many cases it may be preferable to either forego, or minimize the additional terms included in the "correlation" expansions.

2.2. Computational Procedures

The potential energy diagrams of molecules show that in the intermediate scattering region almost inevitably there will be large numbers of electronic states which are strongly coupled in a given process. It is also invariably the case that accurate representations of target eigenstates by means of CI wavefunctions require large numbers of configuration state functions. It therefore appears that the typical size of an accurate intermediate energy calculation will be formidable even leaving aside the considerations discussed at the end of the last section. In the following sections several methods which may improve efficiency and permit these large-scale computations to be performed are discussed.

Single Particle Basis. The construction of a single orthonormal basis of molecular orbitals capable of providing a compact and accurate representation of both a large set of target eigenstates and the scattering wavefunction is one of the most delicate problems in the R-matrix approach. The orbital set includes bound and virtual orbitals which are used to represent the wavefunctions of the N-electron states, ϕ_i. In present implementations these are expanded in terms of Slater functions located on the nuclear centres. These orbitals may in principle be constructed using any one of a number of standard bound state techniques but in practice MCSCF (multiconfiguration self-consistent field) and natural orbital (NO) methods produce virtual orbitals which are less diffuse and give superior target representations. This is illustrated in Table 1 by comparing excitation energies for electronic states of O_2 calculated by valence CI calculations using SCF and NO orbitals.

Table 1. Excitation energies of O_2 electronic states ($R = 2.3a_0$). Comparison of valence CI results using SCF and natural orbitals (NO). Energies in eV.

State	SCF	NO	Estimated[9]
$X^3\Sigma_g^-$	0.0	0.0	0.0
$a^1\Delta_g$	0.93	0.97	0.98
$b^1\Sigma_g^+$	1.44	1.50	1.65
$c^1\Sigma_u^-$	5.54	5.80	6.12
$A^3\Sigma_u^+$	5.86	6.15	6.47
$B^3\Sigma_u^-$	10.90	9.99	9.25

As the calculation of integrals is not computationally expensive the expansion of the continuum orbitals in terms of numerically defined basis functions is preferred because of the additional flexibility they provide with relatively few functions. This is especially useful in the intermediate energy regime where the continuum orbitals may oscillate significantly within the internal region.

Numerically defined basis functions are obtained by solving a model single channel scattering problem

$$\left(\frac{d^2}{dr^2} - \frac{l(l+1)}{r^2} + V(r) + k_i^2\right) u_i(r) = 0 \tag{135}$$

for a spherically symmetric potential $V(r)$ which is usually chosen as the static potential of the target. A discrete set of solutions which is complete for energies less than $k_M^2/2$ is generated for $k_i, i = 1, M$ by enforcing the boundary conditions

$$u_i(0) = 0 \tag{136}$$

at the origin and

$$\left.\frac{a}{u_i}\frac{du_i}{dr}\right|_{r=a} = b_i \tag{137}$$

at the boundary $r = a$. The standard practice has been to choose fixed boundary conditions by taking $b_i = 0, i = 1, \ldots, M$. However Shimamura[10] and Nesbet[11] have used simple models to demonstrate the faster convergence of functions corresponding to arbitrary boundary conditions in which some b_i are chosen nonzero. Faster convergence was also seen in arbitrary boundary condition calculations using Slater basis functions.[12] Studies are underway to verify that similar improvements are realized using a numerically defined arbitrary boundary condition continuum basis constructed by using suitably chosen logarithmic parameters, b_i.

The bound/virtual and continuum sets must be orthogonalized by Schmidt or Lagrange procedures to form a single orthonormal set. The multicentre character of the resultant orbital set ensures that only those continuum orbital angular momenta required asymptotically must be included.

Hamiltonian Construction. The most computationally expensive phase of R-matrix scattering calculations is the construction of the Hamiltonian matrix. This computation determines the range of scattering models which may be treated and therefore will be discussed in more detail. To avoid nonessential complications the usual low-energy R-matrix approach for diatomic targets with a single electron in the continuum is considered in this section.

The current molecular R-matrix scattering programs are based on a bound state configuration interaction method in which the Hamiltonian is generated by first constructing a symbolic representation of the Hamiltonian matrix elements as a set of lists of the form $P, Q, N, \{C_i, L_i, i = 1, N\}$ where

$$H_{P,Q} = \sum_{i=1}^{N} C_i^{P,Q} V(L_i^{PQ}) \tag{138}$$

Each label L_i^{PQ} consists of four orbital sequence numbers needed to uniquely identify an arbitrary one- or two-electron matrix element. The coefficients C and integral labels L are both generated for each Hamiltonian matrix element $H_{P,Q}$. Later in the calculation the lists are combined with calculated matrix elements V, using eq. (138), to form the Hamiltonian matrix.

In general the disk storage required to hold the coefficient lists imposes severe limits on the use of the "formula tape" approach and alternative direct approaches are favoured in large bound-state calculations. However, it is important to recognise that the special structure of scattering calculations imposed by the close-coupling expansion, provides opportunities to use the simple formula-tape approach for relatively large-scale calculations. Two factors are involved. First, from eq. (125), it follows that the target eigenenergies, E_T, appear as additive terms in the diagonal continuum-continuum elements of the scattering Hamiltonian matrix. These contributions need not be evaluated while constructing the scattering Hamiltonian but may be calculated independently and simply added in at a later stage. The burden of calculating the remaining contributions to the continuum-continuum elements is greatly reduced particularly for large target expansions. An analogous technique is proving very successful in atomic R-matrix calculations[13] so the computational savings in molecular calculations are potentially very large.

The second factor is to recognize that continuum-continuum Hamiltonian matrix elements may be constructed from a relatively small set of symbolic formulae which are used as template elements. Templates are constructed by assuming each target is coupled to a continuum electron which is restricted to occupy a single continuum orbital. Elements of the symbolic list corresponding to target contributions may be eliminated during the calculation of template formulae.

The templates are employed with lists of the actual continuum orbital sequence numbers which are required to rapidly create the symbolic formulae needed to form the continuum-continuum elements of the Hamiltonian eq. (138). A computer program implementing this scheme has been written and shown to result in large savings of both disk storage and processor time thereby extending the range of feasible calculations. A more general program for the extended R-matrix approach is under development. There is clearly a great deal of scope for additional improvements by extending these ideas. Implementations appropriate for use on distributed memory parallel computers follow naturally from this approach.

Two important points emerge from this analysis. First, that unmodified bound state techniques for the Hamiltonian construction become increasing inefficient for scattering Hamiltonians involving large CI target expansions. The effort spent in developing scattering programs should be directed at using the close-coupling form and separating out as far as possible the bound-state aspects of the calculation. An analysis along these lines using the graphical methods[14] which are so successful in bound-state calculations may be expected to result in similar advances in scattering techniques.

Second, the template approach outlined above cannot in general be applied to bound-continuum and bound-bound elements of the Hamiltonian. The set of orthogonality terms described previously provides the exception to this statement as these configurations are defined to consist of an N-electron target state coupled to an electron occupying a bound or virtual orbital. If close-coupling and orthogonality configurations are used very efficient procedures are available; if additional optical potential terms dominate these possibilities disappear.

As an adjunct of this second point it is worth noting that scattering calculations are frequently corrected for defects in the representation of the target states by shifting the diagonal Hamiltonian matrix elements into agreement with separate large-scale calculations or with experiment. This procedure is subject to ambiguities relating to the choice of zero energy. The origin of this problem may be related to the inclusion of additional $\mathcal{L}^2$ terms in the expansion of the scattering wavefunction which are not included in the shifting procedure.

External Region Calculations. The R-matrices resulting from the internal region calculation provide the boundary conditions for the solution of the scattering equations in the external region. In the extended R-matrix approach both the incident and one of the valence electrons may be in the external region. The radial coordinates of these electrons then satisfy the relation $r \geq a$ as illustrated in Figure 1. Conversely, if only the incident electron is allowed into this region, the extended R-matrix approach reduces to the usual external region scattering problem described in previous chapters. As the scattering electron is assumed to be distinguishable from the target electrons, exchange is neglected and the scattering equations become a set of coupled second-order ordinary differential equations. Although the number of coupled equations may become large there now exist efficient program packages for obtaining the solution.[15]

The case where two electrons are allowed into the external region may be treated using R-matrix propagation techniques by subdividing the region into subregions as shown in Figure 1. Calculations are carried out in each subregion to determine the R-matrices on the outer edges of the block given the known values on the inner edges.

The most complicated case arises in the diagonal external region blocks where the valence and incident electrons interact with each other and with the residual ionic core. Although each electron in the external region is distinguishable from the core electrons, exchange effects between these two electrons must be taken into account. An orthonormal basis is introduced (typically shifted Legendre functions) and a two-electron R-matrix calculation carried out. A two-dimensional propagator program implementing this algorithm for s-wave scattering is available[16] and more general programs are under development.

The option of taking two electrons in the external region and employing a two-dimensional propagator opens up several important possibilities. First, Rydberg states of the target molecule may be represented without the necessity of extending the size of the internal region thereby increasing the dimension of the internal region Hamiltonian matrix. Second, it may lead to improved descriptions of electron impact ionization. Neither of these possibilities have been explored in detail.

Whichever choice, one or two electrons in the external region, the R-matrices are propagated to distances where it is possible to match to the asymptotic form of the scattering wavefunction and obtain the scattering K-matrix. This matrix may be used to calculate the required cross sections.

The appearance of pseudoresonances in the calculated eigenphase sum was mentioned earlier and methods suggested for possibly reducing or eliminating these spurious effects. Conventional calculations result in large numbers of narrow densely packed pseudoresonances.[1] Their origin has been discussed by Scholz[17] and approaches for averaging the pseudoresonance region of the calculated spectrum to extract useful physical information described.[17-18]

3. MOLECULAR PHOTOIONIZATION

Both lasers and synchrotron radiation sources have been used in recent years to carry out precise and detailed measurements of molecular photoionization and photodissociation processes. For example, vibrationally resolved angular distributions have been obtained at resolutions of a few millivolts over wide energy ranges for a variety of molecules using synchrotron radiation.[19] This data serves as an important check on theoretical models. However, the data also exhibits a variety of resonance phenomena and features which arise from an interplay between the nuclear and electronic degrees of freedom. Because of the complexity of the processes involved, accurate *ab-initio* calculations are required to provide an unambiguous interpretation of the measurements.

General algebraic expressions for molecular photoionization cross sections have been derived for a variety of experimental conditions[20] including situations where the spin of the outgoing electron is observed[21] and for oriented target molecules.[22] Molecular photoionization processes have also been treated using R-matrix theory by Burke.[23] Here only the briefest outline of the formalism is presented as an indication of how the R-matrix techniques described earlier in this volume may be applied to photoionization and photodissociation. For simplicity, it will be assumed that any rotational structure is unresolved and that the target is a diatomic described using an LS coupling scheme. Only single ionization processes will be considered and quantum number labels which are not central to the discussion will be suppressed.

When a photon with energy $h\nu$ exceeding the photoionization threshold strikes an $(N+1)$-electron diatomic molecule AB several processes may occur. The electron may be ejected with orbital angular momentum l leaving a molecular ion in vibronic state $j\nu$ (in general both electronically and vibrationally excited),

$$h\nu + AB(\nu_0) \rightarrow AB^+(j\nu) + e^-, \tag{139}$$

the target may dissociate leading to photodissociation

$$h\nu + AB(\nu_0) \rightarrow A^* + B^* \tag{140}$$

or to dissociative photoionization

$$h\nu + AB(\nu_0) \rightarrow A^* + B^{*+} + e^-(l), \tag{141}$$

(atomic fragments, A and B may be in excited states). Dissociative photoionization will not be considered further here.

The differential cross section for the photoionization of a molecule initially in state $\Psi_0\nu_0$ (vibrational state ν_0) in which the electron is ejected in a direction $\hat{\underline{k}}'$ leaving the ion in state $\Psi_{Ej\nu}$ (vibrational state ν) is given in the dipole length approximation by

$$\frac{d\sigma_{j\nu}}{d\Omega'} = 4\pi^2\alpha\omega\left|\left\langle\Psi^{(-)}_{Ej\nu}(\hat{k}')\left|\hat{\underline{\epsilon}}.\underline{D}\right|\Psi_{0\nu_0}\right\rangle\right|^2. \tag{142}$$

The superscript $(-)$ indicates that the boundary conditions for the final state correspond to incoming spherical waves in all channels and an outgoing spherical wave in channel $j\nu$. The energy of the incident photon is given by ω and α is the fine structure constant. The polarization vector of the photon is given by $\hat{\underline{\epsilon}}$ and the dipole-length operator $\underline{D}$ is given in spherical coordinates by

$$D_\mu = \left(\frac{4\pi}{3}\right)^{\frac{1}{2}}\sum_{i=1}^{N+1} r_i Y_{1\mu}(\hat{\underline{r}}_i') = \left(\frac{4\pi}{3}\right)^{\frac{1}{2}}\sum_{i=1}^{N+1} r_i \sum_{m_\gamma} D^{1*}_{\mu m_\gamma}(\alpha\beta\gamma)Y_{1m_\gamma}(\hat{\underline{r}}_i) \tag{143}$$

where $D^1_{\mu m_\gamma}$ is a rotation matrix,[24] Y_{1m_γ} a spherical harmonic and $\alpha\beta\gamma$ the Euler angles relating the laboratory (primed) and body-fixed (unprimed) coordinates. The component $\mu = 0$ corresponds to linearly polarized light, $\mu = \pm 1$ to circular polarization.

For a gas target of randomly orientated molecules it is necessary to average over the molecular orientations and the differential cross section eq. (142) reduces to the Yang[25] result

$$\frac{d\sigma_{j\nu}}{d\Omega'} = \frac{\sigma_{j\nu}}{4\pi}\left[1 + \beta_{j\nu}P_2(\cos\theta')\right] \tag{144}$$

where $\sigma_{j\nu}$ is the total photoionization cross section, P_2 is a Legendre polynomial and θ' the angle between the incident photon polarization vector and the momentum vector of the outgoing electron.

The asymmetry parameter $\beta_{j\nu}$ may be expressed in terms of the angular momentum transfer between the incident photon and the ejected electron, j_t, as

$$\beta_{j\nu} = 16\pi^3\alpha\omega\sigma^{-1}_{j\nu}\sum_{j_tLL'm_\alpha}(2j_t+1)^{-1}i^{L-L'}e^{i(\sigma_{L'}-\sigma_L)}D^{(-)}_{j\nu L'j_tm_\alpha}D^{(-)*}_{j\nu Lj_tm_\alpha}\Theta(j_tLL') \tag{145}$$

Here σ_l is the Coulomb phase shift and Θ a factor which depends only on the kinematics of the collision. The total cross section for ionization to state $j\nu$ is given by

$$\sigma_{j\nu} = \frac{4}{3}\pi^2\alpha\omega\sum_{Lj_tm_\alpha}\left|D^{(-)}_{j\nu Lj_tm_\alpha}\right|^2. \tag{146}$$

It is seen from eqs. (145) and (146) that the dynamics of the collision are contained in the complex-valued dipole matrix elements $D^{(-)}$. R-matrix calculations are carried out to determine these factors. By taking account of the fact that R-matrix calculations are carried out assuming standing wave boundary conditions, the complex-valued dipole matrix elements may be expressed as linear combinations of real-valued R-matrix dipoles given by

$$D_{j\nu LM_Lm_\gamma} = \left(\frac{4\pi}{3}\right)^{\frac{1}{2}}\left\langle\Psi_{j\nu LM_L}\left|\sum_{i=1}^{N+1} r_iY_{1m_\gamma}(\hat{r}_i)\right|\Psi_{0\nu_0}\right\rangle. \tag{147}$$

The initial normalized bound state $\Psi_{0\nu_o}$ and the final energy normalized continuum state $\Psi_{j\nu}$ are expanded in terms of energy-independent internal region R-matrix states Ψ_k

$$\Psi_{0\nu_0} = \sum_k \Psi_k A_{k0\nu_0} \tag{148}$$

and

$$\Psi_{j\nu} = \sum_k \Psi_k A_{kj\nu}. \tag{149}$$

Both the internal region states Ψ_k and the expansion coefficients are evaluated by R-matrix calculations using the general nonadiabatic formalism[26] described in the chapter by Schneider in which the full Hamiltonian including the nuclear kinetic energy operators is diagonalized. This procedure should, at least in principle, include effects resulting from any coupling between the fixed-nuclei R-matrix states. The formalism takes full account of the coupling between photoionization and photodissociation but the photoionization calculation simplifies if this coupling is weak.

The first stage in the calculation is to obtain the expansion coefficients $A_{k0\nu_0}$ by means of an all-channels-closed scattering calculation. If all channels are closed the solution of the scattering equations amounts to iteratively solving an eigenvalue problem as solutions exist only at energies corresponding to the bound-state energies. The condition for the existence of an eigenstate is obtained by requiring the R-matrix relation eq. (131) be satisfied on the boundary of the internal region and reduces to finding the zeros of a determinant[27]. The possibility of obtaining the initial target state in this way is one of the most significant advantages of the R-matrix treatment of photoionization. Firstly the continuum orbitals in the expansion of the wavefunction are ideal for representing Rydberg or other diffuse states. Secondly both the initial and final states are calculated using an identical basis and the accuracy of the target and residual ion descriptions may be matched.

The wavefunction of the final state is obtained by calculations of electron scattering by the molecular ion. For a dipole interaction and a diatomic target a maximum of three continuum symmetries may contribute to the photoionization. This reduces the computation required to obtain cross sections compared to electron scattering cases. The inclusion of several states in the close-coupling R-matrix expansion of the scattering state generates Rydberg sequences of autoionizing resonances terminating on each excited state. The generalized R-matrices described in the chapter by Schneider represent the result of a full nonadiabatic treatment of the internal region scattering problem and provide the boundary conditions for the solution in the external region for both electron-molecule and atom-atom channels. These solutions define the expansion coefficients, A, of eqs. (148) and (149).

Given the expansion coefficients and transition dipoles between the internal region R-matrix states (calculated using a standard bound-state transition-moment program) the dipoles of eq. (147) may be calculated. Eqs. (145) and (146) are then evaluated to obtain the cross section and asymmetry parameter. Additional technical complications arise if the dipole matrix elements involve significant contributions from the external R-matrix region. The evaluation of the tail contributions and the normalization of the wavefunctions in this case may be performed using methods similar to those used for atomic photoionization.[28]

A general R-matrix computer program implementing the computational scheme outlined above has been written to treat the photoionization and photodissociation of diatomic targets. To date, however, there have been only a limited number of R-matrix photoionization calculations using the fixed-nuclei approximation for H_2,[29-30] N_2 and CO targets. Calculations presently underway for an O_2 target will include a study of photodissociation.

Figure 2 illustrates results for the photoionization of H_2 obtained by Tennyson *et al*[29] using fixed-nuclei R-matrix calculations at internuclear separations of $R = 1.3$ and

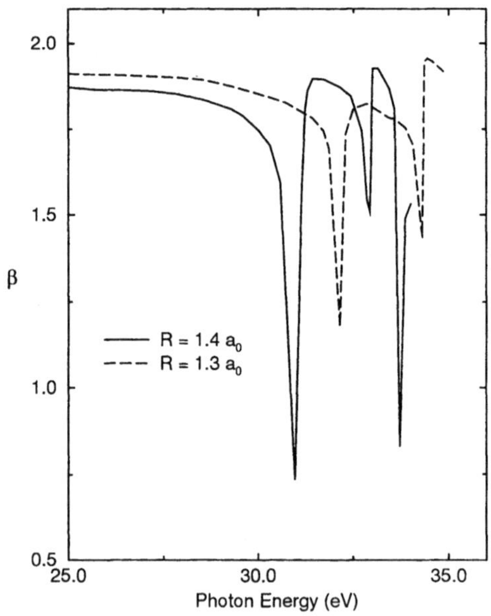

Figure 2. Asymmetry parameter for the photoionization of hydrogen calculated using the fixed-nuclei R-matrix method.[29]

$R = 1.4$ a.u. The calculated asymmetry parameter exhibits strong resonant structure for photon energies between 28 and 35 eV. The structure is expected to be much less pronounced when account is taken of the averaging effects introduced by the vibrational motion but some variation is in fact observed in experimental measurements over this energy range.[31] The resonant structure is a result of doubly excited states of H_2 which manifest as autoionizing resonances in electron scattering by H_2^+. The variation shown in the figure arises from the lowest members of a Rydberg sequence terminating on the $A^2\Sigma_u^+$ electronic state of H_2^+. Similar results have also been obtained by Raseev.[32] The autoionizing resonances in $e^- + H_2^+$ scattering have been the subject of large numbers of theoretical calculations[33] as they are known to play important roles in vibrational excitation[34] and dissociative recombination.[35] More subtle, but theoretically interesting effects, resulting from interference between the resonances have been predicted.[36]

It seems certain that the great success which atomic R-matrix photoionization calculations have enjoyed in recent years will be parallelled in the molecular case in the near future.

3.1. Resonances

As mentioned in the previous subsection a knowledge of the resonances occuring in a particular collision may be sufficient to determine the main features in the cross section for the process. In almost all case this resonance information will provide valuable insight. In this section the limitations in standard fitting procedures are described and alternative methods based on the use of complex-energy R-matrix theory described.

3.2. Resonance Fitting and Definition

Resonances in scattering systems are normally assumed to be a consequence of the incident particle becoming trapped in the interaction region and to result in rapid energy variation or other structure in the observed cross sections. Experimental determinations of resonance parameters therefore depend on fitting scattering data and typically rely on the validity of the Breit-Wigner approximation.[37] This approximation assumes the resonance widths are small compared to the separation between resonances. Correspondingly in theoretical calculations resonances are signalled by the eigenphase sum Δ, given by

$$\det \mathbf{S} = \exp(2i\Delta), \tag{150}$$

increasing rapidly by π radians.[38] The positions and widths of resonances may be determined by least squares fitting the eigenphase sum in a region containing N resonances to functional forms such as

$$\Delta(E) = \sum_{i=1}^{N} \tan^{-1}\left[\frac{\frac{1}{2}\Gamma_i}{(E_i - E)}\right] + \sum_{j=0}^{m} a_j E^j, \tag{151}$$

where the last term in eq. (151) represents a slowly varying background. Breit-Wigner methods are simple and convenient to apply in the case of narrow isolated resonances.[39] In the case of wide or overlapping resonances ambiguities in the fitting procedure and errors in the background determination may both become appreciable. Near thresholds or for wide resonances other functional forms are required. It is therefore useful to consider alternative methods for determining resonance parameters.

Siegert[40] has shown that resonances may be represented mathematically by poles in the scattering matrix $S(E)$ when the energy E is allowed to assume complex values. Physically significant resonances giving rise to structure in experimental cross sections correspond to poles which lie close to the real energy axis. The energy-independent resonance parameters (position, $E_{\rm res}$ and width, $\Gamma_{\rm res}$) are related to the energy of the pole, E_j, by

$$E_j = E^j_{\rm res} + i\frac{1}{2}\Gamma^j_{\rm res} \tag{152}$$

An alternative view has been taken by Fano[41] in which resonances arise when a residual interaction mixes discrete and continuum states of an approximate Hamiltonian of the scattering system. The resonance parameters in this formulation are energy-dependent and are related to the Siegert values at a specific energy which is determined self-consistently.[42]

In the Siegert approach the radial wavefunction for an n-channel electron-molecule collision satisfies the asymptotic form

$$\mathbf{F}(r) \underset{r\to\infty}{\sim} \mathbf{k}^{-\frac{1}{2}}(e^{-\mathrm{i}(\mathbf{k}r-\mathbf{l}\pi/2)} - e^{\mathrm{i}(\mathbf{k}r-\mathbf{l}\pi/2)}\mathbf{S}), \tag{153}$$

where the channel momenta, $\mathbf{k}$, angular momenta, $\mathbf{l}$, are diagonal matrices and the columns of $\mathbf{F}$ represent linearly independent solutions. The total scattering energy is E, E_i^T the target threshold energy in channel i and $k_i^2 = 2(E - E_i^T)$. For the case of two-body channels and short-range interactions the S-matrix is an analytic function of the channel momenta with square-root branch points at each scattering threshold. It may be treated as a single-valued function by introducing cuts from the thresholds and running along the positive energy axis. If the channel momenta are written

$$k_i = |k_i|\, e^{\frac{1}{2}\mathrm{i}\theta_i}, \tag{154}$$

Riemann sheets are determined by the phase-angle ranges $0 \le \theta_i < 2\pi$ and $-2\pi \le \theta_i < 0$. The physical scattering region corresponds to the upper edge of the cut in the physical sheet and unphysical sheets are reached by crossing the cuts.

A pole in the scattering matrix occurs when one of its eigenvalues has a singularity. (See McVoy[43] for a discussion of compound poles).

$$\mathbf{SU} = \mathbf{UD}, \tag{155}$$

where the columns of $\mathbf{U}$ are the eigenvectors and the elements of the diagonal matrix $\mathbf{D}$ are the corresponding eigenvalues. Assuming the i-th eigenvalue is singular at complex energy E_α

$$D_{ii} = \frac{d_{ii}}{E - E_\alpha} + c_{ii} + \mathcal{O}(E - E_\alpha), \tag{156}$$

the ith solution vector $\mathbf{G}_i$ has the asymptotic form

$$\mathbf{G}_i = (\mathbf{FU})_i\,(E - E_\alpha) \underset{E\to E_\alpha}{\sim} -\mathbf{k}^{-\frac{1}{2}}(e^{\mathrm{i}\theta}\mathbf{U})_i d_{ii}. \tag{157}$$

The condition for a resonance is therefore that the components of the ith solution vector $\mathbf{G}_i$ satisfy the Siegert condition

$$\lim_{E\to E_\alpha} G_{ji} \underset{r\to\infty}{\sim} a_{ji}e^{\mathrm{i}k_j r}, \qquad j = 1,\ldots,n. \tag{158}$$

in which there are outgoing waves in all channels. Coefficients, a_{ji}, are complex and non-singular. Complex poles of the S-matrix are required by unitarity to lie on the unphysical sheets.

Several approaches for determining resonance parameters using the analytic properties of the scattering matrix have been suggested.[44-45] In the following subsections an R-matrix method for locating poles in the multichannel scattering matrix will be described.[46]

3.3. Complex energy R-matrix theory

The R-Matrix is a meromorphic function whose poles (and zeros) all lie on the real energy axis.[47] If the R-matrix, eq. (131), is written in a spectrally decomposed form obtained by diagonalizing the internal region Hamiltonian it may be continued into the complex energy plane by allowing the scattering energy, E to become complex valued so

$$R_{ij} = \sum_k \frac{\gamma_{ik}\gamma_{jk}}{E_k - E}, \qquad i,j = 1,\ldots,n, \tag{159}$$

where the reduced width amplitudes, γ_{ik} and the R-Matrix eigenenergies, E_k, are real, but the total energy, E, is complex.

The analytically continued R-matrix provides the boundary conditions for the solution of the external region scattering equations which now must be solved for complex energies and the Siegert asymptotic boundary conditions, eq. (158).

$$\mathcal{F}_{ij}(r) \underset{r\to\infty}{\sim} e^{ik_j r}[\delta_{ij} + a_{ij}\mathcal{O}(r^{-q})], \tag{160}$$

where q is a positive integer. The solutions $\mathbf{F}$ may be written as the linear combination

$$\mathbf{F}(r) = \mathcal{F}(r)\mathcal{C}. \tag{161}$$

where the expansion coefficients, $\mathcal{C}$, are determined by the condition

$$d(E) = \det\left[\mathcal{F} - \mathbf{R}\left(a\frac{d\mathcal{F}}{dr} - b\mathcal{F}\right)\right]_{r=a} = 0. \tag{162}$$

The problem of determining the location of the poles of the S-matrix and consequently of the resonance parameters therefore reduces to finding the zeros of the discrepancy function $d(E)$ on the appropriate Riemann sheet of the complex energy plane. The introduction of complex energy values introduces no significant complications and the solution of the external region scattering equations may be obtained by straight-forward extensions of the standard real-energy procedures.

3.4. Location of zeros

The task of locating the zeros of the discrepancy function may in principle be carried out using Newton's method or similar techniques. The results however are sensitive to the starting point of the search and to the initial step size. Using this type of method it is difficult to ensure that all zeroes which lie within a predetermined region of the energy plane have been located.

Algorithms based on Cauchy's theorem, such as the Delves-Lyness algorithm,[48] provide a remedy for this problem. The integrals round closed contours

$$S_N = \frac{1}{2\pi i}\int_C z^N \frac{f'(z)}{f(z)}\,dz = \sum_{i=1}^{\nu} z_i^N \tag{163}$$

are evaluated by numerical quadrature to determine the sum of powers of the ν zeros of $f(z)$, z_i, $i = 1,\ldots,\nu$, that lie within the contour C. The values of the integrals S_N are then used to define a polynomial

$$P_N(z) = \prod_{j=1}^{N}(z - a_j) = \sum_{j=0}^{N} A_j\, z^{N-j} \tag{164}$$

with zeros, a_j, coinciding with those of the function f. Standard techniques (for example Müller's method,[49]) may be used to extract the location of each of these zeros. To avoid the calculation of the energy derivative of the discrepancy function, variants of the Delves-Lyness method suggested by Carpentier and Dos Santos[50] and by Davies[51] are used. The quadratures are carried out assuming circular contours by the trapezoid rule.

3.5. Electron scattering by O_2^+

Electron scattering by singly-charged oxygen molecular ions provides information on autoionization structure in the photoionization spectrum of molecular oxygen

$$h\nu + O_2 \rightarrow O_2^+ + e^-. \tag{165}$$

To study the low-energy photoionization spectrum the usual R-matrix formulation with a single electron in continuum orbitals is employed. Calculations have been carried out including three electronic states of O_2^+ in the sum over target states ϕ_i in eq. (127). These are the $X^2\Pi_g$, $a^4\Pi_u$ and $A^2\Pi_u$ states. Each target is represented by a valence CI wavefunction providing excitation energies accurate to better than 0.5 eV. Seven channels are included in the external region calculation.

Two scattering symmetries, $^3\Sigma_u^-$ and $^3\Pi_u$, may contribute to dipole transitions from the $X^3\Sigma_g^-$ ground state of O_2. There will be Rydberg series of resonances converging to both of the excited electronic states of O_2^+. The lower members of the series are sufficiently wide and isolated that their parameters may be obtained reliably by fitting the eigenphase sum. Higher members become narrow and closer in energy so reliable results require the complex energy technique. Results obtained using the two methods for $^3\Sigma_u^-$ symmetry are compared in Table 2. As expected there is good agreement for the wider resonances and fitting the eigenphase sum is impractical for the narrow resonances.

Table 2. Resonance positions and widths for $e^- - O_2^+$ scattering in $^3\Sigma_u^-$ symmetry at $R = 2.3a_0$. Comparison of complex energy R-matrix and Breit-Wigner values. Energies in Rydberg.

	R-matrix		Breit-Wigner	
State	Energy, E	Width, Γ	Energy, E	Width, Γ
1	0.2151	1.06(−3)	0.2151	1.06(−3)
2	0.2610	4.24(−4)	0.2610	4.27(−4)
3	0.2809	5.63(−7)		
4	0.2822	1.99(−4)	0.2822	1.99(−4)
5	0.2930	4.80(−7)		
6	0.2937	8.20(−5)	0.2937	8.22(−5)

4. CONCLUSIONS

Although the R-matrix method is successful in describing a wide range of low-energy electron-molecule collisions, in order to include large numbers of accurate target wavefunctions or to treat scattering at intermediate energies new theoretical and computational techniques are necessary. Some of the ideas which may lead to these advances have been explored in this chapter. It seems likely that future work will increasingly focus on these issues and on applications of the extended R-matrix technique to photoionization, electron impact ionization and dissociative processes. Molecular multiphoton processes[52] and double photoionization are two important topics where this R-matrix technique may prove particularly effective. The complex energy R-matrix method emphasizes the fact that detailed insight into scattering processes may be derived from studies of the analytic structure of scattering matrices calculated using *ab-initio* methods such as the R-matrix approach.

Acknowledgements. The author would like to thank Professor P.G. Burke for many very useful discussions on R-matrix theory.

References

[1] P.G. Burke, C.J. Noble and P. Scott, Proc. R. Soc A **410** 341 (1987).
[2] A.L. Smith, Phil. Trans. Roy. Soc. Lond. A **268** 169 (1970).
[3] P.G. Burke and T.G. Webb, J. Phys. B: At. Mol. Phys. **13** L131 (1970).
[4] C. Bloch, Nucl. Phys. **4** 503 (1957).
[5] P.G. Burke, I. Mackey and I. Shimamura, J. Phys. B: At. Mol. Phys. **10** 2497 (1977).
[6] C. Duneczky, R.W. Wyatt, D. Chatfield, K. Haug, D.W. Schwenke, D.G. Truhlar, Y. Sun and D.J. Kouri, Comp. Phys. Commun. **53** 357 (1989).
[7] H. Feshbach, Adv. Phys. (NY) **19** 287 (1962).
[8] I. Bray and A.T. Stelbovics, Phys. Rev. Letts. **69** 53 (1992).
[9] D. Teillet-Billy, L. Malegat and J.P. Gauyacq, J. Phys. B: At. Mol. Phys. **20** 3201 (1987).
[10] I. Shimamura, R-matrix theories, *in:* "Electronic and Atomic Collisions", G. Watel, ed., North-Holland, Amsterdam (1978).
[11] R.K. Nesbet, Phys. Rev. A **6** 2975 (1981).
[12] C.J. Noble, P.G. Burke and S. Salvini, J. Phys. B: At. Mol. Phys. **15** 3779 (1982).
[13] P.G. Burke and V.M. Burke, to be published (1994).
[14] W. Duch, "GRMS or Graphical Representation of Model Spaces", (Lecture Notes in Chemistry, **42**), Springer-Verlag, Berlin (1986).
[15] V.M. Burke and C.J. Noble, Comp. Phys. Commun. *in press* (1994).
[16] M. Le Dourneuf, J.M. Launay and P.G. Burke, J. Phys. B: At. Mol. Opt. Phys. **23** L559 (1990).
[17] T.T. Scholz, J. Phys. B: At. Mol. Opt. Phys. **24** 2127 (1991).
[18] P.G. Burke, K.A. Berrington and C.V. Sukumar, J. Phys. B: At. Mol. Phys. **14** 289 (1981).
[19] M.R.F. Siggel, J.B. West, M.A. Hayes, A.C. Parr, J.L. Dehmer and I. Iga, J. Chem. Phys. **99** 1556 (1993); J.L. Dehmer, P.M. Dehmer, J.B. West, M.A. Hayes, M.R.F. Siggel and A.C. Parr, J. Chem. Phys. **97** 7911 (1992).
[20] A.D. Buckingham, B.J. Orr and J.M. Sichel, Phil. Trans. Roy. Soc. Lond. A **268** 147 (1970); J.C. Tully, R.S. Berry and B.J. Dalton, Phys. Rev. **178** 95 (1968); D. Dill and J.L. Dehmer, J. Chem. Phys. **61** 692 (1974); R.J.W. Henry and L. Lipsky, Phys. Rev. **153** 51 (1967).
[21] N.A. Cherepkov and V.V. Kuznetsov, At. Mol. Clust. **7** 271 (1980); N.A. Cherepkov, Adv. At. Mol. Phys. **19** 395 (1983).
[22] D. Dill, J. Chem. Phys. **65** 1130 (1976).

[23] P.G. Burke, Electron and photon collisions with molecules, *in:* "Collision Theory for Atoms and Molecules", F.A. Gianturco, ed., Plenum, New York (1989); unpublished notes.
[24] M.E. Rose, "Elementary Theory of Angular Momentum", John Wiley, New York (1957).
[25] C.N. Yang, Phys. Rev. **74** 764 (1948).
[26] B.I. Schneider, M. Le Dourneuf and P.G. Burke, J. Phys. B: At. Mol. Phys. **12** L365 (1979).
[27] P.G. Burke and M.J. Seaton, J. Phys. B: At. Mol. Phys. **17** L683 (1984); M.J. Seaton, J. Phys. B: At. Mol. Phys. **18** 2111 (1985).
[28] K.A. Berrington, P.G. Burke, K. Butler, M.J. Seaton, P.J. Storey, K.T. Taylor and Y. Yan, J. Phys. B: At. Mol. Opt. Phys. **20** 6379 (1987); M.J. Seaton, J. Phys. B: At. Mol. Opt. Phys. **19** 2601 (1986); Y. Yan and M.J. Seaton, J. Phys. B: At. Mol. Phys. **20** 6409 (1987).
[29] J. Tennyson, C.J. Noble and P.G. Burke, Int. J. Q. Chem. **XXIX** 1033 (1986).
[30] J. Tennyson, J. Phys. B: At. Mol. Opt. Phys. **19** 4255 (1986).
[31] A.C. Parr, J.E. Hardis, S.H. Southworth, C.S. Feigerle, T.A. Ferrett, D.M.P. Holland, F.M. Quinn, B.R. Dobson, J.B. West, G.V. Marr and J.L. Dehmer, Phys. Rev. A **37** 437 (1988).
[32] G. Raseev, J. Phys. B: At. Mol. Phys. **18** 423 (1985).
[33] L.A. Collins, B.I. Schneider and C.J. Noble, Phys. Rev. **45** 4610 (1992); I. Shimamura, C.J. Noble and P.G. Burke, Phys. Rev. A **41** 3545 (1990); earlier references are listed in these papers.
[34] K. Nakashima, H. Takagi and H. Nakamura, J. Chem. Phys. **86** 726 (1987).
[35] A. Giusti-Suzor, J.N. Bardsley and C. Derkits, Phys. Rev. **28** 682 (1983).
[36] L.A. Collins, B.I. Schneider, C.J. Noble, C.W. McCurdy and S. Yabushita, Phys. Rev. Lett. **57** 980 (1986).
[37] G. Breit and E.P. Wigner, Phys. Rev. **49** 519 (1936).
[38] A.U. Hazi, Phys. Rev. A **19** 920 (1979).
[39] J. Tennyson and C.J. Noble, Comput. Phys. Commun. **33** 421 (1984).
[40] A.J.F. Siegert, Phys. Rev. **56** 750 (1939).
[41] U. Fano, Phys. Rev. **124** 1866 (1961).
[42] M.V. Basilevsky and V.M. Ryaboy, J. Comp. Chem. **8** 683 (1987).
[43] K.W. McVoy, Nuclear resonance reactions and S-matrix analyticity, *in:* "Fundamentals in Nuclear Reaction Theory", A. de Shalit and C. Villi, eds., IAEC, Vienna (1967).
[44] B.I. Schneider, Phys. Rev. A **24** 1 (1981).
[45] D.W. Schwenke, Theor. Chim. Acta. **74** 381 (1988).
[46] C.J. Noble, M Dörr and P.G. Burke, J. Phys. B: At. Mol. Opt. Phys. **26** 2983 (1993).
[47] E.P. Wigner and L. Eisenbud, Phys. Rev. **74** 29 (1947).
[48] L.M. Delves and J.N. Lyness, Math. Comput. **21** 543 (1967).
[49] S.D. Conte and C. de Boor, "Elementary Numerical Analysis - An Algorithmic Approach", McGraw-Hill, New York (1980).
[50] M.P. Carpentier and A.F. Dos Santos, J. Comp. Phys. **45** 210 (1982).
[51] B. Davies, J. Comp. Phys. **66** 36 (1986).
[52] P.G. Burke, P. Francken and C.J. Joachain, J. Phys. B: At. Mol. Opt. Phys. **24** 761 (1991).

THE SCHWINGER VARIATIONAL METHOD

Winifred M. Huo

NASA Ames Research Center, Moffett Field, CA 94035, U.S.A.

1. INTRODUCTION

Variational methods have proven to be invaluable tools in theoretical physics and chemistry, both for bound state problems and for the study of collision phenomena. For collisional problems variational methods can be grouped into two types, those based on the Schrödinger equation and those based on the Lippmann-Schwinger equation. The Hulthén-Kohn[1-3] method belongs to the first type, and their modern development for electron-molecule scattering, incorporating complex boundary conditions, is reported in chapter 1 of this book by Rescigno et al.[4] An offshoot of the Hulthén-Kohn variational method is the variational R-matrix method.[5,6] In chapter 8 of this book Schneider[7] presents a general discussion of the R-matrix method, including the variational R-matrix.

The Schwinger variational (SV) method, which Schwinger introduced in his lectures at Harvard University and subsequently published in 1947,[8] belongs to the second category. The application of the SV method to e-molecule collisions and molecular photoionization has been reviewed previously.[9-12] The present chapter discusses the implementation of the SV method as applied to e-molecule collisions. Since this is not a review of cross section data, cross sections are presented only to serve as illustrative examples.

In the SV method, the correct boundary condition is automatically incorporated through the use of the Green's function. Thus SV calculations can employ basis functions with arbitrary boundary condition. This feature enables the use of an L^2 basis for scattering calculations, and provided the initial motivation for applying the SV method to atomic and molecular physics.[10,13] The initial success led to the development of the iterative Schwinger method[14] which uses single-center expansion techniques and also an iterative procedure to improve the initial basis set. The iterative Schwinger method has been used extensively to study molecular photoionization.[15] For e-molecule collisions, it is used at the static-exchange level to study elastic scattering[16] and coupled with the distorted wave approximation to study electronically inelastic scattering.[17]

The Schwinger multichannel (SMC) method, originally formulated by Takatsuka and McKoy,[18,19] is the first modern computational method for e-molecule collisions explicitly designed to treat the multicenter nature of a polyatomic target. When Gaussian functions are used for the L^2 basis and plane waves for the incoming electron, all two-electron repulsion integrals are calculated analytically. It is also a multichannel method and readily treats electronic inelastic scattering as well as polarization effects in elastic scattering. In addition, integrals involving the Green's function, generally considered the bottleneck in the SV method, can now be computed efficiently and accurately using the insertion method.[20] Alternatively, highly parallel computing can be effectively employed in the numerical integration over the final three-dimensions for this type of integrals.[12] The SMC method has been found to be robust, and has been employed to study molecules of various sizes and with different bonding types. For example, it has been used to study the electronic excitation of [1,1,1] propellane by electron impact,[21] the largest polyatomic molecule studied so far by *ab initio* methods. It has also been used to study electron scattering from the ground state of BeCO, a molecule with a weak van der Waals bond, to simulate scattering from an adsorbate in a physisorbed system.[22] Since the SMC method is the most commonly used SV treatment for e-molecule scattering, the main topic of this chapter will deal with its implementation.

Both the SMC and iterative Schwinger methods calculate fixed-nuclei, body-frame T-matrices. Domcke[23] has demonstrated that nuclear dynamics in e-molecule collisions can be treated efficiently using projection operator formalism and Green's function approach. Results of the fixed-nuclei SMC calculation can be coupled directly with Domcke's treatment of nuclear dynamics. As an example, we present a calculation of vibrational excitations of N_2 by electron impact which combines these two treatments.

2. THE LIPPMANN-SCHWINGER EQUATION AND THE SCHWINGER VARIATIONAL PRINCIPLE

In the integral equation approach, we look for solutions of the Lippmann-Schwinger (LS) equation instead of the Schrödinger equation. Let the Hamiltonian of the electron + molecule system be written as

$$H = \sum_{i=1}^{N+1} \left(-\frac{1}{2}\nabla_i^2 - \sum_{k=1}^{N_p} \frac{Z_k}{|\mathbf{r}_i - \mathbf{R}_k|} + \sum_{j>i}^{N+1} \frac{1}{r_{ij}} \right) + \sum_{k=1}^{N_p} \left(-\frac{1}{2m_k}\nabla_k^2 + \sum_{l>k}^{N_p} \frac{Z_k Z_l}{R_{kl}} \right). \quad (1)$$

Here i, j and k, l sum over electrons and nuclei, respectively; Z_k and m_k denote the charge and mass of the kth nucleus. Also, atomic units will be used unless specified otherwise. Note that H is symmetric and its eigenfunction antisymmetric with respect to electron exchange. Rewriting H as

$$H = H_o + V, \quad (2)$$

with

$$H_o = H_M - \frac{1}{2}\nabla_{N+1}^2, \quad (3)$$

and

$$V = \sum_{i=1}^{N} \frac{1}{|\mathbf{r}_{N+1} - \mathbf{r}_i|} - \sum_{k=1}^{N_p} \frac{Z_k}{|\mathbf{r}_{N+1} - \mathbf{R}_k|}. \quad (4)$$

The use of $N+1$ to label the continuum electron is completely arbitrary. Also, H_M is the molecular Hamiltonian with eigenfunctions Φ_m,

$$H_M \Phi_m = \mathcal{E}_m \Phi_m. \quad (5)$$

The LS equation is given by,

$$\Psi^{(+)} = S + G^{(+)} V \Psi^{(+)}. \tag{6}$$

with

$$\lim_{\delta \to 0} (E - H_o + i\delta) G^{(+)}(E') = \delta(E - E'). \tag{7}$$

and S is the solution of the corresponding homogeneous equation.[1/] Here we use the boundary condition of an incoming plane wave and outgoing spherical waves, which has been incorporated into the Green's function $G^{(+)}$. Due to this feature, it is not necessary to use trial functions which satisfy the correct boundary conditions in solving the LS equation. The Green's function $G^{(+)}$ is expressed in terms of the molecular eigenfunctions and the interaction-free Green's function.

$$G^{(+)}(E) = -\frac{1}{2\pi} \sum_{m=1} |\Phi_m\rangle \frac{exp(ik_m |\mathbf{r}_{N+1} - \mathbf{r}_{N+1}{}'|)}{|\mathbf{r}_{N+1} - \mathbf{r}_{N+1}{}'|} \langle \Phi_m|. \tag{8}$$

where

$$\frac{1}{2} k_m^2 + \mathcal{E}_m = E, \tag{9}$$

The summation m is over the entire molecular spectrum, including continuum states. When $\mathcal{E}_m > E$, k_m becomes imaginary, corresponding to a bound electron. Indeed, including the continuum spectrum of the target allows the outgoing electron to be different from the incoming electron, a necessary feature in order to account for electron exchange.[2/]

Based on the LS equation for $\Psi^{(+)}$, and the corresponding equation for $\Psi^{(-)}$ with outgoing plane wave and incoming spherical waves boundary conditions,

$$\Psi^{(-)} = S + G^{(-)} V \Psi^{(-)}.$$

the Transition matrix, or T-matrix, of a scattering process can be written as

$$T_{mn} = \langle S_m | V | \Psi_n^{(+)} \rangle = \langle \Psi_m^{(-)} | V | S_n \rangle = \langle \Psi_m^{(-)} | V - V G^{(+)} V | \Psi_n^{(+)} \rangle. \tag{10}$$

It is readily seen that a combination of the above expressions

$$T_{mn} = \langle S_m | V | \Psi_n^{(+)} \rangle + \langle \Psi_m^{(-)} | V | S_n \rangle - \langle \Psi_m^{(-)} | V - V G^{(+)} V | \Psi_n^{(+)} \rangle. \tag{11}$$

results in a T-matrix that is stationary with respect to first-order variations in $\Psi^{(+)}$ and $\Psi^{(-)}$, respectively.

$$\delta T_{mn} = \langle S_m | V | \Psi_n^{(+)} + \delta \Psi_n^{(+)} \rangle + \langle \Psi_m^{(-)} | V | S_n \rangle - \langle \Psi_m^{(-)} | V - V G^{(+)} V | \Psi_n^{(+)} + \delta \Psi_n^{(+)} \rangle - T_{mn}$$

$$= 0,$$

1/ We choose to use a homogeneous solution which is the product of the target wave function and the free-particle wave function of the continuum electron, without antisymmetrization between the two. As pointed out in Ref. 24, it is not necessary to antisymmetrize both the initial and final wave functions explicitly. If $\Psi^{(+)}$ is antisymmetric, then the T-matrix element will automatically pick up the antisymmetric part of S. This convention is used here so that the same S will also apply to the SMC equation discussed in Sec. 3.

2/ A different approach to account for exchange is to incorporate the Hamiltonian with permuted indices into the interaction potential, $U = V + \sum_i P_{i,N+1}(H - E)$. See Bransden et al.[25]

and

$$\delta T_{mn} = \langle S_m|V|\Psi_n^{(+)}\rangle + \langle \Psi_m^{(-)} + \delta\Psi_m^{(-)}|V|S_n\rangle - \langle \Psi_m^{(-)} + \delta\Psi_m^{(-)}|V - VG^{(+)}V|\Psi_n^{(+)}\rangle - T_{mn}$$

$$= 0.$$

Equation (11) is referred to as the linear form of the Schwinger variational principle.

In most e-molecule calculations, it is common practice to expand $\Psi_n^{(+)}$ and $\Psi_m^{(-)}$ in terms of a set of antisymmetrized $N+1$-electron trial functions f,

$$\Psi_n^{(+)} = \sum_r b_r(\boldsymbol{k}_n) f_r, \tag{12}$$

and

$$\Psi_m^{(-)} = \sum_s c_s(\boldsymbol{k}_m) f_s. \tag{13}$$

Eq. (11) gives

$$T_{mn} = \sum_r b_r \langle S_m|V|f_r\rangle + \sum_s c_s \langle f_s|V|S_n\rangle - \sum_r \sum_s b_r c_s \langle f_s|V - VG^{(+)}V|f_r\rangle. \tag{14}$$

The requirement that T_{mn} be stationary with respect to first order variations in the expansion coefficients b_r and c_s,

$$\partial T_{mn}/\partial b_r = 0,$$

$$\partial T_{mn}/\partial c_s = 0,$$

gives the following relationships,

$$b_r = \sum_s D_{rs} \langle f_s|V|S_n\rangle, \tag{15}$$

$$c_s = \sum_r \langle S_m|V|f_r\rangle D_{rs}, \tag{16}$$

with

$$(D^{-1})_{rs} = \langle f_r|V - VG^{(+)}V|f_s\rangle. \tag{17}$$

The resulting variational stable expression for T_{mn} is[26]

$$T_{mn} = \sum_{rs} \langle S_m|V|f_r\rangle D_{rs} \langle f_s|V|S_n\rangle. \tag{18}$$

Equation (18) has the advantage that it is independent of the normalization of $\Psi_n^{(+)}$ and $\Psi_m^{(-)}$.

The advantages of the Schwinger variational method have been discussed previously.[9, 10] Because the correct boundary condition is already incorporated into the Green's function and because the variational stable expression of the T-matrix in Eq. (18) is independent of normalization, it allows flexibility in the choice of a basis. Furthermore, the wave function in Eqs. (11) and (18) always appears together with the potential, i.e., $V|\Psi_n^{(+)}\rangle$ or $\langle\Psi_m^{(-)}|V$. Thus a trial basis for $\Psi_n^{(+)}$ needs only to cover the region of space where V does not vanish. This feature allows us to use an L^2 basis to represent $\Psi_n^{(+)}$ when V is a short-range potential.

Another advantage of the SV method is that the Schwinger variational principle is one rank higher than the Kohn variational principle.[27, 28] If the same trial wave function is used, the Schwinger method should give a better converged result. It should be noted that, in comparing the two methods, it is important to use the *same* trial basis because the Kohn method, which requires basis functions with the correct boundary condition, generally uses a trial basis different from the Schwinger method.

A major drawback of the SV method lies in the difficulty in calculating the Green's function matrix elements.[9] The matrix element $\langle \Psi_m^{(-)} | V G^{(+)} V | \Psi_n^{(+)} \rangle$ in Eq. (11) or $\langle f_r | V G^{(+)} V | f_s \rangle$ in Eq. (17) are nine-dimensional integrals if V includes a two-electron potential. This obstacle has recently been removed when it is demonstrated[20] that the insertion technique can be used with efficiency and good accuracy for the evaluation of these matrix elements. This will be discussed in Sec. 3.5. Alternatively, highly parallel computing can be employed for the 3-dimension numerical integration over the wave vector in these matrix elements.

2.1. A Simple Example of Potential Scattering

To illustrate the use of Eq. (18) and to demonstrate the convergence property of the SV method, consider the elastic scattering of an s-wave by a weak potential U. The initial state is described by the Riccati-Bessel function,

$$S = j_0(kr).$$

Since the potential is weak, $j_0(kr)$ can also be used as a reasonable representation for the trial functions of Ψ^+ and $\Psi^{(-)}$. Here the trial functions are of the form

$$\Psi^{(+)} = b_0 j_0(kr),$$

$$\Psi^{(-)} = c_0 j_0(kr).$$

Because the initial boundary condition is described by S in the LS equation, the coefficients b_0 and c_0 need not be set to 1. Instead, they are used as variational parameters and optimized. Note that the use of $j_0(kr)$ as a trial basis is an example of a basis which does not satisfy the correct boundary condition. Otherwise a combination of Riccati-Bessel and Neumann functions should be employed. Because $\Psi^{(+)}$ and $\Psi^{(-)}$ are expanded in a one term expansion, the inverse of the D-matrix, in Eq. (17), is just the inverse of the matrix element itself. From Eq. (18), the SV expression of the elastic T-matrix is given by

$$T_{00} = \frac{\langle j_0(kr)|U|j_0(kr)\rangle \langle j_0(kr)|U|j_0(kr)\rangle}{\langle j_0(kr)|U - UG^{(+)}U|j_0(kr)\rangle}.$$

Since U is a weak potential, we can write

$$\frac{1}{\langle j_0(kr)|U - UG^{(+)}U|j_0(kr)\rangle} = \frac{1}{\langle j_0(kr)|U|j_0(kr)\rangle}\left(1 + \frac{\langle j_0(kr)|UG^{(+)}U|j_0(kr)\rangle}{\langle j_0(kr)|U|j_0(kr)\rangle} + \ldots\right).$$

Substituting the above into the expression for T_{00}, we find

$$T_{00} = \langle j_0(kr)|U|j_0(kr)\rangle + \langle j_0(kr)|UG^{(+)}U|j_0(kr)\rangle + \ldots$$

The first and second terms of T_{00} correspond to the first and second Born terms, respectively.

In Eq. (10) three non-variational expressions for T_{00} are given. They give identical results if the exact $\Psi^{(+)}$ and $\Psi^{(-)}$ are used. Otherwise they give different values for the T-matrix. The first relation,

$$T_{00} = \langle S|U|\Psi^{(+)}\rangle,$$

gives the Born approximation for T_{00}.[3/]

$$T_{00} = \langle j_0(kr)|U|j_0(kr)\rangle.$$

This result is inferior to the SV expression. Using the third relation in Eq. (10),

$$T_{00} = \langle \Psi^{(-)}|U - UG^{(+)}U|\Psi^{(+)}\rangle,$$

T_{00} deviates even further from the variationally stable result,

$$T_{00} = \langle j_0(kr)|U|j_0(kr)\rangle - \langle j_0(kr)|UG^{(+)}U|j_0(kr)\rangle.$$

Now the second Born term is subtracted from, instead of added to, the first Born term. The above exercise clearly illustrates the power of the variational method.

3. THE SCHWINGER MULTICHANNEL METHOD (SMC)

Currently, the most frequently used form of the Schwinger variational principle for e-molecule collisions originates from the work of Takatsuka and McKoy.[18, 19] As seen below, while this formulation is based on the Schwinger variational principle, its equation of motion is obtained by combining the LS and Schrödinger equations. These equations are coupled via the introduction of a projection operator P which projects into the open-channel target space. In the following, we shall first consider some of the properties of Takatsuka's projection operator, then derive the SMC equation. Implementation of the method will then be delineated. Efficient evaluation of the Green's function matrix elements, generally considered a bottleneck in the Schwinger method, will be discussed and extension to include correlated target functions will also be presented.

3.1. The N-Electron Projection Operator P

The projection operator P introduced by Takatsuka and McKoy projects into the open channel target space. Hence it is an N-electron projection operator and different from the Feshbach projection operator,[29] which is a $N+1$-electron operator.

$$P = \sum_{m=1}^{M} |\Phi_m(1,2,\ldots,N)\rangle\langle\Phi_m(1,2,\ldots,N)|, \tag{19}$$

with m summing over all energetically accessible target states. P satisfies the idempotent property,

$$P^2 = P,$$

and commutes with H_o

$$PH_o = H_oP.$$

To illustrate the difference between P and P_{N+1}, the Feshbach projection operator, consider the simplest case when the target wave function Φ_i is expressible in terms of a closed shell, single determinental wave function,

$$\Phi_i(1,2,\ldots,N) = \mathcal{A}\Big\{\phi_1\alpha(1)\phi_1\beta(2)\ldots\phi_{N/2}\alpha(N-1)\phi_{N/2}\beta(N)\Big\}.$$

3/ Here $\Psi^{(+)}$ is not optimized and $b_0 = 1$ is used.

Here α and β denote spin functions and $\mathcal{A}$ the antisymmetrizer. If the wave function of the continuum electron, $g_i(N+1)$, is orthogonal to all the target orbitals, then the $S_z = 1/2$ component of Ψ_i becomes

$$\Psi_i(1,2,\ldots,N,N+1) = \mathcal{A}\left\{\phi_1\alpha(1)\phi_1\beta(2)\ldots\phi_{N/2}\beta(N)g_i\alpha(N+1)\right\}.$$

We have

$$P\Psi_i(1,2,\ldots,N,N+1) = \frac{1}{\sqrt{N+1}}\Phi_i(1,2,\ldots,N)g_i\alpha(N+1), \tag{20}$$

and

$$\langle\Psi_j|HP|\Psi_i\rangle = \frac{1}{N+1}\langle\Psi_j|H|\Psi_i\rangle. \tag{21}$$

On the other hand, the Feshbach projection operator is defined in terms of the $N+1$-electron open channel functions,

$$P_{N+1} = \sum_m |\Psi_m(1,2,\ldots,N,N+1)\rangle\langle\Psi_m(1,2,\ldots,N,N+1)|.$$

where Ψ_m satisfies the correct asymptotic boundary condition. The function Ψ_i is an open channel function in the terminology of the Feshbach projector formalism if g_i satisfies the correct boundary condition. In that case,

$$P_{N+1}\Psi_i = \Psi_i. \tag{22}$$

A comparison of Eqs. (20) and (22) shows how the two projection operators differ. Also note that Takatsuka's projection operator treats Ψ_i as an open channel function as long as it is associated with an energetically accessible target function, regardless of the behavior of g_i at the boundary. The Feshbach projection operator, on the other hand, includes Ψ_i in the open channel space only if g_i has the correct boundary behavior.

It should be noted that Eq. (20) may not be applicable when the orthogonality constraint between the continuum electron function and the target function is relaxed, and/or when the target is described by more sophisticated wave functions. Consider the simple case of $X^1\Sigma_g^+ \rightarrow b^3\Sigma_u^+$ excitation of H_2 by electron impact. Let the wave function for the $X^1\Sigma_g^+$ state be represented by

$$\Phi_X = \mathcal{A}\{1\sigma_g\alpha(1)1\sigma_g\beta(2)\},$$

and the three spin components of the $b^3\Sigma_u^+$ state by

$$\Phi_{b,S_z=1} = \mathcal{A}\{1\sigma_g\alpha(1)1\sigma_u\alpha(2)\},$$

$$\Phi_{b,S_z=0} = \mathcal{A}\{1\sigma_g(1)1\sigma_u(2)\frac{1}{\sqrt{2}}\left[\alpha(1)\beta(2)+\beta(1)\alpha(2)\right]\},$$

$$\Phi_{b,S_z=-1} = \mathcal{A}\{1\sigma_g\beta(1)1\sigma_u\beta(2)\}.$$

It is well established that in calculating this transition, terms of the form

$$\Psi_a = \mathcal{A}\{1\sigma_g\alpha(1)1\sigma_u\alpha(2)1\sigma_u\beta(3)\}, \tag{23}$$

and

$$\Psi_b = \mathcal{A}\{1\sigma_g\alpha(1)1\sigma_g\beta(2)1\sigma_u\alpha(3)\}, \tag{24}$$

should be included in the expansion of the wave function. These are usually called 'penetration' or 'recorrelation terms' because they relax the enforced orthogonality between the continuum and target functions. Let the projection operator P be generated from the four target functions above. We then have

$$P\Psi_a = \frac{1}{\sqrt{3}}\Phi_{b,S_z=1}(1,2)1\sigma_u\beta(3) - \frac{1}{\sqrt{6}}\Phi_{b,S_z=0}(1,2)1\sigma_u\alpha(3). \tag{25}$$

This simple example illustrates how the operation of P depends on the type of functions used.

3.2. The SMC Equation

The projected LS equation is

$$P\Psi_n^{(+)} = S_n + G_P^{(+)}V\Psi_n^{(+)}, \tag{26}$$

The projected Green's function, $G_P^{(+)}$, is defined in the open channel space by,

$$G_P^{(+)} = -\frac{1}{2\pi}\sum_{m=1}^{M}|\Phi_m\rangle\frac{exp(ik_m|\boldsymbol{r}_{N+1}-\boldsymbol{r}_{N+1}'|)}{|\boldsymbol{r}_{N+1}-\boldsymbol{r}_{N+1}'|}\langle\Phi_m|.$$

Multiplying Eq. (26) from the left by V and rearranging, we find

$$(VP - VG_P^{(+)}V)\Psi_n^{(+)} = VS_n \tag{27}$$

Equation (27) describes only the open channel functions, whereas a complete description of $\Psi_n^{(+)}$ also requires the closed channel component. To do this, we use the following identity

$$aP + (1 - aP) = 1. \tag{28}$$

Note that in association with the operation of P, a parameter a, called the projection parameter, is introduced. It is seen from Eqs. (20) and (25) that the operation of P not only removes the antisymmetrization between the target function and continuum orbital, but also generates a constant which multiplies the resulting function. Thus the introduction of a is a logical step. The determination of a will be discussed after the SMC equation is derived.

The closed channel contribution is described by the projected Schrödinger equation,

$$(1 - aP)\hat{H}\Psi_n^{(+)} = 0, \tag{29}$$

with

$$\hat{H} = E - H,$$
$$\hat{H} = \hat{H}_o - V,$$

and

$$\hat{H}_o = E - H_o.$$

Note that

$$\begin{aligned} P\hat{H} &= P\hat{H}_o - PV \\ &= \frac{1}{2}(\hat{H}_oP + P\hat{H}_o) - PV \\ &= \frac{1}{2}(\hat{H}P + P\hat{H}) + \frac{1}{2}(VP - PV). \end{aligned}$$

Equation (29) can be rewritten as

$$[\hat{H} - \frac{a}{2}(\hat{H}P + P\hat{H}) + \frac{a}{2}(PV - VP)]\Psi_n^{(+)} = 0. \tag{30}$$

The SMC equation is obtained by dividing Eq. (30) by a and adding it to Eq. (27),

$$A^{(+)}\Psi_n^{(+)} = VS_n, \tag{31}$$

with $A^{(+)}$ the SMC operator,

$$A^{(+)} = \left\{\frac{1}{2}(PV + VP) - VG_P^{(+)}V + \frac{1}{a}\hat{H} - \frac{1}{2}(P\hat{H} + \hat{H}P)\right\}. \tag{32}$$

The SMC equation[4/] uses the LS equation to describe the open channel component of $\Psi^{(+)}$ and the Schrödinger equation for the closed channel component. Because it includes both the open and closed channel components, it provides a complete solution to the scattering problem.[30]

Using the SMC equation instead of the LS equation, the Schwinger variational expression for the T-matrix is

$$T_{mn} = \langle S_m|V|\Psi_n^{(+)}\rangle + \langle\Psi_m^{(-)}|V|S_n\rangle - \langle\Psi_m^{(-)}|A^{(+)}|\Psi_n^{(+)}\rangle. \tag{33}$$

If the scattering wave function is expressed in terms of $N+1$-electron trial functions f, as in Eqs. (12) and (13), the variational stable expression for T_{mn} is the same as Eq. (18),

$$T_{mn} = \sum_{rs}\langle S_m|V|f_r\rangle D_{rs}\langle f_s|V|S_n\rangle. \tag{34}$$

but with

$$(D^{-1})_{rs} = \langle f_r|A^{(+)}|f_s\rangle. \tag{35}$$

3.3. The Projection Parameter a

The SMC equation is incomplete until a is chosen. A number of ways of determining a have been considered in the literature, and they are described below:

(a) Based on the hermiticity of the principal-value SMC operator. Takatsuka and McKoy[18, 19] and Lima and McKoy[30] argued that the variational stability of T_{mn} requires

$$A^{(+)\dagger} = A^{(-)}. \tag{36}$$

In other words, the principal-valued SMC operator A must be Hermitian. Otherwise, T_{mn} will be unstable with respect to first order variation of either $\delta\Psi_n^{(+)}$ or $\delta\Psi_m^{(-)}$. While Eq. (36) is readily satisfied for trial functions consisting only of L^2 functions, Takatsuka and McKoy noted that, if the trial functions included (shielded) spherical Bessel and Neumann (or Hankel) functions, Eq. (36) is not valid because the matrix element of the kinetic energy operator between Bessel and Neumann functions is non-Hermitian. In this case, they showed that if the projection parameter is chosen to be

$$a = N + 1, \tag{37}$$

4/ Takatsuka and McKoy's original derivation[18,19] used the principal-value SMC operator, A, obtained by replacing $G_P^{(+)}$ with the principal-value Green's function, G_P. Similarly, the wave functions are replaced by those with standing wave boundary conditions. However, most subsequent numerical calculations used $A^{(+)}$.

the following matrix element vanishes in the open channel space,

$$\langle P\Psi_m|\frac{1}{a}\hat{H} - \frac{1}{2}(P\hat{H} + \hat{H}P)|P\Psi_n\rangle = 0.$$

The operator $\frac{1}{a}\hat{H} - \frac{1}{2}(P\hat{H} + \hat{H}P)$ is the only part of the SMC operator which involves the kinetic energy operator. If its matrix element is identically zero in the open channel space, Eq. (36) is guaranteed to hold even for a basis which includes spherical Bessel and Neumann functions.

(b) Based on the stability of the T-matrix with respect to first order variation of a. Huo and Weatherford[31] pointed out that the SMC equation automatically incorporates the proper boundary condition through the Green's function. Thus it is unnecessary to include functions with the proper boundary condition in an SMC basis. Indeed, almost all SMC calculations carried out so far use an L^2 basis. In that case, Eq. (36) is satisfied independent of a. Even if continuum functions are included in the basis, the use of δ-function normalizable continuum functions would again ensure the validity of Eq. (36). (A Neumann function is not δ-function normalizable). Instead, they considered the role of a as a weight factor for the relative contribution between the open and closed channel space. Thus a should be treated as a variation parameter and its optimal choice should be based on the variational stability,

$$\partial T_{mn}/\partial a = 0.$$

If $\Psi_n^{(+)}$ and $\Psi_m^{(-)}$ are expanded in terms of Eqs. (12) and (13), the above condition results in the following expression,

$$\sum_r\sum_s b_r\langle f_r|\hat{H}|f_s\rangle c_s = 0. \tag{38}$$

Equation (38) is to be solved together with Eqs. (15) and (16) so that a can be determined simultaneously with the expansion coefficients b_r and c_s.

Using Eqs. (12) and (13), Eq. (38) can be rewritten as

$$\langle\Psi_n^{(+)}|\hat{H}|\Psi_m^{(-)}\rangle = 0.$$

Thus it will be automatically satisfied if $\Psi_n^{(+)}$ and/or $\Psi_m^{(-)}$ satisfy the Schrödinger equation,

$$\hat{H}\Psi_n^{(+)} = 0,$$

or

$$\hat{H}\Psi_m^{(-)} = 0.$$

Under these circumstances, T_{mn} will be stable for any finite value of a, including $a = N + 1$. This result is related to how the LS and Schrödinger equations are combined in the SMC equation. A true solution should satisfy both equations, with the result independent of how the SMC operator is partitioned. However, in practical calculations we search for a variationally stable T_{mn}. Equation (38), which is identical to the configuration-interaction (CI) equation in electronic structural calculations, is much more amenable to practical calculations than the Schrödinger equation itself. Generally we have found that an iterative solution of a, coupled with the calculation of T_{mn}, adds approximately 10% to the cost of an SMC calculation with fixed a.

(c) Based on supplementing the projected LS equation. An alternate proof for Eq. (37) was provided by Winstead and McKoy.[32] Instead of using the SMC equation, they looked for the variational stability of T_{mn} by using the projected LS equation, Eq. (26), alone.

$$T_{mn} = \langle S_m|V|\Psi_n^{(+)}\rangle + \langle\Psi_m^{(-)}|V|S_n\rangle - \langle\Psi_m^{(-)}|VP - VG_P^{(+)}V|\Psi_n^{(+)}\rangle.$$

While the above expression is not variationally stable, they found that stability can be achieved by adding to the projected LS equation a term $Q\hat{H}$.

$$T_{mn} = \langle S_m|V|\Psi_n^{(+)}\rangle + \langle\Psi_m^{(-)}|V|S_n\rangle - \langle\Psi_m^{(-)}|VP - VG_P^{(+)}V + Q\hat{H}|\Psi_n^{(+)}\rangle, \tag{39}$$

with

$$Q = P - R.$$

The operator R, applied to an antisymmetrized $N+1$-electron function, removes the antisymmetrization between an N-electron function and the one-electron function for the $(N+1)$th electron. Thus we have

$$\langle\Psi_m^{(-)}|R\hat{H}|\Psi_n^{(+)}\rangle = \frac{1}{N+1}\langle\Psi_m^{(-)}|\hat{H}|\Psi_n^{(+)}\rangle.$$

It can be readily shown that Eq. (39) leads to a variational stable expression for T_{mn}, with Eq. (37) for a. However, unlike Eq. (28), the operators P and Q do not span the complete space. In view of the fact that P is an N-electron operator whereas R, and hence Q, is an $N+1$-electron operator, we have

$$P + Q \neq 1.$$

In contrast with the result of Lima and McKoy[30] who used Eq. (28) to prove the completeness of the solutions of the SMC equation, the present approach fails to demonstrate that the solution of the operator equation

$$(VP - VG_P^{(+)}V + Q\hat{H})\Psi^{(+)} = S,$$

is complete.

3.4. Implementation of the SMC Method

The trial wave function $\Psi_n^{(+)}$ in the SMC method is expanded by

$$\Psi_n^{(+)} = \sum_{i=1}^{M}\sum_j b_{ij}^{(n)}\mathcal{A}\{\Phi_i(1\ldots N)\chi_j(N+1)\} + \sum_l b_l^{(n)}\Theta_l(1\ldots N+1), \tag{40}$$

where the index i sums over open channel target functions, j sums over the one-electron basis set χ used to expand the continuum electron function, and l sums over Θ, the antisymmetrized $N+1$-electron configuration-state-function (CSF). The expansion coefficients $b_{ij}^{(n)}$ and $b_l^{(n)}$ are to be determined variationally.

So far, all SMC calculations on molecules have used target functions represented by a Cartesian Gaussian basis. Also, they make use of the fact that the correct boundary conditions have been automatically incorporated in the SMC equation and employ an L^2 basis of Cartesian Gaussian functions to represent the continuum electron. Thus the trial form of $\Psi_n^{(+)}$ contains no information of its behavior at the boundary. Instead, its initial condition is given by S_n and the outgoing wave by $G_P^{(+)}V\Psi_n^{(+)}$ in the SMC equation. For collisions with long range potentials, where an L^2 basis is inadequate, the higher partial wave contributions are determined using a Born closure approximation.[41] In addition, SMC calculations use incoming plane waves, instead of angular momentum waves, for the homogeneous solution S.

$$S_n = \Phi_n(1 \ldots N) exp(i\mathbf{k}_n \cdot \mathbf{r}_{N+1})$$

Thus this is the only method described in this book which uses a linear momentum instead of an angular momentum representation for the T-matrix. A plane wave representation is chosen because the two-electron integral between three Gaussian functions and one plane wave can be evaluated analytically.[13]

Below, we describe the overall organization of an SMC code, the steps for calculating a body-frame fixed-nuclei T-matrix, T_{mn}, the transformation from the body-frame to laboratory-frame, and the calculation of differential cross sections (dcs). Some of the important features in implementing the SMC method are described in more detail in subsequent sections.

(a) Angular quadrature for $\mathbf{k}_m$ and $\mathbf{k}_n$. In the linear momentum representation, T_{mn} is a function of both the magnitude and direction of $\mathbf{k}_m$ and $\mathbf{k}_n$, $T_{mn} = T(\mathbf{k}_m, \mathbf{k}_n)$. The total energy determines the magnitude of the wave vector, see Eq. (9). To simulate the random orientation of a molecule in a gas phase collision, SMC calculations for gas phase e-molecule collisions are performed over angular quadratures of $\mathbf{k}_m$ and $\mathbf{k}_n$. Thus, instead of positing an electron beam with a fixed direction colliding with randomly oriented molecules, we fix a molecule in space and describe its collision with electrons coming from different directions, (θ, φ). Note that molecular symmetry can be employed to reduce the number of quadrature points used. For example, the cylindrical symmetry of a diatomic molecule allows us to to calculate the plane wave contribution from one φ and deduce the contributions from other φ's by rotation. Also, part (f) shows that the number of partial waves retrieved in a partial wave decomposition of $T(\mathbf{k}_m, \mathbf{k}_n)$ depends on the size of quadrature used.

(b) Gaussian basis set. Gaussian basis sets for the calculation of molecular wave functions have been well studied. A variety of basis sets are available, with varying degree of accuracy.[33] All SMC calculations carried out have used bases constructed with the segmented contraction scheme, including Dunning's earlier contracted basis[34] and his correlated consistent contracted basis.[35] For diatomic molecules and polyatomic hydrides, uncontracted bases have also been used. The Gaussian basis for the continuum electron is obtained by augmenting the target basis with even-tempered diffuse functions. Their exponents are determined by

$$\zeta_{i+1}^{\Gamma} = \frac{\zeta_i^{\Gamma}}{\alpha^{\Gamma}}$$

where ζ_i^{Γ} is the exponent of the ith Gaussian of symmetry type Γ and α^{Γ} is a constant, usually between 2 to 3. The initial ζ_i^{Γ} is taken from the most diffuse function of this symmetry type in the target basis. Note that as more diffuse functions are introduced, the basis set is closer to redundancy. Hence, instead of placing diffuse functions at each nuclear center, it becomes advisable to place the most diffuse functions only at the center of mass. Also, if a basis set is truly redundant, the lowest eigenvalue of its overlap matrix is zero. This feature can be used as a quick test for basis set redundancy.

When the Green's function matrix elements are calculated using the insertion technique, an additional basis is needed for the insertion calculation, The choice of this basis, which is critical to the success of the insertion method, will be discussed in Sec. 3.5.

(c) Open and Closed channel functions. The open channel projection operator P is defined to span all energetically accessible target functions. However, in practical calculations it is frequently not feasible to include all open channels defined in this manner, since the size of the calculation will become too large. This problem occurs when the electron energy is close to or larger than the first ionization potential of the target and the full set of Rydberg states, the ionization continuum, and dissociative states become open. Thus a selection process must be made. Generally, a multichannel study is undertaken because a particular set of elastic/excitation processes is of interest. The open channel space should include the initial and final channels for these processes, as well as other open channels which have significant couplings with the channels under study. For example, in a study of valence excitations, Rydberg states are generally neglected because the Rydberg-valence coupling tends to be small.[36] A significant amount of experimentation is required in the selection.

The full Gaussian basis is used to span the continuum electron orbital. This guarantees that the open channel function is invariant when we transform the orbitals to achieve an optimal representation of the closed channel space. Notice that, in the SMC method, terms of the type given in Eqs. (23) and (24), so called penetration terms, belong to the open channel space because they are associated with the open channel target functions and the projection operator P does not annihilate them. In the Kohn method, which employs the Feshbach projector formalism to partition the open and closed channel space, these terms are grouped into the closed channel space instead.

The closed channel functions are chosen so they can describe, in as compact an expansion as possible, the effects of energetically inaccessible channels and/or transient negative ions on the scattering process. For a nonresonant process, the closed channel space contributes to the description of the mutual polarization effects between the target and electron. In the type of trial function used in Eq. (40), this contribution is represented by the use of 1-hole-2-particle configurations obtained by multiplying the dipole-allowed, singly-excited configurations generated from the target function with the continuum orbital. The description "dipole allowed" means the transition dipole moment between the singly excited N-electron CSF and the target function does not vanish. The polarization effect will be accounted for if a complete set of such configurations is included in the trial function.

In practice, however, this will make the calculation quite large, especially if a large one-electron basis is used. Thus truncation of the CSF expansion is necessary. This is achieved by employing natural orbitals derived from bound state calculations. Earlier SMC calculations[37] used natural orbitals from bound state negative ion CI calculations. Because these are bound state calculations, the continuum electron is simulated by a high-lying Rydberg orbital. Nevertheless, the natural orbitals from the CI calculations can be used to provide a shorter expansion for the closed channel functions. A more efficient expansion has recently been proposed by Lengsfield et al.[38] using polarized orbitals, an approach adopted in recent SMC calculations.[22]

For processes involving shape resonances, natural orbitals deduced from a bound state $N+1$-electron CI still provides an optimal representation of the closed channel space. In the cases we have studied so far, the CI wave function suitable for representing the transient negative ion can be readily identified by the presence, in the set of natural orbitals for this state, of an orbital with an occupation number of nearly one and with a strong antibonding valence character. Based on the size of the corresponding CI coefficients, a truncated CSF list can be generated and then used to generate the closed channel configurations. This approach has been employed to obtain an accurate description of the $^2\Pi_g$ resonance in the elastic scattering of N_2,[39] and the shape resonances in the Be atom and small Be clusters.[22]

Polarization effects are also important for resonant channels so one should generally use a combination of the above two strategies. One can either generate a set of polarized orbitals from the natural orbitals excluded from the description of the transient negative ion, or order the set of discarded CSF's by the transition dipole moment between the N-electron function associated with the CSF and the target state. A recent calculation on elastic scattering of CF_4, incorporating polarization effects, is based on the second approach. Some of the results are presented at the end of part (f).

The SMC method can also be used to study Feshbach resonances which arise from the interaction of the open channel with a bound negative ion associated with an excited target state. The closed channel configurations used to represent a Feshbach resonance can be generated readily using the bound state CI techniques described above. An SMC study of a Feshbach resonance in e-H_2 collisions has been reported.[40] However, this calculation used a frozen core approximation for the excited state orbitals instead of CI natural orbitals.

(d) Integrals. Three type of integrals are involved: (1) The matrix elements $\langle \Psi_m^{(-)}|\hat{H}|\Psi_n^{(+)}\rangle$ and $\langle \Psi_m^{(-)}|V|\Psi_n^{(+)}\rangle$. These matrix elements involve integrals between Gaussians which are evaluated using existing quantum chemistry packages. (2) The matrix elements $\langle \Psi_m^{(-)}|V|S_n\rangle$ and $\langle S_m|V|\Psi_n^{(+)}\rangle$. If V is the nuclear attraction potential, it involves integrals between a Gaussian and a plane wave. If V is the two-electron repulsion potential, the integrals are between three Gaussians and a plane wave. Analytical expressions for both one and two-electron integrals have been derived.[13, 42] These integrals are complex and their evaluation are more costly than the Gaussian integrals. For example, the two-electron integrals are an order of magnitude slower than the corresponding Gaussian integrals. Nevertheless, integral packages are available for their calculation. Note that these integrals are calculated over the angular quadrature of $\mathbf{k}_m, \mathbf{k}_n$. (3) The Green's function matrix element. This type of integrals is considered to be the bottleneck in a Schwinger calculation. They will be discussed separately in Sec. 3.5.

(e) Formation of the $A^{(+)}$ matrix and its inversion. Once the integrals are calculated and the open and closed channel configurations are chosen, the gathering of the $A^{(+)}$ matrix elements follows the structure of quantum chemistry codes. The operation of the projection operator P is also straightforward. For SCF target functions, this operation is done directly but for CI target functions, it can be more efficiently performed using density matrices. In Sec. 3.6 we shall consider how SMC calculations avoid a certain type of pseudoresonance when correlated target functions are used.

Due to the use of complex boundary conditions, the $A^{(+)}$ matrix elements are complex. Its inversion is done using matrix inversion routines for complex matrices available in many computational science libraries. The body-frame $T(\mathbf{k}_m, \mathbf{k}_n)$ is then obtained from Eq. (34) using the inverted $A^{(+)}$ matrix, If the projection parameter a is to be determined variationally, then an iterative solution of a is coupled with the calculation of $T(\mathbf{k}_m, \mathbf{k}_n)$ at this step.

(f) Frame transformation and cross section expression. To obtain the differential cross section, we need to transform the body-frame T-matrix into the laboratory-frame. Let (θ_n, φ_n) and (θ_m, φ_m) be the angular coordinates of $\mathbf{k}_n$ and $\mathbf{k}_m$ in the body-fixed frame. We first expand $T(\mathbf{k}_m, \mathbf{k}_n)$ in terms of partial waves of $(\theta_m,$

$\varphi_m)$

$$T(\mathbf{k}_m, \mathbf{k}_n) = \sum_{l,\mu} T_{l\mu}(k_m, k_n, \varphi_n, \theta_n) Y_{l\mu}(\theta_m, \varphi_m). \tag{41}$$

The body-frame partial wave T-matrix is determined by

$$T_{l\mu}(k_m, k_n, \varphi_n, \theta_n) = \int_o^\pi d\theta_m sin\theta_m \int_o^{2\pi} d\varphi_m Y_{l\mu}^*(\theta_m, \varphi_m) T(\mathbf{k}_m, \mathbf{k}_n).$$

Next the body-frame partial wave T-matrix is transformed to a laboratory-frame with the new Z-axis in the direction of $\mathbf{k}_n$. The Euler angles for the frame transformation is $(0, \theta_n, \varphi_n)$.

$$T^L(\theta, \varphi, \theta_n, \varphi_n) = \sum_{l,\mu,\nu} T_{l\mu}(k_m, k_n, \varphi_n, \theta_n) Y_{l\mu}(\theta_m, \varphi_m) D^l_{\mu\nu}(0, \theta_n, \varphi_n).$$

Here (θ, φ) is the solid scattering angle in the laboratory frame. Then $|T^L|^2$ is averaged over θ_n, φ_n to account for the random orientation of the target,

$$\sigma(\theta, \phi) = \frac{1}{16\pi^3} \frac{k_m}{k_n} \int_o^\pi d\theta_n sin\theta_n \int_o^{2\pi} d\varphi_n |T^L(\theta, \varphi, \varphi_n, \theta_n)|^2.$$

The physical cross section is obtained by averaging over the azimuthal angle φ,

$$\sigma(\theta) = \frac{1}{2\pi} \int_o^{2\pi} d\varphi \sigma(\theta, \varphi).$$

Integration of $\sigma(\theta)$ over θ gives the integral cross section. The integral cross section can also be obtained directly in the body-frame,

$$\sigma = \frac{1}{16\pi^3} \frac{k_m}{k_n} \int d\hat{\mathbf{k}}_n \int d\hat{\mathbf{k}}_m |T(\mathbf{k}_m, \mathbf{k}_\mathbf{n})|^2.$$

It should be mentioned that the number of partial waves deducible from Eq. (41) is determined by the size of the angular quadrature used to calculate the body-frame T-matrix for various plane wave orientations. While in principle each plane wave can be decomposed into an infinite sum of angular momentum waves, the finite angular quadrature used in the calculation limits the number of angular momentum waves obtainable from such decomposition. However, as discussed in part (g), the scattering of high l partial waves can be described by perturbation theory and does not require a full scale variational calculation. Thus the limited number of partial waves deduced from Eq. (41) does not present a problem. In most cases, a (8x8) or (10x10) Gaussian-Legendre quadrature for $\hat{k}_m$ and $\hat{k}_n$ give $l_{max} = 6$ for the angular momentum wave.

A recent calculation of e-CF_4 elastic scattering,[43] including polarization effects, serves as an example of an SMC calculation which employs a set of optimized closed channel configurations. An elastic scattering calculation in the static-exchange (SE) approximation has been reported by Winstead et al.[44] who improved an earlier SE calculation by Huo[45] by employing a larger Gaussian basis. They reported an A_1 resonance at $\approx$ 13 eV and a T_2 resonance at $\approx$ 11 eV. The present polarization calculation used a Gaussian basis of 10s6p1d functions at each nuclear center. The target function was described by an SCF function at the experimental equilibrium geometry and an (8x8) angular quadrature for the wave vector was used. The closed channel space was chosen using the procedure described in part (c) above. Thus bound state CI calculations were carried out for negative ions of $^2A_1$, $^2T_{2x}$, $^2T_{2y}$ and $^2T_{2z}$ symmetries and the CI wave

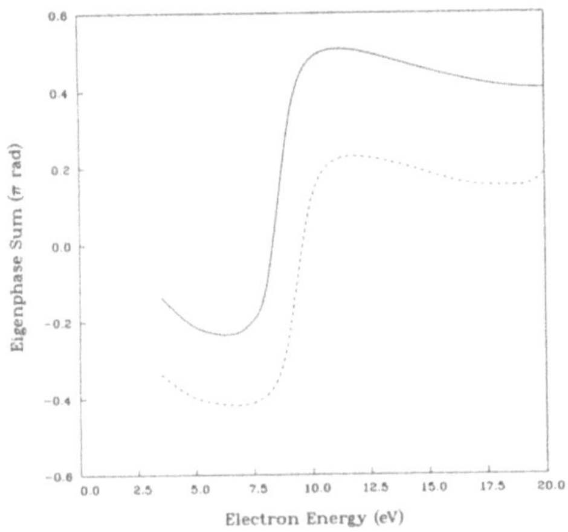

Figure 1. Eigenphase sums for electron-CF_4 scattering, showing resonances in the $^2A_1$ (dashed line) and $^2T_{2x}$ (solid line) partial channels.

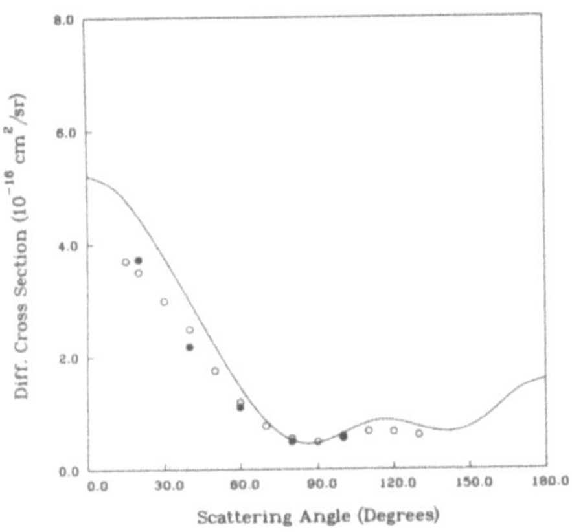

Figure 2. Differential cross section for elastic electron-CF_4 scattering at 8 eV. Solid line, present polarized result, filled circles, experiment of Mann and Linder,[46] and open circles, experiment of Boesten et al.[47]

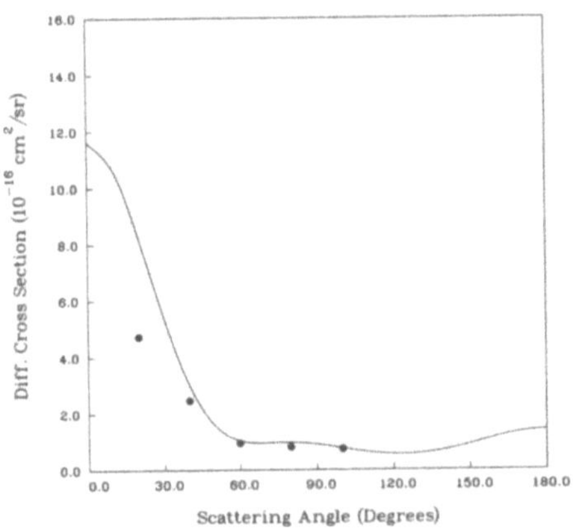

Figure 3. Differential cross section for elastic electron-CF_4 scattering at 11 eV. Solid line, present polarized result, and filled circles, experiment of Mann and Linder.[46]

functions with a strong antibonding character were identified as good candidates to describe the transient negative ions. It was found that the natural orbitals from the $^2A_1$ ion calculation could also provide rather compact representations of the CI functions of the $^2T_{2x}$, $^2T_{2y}$ and $^2T_{2z}$ ions. Thus the $^2A_1$ natural orbitals were used in the scattering calculation and lists of CSF's for the four symmetries were generated from the respective negative ion CI calculations. The lists were augmented to include polarization effects based on the size of the transition moment. They were then used to represent the closed channel functions in the polarized scattering calculation. Figure 1 presents the eigenphase sum of the A_1 and T_{2x} partial channels. The position of the A_1 resonance is lowered to 9.0 eV and the T_2 resonance to 8.6 eV, in much better agreement with experimentally observed features.[46, 47] Figures 2 and 3 present the theoretical differential cross section (dcs) at 8 and 11 eV, together with experimental data.

(g) Born closure for long range potentials. When the interaction potential or transition potential includes long range multiple potentials, such as the dipole potential in elastic scattering or the dipole transition potential in dipole-allowed transitions, the number of partial waves required to converge the cross section, especially in the forward direction, is significantly larger than what can be provided by an SMC calculation with moderate size angular quadrature. On the other hand, due to the large centrifugal barrier for the higher partial waves, they are prevented from reaching the inner region of the target and the scattering tends to be weaker. Thus the Born approximation is expected to be valid for the high l partial waves and can be employed to describe their contributions[48-50] to the cross section. Gibson et al.[41] made use of the following Born closure approximation,

$$\sigma(\theta) = \sigma^{FBA}(\theta) + [\sigma^{SMC}(\theta) - \sigma_{FE}^{FBA}(\theta)]. \tag{42}$$

to include the high l contributions in the dcs. Here FBA denotes the first Born approximation and $\sigma_{FE}^{FBA}(\theta)$ is obtained from a finite expansion of the first Born cross section containing exactly the same number of partial waves as $\sigma^{SMC}(\theta)$. In Eq. (42), the Born value for the low l contributions to the dcs is subtracted out. Thus we use the SMC result for the low l and Born value for the high l contributions to the dcs. In evaluating $\sigma_{FE}^{FBA}(\theta)$, care should be taken so that the same angular quadrature of $\mathbf{k}_m$ and $\mathbf{k}_n$ are used as in the SMC calculation.

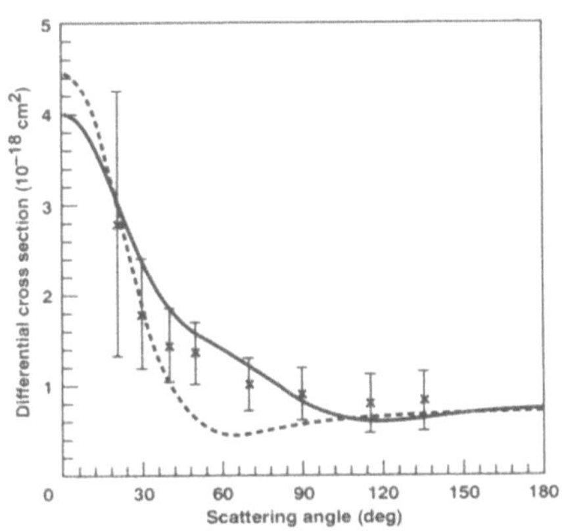

Figure 4. Differential cross section for $X^1\Sigma_g^+ \rightarrow B^1\Sigma_u^+$ transition of H_2 at 15 eV. Solid line, SMC result, dashed line, distorted wave result,[48] and crosses, experimental data of Srivastava and Jensen.[49]

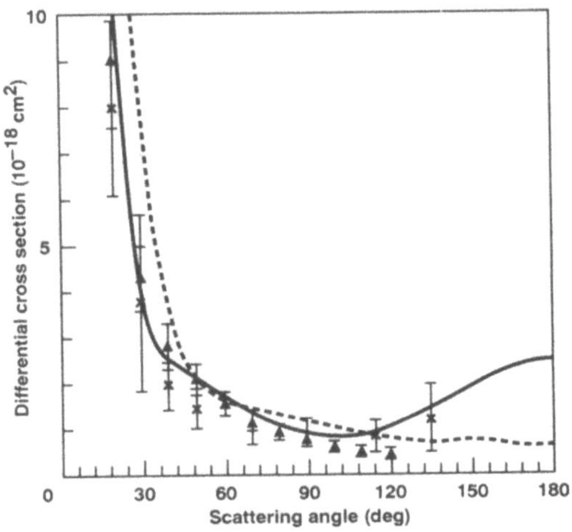

Figure 5. Differential cross section for $X^1\Sigma_g^+ \rightarrow B^1\Sigma_u^+$ transition of H_2 at 20 eV. Solid line, SMC result, dashed line, distorted wave result,[48] crosses, experimental data of Srivastava and Jensen,[49] and triangles, experimental data of Khakoo and Trajmar.[50]

Equation (42) has been applied to a two-channel SMC calculation to study the dipole allowed transition $X^1\Sigma_g^+ \rightarrow B^1\Sigma_u^+$ in H_2. Figures 4 and 5 present the 15 and 20 eV dcs from the SMC calculation with Born closure, the distorted wave calculation of Rescigno et al.,[51] and the experimental data of Srivastava and Jensen[52] and Khakoo and Trajmar.[53]

It should be pointed out that in their implementation of the Born closure approximation to study elastic scattering of polar molecules, Rescigno et al. [4, 54] took a slightly different approach and applied it to the body-frame T-matrix instead of the dcs.

3.5. Evaluation of the Green's Function Matrix Elements

The Green's function matrix element

$$M = \langle \Psi_m^{(-)} | V G_P^{(+)} V | \Psi_n^{(+)} \rangle,$$

can be expressed as

$$M = M_R + iM_I.$$

The imaginary part M_I is given by the residue,

$$M_I = -\frac{i}{8\pi^2} \sum_l k_l \int d\hat{\mathbf{k}}_l \langle \Psi_m^{(-)} | V | S_l \rangle \langle S_l | V | \Psi_n^{(+)} \rangle. \tag{43}$$

where l sums over the poles of $G_P^{(+)}$. Since $G_P^{(+)}$ is a projected Green's function, this is equivalent to a sum over the open channels and k_l is just the wave vector associated with the open channel l. Also, $\langle S_l | V | \Psi_n^{(+)} \rangle$ is the same matrix element used in the expression of the body-frame T-matrix, T_{ln}. The integration over $\hat{\mathbf{k}}_l$ in Eq. (43) can use the same quadrature as the body-frame T-matrix. Consequently, the calculation of M_I involves very little extra work.

The real part of M is given by the principal-value integral,

$$M_R = \frac{1}{(2\pi)^3} \mathcal{P} \int_o^\infty d\epsilon \sum_{l \in open} \frac{k\, U_{mn}}{(E - \mathcal{E}_l) - \epsilon + i\delta}. \tag{44}$$

with

$$U_{mn}(k) = \int d\hat{\mathbf{k}} \langle \Psi_m^{(-)} | V | \Phi_l e^{i\mathbf{k}\cdot\mathbf{r}_{N+1}} \rangle \langle e^{i\mathbf{k}\cdot\mathbf{r}_{N+1}} \Phi_l | V | \Psi_n^{(+)} \rangle, \tag{45}$$

Here $\epsilon = \frac{1}{2}k^2$. Due to the integration over ϵ and $\hat{\mathbf{k}}$, M_R is much more difficult to compute. Its efficient and accurate calculation is an important factor in the development of a practical SMC method. Two approaches to computing this integral are described below.

(a) Direct numerical calculation over a quadrature of $\hat{\mathbf{k}}$ and ϵ. An optimized quadrature is obviously important in this approach. To avoid the rapid variation of the integrand in the ϵ coordinate near the poles of the Green's function, we follow the work of Walters[55] and Heller and Reinhardt[56] and separate the ϵ-integration into two regions, with the poles of the Green's function all located in the first region. The rapid variation in the first integrand is avoided by the addition and subtraction of a term such that the numerical integration can be carried out over a smooth integrand. The ϵ-integral in the first region is rewritten as,

$$I_1 = \sum_{l \in open} \mathcal{P} \int_o^{\epsilon_a} d\epsilon \frac{k\, U_{mn}(k)}{E - \mathcal{E}_l - \epsilon} = \sum_{l \in open} \mathcal{P} \int_o^{\epsilon_a} d\epsilon \left\{ \frac{k\, U_{mn}(k)}{E - \mathcal{E}_l - \epsilon} - \frac{k_l\, U_{mn}(k_l)}{E - \mathcal{E}_l - \epsilon} + \frac{k_l\, U_{mn}(k_l)}{E - \mathcal{E}_l - \epsilon} \right\}, \tag{46}$$

with $\epsilon_a > (E - \mathcal{E}_l)$ for all l. In Eq. (46), the second and third term in the integrand are obtained from the first by replacing $\hat{k}$ and k with $\hat{k}_l$ and k_l. Because $U_{mn}(k_l)$ and k_l are independent of ϵ, the last integral on the right-hand side of Eq. (46) can be evaluated analytically and Eq. (46) is rewritten as,

$$I_1 = \sum_{l \in open} \left\{ k_l\, U_{mn}(k_l)\, ln \left(\frac{\epsilon_l}{\epsilon_a - \epsilon_l} \right) + \mathcal{P} \int_o^{\epsilon_a} d\epsilon \left[\frac{k\, U_{mn}(k)}{E - \mathcal{E}_l - \epsilon} - \frac{k_l\, U_{mn}(k_l)}{E - \mathcal{E}_l - \epsilon} \right] \right\}.$$

Since the first and second term in the ϵ-integand above have poles at the same place, the large changes in the integrand near the pole region are canceled out and the integrand remains smooth, suitable for numerical integration.

To demonstrate the convergence of the numerical quadrature scheme, we calculated the elastic scattering of CO in the static-exchange approximation, using an SCF target at the experimental equilibrium geometry. A Gaussian basis, with 5s3p1d contracted functions at each nuclear center, represented the continuum electron. The SMC calculation included up to $l_{max} = 6$. Born closure was not included since our purpose was to demonstrate the convergence of the $VG_P^{(+)}V$ quadrature, and the introduction of a constant correction term from the Born closure would not affect our conclusions. Also, all calculations used the Gauss-Legendre quadrature.

Figure 6 presents the partial integral cross section for e-CO elastic scattering. A (20x20) quadrature was used for the angular integration and a 20-point quadrature for the second region ϵ integral. Three different quadratures were used for the first ϵ region, 20, 32, and 48 points. The three cross section curves virtually coalesce, indicating a well converged result.

The cross section is much more sensitive to the quadrature used in the second ϵ integration. Figure 7 shows the cross sections obtained using 20, 32, and 48 quadrature points for the second region integration and keeping the other quadratures constant. Note that the peak of the cross section curve at $\approx$ 3.5 eV, arising from the $^2\Pi$ shape resonance in CO, is shifted to a lower energy when a larger quadrature is used. Also, the 32- and 48-point calculations have similar resonance positions, but the cross sections themselves differ by $\approx 10\%$ at the peak.

Sensitivity to the angular quadrature was also studied. Figure 8 presents the results for the variation in the θ quadrature, keeping all other integration parameters constant. While the 32- and 48-point integrations give substantially similar results, the 20-point quadrature has not yet converged. In particular, note that the maximum of the cross section, at $\approx$ 3.5 eV energy, is shifted to a higher energy when a larger quadrature is used.

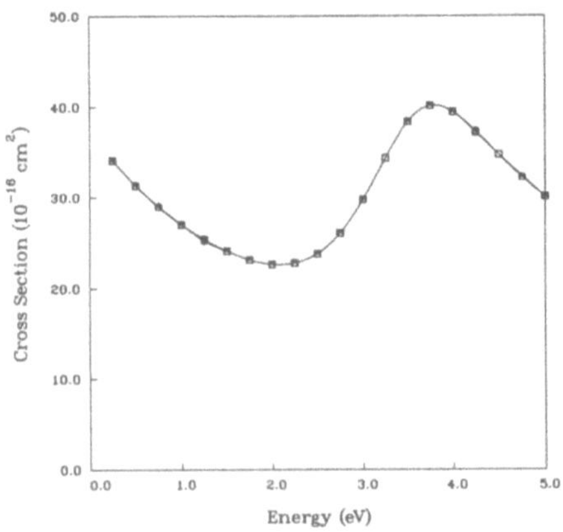

Figure 6. Partial integral cross section (l_{max}=6) for e-CO elastic scattering in the static-exchange approximation. A (20x20) quadrature is used for the angular integration in the $VG_P^{(+)}V$ term and 20-point for the second region ϵ-integration. The quadrature size for the first region ϵ-integration is: circle, 20 points, square, 32 points, and triangle, 48 points.

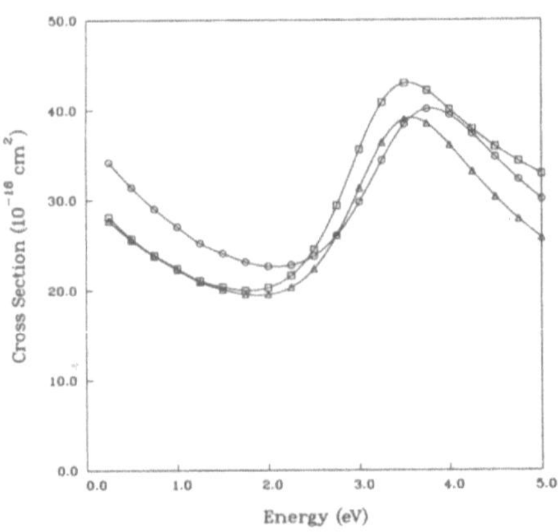

Figure 7. Same as Fig. 6. A (20x20) quadrature is used for the angular integration and 20-point for the first region ϵ-integration. The quadrature size for the second region ϵ-integration is: circle, 20 points, square, 32 points, and triangle, 48 points.

When a bigger basis set is used for the target function, the demand for a larger angular quadrature becomes higher, particularly if the basis set is uncontracted and has very large exponents. Also, the second ϵ integral shows more sensitivity to the size of the angular quadrature. These features suggest that the off-shell contribution to the $VG_P^{(+)}V$ term, coming from the high energy region of the target spectrum, may be responsible for the need of large quadratures. In the above example, a converged result requires over forty thousand quadrature points.

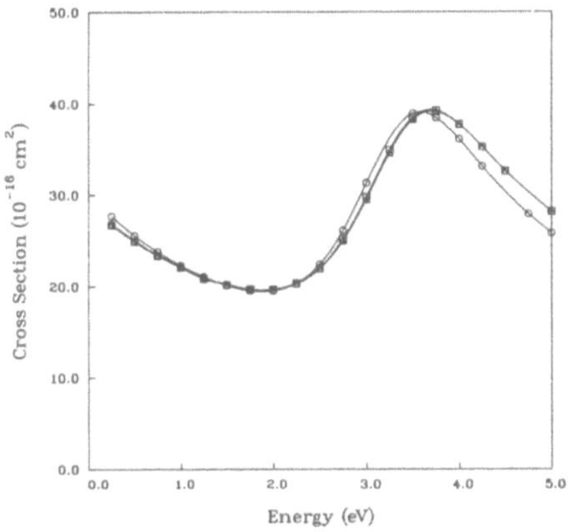

Figure 8. Same as Fig.6. A 20-point quadrature is used for the first region ϵ-integration, 48 for the second, and 20 for the φ integration. Circle, 20-point θ quadrature, square, 32 points, and triangle, 48 points.

The direct numerical integration approach was first investigated by Lima et al.[57] To expedite the integral calculation, Winstead et al.[58] have successfully adapted the integral calculation and the three-index transformation[5/] to the highly parallel computers. Parallel computers are particularly well suited for this purpose because each integral calculation is independent of others. SMC calculations using this technique have been carried out on many polyatomic molecules.[12]

(b) Insertion using a Gaussian basis. Provided that a Gaussian basis satisfies the approximate relation

$$\sum_a |p_a\rangle\langle p_a| \approx 1.$$

the free-particle Green's function can be projected into this basis, called the insertion basis. The principal-value Green's function matrix element is now written as

$$M_R = \frac{1}{(2\pi)^3} \sum_{l \in open} \sum_a \sum_b \langle \Psi_m^{(-)} | V | \Phi_l \, p_a(\mathbf{r}_{N+1}) \rangle \langle \Phi_l \, p_b(\mathbf{r}_{N+1}) | V | \Psi_n^{(+)} \rangle$$

$$\times \mathcal{P} \int_o^\infty d\epsilon \int d\hat{\mathbf{k}} \, k \, \frac{\langle p_a(\mathbf{r}_{N+1}) | e^{i\mathbf{k}\cdot\mathbf{r}_{N+1}} \rangle \langle e^{i\mathbf{k}\cdot\mathbf{r}_{N+1}} | p_b(\mathbf{r}_{N+1}) \rangle}{(E - \mathcal{E}_l) - \epsilon + i\delta}. \tag{47}$$

Both ϵ and $\hat{\mathbf{k}}$ integrals are expressed in closed form.[59] Furthermore, calculations of these Gaussian integrals are an order of magnitude faster than the three-Gaussian-one-plane wave integrals in Eq. (45).

It is important to recognize that in Eq. (47), an L^2 basis is used in the evaluation of an integral over the Green's function, but not to represent the Green's function itself. Just as an L^2 basis can be used to represent a trial function in a Schwinger variational calculation because the trial function is always multiplied by V in the expression of the T-matrix, in the present case the free-particle Green's function is integrated over $V|\Psi^{(+)}\rangle$, with the trial $\Psi^{(+)}$ itself expressed in terms of another L^2 basis. Thus, the long range oscillatory part of the Green's function does not contribute to the integral. In this respect, the application of an L^2 basis to evaluate M_R is analogous in spirit to

[5/] The fourth index is the plane wave.

Schneider's calculation[60] of the Schwinger separable potential by projection onto an L^2 basis.

The insertion technique was used in early SMC calculations when large, uncontracted Gaussian bases were used for the scattering basis, with the insertion basis usually composed of the scattering basis supplemented by additional Gaussians. Good convergence behavior was observed.[39] The change to the direct numerical integration technique by Lima et al.[57] was motivated by the convergence problems they encountered. Such problems appeared to be related to choosing a suitable insertion basis.

In view of recent progress made in the field of quantum chemistry in developing flexible Gaussian bases,[33] and the efficiency in the computation of Gaussian integrals themselves, we re-investigated the insertion technique.[20] The choice of a suitable basis was guided by the convergence characteristics observed in the numerical integration approach, namely that the off-shell high energy contribution to M_R seems to play an important role in the convergence. To describe this type of contribution, we included Gaussians with large exponents. Also, a multicentered Gaussian basis readily simulates the dependence on the angular quadrature. When the insertion basis was chosen using these guidelines, good convergence was observed. As an example, we studied the e-CO static-exchange problem using the same scattering basis as discussed above. The insertion basis is obtained by supplementing the scattering basis by 10s18p9d Gaussians at each nuclear center and 16s5p9d Gaussians at the center of mass.[6/] Figure 9 compares the cross sections obtained using the insertion method versus the direct numerical integration approach. Good agreement is observed. The largest discrepancy between the two results, less than 3%, is found at the low energy region. In the $^2\Pi$ shape resonance region, the two curves almost coalesce.

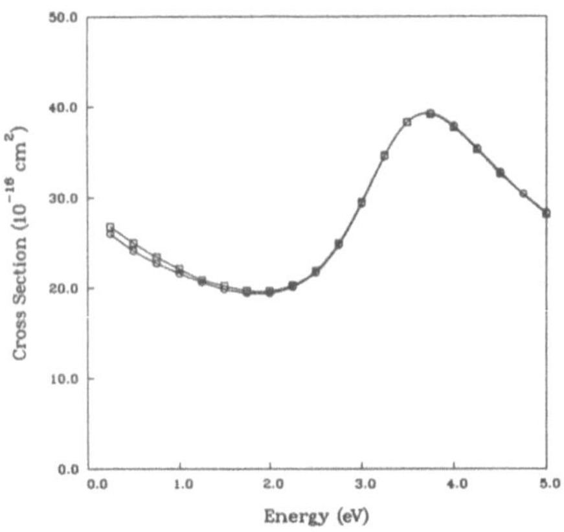

Figure 9. Partial integral cross section (l_{max}=6) for e-CO elastic scattering in the static-exchange approximation. Square, calculated using a direct numerical integration, and circle, calculated with the insertion technique. The numerical integration uses a (32x32) angular quadrature, 20-point for the first ϵ integration, and 48-point for the second. The insertion basis is described in the text.

While the insertion basis used for the CO calculation appears to be very large, it should be pointed out that the insertion basis has 238 Gaussian functions and the total number of Gaussian integrals used is 15.2 million. The numerical integration calculation

[6/] Not all components of the d functions were used. Also, a smaller size p_z basis was used at the nuclear center.

requires more than 2.78 billion three-Gaussian-one-plane wave type integrals.[7/] Taking into account the difference in the cycle time required to calculate the two types of integrals, the insertion technique requires orders of magnitude less CPU and storage resources.

Recently, Bettega et al.[61] applied the pseudopotential method and removed the core electrons from the SMC calculations. This raises an interesting question. Since the off-shell high energy contribution to M_R that we observed originate, at least partially, from the core electrons, the use of pseudopotentials may lessen the demand for large quadratures in the numerical integration technique and the demand for tight Gaussians in the insertion technique. This aspect remains to be tested.

3.6. Correlated Target Function in the SMC Method

Because molecular bonding changes with nuclear geometry, multiconfiguration wave functions are required to provide a consistent description of the target over a range of geometries. Similarly, the description of several electronic states with a common set of orbitals, a practice frequently used in the study of electronic excitations by electron or photon impact, requires the use of multiconfiguration target functions. Thus correlated target functions are used with some frequency in *ab initio* studies of e-molecule collisions.[62–65]

An ubiquitous feature in the use of multiconfiguration target functions is the presence of pseudoresonances at intermediate energies.[62, 65] Lengsfield and Rescigno[62] attributed these resonances to the fact that multiconfiguration target functions can introduce certain terms in the $N+1$-electron wave function which are associated with excited states of the target excluded from the open channel space. Pseudoresonances appear at an energy where the excited target state becomes open. Consider the example of a two-configuration function for the ground state of H_2[8/]

$$\Phi_X = a_1\phi_1 + a_2\phi_2,$$

with

$$\phi_1 = \mathcal{A}\{1\sigma_g\alpha(1)1\sigma_g\beta(2)\},$$
$$\phi_2 = \mathcal{A}\{1\sigma_u\alpha(1)1\sigma_u\beta(2)\}.$$

Here Φ_X is the simplest function that can describe the dissociation of H_2 properly.

Two $N+1$-electron CSF's are used to relax the orthogonality between the target and the continuum electron,

$$\Theta_a = \mathcal{A}_{N+1}\Big\{\Phi_X(1,2)1\sigma_g\alpha(3)\Big\} = a_2\mathcal{A}_{N+1}\Big\{\phi_2(1,2)1\sigma_g\alpha(3)\Big\}.$$

and

$$\Theta_b = \mathcal{A}_{N+1}\Big\{\Phi_X(1,2)1\sigma_u\alpha(3)\Big\} = a_1\mathcal{A}_{N+1}\Big\{\phi_1(1,2)1\sigma_u\alpha(3)\Big\}.$$

Here $\mathcal{A}_{N+1}$ antisymmetrizes electron 3 with electrons 1 and 2. Note that Θ_a includes the terms $1\sigma_g\alpha(1)1\sigma_u\alpha(2)1\sigma_u\beta(3)$ and $1\sigma_g\alpha(1)1\sigma_u\beta(2)1\sigma_u\alpha(3)$. These terms are associated with the $b^3\Sigma_u^+$ target state, with the $1\sigma_u$ orbital representing the continuum electron. In the Kohn calculation it was found that if the $b^3\Sigma_u^+$ state was not included in the open

7/ As a result of symmetry operations, in both cases the actual number of distinct integrals to be calculated are less. However, the ratio of the the two numbers will be approximately the same when symmetry is taken into account.

8/ This example was originally used by Lengsfield and Rescigno.[4, 62]

channel space, a pseudoresonance would occur around the energy where this channel became open. Similar pseudoresonances were also observed in the R-matrix calculation of the electronic excitations of H_2.[65] The strategy used in the Kohn method to eliminate such terms has been discussed.[4, 62]

Due to the use of the N-electron projection operator, the SMC method treats correlated target functions differently. As seen in Sec. 3.1, the N-electron projection operator selects the terms that are associated with the target function from an antisymmetrized $N+1$-electron CSF, resulting in an un-antisymmetrized product of the N-electron target function and an one-electron orbital describing the continuum electron. For a multiconfiguration target function, this operation has additional consequences. As an illustration, consider the same two-configuration wave function. The projection operator P is given by

$$P = |\Phi_X\rangle\langle\Phi_X|.$$

The operation of P on Θ_a gives,

$$P|\Theta_a\rangle = \frac{a_2}{\sqrt{3}}\Phi_X(1,2)1\sigma_g\alpha(3).$$

Thus the operation of P recovers the ϕ_1 component in Θ_a, originally eliminated by antisymmetrization. In this manner, the projector P removes the $b^3\Sigma_u^+$ character from Θ_a, and $P\Theta_a$, unlike Θ_a itself, consists only of the open channel target.

In an SMC calculation, a pseudoresonance usually appears when the denominator in the expression of the T-matrix becomes particularly small at a certain energy. Because Θ_a includes a term that behaves like a open channel function, the matrix element

$$\langle\Psi^{(-)}|\hat{H}|\Theta_a\rangle,$$

as a function of energy E goes through a minimum around the energy when that channel becomes open. On the other hand, $P\Theta_a$ does not have any $b^3\Sigma_u^+$ character, and

$$\langle\Psi^{(-)}|\hat{H}P|\Theta_a\rangle,$$

does not exhibit the same minimum.

Based on this analysis, an SMC calculation is expected to perform differently from other methods when a multiconfiguration target is used. Figure 10 presents an SMC study of the ${}^2\Sigma_g^+$ channel elastic scattering of H_2, using the static-exchange approximation. The calculation was carried out at $R = 4.0\ a_o$, using a Gaussian basis of 6s6p at H and 4s4p at the center of mass, The CI coefficients of the two-configuration target function are 0.8785 and -0.4778. As seen in Fig. 10, the SMC cross section curve is smooth over the energy range of 0.25 to 21 eV. The pseudoresonance found in other methods, arising due to the term Θ_a, is absent here. Thus it appears that the role of the projection of P on Θ_a is to associate Θ_a with the correct open channel, and thereby removes the spurious resonance.

The above results suggests that the N-electron projection operator has additional advantages besides providing a convenient treatment for exchange. The use of correlated target function in the SMC method is relatively recent. Only a few studies have been made using a multiconfiguration target function.[22] Future studies should explore this aspect further.

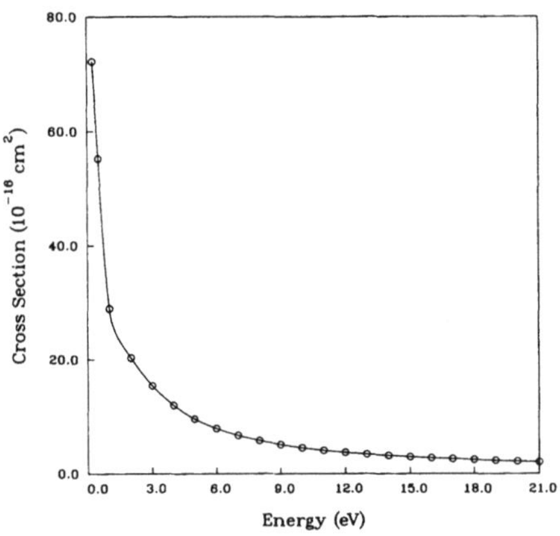

Figure 10. Integral cross section for e-H_2 $^2\Sigma_g^+$ channel elastic scattering in the static-exchange approximation, calculated using a two-term CI function for the target

4. THE USE OF SMC RESULTS IN THE STUDY OF VIBRATIONAL EXCITATIONS

All SMC calculations use the fixed-nuclei approximation. However, the fixed-nuclei SMC results can be readily used as a starting point in the application of the nuclear dynamics treatment of Domcke. Domcke's treatment, which is based on the projection operator formalism, has been reviewed recently,[23] and it will not be repeated. Here we consider an example of combining the two approaches in the study of the resonant enhanced vibrational excitation of N_2 by electron impact. Because the $^2\Pi_g$ resonance in N_2 is a pure d-wave resonance, we rewrite the T-matrix for the l=2 partial wave in terms of the K-matrix,

$$T_{l=2}(E,R) = \frac{K(E,R)}{1 - iK(E,R)},$$

where R is the internuclear distance. The K-matrix element is related to the width $\Gamma(E,R)$ and shift $\Delta(E,R)$ functions by

$$K(E,R) = -\frac{\Gamma(E,R)}{2[E - \epsilon_d(R) - \Delta(E,R)]}.$$

Here ϵ_d is the difference between the target and negative ion potential curves. The shift function is given by the principal value integral,

$$\Delta(E,R) = \frac{1}{2\pi}\mathcal{P}\int dE' \frac{\Gamma(E',R)}{E - E'},$$

and the width function can be expressed in terms of the potential U_{dE} which Domcke called the entry amplitude,

$$\Gamma(E,R) = 2\pi |U_{dE}|^2.$$

An analytical fit of U_{dE} can be obtained using the K-matrix determined from the SMC calculation. Notice that U_{dE} is a nonlocal potential, depending on both R and E.

The calculation on N_2 has been reported previously.[39, 66] In applying Domcke's method, we used a Schwinger-type separable potential approach to evaluate the T-matrix element. A complete set of the bound vibrational wave function of the transient negative ion, obtained using a numerical solution of the 1-d schrödinger equation, was used in the separable potential calculation. The negative ion curve was determined empirically by Berman et al.[67]

The results of this calculation compared well with experiment, and the comparison will not be repeated. Here we consider an example of applying our result to an experimentally inaccessible region. In very high temperature plasmas, as in the flow field of a high-speed space vehicle upon entry to the earth's atmosphere, molecules are very hot ro-vibrationally and it is difficult to duplicate such conditions in the laboratory. Figure 11 presents the $v = 0 \rightarrow 1$ and $v = 3 \rightarrow 4$ vibrational excitation cross sections of N_2 by electron impact. While many experimental measurements on transitions from $v = 0$ have been done, there has been only one reported measurement with the initial state at the $v = 1$ level.[68] Due to the difficulties involved in preparing a significant quantity of N_2 at the $v = 3$ level, no measurement has been reported for excitations from this level and theoretical data is the only available source for these cross sections. It is seen from Fig. 11 that the resonance structures for the $v = 3 \rightarrow 4$ transition is more extended both at high and low energies in comparison with the $v= 0 \rightarrow 1$ transition, and not as strongly peaked. The calculation assumed the molecule was initially at $J = 50$, with the centrifugal barrier in the vibrational potential determined by this J value. Note that our calculation did not include rotation in the dynamics and the effect of rotation only entered through the centrifugal term.

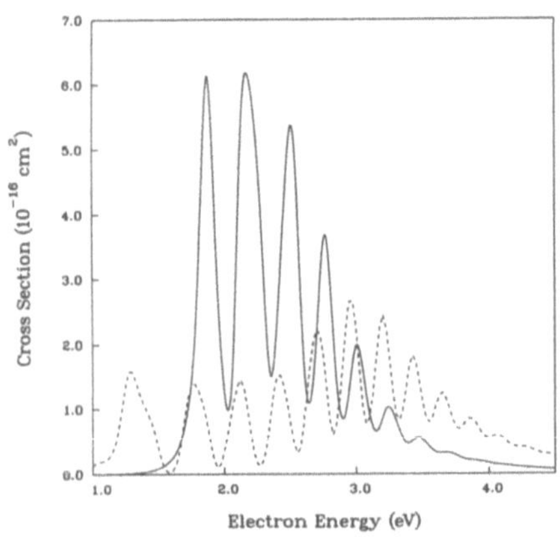

Figure 11. N_2 vibrational excitation cross sections by electron impact. Solid line, $v = 0 \rightarrow 1$, dashed, $v = 3 \rightarrow 4$. The calculation assumes the molecule is at $J = 50$.

Figure 12 presents the $v = 0 \rightarrow 1$ excitation cross sections for N_2 at J=0, 50, and 150. Again rotational effects contribute only through the centrifugal barrier of vibrational motion. Because the differences between the potential curves of the neutral and transient negative ion, the large centrifugal barriers at $J = 50$ and 150 have differential effects on the vibrational excitation. The resonance structure in the cross section moves to lower energies and narrows as J increases.

Future studies that couple vibrational and rotational motion at high J will extend our understanding of the collisional processes involving ro-vibrationally hot molecules.

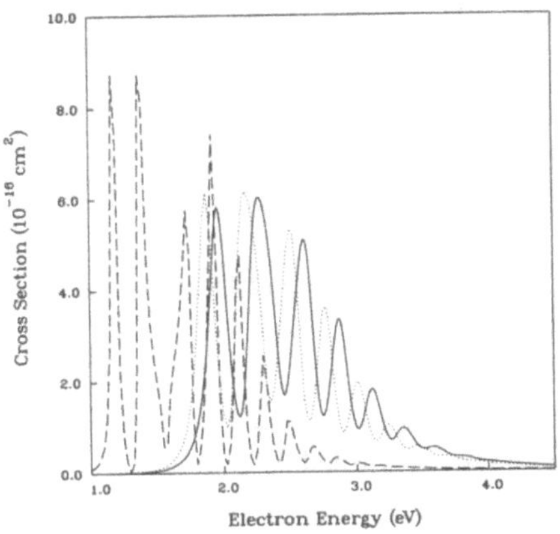

Figure 12. N_2 $v = 0 \rightarrow 1$ vibrational excitation cross section by electron impact. Solid line, $J = 0$, dotted, J=50, and dashed, J= 150.

5. SUMMARY

The SMC method, a modified version of the SV method, has proven to be a versatile tool in the study of e-molecule collisions. The SMC method has been successfully applied to the study of technologically important electron collision processes involving polyatomic targets. This application will continue as the SMC method is further developed and refined. The present discussion of its implementation also serves to illustrate some of the physical basis in its formulation, and provides a foundation for the future development of the method.

References

[1] L. Hulthén, Kgl. Fysiograf. Sälbkap. Lund. Fröh. **14**, 257 (1944).
[2] W. Kohn, Phys. Rev. **74**, 1763 (1948).
[3] S.I. Rubinow, Phys. Rev. **96**, 218 (1954).
[4] T.N. Rescigno, C.W. McCurdy, A.E. Orel, and B.H. Lengsfield III, "The Complex Kohn Variational Method," chapter 1 in this book.
[5] J.L. Jackson, Phys. Rev. **83**, 301 (1951).
[6] R.K. Nesbet, *Variational Methods in Electron-Atom Scattering Theory*, Plenum Press, New York (1980).
[7] B.I. Schneider, "An R-Matrix Approach to Electron Molecule Collisions," chapter 8 in this book.
[8] J. Schwinger, Phys. Rev. **56**, 750 (1947).
[9] R.R. Lucchese, K. Takatsuka, and V. McKoy, Phys. Rep. **131**, 147 (1986).
[10] D.K. Watson, Adv. At. Mol. Phys. **25**, 221 (1988).
[11] M.A.P. Lima, T.L. Gibson, L.M. Brescansin, V. McKoy, and W.M. Huo, "Studies of Elastic and Electronically Inelastic Electron-Molecule Collisions," in *Swarm Studies and Inelastic Electron-Molecule Collisions*, ed. L.C. Pitchford, B.V. McKoy, A. Chutjian, and S. Trajmar, Springer-Verlag, New York (1987), pp 239-264.
[12] C. Winstead and V. McKoy, "Studies of Electron-Molecule Collisions on Highly Parallel Computers," in *Modern Electronic Structure Theory Vol. 2*, ed. D. Yarkony, World Scientific, Singapore (1994).

[13] D.K. Watson and V. McKoy, Phys. Rev. A **20**, 1474 (1979).
[14] R.R. Lucchese, G. Raseev, and V. McKoy, Phys. Rev. A **25**, 2572 (1982).
[15] See, for example, G. Bandarage and R.R. Lucchese, Phys. Rev. A **47**, 1989 (1993); M.-T. Lee, K. Wang, and V. McKoy, J. Chem. Phys. **97**, 3108 (1992).
[16] M.-T. Lee, M.M. Fujimoto. S.E. Michelin, L.E. Machado, and L.M. Brescansin, J. Phys. B. **25**, L505 (1992).
[17] M.-T. Lee, S.E. Michelin, L.M. Brescansin, G.D. Meneses, and L.E. Machado, J. Phys. B. **26**, L477 (1993).
[18] K. Takatsuka and V. McKoy, Phys. Rev. A **24**, 2473 (1981).
[19] K. Takatsuka and V. McKoy, Phys. Rev. A **30**, 1734 (1981).
[20] W.M. Huo and J.A. Sheehy (to be published).
[21] C. Winstead, Q. Sun, and V. McKoy, J. Chem. Phys. **97**, 9483 (1992).
[22] W.M. Huo and J.A. Sheehy, "Theoretical Study of Electron Scattering by Small Clusters and Adsorbates," in *Electron Collisions with Molecules, Clusters, and Surfaces*, ed. H. Ehrhardt and L.A. Morgan, Plenum, New York (1994), pp 171-182.
[23] W. Domcke, Phys. Rep. **208**, 97 (1991).
[24] R.G. Newton, *Scattering Theory of Waves and Particles*, Springer-Verlag, New York (1982).
[25] B.H. Bransden, R. Hewitt, and M. Plummer, J. Phys. B **21**, 2645 (1988).
[26] S.K. Adhikari and I.H. Sloan, Phys. Rev. C **11**, 1133 (1975).
[27] K. Takatsuka, R.R. Lucchese, and V. McKoy, Phys. Rev. A **24**, 1812 (1981).
[28] J.T. Taylor, *Scattering Theory*, R. E. Krieger Publishing, FL (1983), pp. 274-279.
[29] H. Feshbach Ann. Phys. **5**. 357 (1958); *ibid* **19**, 287 (1962).
[30] M.A.P. Lima and V. McKoy, Phys. Rev. A **38**, 501 (1988).
[31] W.M. Huo and C.A. Weatherford, Bull. Am. Phys. Soc. **36**, 1265 (1991).
[32] C. Winstead and V. McKoy, Phys. Rev. A **47**, 1514 (1993).
[33] T. Helgaker and P.R. Taylor, "Gaussian Basis Sets and Molecular Integrals" in *Modern Electronic Structure Theory Vol. 2*, ed. D. Yarkony, World Scientific, Singapore (1994).
[34] T.H. Dunning, J. Chem. Phys. **53**, 2823 (1970).
[35] T.H. Dunning, J. Chem. Phys. **90**, 1007 (1989), and D.E. Woon and T.H. Dunning, J. Chem. Phys. **98**, 1358 (1993).
[36] As an example of valence excitation calculations which neglected Rydberg states in the open channel configurations, see Q. Sun, C. Winstead, V. McKoy, J.S.E. Germano, and M.A.P. Lima, Phys. Rev. A **46**, 2462 (1992).
[37] W.M. Huo, M.A.P. Lima, T.L. Gibson, and V. McKoy, Phys. Rev. A **36**, 1642 (1987).
[38] B.H. Lengsfield, T.N. Rescigno, and C.W. McCurdy, Phys. Rev. A **44**, 4296 (1991).
[39] W.M. Huo, T.L. Gibson, M.A.P. Lima, and V. McKoy, Phys. Rev. A **36**, 1632 (1987).
[40] A.J.R. da Silva, M.A.P. Lima, L.M. Brescansin, and V. McKoy, Phys. Rev. A **41**, 2903 (1991).
[41] T.L. Gibson, M.A.P. Lima, V. McKoy, and W.M. Huo, Phys. Rev. A **35**, 2473 (1987).
[42] N.S. Ostlund, Chem. Phys. Letters, **34**, 419 (1975).
[43] W.M. Huo and J.A. Sheehy (to be published). See also J.A. Sheehy and W.M. Huo, "Low-Energy Elastic Electron Scattering from Carbon Tetrafluoride" in ICPEAC

Abstracts Vol. I, ed. T. Andersen, B. Fastrup, F. Folkmann, H. Knudsen, (1993), p. 259.
[44] C. Winstead, Q. Sun, and V. McKoy, J. Chem. Phys. **98**, 1105 (1993).
[45] W.M.Huo, Phys. Rev. A **38**, 3303 (1988).
[46] A. Mann and F. Linder, J. Phys. B **25**, 545 (1992).
[47] L. Boesten, H. Tanaka, A. Kobayashi, M.A. Dillon, and M. Kimura, J. Phys. B **25**, 1607 (1992).
[48] D.W. Norcross and N.T. Padial, Phys. Rev. A **25**, 226 (1982).
[49] S. Chung and C.C. Lin, Phys. Rev. A **17**, 1874 (1978).
[50] A.W. Fliflet and V. McKoy, Phys. Rev. A **21**, 1863 (1980).
[51] T.N. Rescigno, C.W. McCurdy, Jr., and V. McKoy, J. Phys. B **7**, 2396 (1974).
[52] S.K. Srivastava and S. Jensen, J. Phys. B **10**, 3341 (1977). See S. Trajmar, D.F. Register, and A. Chutjian, Phys. Rep. **97**, 219 (1983) for renormalization of this data.
[53] M.A. Khakoo and S. Trajmar, Phys. Rev A **34**, 146 (1986).
[54] T.N. Rescigno, B.H. Lengsfield, C.W. McCurdy, and S.D. Parker, Phys. Rev. A **45**, 7800 (1992).
[55] H.J.R. Walters, J. Phys. B **4**, 437 (1971).
[56] E.J. Heller and W.P. Reinhardt, Phys. Rev. A **7**, 365 (1973).
[57] M.A.P. Lima, L.M. Brescansin, A.J.R. da Silva, C. Winstead, and V. McKoy, Phys. Rev. A **41**, 327 (1990).
[58] C. Winstead, P.G. Hipes, M.A.P. Lima, and V. McKoy, J. Chem. Phys. **94**, 5455 (1991).
[59] D.A. Levin, A.W. Fliflet, M. Ma, and V. McKoy, J. Comp. Phys. **28**, 416 (1978).
[60] B.I. Schneider, Phys. Rev. A **31**, 2188 (1985).
[61] M.H.F. Bettega, L.G. Ferreira, and M.A.P. Lima, Phys. Rev. A **47** 1111 (1993).
[62] B.H. Lengsfield III and T.N. Rescigno, Phys. Rev. A **44**, 2913 (1991).
[63] T.N. Rescigno, B.H. Lengsfield III, and A.E. Orel, J. Chem. Phys. **99**, 5097 (1993).
[64] C.J. Gillan, O. Nagy, P.G. Burke, L.A. Morgan and C.J. Noble, J. Phys. B **20**, 4585 (1987).
[65] S.E. Branchett and J. Tennyson, Phys. Rev. Letts. **64**, 2889 (1990).
[66] W.M. Huo, V. McKoy, M.A.P. Lima, and T.L. Gibson, "Electron-Nitrogen Molecule Collisions in High-Temperature Nonequilibrium Air," in *Thermalphysical Aspects of Re-entry Flows*, ed. J.N. Moss and C.D. Sott, AIAA, New York (1986), pp. 152-196.
[67] M. Berman,H. Estrada, L.S. Cederbaum, and W. Domcke, Phys. Rev. A**28**, 1363 (1983).
[68] S.F. Wong, J.A. Michejda, and A. Stamatovic, unpublished data.

INDEX

MIX
Papier aus verantwortungsvollen Quellen
Paper from responsible sources
FSC® C105338

If you have any concerns about our products,
you can contact us on
ProductSafety@springernature.com

In case Publisher is established outside the EU,
the EU authorized representative is:
Springer Nature Customer Service Center GmbH
Europaplatz 3, 69115 Heidelberg, Germany

Printed by Libri Plureos GmbH
in Hamburg, Germany